Mathematical Methods in Artificial Intelligence

IEEE Computer Society Press

SELECTED TITLES

Stiquito: The Design and Implementation of Nitinol-Propelled Walking Robots
    James M. Conrad

Mathematical Methods in Artificial Intelligence
    Edward A. Bender

Organizational Intelligence: AI in Organizational Design, Modeling, and Control
    Robert W. Blanning and David R. King

Mathematical Methods in Artificial Intelligence

Edward A. Bender

IEEE Computer Society Press
Los Alamitos, California

Washington • Brussels • Tokyo

Library of Congress Cataloging-in-Publication Data

Bender, Edward A., 1942–
    Mathematical methods in artificial intelligence / Edward A. Bender.
        p.   cm.
    Includes bibliographical references and index.
    ISBN 0-8186-7200-5
        1. Artificial intelligence—Mathematics.   2. Expert systems
(Computer science)   I. Title.
Q335.B45   1996
006.3' 01 ' 5—dc20

95-24708
CIP

IEEE Computer Society Press
10662 Los Vaqueros Circle
P.O. Box 3014
Los Alamitos, CA 90720-1264

IEEE Computer Society Press Order Number BP07200
Library of Congress Number 95-24708
ISBN 0-8186-7200-5

Additional copies may be ordered from:

IEEE Computer Society Press	IEEE Service Center	IEEE Computer Society	IEEE Computer Society
Customer Service Center	445 Hoes Lane	13, Avenue de l'Aquilon	Ooshima Building
10662 Los Vaqueros Circle	P.O. Box 1331	B-1200 Brussels	2-19-1 Minami-Aoyama
P.O. Box 3014	Piscataway, NJ 08855-1331	BELGIUM	Minato-ku, Tokyo 107
Los Alamitos, CA 90720-1264	Tel: +1-908-981-1393	Tel: +32-2-770-2198	JAPAN
Tel: +1-714-821-8380	Fax: +1-908-981-9667	Fax: +32-2-770-8505	Tel: +81-3-3408-3118
Fax: +1-714-821-4641	mis.custserv@computer.org	euro.ofc@computer.org	Fax: +81-3-3408-3553
Email: cs.books@computer.org			tokyo.ofc@computer.org

Technical Editor: Jon Butler
Production Editor: Lisa O'Conner
Copy Editor: Emily Thompson
Cover Design: Christa Schubert
The limerick on page 104 is from the *Canadian Artificial Intelligence Newsletter,* #9, Sept. 1986.
Reprinted with permission of CSCSI.

Printed in the United States of America by BookCrafters

The Institute of Electrical and Electronics Engineers, Inc

Contents

Preface

If AI is ever to become a respectably hard science,
then a firm, formal basis is needed.

—Derek Partridge (1991)

Philosophy

Teaching an AI course presents a problem. The field is so broad that an attempt to cover most of it is bound to result in a fairly shallow survey course. Nevertheless, it is important to discuss some important tools of AI in some depth. The tools can be roughly divided into three types: tools for implementing a plan (e.g., Lisp, microprocessors), tools for designing a plan (e.g., algorithms), and tools for designing tools.

A hands-on approach based on implementing plans is often pursued in computer science. Unfortunately, toy AI problems are of limited pedagogical use while real AI problems are often on such a scale that programming only one of them is a major project. Moreover, a hands-on approach often gives students the ability to implement some plans without giving them the ability to understand or develop the tools on which such plans are based.

As a result, I believe it's critical to focus on tools for designing AI tools. Since AI is a young science, we must to some extent anticipate what these tools will be. Mathematics has been the major tool designing tool in the sciences; therefore, I am persuaded that AI will not be an exception.

Possible Courses

It's popular to say that a book does not require much formal mathematics, but does require some mathematical maturity. That's true here. Much of the material can, in theory, be read and understood with no more background than high school algebra. In practice, however, students need more than this or they will be overwhelmed by the need to think mathematically. Furthermore, after the first ten chapters, some background in calculus is needed.

This text can be used for an introductory course in AI for upper-division or graduate students who have had a standard lower-division calculus course. Many courses are possible, depending on the time available, the capabilities of the students, and the interests of the instructor. All courses should include at least Chapter 1 and most of Chapters 2, 3, 5, and 10. Possible supplements are (a) logic from Chapters 4 and/or 6; (b) neural nets from Chapter 11; (c) probability and its uses from Chapters 7, 8, and 9; and (d) material from Chapter 14. The more mathematical exercises and proofs can be emphasized or deemphasized as circumstances dictate.

This text can also be used for a second course in AI for students interested in AI research. Many monographs and research papers are inaccessible to such students because they assume a mathematical background not provided by standard AI courses. The mathematics in this text helps bridge that gap.

Instructors may obtain a TeX diskette containing solutions to many of the exercises from Computer Society Press.

The following diagram illustrates some dependencies among all chapters but the last, ranging from weak (dotted lines) to nearly complete dependency (solid lines). Dashed lines indicate that only some sections are essential. More details on dependencies are found at the end of each chapter introduction.

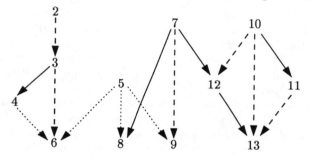

Acknowledgments

Various people have helped with this text. I'd particularly like to thank my students whose confusions and misunderstandings uncovered poorly written passages, my colleagues Frederic Bien, Te C. Hu, Fred Kochman, Alfred Manaster, Jeff Remmel, and Malcolm Williamson for their suggestions, my copy editor Emily Thompson for her ample, apposite use of red ink, and Computer Society Press for their helpful editorial assistance.

Dear Student

History teaches that new technology will require new mathematics.
... The question is: Which mathematics to use?
—Monique Pavel (1989)

Many introductory AI texts give the impression that AI is a collection of heuristic ideas and data structures implemented in Lisp and Prolog. The prognosis for such a discipline would be grim. Fortunately, AI researchers use mathematics and are developing new tools. Unfortunately, most of what you need is found in monographs and research articles—inappropriate material for a beginning course. This text is my attempt to fill the gap.

Since some of the mathematics used in AI is not part of a standard undergraduate curriculum, you'll be learning mathematics and seeing how it's used in AI at the same time. As with most mathematically oriented texts, this one isn't easy. I've written the next couple of pages to help you through it. Please read them.

Goals

In this text I hope to introduce you to some mathematical tools that have been important in AI and to some of their applications to the design of algorithms for AI. Since expert systems (broadly interpreted) comprise a large part of AI and have been the main focus of mathematically based tools, I have centered the book around the expert system idea.

As a result of studying this text, you should be in a much better position to read the technical literature in AI and should be able to easily fill in gaps in coverage by reading one of the more broadly based survey texts.

Reading Mathematics

Many people learn mathematics the way I learned history in high school. The exams contained two columns and the goal was to match each date in column A with one of the persons, places, and events in column B. Being lazy, I learned the "whens" of history but never the "whys." I missed a whole world of ideas.

When mathematics is taught and learned by rote, students miss a world of ideas. Mathematics should be learned as an aid to thinking, not as a replacement for it. Learning mathematics is a skill that's seldom taught. If, like many students, you haven't mastered it, the following comments should be helpful.

The key is to work on understanding—not on memorization. How can you do this?

Let's begin with definitions. Whenever you meet a new concept, develop an understanding of it by relating it to ideas you already know and by looking at what it means in specific cases. For instance, when learning what a polynomial is, look at specific polynomials; when learning what continuity is, see what it means for a specific function like x^2. *The importance of understanding the general through the specific cannot be overemphasized*—even by using italics. The discussions and examples that immediately precede and follow definitions are often designed to foster understanding. If a definition refers to an earlier, unclear concept, stop! If you proceed, you may end up wandering aimlessly in a foggy landscape filled with shadowy concepts and mirages. Go back and improve your understanding of the earlier concepts so that they're practically solid objects that you can touch and manipulate. Finally, ask yourself why a definition has been introduced: What is the important or useful concept behind it? You may not be able to answer that question until you've read further in the text, but you can prepare your mind to recognize the answer when you see it.

What about theorems? The comments for definitions apply here, too: Look at specific examples, try to relate the theorem to other things you know, ask why it's important. Be sure you're clear on what the theorem *claims* and on what its words *mean*. In addition, attempt to see why the result seems reasonable before you read the proof. Reading and understanding the proof is the last step. If the proof is long, it may be helpful to make an outline of it. But don't mistake the ability to reproduce a proof for understanding. That's like expecting a photograph to understand a scene. There are better tests of understanding: Do you see where all of the assumptions are used? Can you think of a stronger conclusion than that in the theorem? If so, can you see why the stronger conclusion is not true, or at least why the proof is insufficient to establish the stronger conclusion?

Examples play a key role in mathematics. In practically every mathematics text, they fall into three categories.

- The type that aren't in the text: They're the ones you create by following the preceding advice.

- The obvious ones that are labeled "example" in the text. They're usually illustrations of definitions, algorithms, or theorems. Sometimes they develop related ideas.

- The type that comes from homework problems: These examples are the *solutions* to the problems you do yourself, not the problems themselves.

If you neglect any of these three types of examples, your mathematical text will be most useful to you as a doorstop.

Navigation Aids

Here's some information to help you navigate this text.

- A chapter introduction usually tells what's in the chapter, why it's there, and how the chapter is laid out. The overview it provides will help you organize the chapter in your mind.

- Numerous quotations highlight ideas and controversies, offer insight, provoke thought, and perhaps provide comic relief.

- Starred material either is more difficult than the text in which it is embedded or is peripheral.

- A remark that's somewhat off the track may appear as an Aside set in smaller type. Asides can be skipped without losing the thread of the discussion.

- There are four types of exercises. Here they are in order of difficulty.

  - Some exercises are lettered and some numbered; for example, 2.4.A versus 2.4.1. The purpose of lettered exercises is to make sure you absorbed the basic ideas. Their solutions can be found by rereading the preceding material. *You should do all lettered exercises.* It's often necessary to know the answers to these exercises before reading further.

  - A few numbered exercises are there to be sure you've picked up basic ideas that are needed soon. The answer to such an exercise is given immediately after the exercise section. *You should do all these exercises, then read the solutions.* If you've made an error, study the section further or ask for help. It's important to understand how to do these exercises before reading further.

— The solutions to most exercises are neither very simple nor very difficult. Many you'll be able to do. Ask for help on those that baffle you.

— Starred exercises are ones that I consider difficult or that refer to starred material.

A few exercises don't have just one right answer. They may ask for your opinion or they may ask for you to construct an example of something. If an exercise asks for a proof, use full sentences. Read your proof aloud—it'll help you catch mistakes and incoherent thinking.

Enjoy your exploration of AI and its mathematical foundations.

Sincerely,
Ed Bender

1

First Things

Anyone teaching a course [on AI] ... will have to decide what artificial intelligence is, even if only because inquiring minds want to know.
—Stuart Russell and Eric Wefald (1991)

"Can machines think?" [is] as ill-posed and uninteresting as "Can submarines swim?"
—Edsger W. Dijkstra (ca 1970)

Our minds contain processes that enable us to solve problems we consider difficult. "Intelligence" is our name for whichever of those processes we don't yet understand
—Marvin Minsky (1985)

Introduction

From golems to androids, manmade intelligences have been a dream and night-mare of mankind for centuries. In the 1950s, electronic brains led to the birth of the science of artificial intelligence. Will AI, as the field is commonly called, fulfill its promise to convert mankind's fantasies into reality?

We'll begin exploring the nature of AI by examining its goals, tools, and accomplishments, and some of the debates it has engendered. Such an exami-nation should give us a revealing picture of the current state of this promising discipline.

Next, we'll discuss the why's and wherefore's of this text. Why the em-phasis on mathematics? What do future chapters hold in store?

The final sections introduce two important topics: the computation prob-lem and expert systems. The computation problem permeates AI, but is not always evident. Meeting it face to face now is important because overlooking its presence is a ticket to disaster. Expert systems provide a unified way of viewing most of AI.

1.1 Delimiting AI

We'll examine AI from three different viewpoints, or "coordinates":

$$\text{goals,}$$
$$\text{methods or tools and} \tag{1.1}$$
$$\text{achievements and failures.}$$

For example, your goal may be to understand what AI is all about; your method, talking to AI researchers; and your achievement, a new overview of AI. Some parts of AI (such as machine learning) are primarily defined by goals, others (such as neural networks) primarily by methods. Achievements and failures give information on how a field has progressed.

Some Goals of AI

When you read the following list, interpret words like "reasoning" and "understanding" as referring to the *results*, not the *methods*. In other words, focus on a program's output rather than its algorithm. (Mimicking human algorithms is a concern of cognitive science, not AI.)

- Reasoning: Given some general knowledge together with some specific facts, deduce certain consequences. For example, given knowledge about diseases and symptoms, diagnose a particular case on the basis of information about the symptoms. The most difficult type of reasoning is based on what people call "common sense."

- Planning: Given (a) some knowledge, (b) the present situation, and (c) a desired goal, decide how to reach the goal; that is, use *goal-directed reasoning*. How do planning and reasoning differ? They overlap, but, roughly, reasoning seeks the answer to What?; planning seeks the answer to How? For example, "What sort of student am I?" versus "How can I be an A student?"

- Learning: Acquiring knowledge (learning) is a central issue since knowledge must be acquired before it can be used. In some situations, it is feasible to build knowledge into a system. In others, it is infeasible or undesirable. Then we want a system that can repeatedly extend its knowledge base in a coherent fashion by acquiring new facts and integrating them with previous knowledge, often by some process of abstraction. For example, if a system is exposed to various examples of chairs, how can it abstract the concept "chair"?

The previous goals are rather general in nature and are relevant to many parts of AI. We now look at goals that may be viewed as more specific. Although they draw on results in the previous areas, they are very much separate parts of AI with their own tools and problems.

- **Language Understanding and Use**: Obviously, this relies heavily on reasoning and learning, but it deserves a separate category. "Common sense" plays an important role in language. Unfortunately, common sense is an extremely elusive topic that appears to require a considerable knowledge base. Attempts to understand spoken language must face additional complications.

- **Processing Visual Input**: Vision is only one type of sensory input that must be processed, but it is by far the most complex. Abstracting useful information from visual input is proving very difficult.

- **Robotics**: Robotics must marry AI with engineering. In all but the simplest industrial settings, reality is dauntingly complex. The AI techniques used in robotics must produce results in real time and, for an autonomous robot, must not require excessive computer power.

Some Tools of AI

Knowledge about knowledge is the focus of AI. Knowledge is given either *declaratively*—in declarative statements—or *procedurally*—by procedural rules. Specific knowledge tends to be represented declaratively and general knowledge procedurally. (The situation is not this cut-and-dried, but the distinction is still useful.) Declarative knowledge is stored in what is called a *knowledge base*. Knowledge about knowledge provides tools for interacting with the knowledge base:

- **Knowledge Organization Tools**: Data structures and algorithms facilitating the organization of the knowledge base.

- **Knowledge Manipulation Tools**: Methods for extracting new knowledge from the knowledge base; for example, reasoning and planning.

- **Knowledge Acquisition Tools**: Methods for incorporating new knowledge into the knowledge base or modifying the tools ("learning"). The border between acquisition and manipulation is fuzzy.

In many areas of computer science, algorithms are primary and data structures are secondary. In contrast, knowledge representation is a central problem in AI. The form of declarative knowledge (the data structures) limits what we can state and the ease with which we can manipulate it. AI's declarative knowledge is seldom considered "just data." Thus, the tools of AI could be thought of in terms of what can be incorporated in the data structures. Here are some data structures and where to find them and their tools.

- **Limited Structure:** Relatively unstructured search spaces are attractive because they impose few restrictions. Sadly, the lack of structure makes computations overwhelming for all but the simplest problems (Chapters 2 and 13).

- **Mathematical Logic:** Mathematical logic allows us to represent facts about the world in a form that can be manipulated (Chapters 3–6).

- **Logic-like Representations:** Representational awkwardness and other handicaps motivated some researchers to seek alternatives. Some approaches, such as rule-based systems and semantic nets, can be recast in the framework of logic (Chapter 6). Other approaches, such as reasoning by analogy as in case-based reasoning, use other methods and are lightly touched upon in Chapter 14.

- **Numerical Information:** Numerical information can play a central role in describing uncertainty about the world, as in "a 40% chance of rain" (Chapters 8 and 9).

- **Nonsymbolic Structures:** The previous structures are designed to represent and manipulate information symbolically. A growing number of researchers have questioned this approach (Chapters 10, 11, and 13).

On another level, we could say that the tools of AI are those things that provide the basis for creating the knowledge tools. They tend to fall into four areas:

- **Hardware:** AI makes heavy use of computers for developing and testing ideas. Some parts can benefit from special-purpose devices.

- **Software:** AI's large software systems make it an important developer and consumer of programming tools.

- **Mathematics:** Some parts of mathematics have proven useful in AI. (Every formal manipulation of concepts is a part of mathematics.)

- **Heuristics:** Sometimes called "rules of thumb," heuristics are empirical principles. Heuristics may use mathematics, but often do not.

What Has AI Given the World?

Many of AI's contributions contain no AI: They are simply tools that were developed to aid AI research. We'll begin with these spinoffs and move on to results that do contain some AI. There is no consensus on where to draw a line between contributions containing little or no AI and those with significant amounts. Many draw the line just before or just after "game-playing programs."

- **Timesharing**: Much early work on timesharing was done by MIT's project MAC—a dual acronym meaning either "machine-aided cognition" or "multiple-access computing." (Some wags called it "man against computer.")

- **Windows and Graphical User Interfaces**: These were developed at Xerox's Palo Alto Research Center to provide easier computer access for AI researchers.

- **Programming Paradigms**: These include
  - constraint propagation (now used in spreadsheet programs),
  - object-oriented programming,
  - functional programming (the basis of Lisp), and
  - logic, or declarative, programming (the basis of Prolog).

- **Fuzzy Controllers**: "Fuzzy logic" leads to more stable and flexible means of regulating machines.

- **Game-Playing Programs**: Game playing was a favorite topic in the early years of AI research. By 1995, the best artificial chess player could beat all but the best human players. Backgammon programs achieved a similar level: One beat the world champion because lucky rolls of the dice compensated for somewhat inferior play.

- **Expert Systems**: Commercial expert systems have been proliferating in recent years and many businesses are using special-purpose software to write expert systems for in-house use.

- **Natural Language Interfaces**: A limited ability to understand natural language is providing friendlier user interfaces for some programs.

- **Dictation Systems**: Systems able to transcribe speech have begun to appear on the market. So far, vocabulary and speed are rather limited.

The flip side of achievement is failure—the skeleton in the closet. Here are three of them.

- **Wild Optimism**: The seeds of a variety of failures were planted in the 1950s—the early, heady years of AI when almost everything was "just around the corner." In the 1980s, a minor relapse into unbridled optimism was caused by the rebirth of neural networks—a methodology inspired by the highly interconnected, self-modifying nature of biological neural systems.

- **Game-Playing Programs:** Many hoped that studying games would lead to significant progress in AI. The rate of return has been low, however—perhaps because competitions tend to focus on immediate improvement rather than new ideas.

- **Ad Hoc Developments:** Much research has been based on ad hoc methods rather than solid foundations. People argue about whether seat-of-the-pants design is inherent in the subject matter of AI or just a passing stage. Advocates of ad hoc methods are called *scruffies*; advocates of theoretical methods are called *neats*.

Results versus Methods: Cognitive Science

> *Artificial intelligence is an invention.*
> *In contrast, a theory of human intellect is a discovery.*
>
> —Morton Wagman (1991)

For some, a major goal of AI is the construction of an artificial intelligence having human-level abilities. Progress has certainly been made, but the goal is still far away, if not impossible. Other people are concerned with the methods *humans* use to achieve their abilities. As noted earlier, these people are *cognitive scientists*.

Like AI, cognitive science is an umbrella field related to "intelligence." Cognitive science, which includes topics like cognition and consciousness, seems to be striving to achieve many of the goals listed for AI. Unlike AI, cognitive science focuses on learning how human minds achieve such goals rather than on creating artificial methods for achieving them. There are a variety of introductions to cognitive science; for example [14] and [49].

Allen Newell [32] was a major advocate for developing unified theories of cognition that can be tested and expressed through programs. He argues cogently that both cognitive science and AI will profit from such attempts in the short term, but will go their separate ways in the long term [32, p. 57]. To see why this might be so, consider a crude parallel—a slightly fictional history of flight. Cognitive science corresponds to understanding bird flight, and AI to creating artificial flight. Understanding and adapting some aspects of bird flight informed the early development of artificial flight. Conversely, attempts at artificial flight provided tests for the understanding of bird flight. Major progress required an understanding of the principles of aerodynamics, at which point the methods employed by birds were no longer relevant. (Actually, studying flying fish may have been more productive for early attempts at flying.)

Exercises

1.1.A. What are some goals of AI?

1.1.B. What are some general tools of AI?

1.1.C. What are some contributions of AI?

1.1.D. What is the difference between AI and cognitive science?

1.2 Debates

> *There is nothing which is not the subject of debate, and in which men of learning are not of contrary opinions. The most trivial question escapes not our controversy, and in the most momentous we are not able to give any certain decision.*
>
> —David Hume (1740)

> *Consciousness is a subject about which there is little consensus, even as to what the problem is. Without a few initial prejudices one cannot get anywhere.*
>
> —Francis Crick (1994)

One of the ongoing debats in AI is the definition of the AI field itself. Actually, the variety in the field probably makes it impossible to give a concise definition that is neither too broad nor too narrow. To see why this is so, try the much simpler problem of defining what is meant by a sport. Your definition should include bowling and recreational cycling, but not chess or dancing.

Here are three debates that provide some insights about AI.

Consciousness and Intelligence

A better understanding of cognitive science topics like intelligence and consciousness could benefit AI research. Thus we'll look briefly at these debates, even though they do not belong in AI.

A question like "Can machines think?" is difficult. We often start from the premise that we understand what this question means when, in fact, ongoing debates show that we have not yet figured out what we're talking about. Even the first step—agreeing on the definition of "intelligence"—has not been taken. Some people believe that the most famous proposed test for machine intelligence, the Turing test, should be regarded as a definition of intelligence. Other people disagree. (See Exercise 1.2.1.)

Perhaps thought and intelligence are the wrong issues to address. Instead, consciousness may be a more fundamental issue. We seem to know less about this subject than some experts would like to believe. The study of consciousness belongs to philosophy, psychology, and cognitive science. If this line of study appeals to you, you may find the books by Churchland [8], Dennett [12], and Moody [31] of interest.

The range of beliefs (or hopes) regarding intelligence and consciousness is quite broad.

- At one extreme, strong AI supporters maintain that it is possible to create an intelligent, conscious machine and that something like the Turing test (Exercise 1.2.1) is adequate to determine if the machine is intelligent and conscious. One expression of this is the *physical symbol hypothesis* of Newell and Simon [33]. They define a physical symbol system to be something that is capable of manipulating physical patterns (such as data in a computer or strengths of connections among neurons) and hypothesize that such a system is necessary and sufficient for implementing general intelligent behavior.

- At the other extreme are those who maintain either (a) that intelligent, conscious behavior has a nonphysical component (as in Cartesian dualism) or (b) that it involves something inherently biological. These people conclude machines will never achieve such behavior.

Given the current state of AI, researchers need not worry about such issues any more than the Wright brothers needed to worry about the sound barrier.

Symbols versus Connections

The symbols versus connections debate might also be described as "intelligence by design" versus "intelligence as an emergent property."

The traditional approach to AI has been symbolic; that is, knowledge is represented at a symbolic level comprehensible to us. The strict symbolic viewpoint is that the way to make real progress in AI is through the development of powerful data structures and algorithms for the representation and manipulation of knowledge on a symbolic level. Most defenders of this view believe the symbolic approach mimics conscious human reasoning. The choice of a symbolic framework has been debated. Some want to base the symbolic approach on mathematical logic; others insist that numerical methods should play a central role.

There has recently been a revival of the connectionist approach. Like much of AI, this approach was born amidst the rosy predictions of the 1950s. It nearly disappeared in 1969 after Minsky and Papert [30] emphasized the limitations of the methods then available. Interest blossomed anew in the 1980s. Since then, considerable research has been done using simulations of networks of simple interconnected processors, that is, *neural networks*. Strict connectionists believe that one should design complex networks of simple processors and then train these networks. Intelligence, they maintain, will emerge as a consequence, but won't be found in the parts of the network separately. This is the sort of internal representation and manipulation of knowledge that the human brain apparently uses on the physiological level, with neurons as processors.

Which approach is better? The answer may depend on the application. It may be best to combine the approaches—people are experimenting with hybrid systems. At any rate, it's too soon to tell.

The Role of Theory

The word "theory" encompasses mathematics as well as such things as the theory of general relativity. It does not include simple facts and rules of thumb based on them. For example, the commonsense advices "get a good night's sleep before an exam" is not a theory. It's a heuristic rule based on personal observation. What, then, is the practical relevance of theory for AI?

"The theory *is* the program" view of some nontheorists is at one extreme. This attitude should not be confused with the idea that computer programs in AI (should) play the role of experiments—no one claims that a theory is an experiment. In contrast, "The theory is the program" means you may ask how well the program works but you can't ask for a foundation on which the program is based.

At the other extreme is the ultra-logicist claim that, ultimately, AI will succeed by employing a theoretically justified system of symbolic reasoning.

Naturally, most researchers' beliefs lie between these two extremes. The issue then is "What is the best blend between heuristics and theory?" The answer to this question depends on the researcher, on the subject, and on its state of development: On the researcher, because abilities vary from person to person; on the subject, because simpler areas are more easily fit into a theoretical framework; and on the state of development because mathematics is gradually making greater inroads into various areas of AI.

*Exercises

To the student: These exercises are likely to be time-consuming. Most instructors (myself included) won't assign any because of time pressure. Read them anyway—they provide food for thought.

To the instructor: See above.

1.2.1. The *Turing test* [51]: An evaluator E is allowed access to two subjects W (a woman) and X (not a woman) only through a remote terminal. The experimenter tells E that exactly one of W and X is a woman and instructs E to determine which it is by whatever method E wishes using the remote terminal. Each of W and X attempts to react like a woman when responding to E's questions. If E decides that X is a woman, then E has been deceived. By averaging over many E's, W's, and X's, we can obtain a success rate for deception. In particular, we can compute the deception rate when X is a man—the deception rate for men. We can also compute the deception rate for a computer program. In the Turing test, the program is declared to possess intelligence if its deception rate is at least as great as the deception rate for men.

(a) Consider the following statement: "The Turing test is based on the idea that the ability to misrepresent oneself is a measure of intelligence." Do you agree? Why? If you agree with it, do you think that ability is a measure of intelligence? Why?

(b) In some statements of the Turing test, the deception rate for a computer program is simply required to exceed some value. Which version do you think is better? Why?

(c) Suppose a species as intelligent as humans were found. (Intelligence in this sentence does not refer to the Turing test, but to a "commonsense" assessment.) Do you think such aliens could pass the Turing test? Why?

(d) Given the existence of an intelligent alien species, suggest and defend a less species-biased test for computer intelligence.

(e) As stated, passing the Turing test depends on the computer's possessing extensive knowledge of the nature of human beings, both physical and psychological, as well as their culture, history, literature, and so forth. Suggest and defend modifications of the Turing test that would reduce the need for such knowledge. To what extent can such a need be eliminated without affecting the validity of the test?

(f) More generally, can you formulate a better test?

1.2.2. Suppose we are considering cognitive skills, learning abilities, or some other human skill that is relevant to AI. Imagine a three-sided debate:

1. The (nearly) best way to achieve this skill has been found by evolution.

2. By reason and experiment, we'll be able to improve considerably on human skills.

3. Neither of the two previous views is correct.

Come to class prepared to carry out such a debate. (You may be assigned a particular viewpoint to defend.)

1.2.3. For each of the three sides of the debate in the previous exercise, describe the implications for AI work on a particular skill if the side is correct.

1.2.4. Newell [34, p. 19] lists a variety of things a mind is able to do, many of which are reproduced below. Which of these abilities do you think a computer program should have in order to deserve being considered a major AI project? Explain your choices.

Hint. There is a wide latitude for acceptable answers, but you may have to decide what *you mean by AI* in order to answer.

(a) Behave flexibly as a function of the environment

(b) Exhibit adaptive (rational, goal-oriented) behavior

(c) Operate in real time

(d) Operate in a rich, complex, detailed environment

- Perceive an immense amount of changing detail
- Use vast amounts of knowledge
- Control a motor system of many degrees of freedom

(e) Use symbols and abstractions

(f) Use language, both natural and artificial

(g) Learn from the environment and from experience

(h) Acquire capabilities through development

(i) Operate autonomously, but within a social community

(j) Be self-aware and have a sense of self

1.3   About This Text

> *The paradox is now fully established that the utmost abstractions are the true weapons with which to control our thought of concrete fact.*
>
> —Alfred North Whitehead (1925)

> *Understanding in mathematics cannot be transmitted by painless entertainment any more than education in music can be brought by the most brilliant journalism to those who have never listened intensively. Actual contact with the content of living mathematics is necessary.*
>
> —Richard Courant (1941)

> *Teach nothing that pupils can teach themselves.*
>
> —Amos Bronson Alcott (1799–1888)

Mathematics is the term we use to describe the process of symbolically deducing conclusions from conceptual assumptions, whether these be the axioms of geometry, the laws of physics, or the assumptions in economics' utility theory. Mathematics with bad assumptions is useless; with good assumptions, it is a wonderful tool.

Heuristics is the term we use to describe empirical principles and techniques, such as "Avoid the use of GOTO," "The best offense is a good defense," and "graphical user interfaces." Good heuristics whose limits are well understood are very useful.

Programming is a means of testing ideas, creating tools, and generating information that may spark new research. Because of AI's complexity, we often use special languages (most notably Lisp and Prolog) or simulator packages (especially for neural nets).

Programming, heuristics, and mathematics are all important in AI.

Because AI is a large field, textbook authors must make choices. Most authors emphasize heuristics and relatively simple programming exercises. Since writing large programs and studying mathematics are time-consuming, this approach allows the broadest coverage of topics. After learning some Lisp or Prolog and taking a course that involves a large programming project, you should be able to study AI programming methods and write such programs. On the other hand, it's much harder to study mathematics on your own.

My goal is to provide an introductory AI course based on the most important mathematics and its applications. To keep the length manageable, material must be cut. My algorithm is simple: Focus on important AI topics that involve the most broadly applicable mathematics and cut back on others. What does that leave? The main mathematical tools for representing and manipulating knowledge symbolically are (a) various forms of logic for qualitative knowledge and (b) probability and related concepts for quantitative knowledge. The main tools for manipulating knowledge nonsymbolically, as in neural nets, are optimization methods and statistics. I've organized that material as follows.

- **Trees and Search**: Since search plays a central role in AI, elementary aspects of search trees are discussed in Chapter 2. Some additional aspects of search are briefly discussed in Chapter 13 after the necessary probability theory has been introduced in Chapter 12.

- **Classical Mathematical Logic**: First-order predicate logic, the starting point for the use of logic in AI, is presented in Chapters 3 and 4. Prolog is introduced to show how the concepts and results can be implemented in a programming language. The reasoning engine in Prolog combines a search strategy with a deductive method from logic. (You won't, however, learn how to program in Prolog from this brief introduction.)

- **Uncertainty in Reasoning**: AI systems based on classical mathematical logic have various shortcomings. Among these are the following:

  - We can't easily allow for general rules that have exceptions (for example, "mammals have legs" and "whales are mammals without legs").

- We can't allow for uncertain statements (for example, "When the barometer is falling, it often rains by the following day.")

Qualitative approaches based primarily on extending logic are discussed in Chapter 6. Quantitative approaches are discussed in Chapters 8 and 9 after the necessary probability theory has been introduced in Chapter 7.

- Automatic Classification: An alternative to incorporating knowledge-based rules into expert systems is to design programs that develop their own "rules" from examples. These are called *pattern classifiers* and are discussed in Chapters 10, 11, and 13. After discussing neural nets and optimization in Chapter 11, I digress to introduce some probability, statistics, and information theory in Chapter 12. This is applied to neural nets and decision trees in Chapter 13.

- Other Things: The previous material omits important areas of AI. One is robotics, in which sensory-input processing (especially vision) and motion planning involve considerable mathematics. Another is language, where linguistics and speech processing use mathematics. The final chapter contains brief introductions to these omissions and to some less mathematical topics so that you'll have a bit of background and some references for further study.

1.4   The Computation Problem in AI

Any program that will successfully model even a small part of intelligence will be inherently massive and complex. Consequently artificial intelligence continually confronts the limits of modern computer-science technology.

—J. Michael Brady, Daniel G. Bobrow, and Randall Davis (1993)

Designing algorithms is a central problem in almost any computer oriented field, and AI is no exception. Unfortunately, algorithms are particularly troublesome in AI. Three reasons for this are as follows.

- Complexity: Problem complexity makes designing and implementing algorithms difficult.

- Time: Algorithms frequently explore potential solutions in the course of searching for an acceptable one. For problems of realistic size, a simple search process may take too long because of *combinatorial explosion*— a rapid growth in the number of possible solutions. Unfortunately for algorithm design in AI,

Very rapid growth is typical in AI problems.

- **Impossibility:** It may be impossible to design an algorithm for the given problem. Here's a specific example. We would like to design an algorithm that takes as input (a) a computer program in some suitable language and (b) some data for the program. The algorithm must determine if the program will stop or run endlessly—the *halting problem*. In designing the algorithm, we imagine an abstract computer having infinite storage. (Of course, compromises will have to be made when we get around to implementing the algorithm.) In a classic paper in 1935, Turing *proved* that *no such algorithm can exist*. Thus, the problem is impossible.

Here are some ways of dealing with these problems.

- **Find a much better algorithm:** This is the ideal solution. Unfortunately, we often cannot find a much better algorithm.

- **Settle for an algorithm that sometimes fails:** These are algorithms that sometimes fail either by stopping with no solution or, worse, by giving an incorrect solution. It's possible to create such an algorithm by imposing a time limit on another algorithm. For example, the famous simplex algorithm in linear programming has a very bad worst-case time and a very good average-case time [6]. Thus, an intelligently designed time limit would lead to a solution in most cases. For this approach to be useful, we must know from theory or experience that failure is relatively rare.

- **Settle for an approximate solution:** Such a solution is often good enough. Simon coined the term *satisficing* for finding a good enough solution. Sometimes, obtaining good approximate solutions may be as difficult as the original problem.

- **Replace the problem with an easier one:** Solving the easier problem may produce useful results. Also, exploring the easier problem may lead to ideas for the original problem.

- **Give up:** No comment.

All of these approaches are used in AI, often in combination. Inventing compromise algorithms is a tricky, creative business. Mathematics may help in inventing and assessing compromises, but is seldom sufficient. The final weighing of gains and losses in a compromise is a value judgment based on your goals.

How much time should an algorithm be allowed to take? More time often means a better result. On the other hand, speed of response is important; for example, a user is less likely to use a sluggish expert system than a quick one. Figure 1.1 illustrates this idea. Unfortunately, the information needed to construct the curves in the figure is seldom available. In this case, an *anytime algorithm* can be quite useful. This is an algorithm that can be interrupted at any time to obtain an approximate answer. Here's a simple example of such an algorithm. Suppose we know that f is continuous on the interval $[a, b]$, that $f(a) < 0$, and that $f(b) > 0$. We want to obtain an estimate for an

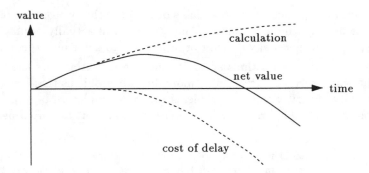

Figure 1.1   The typical effect of response time. The vertical scale measures value in some unspecified manner. The upper dashed curve shows how the value of a result varies with computation time. It ignores the costs of time delay. The lower dashed curve shows the cost of delay in response time. The middle curve, which combines the two, shows the net value of the response. Computation should end at the middle curve's maximum: Even though more calculation would give a better result, the cost of delay outweighs the gain.

$x \in (a, b)$ such that $f(x) = 0$. Simply repeat the following two steps: Let $c = (a + b)/2$. If $f(c) \leq 0$, let $a = c$; otherwise, let $b = c$. Whenever the algorithm is interrupted, it returns the estimate c for x.

NP-Hard Problems

Theoretical computer scientists consider an algorithm to be fast if its running time can be bounded by a polynomial in the number of bits needed to express the input and the output. This means that, even in the worst case, the algorithm is reasonably fast on very large problems. It says nothing about average running time. Indeed, it may be very difficult to define an average running time since it may be unclear what to average over.

In the theory of algorithms, a certain class of problems is called *NP-complete*. Hundreds of problems of interest to computer scientists have been shown to be NP-complete. It has been proved that either a fast algorithm exists for *all* NP-complete problems, or no fast algorithm exists for *any* NP-complete problem. Since no fast algorithm has been found after many years of research, it seems unlikely that any exists.

An *NP-hard* problem is one that is at least as difficult as an NP-complete problem. Even when a problem is NP-hard, there may well be an algorithm that works well on the situations that arise in actual usage—that is, the worst cases simply don't arise in practice. (Of course, as soon as you decide this and release your program to the world, Murphy's law dictates that someone will come up with a use where the worst cases occur.) In other words, the relevant time is the average running time over inputs that will actually occur. Unfortunately, this time is usually difficult or impossible to determine theoretically.

Polynomial time algorithms and NP-complete algorithms are the bottom levels of a whole series of increasingly more difficult problems that are studied in complexity theory. Some AI problems are NP-complete. Many more are even more difficult. As a result, compromises of some sort are often needed.

Aside. Here's a technical note for those who want to know a bit more about NP-complete. Let $|y|$ denote the number of bits needed to describe y. We say that an algorithm is (at most) "g time" if the running time of the algorithm with input x is bounded by $g(|x|)$.

Suppose we want to determine whether certain things in a set S have some property F. This is called a *recognition problem*. A recognition problem is in the class P if there exists a polynomial time algorithm that can determine if $F(x)$ is true or false. For example, S could be the positive integers and F could be "composite" (not a prime). In this case, $F(x)$ is true if and only if x is not a prime. No polynomial time algorithm is known for this example.

It may be much easier to verify that $F(x)$ is true for a given x if we're given some additional information. This added information is called a *certificate*. Thus, a certificate could change a hard problem into an easy one. (Of course, it might be *very hard* to create such a certificate.) Note that this makes no provision for verifying that $F(x)$ is false. For the composite number example, a certificate $c(x)$ for x could be a factor of x. To verify that $F(x)$ is true, all we need to do is check that $x/c(x)$ is an integer between 1 and x.

A certificate-checking algorithm is NP if it is polynomial time and $|c(x)|$ is bounded by a polynomial in $|x|$. "NP" stands for "nondeterministic polynomial." It should be clear how *polynomial* applies to the definition, but where does *nondeterministic* come in? An algorithm that makes lucky guesses could do the hard part—that is, create $c(x)$ in polynomial time by guessing. Guessing is a nondeterministic process. Combining this with the certificate-checking algorithm gives a nondeterministic polynomial time algorithm for $F(x)$. A recognition problem is in the class NP if there exists an NP algorithm for it.

Suprisingly, there exists a class of "hardest" recognition problems in NP. These are the NP-complete problems. In what sense are they hardest? Suppose we have a g time algorithm for a problem. We say another problem is no harder than this if it has a $g(p)$ time algorithm for some polynomial p. This says that, to within a polynomial adjustment, all NP-complete problems have the same running time bound and no NP recognition problem has a larger bound. Let's put this another way. Suppose that $\mathcal{P}$ is a recognition problem that has a polynomial time certificate checking algorithm and let g be the running time for the best possible noncertificate algorithm for some NP-complete problem. Then the best noncertificate algorithm for $\mathcal{P}$ is at most $g(p)$ time for some polynomial p.

Since a polynomial time algorithm can check if $F(x)$ is true in polynomial time even without a certificate, any problem in the the class P is contained in the class NP. It's not known if the two classes are equal; however, this seems very unlikely. Why? NP-complete problems have been studied extensively and no polynomial time algorithm has been found. (On the other hand, it hasn't been proven that a poly-nomial time algorithm cannot exist.)

A problem is NP-hard if it is at least as hard as an NP-complete problem. An NP-hard problem need not be a recognition problem.

Goals, Difficulties, and Compromises

Computation problems often force compromises—we saw some possible ones earlier. In fact, compromise is a pervasive aspect of AI. Being aware of this will help your understanding and creativity, so develop the habit of asking the following questions:

> What are the goals?
> What are the difficulties?
> What are the compromises?

Try going back to the previous section and picking out the goal(s), problem(s), and compromise(s) involved in my writing of this text.

Exercises

1.4.A. Why are algorithms particularly troublesome in AI?

1.4.B. What are some ways of dealing with the problems to which algorithms in AI often lead?

1.4.C. Roughly speaking, what are NP-complete and NP-hard problems?

1.4.D. Why may it not be too important that a problem is NP-hard?

1.4.1. Prove that the anytime algorithm for finding a solution to $f(x) = 0$ has the following three properties. Assume that there is no roundoff error in the computations.

  (i) There is always such an $x \in [a, b]$.

  (ii) After n iterations, the length of the interval $[a, b]$ is 2^{-n} times its original length.

  (iii) No matter how close to a solution of $f(x) = 0$ we want to be, we can get that close if we allow the algorithm to run long enough.

1.5   Expert Systems

"You really are an automaton—a calculating machine," I cried.
"There is something positively inhuman in you at times."
—Arthur Conan Doyle (Watson to Holmes) (1889)

Expert system research has, by general consensus, not been as
successful as its most vehement proponents still claim and it is
open to us to wonder just why. My view is that it is due to the
divergence between formalised rule and the social nature of being
an expert.
—Philip Leith (1990)

Definition 1.1   Expert System

As a rough working definition, an *expert system* for some special field is
an artificial system that

- exhibits abilities in that field,
- accepts input regarding a specific problem,
- delivers advice, actions, or something similar as its output, not just
  organized data, and
- uses domain-specific knowledge.

The traditional AI definition was more restrictive. It required that the expert system obtain its results by a process akin to abstract reasoning, that
it be able to explain how it reached its conclusions, and that it exhibit abilities at least comparable to those of a human being. The abstract reasoning
requirement probably arose from a combination of the desire for explanations
and an intellectual prejudice concerning how AI should be done. The desire
for explanations was based on the observation that people often insisted on
checking the computer's "reasoning" before accepting its conclusions. Finally,
if the system was not at least as good as a human in its area of expertise, no
one would use it.

The broader definition given here allows for expert systems that are
not based on a symbolic manipulation of data, for example, neural nets.
It also allows for systems in areas where humans exhibit little if any conscious reasoning, for example, in processing visual input. Finally, it allows
for useful systems that are less capable than humans but are still valuable

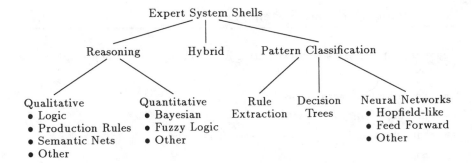

Figure 1.2 Possible engines for expert system shells. "Other" signifies the most blatant omissions. "Rule Extraction" develops input for "Reasoning" systems. "Hybrid" refers to systems that use more than one method, a practice that is becoming more common.

---

in research or applications, for example, natural language processing systems.

Closely related to the notion of an expert system is that of an *expert system shell*, which is important in the development of commercial systems. Roughly speaking,

An expert system shell is to an expert system

as

a compiler or interpreter is to a program.

In this analogy, program statements correspond to domain-specific knowledge, which is often expressed declaratively either in rules or in examples. Just like interpreters and compilers, expert system shells tend to fall into two categories:

- Rule-based knowledge is normally used at run time in a symbolic reasoning process.

- Example-based knowledge is normally used at compile time in a pattern classification process.

Figure 1.2 illustrates some of the possibilities for expert system shells.

Constructing an Expert System

There are various steps to constructing an expert system. One possible break-down is

- selection of a tractable problem,
- selection of an appropriate shell,
- acquisition and preparation of knowledge, and
- testing.

Actually blending and feedback take place among the steps. For example, we might defer the shell choice until we have acquired some knowledge, or, in the process of testing, we might decide that the knowledge base is inade-quate.

Step 1. The Problem: To begin with, you must have a "good" problem, that is, one for which an expert system is likely to be useful. How can you tell if this is the case?

Best performance is usually obtained by choosing a narrow subject; that is, one in which the knowledge base is well delimited. In particular, we should *avoid problems that involve "common sense."* Some AI problems, like natural language understanding, are plagued by the need for common sense.

Performance has generally been disappointing in areas where evolution has apparently led to some "hard wiring" in human brains. The foremost example is expert systems related to vision. Successful expert systems in such areas have generally been limited to *very* specific problems such as identifying handwritten Zip codes.

Step 2. Shell Selection: Once you've clearly stated the problem and gained some understanding of the field, you should choose a method of implementa-tion. You can program a system from scratch, but using an appropriate shell is usually much more efficient.

Step 3. Knowledge Acquisition: It is generally difficult to obtain accurate in-formation from experts—they misstate the rules they use, forget important factors, contradict themselves (and each other), and estimate numerical val-ues poorly. Furthermore, their knowledge is probably poorly organized for use in an expert system shell. Although the art of obtaining information from experts is important in building an expert system, we won't study it.

In some areas, such as vision processing, experts do not use conscious methods. In this case, you must either attempt to discover rules yourself or you must abandon rules and create the expert system from a collection of "typical" examples.

Step 4. Testing: Testing is often referred to as validation. You can expect problems that will send you back to Step 3 repeatedly. It is hard to decide when a system has finally passed the testing phase and is ready for use. In the first place, users typically come up with situations that software designers did not anticipate. Second, we often do not expect 100% success, so it is hard to judge the failures we observe. In this case, systems that provide explanations are quite helpful—the reasons given for a wrong answer can help us decide if we want to attribute it to a design error or to a limitation that we cannot (or do not want to) overcome.

Examples of Expert Systems

To give you a more concrete appreciation, I'll briefly discuss a few of the many expert systems that have been written. My choices were motivated by a desire for breadth not by the commercial success, if any, of the system. Inevitably, the brevity of the descriptions has led to some distortion.

LOGIC THEORIST (1956)
In the early years of the twentieth century considerable effort was devoted to providing a solid foundation for mathematics. The most massive attempt was Whitehead and Russell's *Principia Mathematica* (1910).

Aside. This search for a solid foundation is one of the modern impossibility problems. Its impossibility was proved by Gödel in 1931. He showed that any system based on the usual methods of logical reasoning and arithmetic must contain true theorems that could not be proved within the system ("incompleteness"). In another paper, he showed that the real numbers could not be completely specified in such a system ("independence of the continuum hypothesis"). The classical impossible problems are the trisection of the angle, the doubling of the cube, and the squaring of the circle. The Renaissance impossible problem is the solving of the general fifth-degree equation by radicals. The proof that π is transcendental established the impossibility of squaring the circle. Galois theory was used to establish the impossibility of the other three problems.

Since the framework provided by Whitehead and Russell allows theorems to be proved with no "understanding" of the concepts, it's a reasonable candidate for symbolic manipulation by computer. Newell, Simon, and Shaw took up this task and produced LOGIC THEORIST.

As you've undoubtedly discovered, it's not always clear what steps must be taken to prove a theorem. Because of this, LOGIC THEORIST used an ad hoc trial-and-error method. An essential part of a trial-and-error method is deciding what to try. The program used two approaches:

- Suppose the goal is to prove Z and we have an axiom or theorem that says, "If A is true, then Z is true." We can attempt to prove A.

- Suppose the goal is to prove that "If A is true, then Z is true." We can try to find M such that one of the two statements

  "If A is true, then M is true."   and   "If M is true, then Z is true."

  is either an axiom or a theorem. Then we try to prove the other of the two statements.

This method is related to the reasoning engine used in the Prolog language. Unlike LOGIC THEORIST, Prolog has a theoretical foundation that provides power and clarifies its limits. On the other hand, the construction and arrangement of Prolog statements are more critical to its success.

Few, if any, researchers claim that the formal methods of LOGIC THEORIST and Prolog are used in day-to-day human reasoning. Nevertheless, a large number of researchers believe that extensions of these ideas will prove adequate for much of the reasoning needed in AI.

Mathematical logic and Prolog are discussed in Chapters 3, 4, and 6.

MYCIN (1972)

Beginning in 1972 at Stanford, Shortliffe and others developed MYCIN, which is one of the best known expert systems. In its area of competence, MYCIN was able to diagnose illnesses as well as or better than most physicians. It was also able to explain how it reached its conclusions. Nevertheless, it never received more than token acceptance from the medical community.

MYCIN is a *rule-based system* with *uncertainty*. In real life, many rules are not certain. An example of such a rule is "If you do not study, then you will get a bad grade." However, you might happen to be lucky, so the rule may be valid only 95% of the time. A typical MYCIN rule has the form

If the result of test A is R_A and ... and the result of test Z is R_Z, then there is evidence that the disease organism is D.

Included with the rule is a numerical value in the interval $[-1, +1]$, called a *certainty factor* (CF). This value is intended as a measure of the strength of the rule's conclusion, given that its hypotheses are satisfied. In particular

$$CF = \begin{cases} +1, & \text{given the evidence, } D \text{ is certainly correct;} \\ -1, & \text{given the evidence, } D \text{ is certainly wrong;} \\ 0, & \text{the evidence gives no information about } D. \end{cases}$$

The meaning of intermediate values is not so clear.

The MYCIN reasoning engine proceeds from diagnostic evidence toward causes, eventually producing certainty factors for various diagnoses. In the process, certainty factors are combined using an ad hoc rule. More recently, certainty factors have been given a probabilistic interpretation, and Bayesian nets have provided less ad hoc (but more complex) methods for combining certainty factors.

Although numerical methods like that used in MYCIN provide ways of incorporating uncertainty into reasoning, it is unlikely that human reasoning is based on such processes—people are notoriously poor at assigning numerical values to evidence. On the other hand, the use of numerical methods might lead to AI systems that reason more accurately than humans do.

Bayesian nets and certainty factors are discussed in Chapter 8.

NETtalk (1986)

DECtalk is a complicated rule-based system for converting written English to spoken English. Sejnowski and Rosenberg developed the *neural network* NETtalk to do the same thing. In contrast to DECtalk, NETtalk contains no rules. Instead, it contains about 100 interconnected units (neurons). The network was given paired samples of written and spoken English from which it trained itself, using a process that adjusts the strengths of the connections between the neurons. NETtalk uses the seven most recent text symbols (letters, punctuation, and spaces) to drive a digital speech synthesizer.

Networks have trained themselves for a variety of tasks. In contrast to the more cognitive approaches used in the other examples, networks have no cognitive information built in. Researchers believe that neural networks mimic somewhat the low-level behavior of biological networks of neurons. As a result, they believe that this approach may hold the key to designing AI systems that have some of the capabilities of biological systems.

Neural nets are discussed in Chapters 11 and 13.

DEEP THOUGHT (1990)

Game-playing programs were a favorite research area in the early years of AI. For various reasons, research interests have since moved in other directions, but the area has not been completely abandoned.

Chess is the most actively researched game. Programs are available that will easily beat average players. Thanks to faster processors, special-purpose devices, and improvements in programs, the top silicon-based players are now nearly as good as the top human players.

DEEP THOUGHT, by Hsu, Anantharaman, Browne, Campbell, and Nowatzyk uses special-purpose hardware to search the possibilities for several moves into the future. The quality of each possible position is evaluated and, sometimes, further search is carried out. DEEP THOUGHT's strength lies in the depth to which it can search. In contrast, Nitsche's MEPHISTO searches less and spends more time assessing the positional aspects of the situation. DEEP THOUGHT plays at or near the grandmaster level and MEPHISTO plays at a slightly lower level.

Search plays a major role in AI, but brute-force search has very limited application owing to combinatorial explosion. Some researchers believe that combining search techniques with "heuristic evaluations" and "methods of abstraction" will prove important in some parts of AI.

Search is discussed in Chapter 2 and briefly in Chapter 13.

CHATKB (1992)

Hekmatpour and Elkan developed CHATKB is an expert system to aid users of certain VLSI design tools. The rapid acceptance of this system is in marked contrast to that of others such as MYCIN. The difference may be due to the fact that CHATKB users are already using computers on a regular basis for other high-level activities such as CAD.

When faced with a user problem, CHATKB determines the category to which it belongs. This is done by an iterative questioning process similar to the game of Twenty Questions. Such processes are called *decision trees*. Nonautomated decision trees have been used for many years in natural history field guides for classifying plants and animals.

Each category contains a data base of previously analyzed problems. CHATKB finds the closest matching problem in the data base for the current problem's category. It then presents that problem and its solution to the user. If the user rejects this solution, CHATKB presents the second best match, and so on. Matching in this manner is a form of *case-based reasoning*.

Some researchers believe that this type of dichotomous approach—classify then look for similar cases—is typical of higher level day-to-day human reasoning. Consequently, they expect some such method to play an important role in the design of intelligent systems.

Decision trees are discussed in Chapter 13. Case-based reasoning is mentioned very briefly in Chapter 14.

Exercises

1.5.A. What is an expert system?

1.5.B. What are the steps in building an expert system?

Notes

Crevier [10] has written an informative, lively, nontechnical book on the history of AI based on his own background and on extensive interviews with major researchers. He brings the participants to life and accurately explains important concepts and achievements in layman's terms. You would probably enjoy it.

If my brief treatment in Section 1.1 left you dissatisfied, you may wish to look at other AI texts such as those by Charniak and McDermott [7], Firebaugh [15], Ginsberg [18], Rich and Knight [39], Russell and Norvig [41], and Winston [53].

For a discussion of topics that I've slighted, or for a less mathematical discussion of those I've covered, consult some of the available textbooks and surveys. The texts by Dean, Allen, and Aloimonds [11], Charniak and Mc-Dermott [7], Firebaugh [15], Ginsberg [18], Rich and Knight [39], Russell and Norvig [41], and Winston [53] are all broad-based introductions to AI, but the discussions of neural networks may be limited. Of these, I particularly recommend Ginsberg's and Russell and Norvig's texts. More mathematical, but less broad, are those by Dougherty and Giardina [13], Laurière [27], and Shinghal [45].

Survey and expository articles can sometimes be found in journals and in conference proceedings. Some journals, such as *Artificial Intelligence: An International Journal*, publish special issues containing several such articles. In addition, handbooks and surveys such as [3] and [46] and books such as [38] appear from time to time. For briefer discussions, there is the encyclopedia [44].

Reading original sources in any field is often a good idea, but it can be daunting because the authors usually assume readers are researchers with the necessary background. A solution is provided by annotated collections, for example, Morgan Kaufmann Publishers' *Readings in ...* books, such as [9]. A source in neurocomputing is [2].

I mentioned functional programming and logic programming as two of the paradigms that AI brought to computer science at large. Logic programming as implemented in Prolog will be discussed in Chapters 3 and 4 to help your understanding of logic. Functional programming ideas were implemented by McCarthy in Lisp. The theoretical foundation is provided by the *lambda calculus*, developed primarily by the logicians Church and Kleene. I won't be discussing these topics, but some AI texts have a Lisp-based introduction to the subject, which is good if your focus is understanding Lisp. On the other hand, MacLennan [28] discusses the general methodology of functional programming from both a concrete and an abstract viewpoint without relying on Lisp. For texts on Lisp and Prolog, see the notes at the end of Chapter 3.

The discussion about what constitutes AI continues. Almost any textbook will begin with a discussion of what AI is about and articles appear from time to time in journals and magazines; see, for example, [43]. The nature of AI and other topics of debate appear in the essays edited by Graubard [19]. These were written for a general audience. The essays in Partridge and Wilks [36] and in volume 47 of *Artificial Intelligence* (nos. 1–3, Jan. 1991) are more technical. Material on the connectionist versus symbolic debate can be found in [37] and in [40].

AI frequently employs complex nonlinear feedback systems. Their behavior is often counterintuitive—at least until extensive experimentation leads to the development of a new intuition. For the simplest such systems, mathematical control theory has produced some theoretical results. Forrester has explored complex systems by simulating corporations, cities [16], and the entire world. Many other people have simulated complex systems and attempted to obtain heuristic principles and theoretical results. Progress has been painfully slow. It is quite possible that this area will remain intractable, but giving up now would be extremely premature.

Turing's proof of the impossibility of the halting problem depends on the concept of finite automata. You can find a proof in Bender and Williamson's text [4, pp. 178–179], any book on automata theory, or some texts on discrete mathematics. The ideas relating to Figure 1.1 are discussed more thoroughly by Russell and Wefald [42, Ch. 1]. For further discussion of NP-completeness, see the texts by Papadimitriou and Steiglitz [35] and Wilf [52] or see the article [20]. Garey and Johnson's [17] classic book on the subject lists many NP-complete problems, but the list is now *much* longer.

A discussion of expert systems can be found in many AI texts. There are also books devoted to expert systems. These include Stefik's [48] extensive introduction; the text by Jackson [24], which covers a large part of the material found in a standard AI course; the text by Lucas and van der Gaag [26], which treats fewer topics but in greater depth; and the collection [50], which discusses a variety of applications, devoting a few pages to each. The chapter discussion passed quickly over the difficult problem of knowledge acquisition. Many techniques, problems, and specific examples are discussed in [25].

Biographical Sketches

John McCarthy (1927–)

Born in Boston, he received a bachelor's degree from Caltech and a doctorate from Princeton, both in mathematics. He received the 1971 Turing Award.

McCarthy named the field; he invented the name "artificial intelligence" when writing the proposal for the first AI conference. In 1957, he and Minsky founded the Artificial Intelligence Group at MIT. While there, McCarthy invented timesharing and Lisp. In 1963, McCarthy moved permanently to Stanford, where he founded and directed SAIL (Stanford Artificial Intelligence Laboratory). The MIT and Stanford groups have had a profound influence on AI for many years.

His major concern has been understanding "commonsense" reasoning so that it can be used in AI. As a result, he's focused on achieving a fundamental understanding of knowledge and has advocated a publicly accessible knowledge base for common sense.

Many interesting stories about McCarthy can be found in the biography by Hilts [21, pp. 197–287]. The sketch by Israel [23] provides more technical information.

Marvin Minsky (1927–)

Born in New York City, he studied at Harvard and Princeton, receiving a doctorate in mathematics. As a postdoc at Harvard, he designed the first confocal microscope, a device which is now quite important in optical microscopy. Minsky received the 1969 Turing Award.

In 1957, McCarthy and Minsky founded MIT's Artificial Intelligence Group, where he has continued to inspire excellent thesis research in a variety of areas including

- MACSYMA (the forerunner of Maple and Mathematica),
- analogical reasoning (A is to B as C is to which of the following?),
- language comprehension, and
- robot vision.

Minsky introduced the idea of "frames," which are used in AI and, more recently, in object-oriented programming languages. In 1969, Minsky and Papert dealt a blow to perceptrons—a type of neural network—by proving that they were quite limited [30].

Recalling his student days, Minsky remarked that "The problem of intelligence seemed hopelessly profound. I can't remember considering anything else worth doing." [5, p. 77]. His career has focused on learning what computers are capable of doing on nonarithmetic problems.

Bernstein's [5, pp. 9–128] biographical account contains extensive quotations from Minsky.

Allen Newell (1927–1992)

Born in San Francisco, he received a bachelor's degree in physics at Stanford and began a doctorate in (pure) mathematics at Princeton. Concerned about a lack of breadth, he left Princeton for RAND where he met Herbert Simon. Newell received a doctorate in industrial administration under Simon at Carnegie Tech (now Carnegie-Mellon University), where he became a professor. Newell and Simon received the 1975 Turing Award.

He and Simon began a long and fruitful cooperation in 1955 when, with J. C. Shaw, they designed the list-processing language IPL and used it to write the LOGIC THEORIST, a program that was able to prove results found in Russell and Whitehead's *Principia Mathematica*. As a result, Newell, Shaw, and Simon are often called the parents of AI. The realization that computers are more than just rapid arithmetic calculators—that they can be used to manipulate symbols—was an important observation at the time and is now taken for granted.

In 1956, Newell, Simon, Chomsky, McCarthy, Minsky, and others launched cognitive science at a conference at MIT.

The focus of Newell's career has been the formalization of problem solving and complex task performance by human beings. The scope of this undertaking has grown over the years, moving from attempts to model the performance in specific cognitive areas to a drive to model the entire cognitive process. This has culminated in SOAR, a blend of AI and cognitive psychology. Theories of how humans solve problems provide the motivation for this ongoing programming project whose aim is to simulate significant aspects of human cognition.

More information about Newell and SOAR can be found in [29] and about his interaction with Simon in [47].

Herbert A. Simon (1916–)

Born in Milwaukee, he studied at the University of Chicago, where he received a doctorate in political science. In his autobiography [47, p. 85], he relates that by this time he "had made a modest beginning in mathematics, a basis for subsequent self-instruction." Most of his career has been spent at Carnegie-Mellon University (CMU). Newell and Simon received the 1975 Turing Award. Simon received the 1978 Nobel Prize in Economics.

After being involved in the establishment of the CMU Graduate School of Industrial Administration, he began his shift to AI and cognitive psychology in 1955. He contributed to the establishment of CMU's fruitful interdepartmental computer science program.

As the preceding biographic sketch mentioned, Simon and Newell worked jointed for many years. But, unlike Newell, Simon has continued to focus on more limited problem-solving simulations rather than on the entire cognitive process.

Much of Simon's career has focused on the implications of "bounded rationality" in economics and cognitive science. Traditional economics postulates a very knowing and rational man; he has complete knowledge of all relevant factors, including the details of his own preferences, and is able to carry out any amount of reasoning (and computation). In the early 1950s, Simon broke with this tradition and postulated *bounded rationality*—incomplete knowledge of factors and preferences and limited reasoning abilities.

Simon's autobiography [47] is part of the Alfred P. Sloan Foundation Series—a growing collection of generally excellent autobiographies by prominent contemporary scientists.

Alan M. Turing (1912–1954)

Born in London, he took his degrees in mathematics at Cambridge, where he remained until joining the British war effort in 1938 as their first cryptanalyst. There he played a major part in setting up the system for routinely decoding the German Enigma code. After World War II, Turing spent time at the National Physical Laboratory and at Manchester.

The ACM's Turing Award is named after him, as are Turing machines and the Turing test of Exercise 1.2.1 (p. 10). (The Turing Award lectures through 1985 are collected in [1].) Turing machines illustrate Turing's focus on logic and computation. A Turing machine is an elegantly simple abstract computer. Using these simple computers, he showed that the halting problem for computer programs has no computable solution. This was done in 1935, before the birth of the electronic computer. Using the lambda calculus, Church also showed the existence of well defined noncomputable functions. This nonexistence result has implications for first-order logic, which is the subject of Chapters 3 and 4.

Hodges [22] has published a thorough nontechnical biography of Turing. For more information on Turing machines, consult a text on automata theory.

References

1. *ACM Turing Award Lectures. The First Twenty Years: 1966 to 1985*, ACM Press, New York (1987).

2. J. A. Anderson and E. Rosenfeld (eds.), *Neurocomputing. Foundations of Research*, MIT Press, Cambridge, MA (1988).

3. A. Barr, P. R. Cohen, and E. A. Feigenbaum (eds.), *The Handbook of Artificial Intelligence*, Vols.1–4, Morgan Kaufmann, San Mateo, CA, and Addison-Wesley, Reading, MA (1981, 1982, 1989).

4. E. A. Bender and S. G. Williamson, *Foundations of Applied Combinatorics*, Addison-Wesley, Reading, MA (1991).

5. J. Bernstein, *Science Observed*, Basic Books, New York (1982).

6. K. H. Borgwardt, *The Simplex Method. A Probabilistic Analysis*, Springer-Verlag, Berlin (1987).

7. E. Charniak and D. McDermott, *Introduction to Artificial Intelligence*, Addison-Wesley, Reading, MA (1985).

8. P. M. Churchland, *Matter and Consciousness*, rev. ed., MIT Press, Cambridge, MA (1988).

9. A. Collins and E. E. Smith (eds.), *Readings in Cognitive Science*, Morgan Kaufmann, San Mateo, CA (1988).

10. D. Crevier, *AI: The Tumultuous History of the Search for Artificial Intelligence*, BasicBooks, New York (1993).

11. T. Dean, J. Allen, and J. Aloimonds, *Artificial Intelligence Theory and Practice*, Benjamin/Cummings, Redwood City, CA (1994). Includes discussions of time and space complexity of AI algorithms.

12. D. C. Dennett, *Consciousness Explained*, Little, Brown, Boston (1991).

13. E. R. Dougherty and C. R. Giardina, *Mathematical Methods for Artificial Intelligence and Autonomous Systems*, Prentice Hall, Englewood Cliffs, NJ (1988).

14. M. W. Eysenck and M. T. Keane, *Cognitive Psychology: A Student's Handbook*, Lawrence Erlbaum Associates, Hillsdale, NJ (1990).

15. M. W. Firebaugh, *Artificial Intelligence: A Knowledge-Based Approach*, Boyd and Fraser, Boston (1988).

16. J. W. Forrester, *Urban Dynamics*, MIT Press, Cambridge, MA (1969).

17. M. R. Garey and D. S. Johnson, *Computers and Intractability. A Guide to the Theory of NP-Completeness*, W. H. Freeman, New York (1979).

18. M. Ginsberg, *Essentials of Artificial Intelligence*, Morgan Kaufmann, San Mateo, CA (1993).

19. S. R. Graubard (ed.), *The Artificial Intelligence Debate. False Starts, Real Foundations*, MIT Press, Cambridge, MA (1988). Reprinted from *Daedalus* **117** (1988).

20. J. Hartmanis, Overview of computational complexity theory, *Proceedings of the Symposia in Applied Mathematics* **38** (1989) 1–17.

21. P. J. Hilts, *Scientific Temperaments. Three Lives in Contemporary Science*, Simon and Schuster, New York (1982).

22. A. Hodges, *Alan Turing: The Enigma*, Simon and Schuster, New York (1983).

23. D. J. Israel, A short sketch of the life and career of John McCarthy. In V. Lifschitz (ed.), *Artificial Intelligence and Mathematical Theory of Computation: Papers in Honor of John McCarthy*, Academic Press, Boston (1991).

24. P. Jackson, *Introduction to Expert Systems*, 2d ed., Addison-Wesley, Reading, MA (1990). This is considerably expanded from the first edition.

25. A. L. Kidd (ed.), *Knowledge Acquisition for Expert Systems*, Plenum Press, New York (1987).

26. P. Lucas and L. van der Gaag, *Principles of Expert Systems*, Addison-Wesley, Reading, MA (1991).

27. J.-L. Laurière, *Problem Solving and Artificial Intelligence*, Prentice Hall, Englewood Cliffs, NJ (1990). Translated from the 1987 French edition by J. Howlett.

28. B. J. MacLennan, *Functional Programming. Practice and Theory*, Addison-Wesley, Reading, MA (1990).

29. J. A. Michon and A. Akyürek (eds.), *SOAR: A Cognitive Architecture in Perspective*, Kluwer, Dordrecht (1992)

30. M. L. Minsky and S. Papert, *Perceptrons: An Introduction to Computational Geometry*, MIT Press, Cambridge, MA (1969). It has been reprinted with some additional discussion as *Perceptrons: An Introduction to Computational Geometry, Expanded Edition*, MIT Press, Cambridge, MA (1988).

31. T. C. Moody, *Philosophy and Artificial Intelligence*, Prentice Hall, Englewood Cliffs, NJ (1993).

32. A. Newell, Unified theories of cognition and the role of Soar. In [29], 25–79.

33. A. Newell and H. A. Simon, Computer science as empirical inquiry: Symbols and search, *Communications of the ACM* **19** (1976) 113–126. This is their 1975 ACM Turing Award Lecture.

34. A. Newell, *Unified Theories of Cognition*, Harvard University Press, Cambridge, MA (1990).

35. C. H. Papadimitriou and K. Steiglitz, *Combinatorial Optimization: Algorithms and Complexity*, Prentice Hall, Englewood Cliffs, NJ (1988).

36. D. Partridge and Y. Wilks (eds.), *The Foundations of Artificial Intelligence. A Sourcebook*, Cambridge University Press, Cambridge, Great Britain (1990).

37. S. Pinker and J. Mehler (eds.), *Connections and Symbols*, MIT Press, Cambridge, MA (1988). Reprinted from *Cognition: International Journal of the Cognitive Sciences* **28** (1988).

38. Z. W. Ras and M. Zemankova, *Intelligent Systems. State of the Art and Future Directions*, Ellis Horwood, New York (1990).

39. E. Rich and K. Knight, *Artificial Intelligence*, 2d ed., McGraw-Hill, New York (1991). This is a major revision of Rich's first edition.

40. D. E. Rumelhart, J. L. McClelland, and the PDP Research Group, *Parallel Distributed Processing. Explorations in the Microstructure of Cognition. Volume 1: Foundations*, MIT Press, Cambridge, MA (1986).

41. S. Russell and P. Norvig, *Artificial Intelligence. A Modern Approach*, Prentice Hall, Englewood Cliffs, NJ (1994).

42. S. Russell and E. Wefald, *Do the Right Thing: Studies in Limited Rationality*, MIT Press, Cambridge, MA (1991).

43. R. C. Schank, Where's the AI? *AI Magazine* **12** (1991) 38–49.

44. S. C. Shapiro (editor-in-chief), *Encyclopedia of Artificial Intelligence*, 2d ed., John Wiley and Sons, New York (1992).

45. R. Shinghal, *Formal Concepts in Artificial Intelligence*, Chapman and Hall, London (1992).

46. H. E. Shrobe and the American Association for Artificial Intelligence (eds.), *Exploring Artificial Intelligence: Survey Talks from the National Conferences on Artificial Intelligence*, Morgan Kaufmann, San Mateo, CA (1988)

47. H. A. Simon, *Models of My Life*, Basic Books (1991).

48. M. Stefik, *Introduction to Knowledge Systems*, Morgan Kaufmann, San Mateo, CA (1995).

49. M. A. Stillings, N. H. Feinstein, J. L. Garfield, E. L. Rissland, D. A. Rosenbaum, S. E. Weisler, and L. Baker-Ward, *Cognitive Science: An Introduction*, MIT Press, Cambridge, MA (1987).

50. E. Turban and P. R. Watkins (eds.), *Applied Expert Systems*, North-Holland, Amsterdam (1988).

51. A. M. Turing, Computing machinery and intelligence, *Mind* **59** (1950). Reprinted in [9 pp. 6–19].

52. H. S. Wilf, *Algorithms and Complexity*, Prentice Hall, Englewood Cliffs, NJ (1986).

53. P. H. Winston, *Artificial Intelligence*, 3d ed., Addison-Wesley, Reading, MA (1992). This is the first edition containing material on neural nets.

2

Trees and Search

Introduction

In AI's youth, researchers hoped that much could be achieved by using very general methods—so-called *weak methods*, as opposed to methods that make significant use of particular knowledge in the problem area. General search procedures are the most important weak methods in AI. Although knowledge-intensive approaches are usually favored now, basic search ideas continue to play a major role.

Some Examples

Search is something everyone does. Consider some examples from everyday life.

- Simple Search: Also called *brute-force search* and *British Museum search*, this simple method blindly looks everywhere until it finds a solution. This is the method I might use if I've misplaced my keys somewhere at home, but have no idea where. It's impractical for many AI problems because there are too many places to look. To avoid simple search, I need some sort of additional information about the space being searched.

- Heuristic Search: I can improve on the brute-force key-finding algorithm by using additional information called a *heuristic*. For example, it's unlikely, but possible, that I put my keys in a drawer. Thus, I'll give a higher priority to searching on surfaces than to searching in drawers. Of course, if I don't find them on a surface, I may end up looking in drawers after all. A heuristic just helps me organize my search—it's not a sure thing.

- **Pruning**: This is a method that allows me to eliminate possibilities. As I walk into the den, I see my briefcase in plain sight—just where I left it before going to the university this morning. Thinking for a second, I realize that because I misplaced my keys *after* coming home and because this is the first time since coming home that I've seen my briefcase, my keys can't possibly be in the den. Thus I "prune" the den; that is, I don't search it.

- **Chess and Other Games**: Sometimes it's impossible to conduct a full search. This is typical in games like chess, where the set of all possible ways the game can proceed from its present position to its final end is too large to examine. Players select moves because they "look good." Thus they must have some method of looking at board positions and deciding which ones appear better. In other words, a player has a heuristic for evaluating positions. To write a chess-playing program, we also need a similar heuristic. An evaluation heuristic alone is not enough: It tells us which lines of play to explore further, but it doesn't tell us how far ahead to look or how many lines of play to explore.

- **Doing Homework**: Solving homework problems often involves search. The method a student uses depends on the nature of the homework and on the student. It may range from simple search—look for a formula in the chapter that fits—to sophisticated applications of heuristics and pruning to even more sophisticated planning methods. In fact, the mastery of more sophisticated methods often distinguishes those who understand a course from those who just barely make it through. Researchers have a poor understanding of these methods. Consequently we don't know how to teach them well or how to do a good job of building them into AI programs.

- **Evolution**: Living organisms are quite complicated. Large random changes in an organism are almost certain to be fatal—so much so that such changes would not be able to form the basis for evolution. One key to evolution's ability to work is the fact that nearby states (such as mutations and crossovers) often involve relatively small changes in the structure and function of the organism. Another key is the recombination of genetic material through mating or chromosome interchange. This allows useful changes to be combined.

All these ideas except evolution are important in classical AI search. Evolution has appeared as a tool for AI under the name genetic algorithms. Let's postpone our discussion of genetic algorithms, planning, and those aspects of search that require probability theory. The other aspects of search are discussed in this chapter.

Chapter Overview

We begin by introducing some notions from combinatorics, namely directed graphs and ordered trees. Next, induction and recursion are reviewed. This provides the concepts for converting a search space into a search graph and a search (or decision) tree. The remainder of the chapter is devoted to search trees. Naturally, it begins with the three basic methods of simple search: breadth-first, depth-first, and iterative-deepening.

The blindness of simple search limits its usefulness. In order to guide our search, we need some sort of crystal ball to estimate the quality of a state (vertex in the search tree). The crystal ball is called a heuristic and forms the basis of heuristic search. The efficiency and success of the search depend heavily on the nature of the heuristic. Designing a heuristic for a *specific* problem is often a knowledge-intensive art and thus will not be discussed here.

All other sections of the chapter deal with searching until a goal state is found. In AI, this is often impossible or undesirable. The last section takes up the important topic of partial search. Ideas in this area are based on heuristic and iterative-deepening search.

Games are more fruitfully viewed in terms of a somewhat different search tree structure called an AND/OR tree. Alpha-beta pruning is the classic method for reducing the amount of work in searching AND/OR trees. Since this algorithm is often misunderstood, we'll develop it in stages.

Much of the material on search can be viewed from a recursive viewpoint. In this chapter, recursion will appear in the definition of trees, in search algorithms, and in proofs. Recursion and induction are reviewed in Section 2.2. If you find recursive methods natural, you can skim or even skip this material. If you find recursive methods difficult, don't skim. You'll only find yourself in trouble later because the definitions and methods in logic rely heavily on recursion and because many proofs in later chapters rely on recursion and induction. In fact

> | Recursive methods permeate much of AI. |

Prerequisites: Since this is the first mathematical chapter, no earlier material is needed.

Used in: The terminology on graphs in the first section is referred to in various other chapters. Recursion, which is discussed in Section 2, plays a central role in AI. Thus, you should read at least the first two sections of this chapter. The material on Prolog in Chapters 3 and 4 requires the material on depth-first force search in Section 4. Parts of Chapter 13 use material from this chapter.

2.1 Graphs and Trees

The term *graph* is used to describe two completely different concepts. The concept we do *not* want is the one associated with the graph of a function. Instead, we want the combinatorial concept, which is frequently used in computer science. The graphs of greatest interest to us are directed graphs and ordered trees. You've probably seen trees, at least informally, in other courses. Figures 2.2 (p. 38) and 2.4 (p. 50) contain some examples. In this section, we'll develop a precise, recursive definition. After defining these concepts, we'll use them to discuss recursion and to provide a framework for search.

Unfortunately, there's a lot of terminology to define, so let's use this section to define some important terms. If we define them as they're needed, we'd break up the discussion—and they'd be harder to locate later. You may not remember it all at first, but you can always refer to this section while reading the rest of the chapter.

Remember that, for two sets A and B, the *Cartesian product* $A \times B$ consists of all ordered pairs of the form (a, b), where $a \in A$ and $b \in B$.

Definition 2.1   Directed Graphs (Digraphs)

Let V be a set. A *directed graph*, or *digraph*, is V together with a subset E of $V \times V$. We refer to V as the *vertices* of the digraph and to E as the *edges* of the digraph. We denote the digraph by (V, E). If (v, w) is an edge, we call it an edge from v to w, call v its *tail*, and call w its *head*.

A digraph is represented pictorially as follows. The vertices $v \in V$ are indicated by their names, possibly circled. An edge $(u, v) \in E$ is represented by a line or a curve connecting the representations of u and v, with an arrowhead indicating the direction from u to v. I'll usually abuse terminology and refer to such a representation as if it were the digraph. Since nothing matters for a digraph except V and E, the shapes and positions of the representations of the vertices as well as the shapes and crossings of the curves representing the edges are irrelevant. Figure 2.1 contains pictorial representations of two digraphs.

Definition 2.2   Paths and Cycles

Let $\mathcal{D} = (V, E)$ be a digraph. If $v_1, v_2, \ldots, v_n$ are distinct vertices of $\mathcal{D}$ and $(v_i, v_{i+1}) \in E$ for $1 \leq i < n$, we call $v_1, \ldots, v_n$ a *directed path* in $\mathcal{D}$. If we require only that either (v_i, v_{i+1}) or (v_{i+1}, v_i) be an edge, we have an *undirected path*. If $v_1, \ldots, v_n$ is a directed path and $(v_n, v_1) \in E$, we call $v_1, \ldots, v_n, v_1$ a *directed cycle* in $\mathcal{D}$. (An *undirected cycle* is defined similarly, but we must rule out repeated edges to avoid the triviality of a two-vertex undirected cycle.)

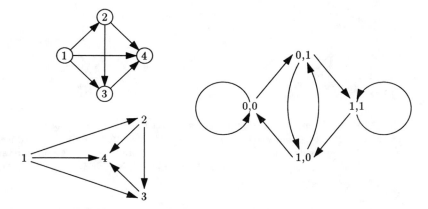

Figure 2.1   Pictorial representations of two directed acyclic graphs (DAGs). The left-hand side shows two representations of a DAG that has $V = \{1, 2, 3, 4\}$ and $(u, v) \in E$ whenever $u < v$. In the right-hand digraph, imagine a sequence of $d_1, d_2, \ldots$ of zeros and ones. Each vertex u corresponds to two adjacent digits, say d_{i-1}, d_i. By choosing the correct edge (u, v) leading out of u, we can move to d_i, d_{i+1}. For example, given the sequence 0,1,1,1,0,0, we would start at 0,1, go to 1,1, then to 1,1 again, then to 1,0, and finally to 0,0.

It's quite simple to define the notion of a rooted tree in terms of a directed graph: It's a directed graph (a) having no undirected cycles and (b) having a special vertex r (called the *root*) such that there is a directed path from r to every other vertex of the graph. One often needs an ordering for the edges coming out of each vertex. Such a tree is called an ordered tree.

Here's a geometric visualization of ordered trees: The topsy-turvy convention used by mathematicians and computer scientists usually places the root at the top of the picture. For any vertex v, we draw the edges $(v, x_1), (v, x_2), \ldots$ in a downward direction from v and in a left-to-right order according to their ordering. This is the reason why rooted trees with the edges from each vertex ordered are also called rooted *plane* trees—drawing a rooted tree in the plane automatically gives a left-to-right ordering. Some ordered trees are shown in Figure 2.2(a).

In many cases, it's helpful to have a recursive definition of ordered trees. Definition 2.3 gives us one. Although it's long, it's not really complicated if you keep the geometric picture in mind. Part (b) corresponds to the following picture. Draw the trees $T_1, \ldots, T_k$ so that their roots $r_1, \ldots, r_k$ lie in a line from left to right. Place a new vertex r above the r_i's and draw lines connecting r to each of the r_i's. See Figure 2.2(b).

(a)   (b)

Figure 2.2 Pictorial representations of (a) five ordered trees and (b) the recursive construction step. The topmost vertex of each tree is its root. All are distinct ordered trees, but the two five-vertex trees represent the same rooted tree. To avoid clutter, the direction of the edges, which is always downwards, is not indicated. The new edges in the recursive construction are dotted.

Definition 2.3   Ordered Tree

An *ordered tree* consists of four things: a set V of vertices, a root $r \in V$, a set E of directed edges, and an ordering of the edges out of each vertex. It is defined recursively as follows:

(a) If $V = \{r\}$ and $E = \emptyset$, calling r the root gives an ordered tree.

(b) Suppose that, for $1 \le i \le k$, T_i is an ordered tree with root $r_i \in V_i$ and edges E_i which have been partially ordered. Suppose that the V_i's are disjoint; that is, $V_i \cap V_j = \emptyset$ whenever $i \ne j$. Suppose further that $r \notin V_1 \cup \cdots \cup V_k$. The following defines an ordered tree.

$$V = \{r\} \cup \left(\bigcup_{i=1}^{k} V_i \right),$$

$$\text{root} = r, \tag{2.1}$$

$$E = \bigcup_{i=1}^{k} \left(\{(r, r_i)\} \cup E_i \right)$$

and the orderings of the edges consist of the orderings of the edges of the T_i's together with

$$(r, r_1) < \cdots < (r, r_k).$$

If v is a vertex in an ordered tree and there are no edges of the form (v, x), we call v a *leaf* or *terminal vertex*.   If we ignore the ordering of the edges, we obtain simply a *rooted tree*. The r_i's are called the *sons* or *children* of r.

Aside. There's one problem with this definition for ordered trees: It doesn't allow us to construct infinite trees. I want to allow trees that may have infinitely long paths; but rather than give a definition, I'll assume that the concept is clear enough without one. Of course, I don't really *want* infinite trees; but they do occur and can cause problems for programs.

Decision trees are a particular manifestation of ordered trees. They provide a useful way of thinking about search trees and are important in designing expert systems that automatically classify data. The study of automatic classification starts in Chapter 10.

Definition 2.4   Decision Tree

An ordered tree is viewed as a *decision tree* as follows. If v is a nonleaf vertex and $(v, x_1), \ldots, (v, x_n)$ are the edges from v, we call $x_1, \ldots, x_n$ the *possible decisions* at v. We refer to x_i as the *i*th *choice* at v and call x_{i+1} the *next choice* after x_i.

Abstractly, decisions at a vertex need not be ordered; however, in practice, they are ordered because a program studying a decision tree must go through the decisions at a vertex in some order.

Aside. The meaning of "tree" without a modifying adjective varies from one discipline to another and even within a discipline. In computer science, "tree" often means "rooted tree" and sometimes means "ordered tree." In mathematics, "tree" often means "free tree," which is essentially a tree without a root. This confusion won't affect us because the only trees we'll need will be ordered trees and decision trees.

Exercises

At the beginning of many exercise sections, you'll find exercises that are lettered rather than numbered. Their purpose is to make sure you understand the basic ideas in the text. Their solutions can be found by rereading the preceding material.

2.1.A. Define and give an example of a digraph. An ordered tree. A decision tree.

2.1.1. (*Answer follows*) Every workday I'm faced with some decisions. First, I must decide if I think the weather may be bad. If so, I must decide whether to take an umbrella or other rain gear. If not, I must decide whether to cycle in or drive in. (In bad weather, I'll certainly drive.) If I decide to drive, I must decide whether to leave home early or late. Draw a decision tree and label the vertices according to the states they represent.

Answers

2.1.1. To avoid your inadvertently seeing the answer, I'll list the edges instead of drawing the tree. Let b and g refer to weather, u and o to rain gear, c and d to transportation and e and l to timing. The edges are (r, b), (r, g), (b, u), (b, o), (u, e), (u, l), (o, e'), (o, l'), (g, c), (g, d), (d, e''), and (d, l''), where I've used primes to distinguish different vertices with the same labels.

2.2   A Review of Recursion and Induction

> *When [recursion] is first presented, students often react as if they had just been exposed to some conjurer's trick rather than a new programming methodology. Given that reaction, many students never learn to apply recursive techniques and proceed to more advanced courses unable to write programs which depend on the use of recursive strategies.*
>
> —Eric S. Roberts (1986)

The notion of recursion appears in definitions, algorithms and proofs. A *recursive definition* normally consists of two parts:

- a description of the simplest items and

- a description of how to *build up* more items from other items.

This approach appears in the recursive definition of ordered trees (Definition 2.3). A *recursive algorithm* usually has a similar pattern:

- an algorithm for treating the simplest cases and

- an algorithm for treating the present case based on the results for simpler cases.

Example 2.1   Two Recursive Ordered-Tree Algorithms

Suppose we want to list the vertices in an ordered tree. Given an ordered tree T, let $r(T)$ be the root of T, let $k = k(T)$ be the number of ordered trees that were joined to r in the recursive step to form T, and let $T_1, \ldots, T_k$ be those ordered trees. Here's an algorithm:

```
List(T)
  Output r(T).
  If r(T) is not a leaf, then
      For i = 1,...,k, List(Tᵢ).
  End if.
End.
```

Now suppose that we want to "parse" an arithmetic expression that is built using the binary operations of $+$, $-$, $\times$, and $/$. Given an expression, we want the function `last_op` that returns the last operation needed to perform the calculation. For example, the last operation in $(a+1) \times (a-b)$ is the multiplication. We also want functions `left` and `right` to return the expression to the left and right of an operation.

```
Parse(exp)
  If exp contains no operation, then
      Return an ordered tree with one vertex, labeled exp.
  Else
      op = last_op(exp).
      T₁ = Parse(left(exp,op)).
      T₂ = Parse(right(exp,op)).
      Return the ordered tree built from T₁ and T₂
          with root labeled op.
  End else.
End.  ■
```

Inductive, or recursive, proofs follow the same format as definitions and algorithms.

Example 2.2   A Recursive Ordered-Tree Proof

Let's prove that every ordered tree contains one more vertex than edge. Let $v(T)$ and $e(T)$ be the number of vertices and edges, respectively, of the ordered tree T. It's now simply a matter of using Definition 2.3. In part (a), $v(T) = 1$ and $e(T) = 0$, so the result is true. In part (b), it's simply a matter of applying the counting functions v and e to (2.1):

$$v(T) = 1 + \sum_{i=1}^{k} v(T_i) \quad \text{and} \quad e(T) = \sum_{i=1}^{k} \left(1 + e(T_i)\right).$$

Since the construction is recursive, we may assume that $v(T_i) = 1 + e(T_i)$, and so it follows from the equations above that $v(T) = 1 + e(T)$. This completes the proof. ■

It may almost seem that we got something for nothing: Why can we assume that $v(T_i) = 1 + e(T_i)$? Don't we need to find some n to induct on? The answer is yes and no. Let's look at proof by induction and recursive definitions more closely.

In the simplest form of proof by induction, we have some statement that depends on a positive integer n, call it $\mathcal{A}(n)$. Proof by induction then consists of two parts: First, show that $\mathcal{A}(1)$ is true. Second for $n > 1$, show that, whenever $\mathcal{A}(n-1)$ is true, so is $\mathcal{A}(n)$. In a slight generalization of this, the second step is replaced by another step: For $n > 1$, show that, whenever $\mathcal{A}(m)$ is true for $m < n$, so is $\mathcal{A}(n)$.

You should've seen all this in previous mathematics courses. However, you may not have seen a proof that induction is a valid method of proof. Here's one possible argument. Suppose that $\mathcal{A}(n)$ is false for some value of n. Then there is a smallest value of n for which it is false, say $n = n_0$. We cannot have $n_0 = 1$ because the induction proof showed that $\mathcal{A}(1)$ was true. On the other hand, we cannot have $n_0 > 1$ because (i) $\mathcal{A}(m)$ is true for $m < n_0$ by the definition of n_0 and (ii) the induction proof showed that then $\mathcal{A}(n_0)$ must be true.

The key point in this argument is that, if $\mathcal{A}(n)$ is not always true, then there must be an n_0 where it is false, but for every case smaller than n_0 it is true. Now, don't limit your thinking of "smaller than" to the integers. In particular, look at the recursive definition of ordered trees. We can think of the trees $T_1, \ldots, T_k$ that are used to construct T as all being smaller than T. The smallest case (corresponding to $n = 1$) is the ordered tree consisting of a single vertex. In this sense of smaller, we can have two trees, neither of which is smaller than the other, for example, ⋀ and ⋀. This does not cause any difficulty for the proof that induction works. We can even have several "smallest" cases—although we had only one here.

Admittedly, the previous discussion is rather sketchy. The important point is that a recursive definition (or algorithm) *automatically* provides the framework for proofs by induction. This is important, so here it is again.

> **Principle:** Suppose a concept is defined recursively and suppose that you want to prove something about the concept. Almost certainly, an inductive proof will be required. The simplest inductive proof will probably be based on upward recursion paralleling the definition.

(2.2)

Following this principle will lead to clearer inductive proofs and will help you find them more rapidly. Of course, there are situations in which no recursive algorithm or definition is at hand. In that case, the principle is useless in creating an inductive proof.

Here's a more complicated proof for ordered trees to illustrate proof by induction again. The result is almost intuitively obvious, but it's not clear how to prove it.

Example 2.3   Another Recursive Ordered-Tree Proof

Claim: In an ordered tree, there is exactly one undirected path between any two distinct vertices.

To prove this, note that it is trivially true for the single-vertex case—there aren't two distinct vertices. However, it will be convenient to think of a single vertex r as a path from r to itself containing no edges. Then the condition that the vertices be distinct can be removed.

Now let u and v be the vertices and suppose that T is constructed from r and the ordered trees $T_1, \ldots, T_k$. How can we possibly use induction here? We must somehow reduce the study of paths to the study of paths in the individual T_i's. The key to doing this is the observation that the only edge between T_i and the rest of the tree is (r, r_i). Another useful remark is that no path can contain r twice or r_i twice.

To prove the inductive step, there are three separate types of paths to consider.

- $u, v \in V_i$: By the previous paragraph, any path between u and v not lying wholly in T_i would contain r_i twice, a contradiction. Therefore, the only possible paths are wholly in T_i and, by induction, there is exactly one such path.

- $u = r$, $v \in V_i$: Any such path cannot enter another tree T_j because it must leave it, thereby using the edge (r, r_j) twice. Thus any such path consists of (r, r_i) and a path between r_i and v wholly in T_i. Again, there is exactly one such by induction.

- $u \in V_i$, $v \in V_j$, $i \neq j$: Any such path must consist of a path wholly in T_i from u to r_i, the edge (r, r_i), the edge (r, r_j), and a path wholly in T_j from r_j to v. Again, uniqueness follows by induction.

This completes the proof. ■

The principle (2.2) can be adapted to help develop recursive algorithms: Notice how the recursive form of the algorithms in Example 2.1 follows that of the definition for ordered trees.

Exercises

I hear and I forget. I see and I remember. I do and I understand.

—Chinese Proverb

2.2.A. What is the structure of a recursive definition? A recursive algorithm?

2.2.B. What is the usual relationship between recursive concepts and proofs that involve them?

2.2.1. A binary rooted (or ordered) tree is a rooted (or ordered) tree in which, for every $v \in V$, the number of edges with tail v is either 0 or 2.

    (a) Prove that all binary ordered trees are obtained by changing the definition of ordered trees so that $k = 2$.

    (b) Prove that the number of vertices in a binary ordered tree is one less than twice the number of leaves.

2.2.2. A decision tree in which all decisions are yes/no is a binary rooted tree. (See previous exercise.) Let $d(v)$ be the number of decisions needed to reach the leaf v. Prove that the sum of $2^{-d(v)}$ over all leaves v equals 1.

2.2.3. A directed graph is called *acyclic* if it has no directed cycles. (In this exercise you're asked to prove a property of such graphs which is useful in Chapter 8.) Let V be a finite set, let (V, E) be an acyclic directed graph, and let $n = |V|$.

    (a) Prove that there is some vertex v_n such that there is no edge of the form (v_n, w); that is, for some $v_n \in V$, we have $(v_n, w) \notin E$ for all $w \in V$.

    (b) Prove, by induction on n using part (a), that the vertices V can be ordered $v_1, \ldots, v_n$ so that there is no edge (v_i, v_j) with $i > j$.

2.3 Problem Spaces and Search Trees

> *Rational search within a problem space is not possible until the space itself has been created, and is useful only to the extent that the formal structure corresponds effectively to the situation. It should be no surprise, then, that the area in which artificial intelligence has had the greatest difficulty is in the programming of common sense.*
>
> —Terry Winograd and Fernando Flores (1986)

The concept of a problem space provides the starting point for defining a search tree. A problem space is where the search is carried out by moving from one state to another using operators in the problem space. We associate a directed graph with the search space. Then we convert the graph into a decision tree—the search tree. This tree provides the framework for our investigations in this chapter.

Definition 2.5   Problem Spaces and Goals

A *problem space* consists of a set S of *states* and a set $\mathcal{F}$ of *operators*. An operator is a function whose domain is a subset of S and whose range is S. If $s \in S$, $f \in \mathcal{F}$, and $f(s)$ is defined, then $f(s)$ is a state that can be reached directly from s. *Goals* are a set $\mathcal{G} \subset S$. A *search procedure* is a method of looking for one or more goals by moving around in the problem space using $\mathcal{F}$.

Let $\mathcal{F}(s)$ be the set of all $t \in S$ such that $f(s) = t$ for some $f \in \mathcal{F}$. In other words, $\mathcal{F}(s)$ are the vertices that can be reached directly from s. By *expanding* s, we mean generating $\mathcal{F}(s)$ somehow.

Don't be put off by the term *operator*; it just tells us one way to get from some states to other states.

$$\boxed{\text{From now on, we assume that } \mathcal{F}(s) \text{ is finite for all } s.}$$

Definition 2.6   Search Graph

A directed graph $\mathcal{D} = (S, E)$ can be associated with a problem space. The vertices are the states S. We have $(s,t) \in E$ if and only if $t \in \mathcal{F}(s)$ (that is, $f(s) = t$ for some $f \in \mathcal{F}$). This graph is called a *search graph*. In search, we are given a vertex $r \in S$. Search techniques start at r and traverse the edges of $\mathcal{D}$ in an attempt to reach a vertex that is a goal state.

Imagine that we are standing at a vertex on a picture of the search graph. To carry out a search, we must select one of the edges leading out from our present vertex. In other words, we must make a *decision*. This leads to the idea of associating a decision tree with the problem space.

Definition 2.7   Search Tree

A *search tree* is associated with a problem space as follows. Let $(S, \mathcal{F})$ be a problem space with goals $\mathcal{G} \subseteq S$ and a starting state $r \in S$. Order the elements of each $\mathcal{F}(s)$ somehow. The search tree will be a decision tree whose vertices will be labeled using the elements in S. Vertices can have the same labels; that is, an element in S might be used to label many vertices of the ordered tree. The root is labeled with r. Suppose that we have constructed a vertex v in the search tree with label s and that s is not a goal. Order the elements $\mathcal{F}(s)$ and let $k = |\mathcal{F}(s)|$. Construct k new vertices x_i and k edges (v, x_i). Label x_i with the ith element of $\mathcal{F}(s)$.

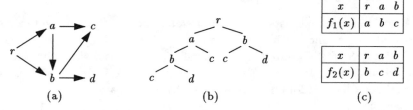

Figure 2.3 (a) The search graph and (b) a search tree corresponding to the problem space $S = \{r, a, b, c, d\}$. The operators in $\mathcal{F}$ might have been defined in many ways. One possibility is $\mathcal{F} = \{f_1, f_2\}$ as given in the tables shown in (c). The decision tree has a branching factor equal to 2.

We could also have derived the search tree from the search graph. (You should see how to do this.) Figure 2.3 shows a simple search graph and a corresponding search tree.

Because a vertex label may appear many times in a search tree, the tree may be much larger than the search graph. It may even be infinite. For example, if the search graph contains the edges (s, t) and (t, s), then the sequence of decisions s to t to s to t to s to ... produces an infinite path. Yes, it is a path: Although s and t appear many times in the path, each appearance corresponds to a new vertex in the search tree. The names s and t are simply labels associated with vertices—they are not actually the vertices of the search tree. Nevertheless, people often abuse terminology and refer to the vertex s.

Since a search tree associated with a search graph can be much bigger than the graph, why should we use a decision tree instead of a graph? Paradoxically, using a decision tree usually requires much less storage.

> The entire search tree or search graph is not stored. Instead, we generate information as needed and store only what may be needed for future use.

In using a decision tree, we might retain only the path from the starting state r to the current vertex. In a search graph, we need to remember which vertices the search has visited. If we did not, we would allow vertices to be revisited and this essentially amounts to converting (part of) the search graph to a decision tree.

Definition 2.8   Depth and Branching Factor

Let a decision tree be given. The number of decisions needed to reach a particular vertex in the decision tree is its *depth*. In particular, the depth of the root r is 0. Alternatively, the depth of a nonroot vertex is the number of edges in the path from the root to the vertex.

If each nongoal vertex of the decision tree has b possible decisions, we call b the *branching factor* of the tree.

The branching factor plays a crucial role in analyzing search algorithms. It's possible to formulate a notion of (average) branching factor for more general decision trees, but it's tricky to define and use. Although results that refer to a branching factor will be rather useless as exact statements—such branching regularity is seldom seen—they are suggestive of what happens in general.

Exercises

2.3.A. What is a search tree? Explain the idea informally without invoking the concept of a problem space or a search graph.

2.3.B. Define depth, branching factor, and expanding a vertex.

2.3.1. Suppose that a search graph is finite. Find a necessary and sufficient condition on the search graph for the corresponding search tree to be finite, too.

2.3.2. Suppose we have a search tree with branching factor b and suppose that every goal has depth exceeding n.

    (a) Show that there are at most b^k vertices of depth k and that this is exactly the number when $k \leq n$.

    (b) Show that there are at most $(b^{k+1} - 1)/(b - 1)$ vertices of depth at most k and that this is exactly the number when $k \leq n$.

2.3.3. The game of fox and hounds is played on a checkerboard. There are four hounds, which are placed along the bottom row on the black squares. There is one fox, which is placed on any black square in the top row. Placing the fox is the first move of the player called FOX. Thereafter, players alternate moves. FOX must move the fox one square diagonally in any direction. The player called HOUNDS must move whatever hound he desires one square diagonally upward. If the fox reaches one of the squares along the bottom row, FOX wins; otherwise HOUNDS wins. A state is a board position together with an indication of who is to move. An operator produces a single move.

    This game has too many states to conveniently list by hand. Therefore, we will look at the case of only two hounds on a 4 × 4 board.

    (a) Draw that portion of the search graph up through the time when FOX has responded to HOUNDS's first move. (This is three moves: FOX places piece, HOUNDS moves, and FOX moves.)

    (b) What is the portion of the decision tree associated with the portion of the search graph in the preceding part? How many vertices are there at depths 0, 1, 2, and 3?

    (c) If the previous portion is extended by one more move, how many vertices will be in the search graph? the search tree?

2.3.4. In the previous exercise, order choices of the form $F \rightarrow i$ according to i and choices of the form $Hi \rightarrow j$ according to i and then according to j; that is, as the two-digit numbers ij are ordered. Order a vertex based on the move that leads to it. Thus, the ordering of $\mathcal{F}(s)$ if s has depth 1 is based on the move ordering $H7 \rightarrow 5$, $H8 \rightarrow 5$, $H8 \rightarrow 6$. Give the ordering of $\mathcal{F}(s)$ for all vertices of depth 2 in the tree. Do the same for all vertices of depth 3.

2.4   Three Simple Search Methods

Seek not out the things that are too hard for thee,
neither search the things that are above thy strength.

—The Apocrypha

We want to examine the search tree until a goal is found. One approach is to examine all vertices at depth 1 by expanding the root, then all at depth 2 by expanding those at depth 1, and so on until a goal is found. This is *breadth-first search*. A big drawback of this method is the amount of storage required. *Depth-first search* avoids excessive storage by going to the opposite extreme. It follows paths downward in the search tree until it finds a goal or until it must go back because a dead end is reached. A big drawback of depth-first search is that it may go much deeper than needed, perhaps even running forever down a path that never ends. *Iterative-deepening search* is a hybrid of the two methods. It saves storage by doing a depth-first search, but it goes to a limited depth D_1. If it fails, the depth limit is increased to $D_2 > D_1$ and the search is repeated. If it still fails, the search is repeated with a depth limit of $D_3 > D_2$ and so on until a goal is found. It's similar to a breadth-first search because it first examines all vertices at depth at most D_1, then all at depth at most D_2, and so on until a goal is found.

Breadth-First Search

Suppose we want to find the goal of least depth in the search tree—in other words, that goal which can be reached by the fewest decisions starting at the root. We can do this in a fairly simple manner.

Algorithm 2.1   Breadth-First Search

If r is a goal state, publish r and stop. Otherwise

1. Create a list Λ containing the entry r.

2. Remove the first entry $s \in \Lambda$ from Λ. (If there is no entry to remove, the search has failed.)

3. For all $t \in \mathcal{F}(s)$, if t is a goal state, publish t and stop. Otherwise, add t to the end of the list.

4. Go to 2.

When the algorithm is executed, it first replaces the root by all vertices of depth 1. Next, the vertices of depth 1 are replaced one at a time by the vertices of depth 2. The algorithm proceeds in this manner through the decision tree. It expands all vertices at a given depth before moving deeper. Pictorially, it moves across the breadth of the tree whenever possible. The need to store all of Λ is a serious limitation of breadth-first search. (See Exercise 2.4.3 (p. 51).)

Of course, in using the algorithm, we would probably put more than just the vertex on the list. For example, we'd probably want to know the sequence of decisions that lead from the root to the vertex.

Aside. For those readers familiar with the terminology, our list is called a *FIFO* (first-in/first-out) list or a *queue*.

There's an inefficiency in Algorithm 2.1. Suppose that state t was placed on the list at some time. Since states can appear more than once in the search *tree*, we may be instructed to place t on the list at some later time. If we do so, we'll cause duplicate work. By never placing t on the list twice, we can avoid such duplication. In effect, this converts the algorithm from one that traverses the search tree to one that traverses the search graph. Here's the modified algorithm

Algorithm 2.2   More Efficient Breadth-First Search

If r is a goal state, publish r and stop. Otherwise

1. Create a list Λ containing the entry r. Mark all vertices except r as unvisited.

2. Remove the first entry $s \in \Lambda$ from Λ. (If there is no entry to remove, the search has failed.)

3. For all $t \in \mathcal{F}(s)$, if t is a goal state, publish t and stop. Otherwise, if t is unvisited, add t to the end of the list and mark it as visited.

4. Go to 2.

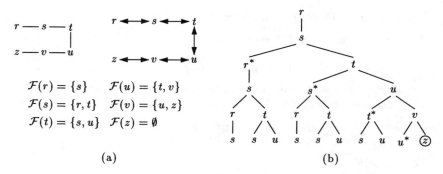

(a)   (b)

Figure 2.4 A trivial maze: (a) Its search graph and (b) part of its search tree. Asterisks indicate vertices not added to Λ by Algorithm 2.2.

Keeping a list of *all* vertices to mark as visited/unvisited may be impractical. Instead, one could keep track of them by dividing Λ into two parts—removed/unremoved—and checking Λ before adding t.

Example 2.4   A Trivial Maze

Figure 2.4 shows a trivial maze. It has six junctions (vertices) labeled r, s, t, u, v, and z. We start at r and attempt to reach the goal z. Although this example is absurdly simple, it allows us to see several of the ideas in this chapter in action. The figure also includes the search graph. Those vertices in the search tree having depth at most 5 are also shown. The goal closest to the root has depth 5.

Simple breadth-first search (Algorithm 2.1) generates all thirteen vertices of depth less than 5. How many of the ten vertices at depth 5 must be generated depends on the order in which vertices at depth 4 are expanded. If it's done alphabetically, the total number generated is 23.

The more efficient Algorithm 2.2 generates nine or ten vertices: Those vertices marked with an asterisk have been encountered previously and so are not be added to Λ. Whether nine or ten are generated depends on whether z or u is generated first in $\mathcal{F}(v)$. ■

Exercises

2.4.A. Give an algorithm for breadth-first search.

2.4.1. In this exercise you prove that Algorithm 2.1 and Algorithm 2.2 are correct; that is, prove that each always finds the way to reach the goal which requires the fewest decisions. (It's easy to convince yourself that the algorithm works, but it's a bit tricky to *prove* that it works.)

(a) Show that if Algorithm 2.2 is correct then so is Algorithm 2.1, and that they find the same path to a goal.

(b) Change Algorithm 2.2 so that it doesn't stop at a goal state. Show that every vertex eventually appears on the Λ of the modified algorithm.

(c) For the modified algorithm, prove that Λ contains some initial list of vertices at some depth d and a (possibly empty) final list of vertices at depth $d + 1$.

(d) For the modified algorithm, conclude that depth is nondecreasing as vertices are removed from Λ.

(e) Prove that the original algorithm works.

2.4.2. In this exercise, you are asked to consider the amount of storage required by breadth-first algorithms when the decision tree has branching factor b.

(a) Assuming no dead ends, show that the length of the list Λ is always increasing.

(b) Assuming no dead ends, show that when all the vertices of depth $k - 1$ have been expanded, the list Λ in Algorithm 2.2 contains b^k entries.

(c) Assuming no dead ends or repeated vertices in the search tree, show that when all vertices of depth $k - 1$ have been expanded, Algorithm 2.2 will need to retain information concerning

$$1 + b + b^2 + \cdots + b^k = \frac{b^{k+1} - 1}{b - 1} \text{ vertices.}$$

2.4.3. *Amdahl's Law* states that a computer that can do about N instructions per second has roughly N bytes of memory. This law holds for many computers from original personal computers to supercomputers. Suppose we are given an Amdahl Law computer and that expanding a vertex does not require much computation. What can you say about the running time of Algorithm 2.1, given that it reaches a goal without running out of memory?

You should at least read and think about the following exercise since it is referred to in Section 2.5.

2.4.4. Sometimes the number of decisions is not the appropriate measure to use in looking for a goal. For example, suppose we have a map of highways with mileages between highway intersections. The goal is to find the shortest route between intersections A and B.

(a) Explain how a search graph can be constructed where the vertices are intersections and each edge has a "cost" equal to the number of miles.

(b) Show how this leads to a search tree with (a) the root labeled A, (b) all goals labeled B, (c) a cost for each edge equal to mileage, and (d) the aim of finding that path from the root to a B for which the sum of the edge weights is a minimum. We can associate with each vertex v a cost $C(v)$ that equals the sum of the edge weights on the path from the root to v. (The path is unique since we are in a tree.)

(c) We can generalize the previous part to a search tree in which each vertex v has an associated cost $C(v)$ and costs increase as we move downward. Modify the breadth-first algorithm to produce a "best-first" algorithm that finds the least-cost goal.
Hint. Always remove the vertex with the least cost from Λ and do not check whether a vertex is a goal until you remove it from Λ.

(d) Prove that the algorithm you have given does in fact find the least cost goal.

The remaining exercises deal with a variant of breadth-first search called *meet-in-the-middle* or *bidirectional search*. They may be omitted. Suppose that we have both the starting state r and the desired goal state z. The problem is to find a path in the search graph from r to z. Since z is known, we could branch backward from z instead of forward from r—or we could do both and "meet in the middle." To begin with, assume that there is a branching factor. Actually, there are two, a forward branching factor b and a backward branching factor β.

2.4.5. Show that if we start with either r or z and branch, it is best to begin with r if $b < \beta$ and with z if $b > \beta$.

2.4.6. Suppose that $b = \beta$ and the depth of z in the search tree is $d = 2\delta$ where δ is an integer. We can branch forward from r to depth δ and backward from z the same distance. Suppose $\mathcal{V}(r)$ and $\mathcal{V}(z)$ are the sets of vertices reached by the two branchings—each vertex with information on how it was reached.

(a) Explain how a path from r to z can be found by looking at $\mathcal{V}(r) \cap \mathcal{V}(z)$.

(b) Show that the number of vertices generated is roughly twice the square root of the number generated if we simply do a breadth-first search starting at r.

2.4.7. Design a bidirectional algorithm that does not require that the depth of z be known. As in the previous exercise, you may assume that $b = \beta$.

*2.4.8. In this exercise, we look at how the previous results can be modified for the case in which $b \neq \beta$. Actually, the final algorithm does not even assume a branching factor.

(a) Suppose that the depth of z is d and that we branch forward from r to a depth x and backward from z to that depth. Show that the number of vertices generated is

$$\frac{b^{x+1} - 1}{b - 1} + \frac{\beta^{d-x+1} - 1}{\beta - 1}.$$

(b) Show that the number of vertices generated is near its minimum when x is chosen so that $b^x \approx \beta^{d-x}$.

(c) Using the previous result as a hint, design an algorithm for bidirectional search given r and z, but no depth or branching factor.

Depth-First Search

Depth-first search avoids the storage bottleneck that plagues breadth-first search. Unfortunately, simple depth-first search can fail to terminate or to find a "good" solution. To overcome these defects, we look at an important modification called iterative-deepening. As its name suggests, *depth-first search* is a search procedure which moves downward in the search tree whenever possible. To implement the search, we must keep track of the path from the root to the current vertex. We need not retain any more vertices than that.

Algorithm 2.3   Depth-First Search

Let a potential search tree be given; that is, we know how to expand any given vertex, but the expansion may not have been carried out ahead of time. The following algorithm starts at the root r and attempts to produce a goal. To use the algorithm, execute `Depth_First(r)`.

```
Depth_First(v)
    If v is a goal, publish v and stop all execution.
    Expand v.
    Set i = 1.
        While there is an ith decision x_i at v
        Call Depth_First(x_i).
        Set i = i + 1.
    End While.
    Return.
End.
```

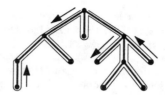

Figure 2.5 An arbitrary search tree with edges shown in heavy lines. The route of an ant doing a depth-first traversal is shown by a thin line. Arrows indicate the direction of the ant's movement.

Example 2.5   Applying Depth-First Search

Let's see how the algorithm works on Figure 2.3 (p. 46). Assume that the ordering of $\mathcal{F}(s)$ is alphabetic and that d is the goal. We begin by calling **Depth_First**(r). Since r is not a goal and the first decision at r is a, the algorithm calls **Depth_First**(a). This calls **Depth_First**(b), which then calls **Depth_First**(c). Now, c is not a goal, but the **while** loop terminates immediately because there are no decisions available. Control is then returned to the calling procedure **Depth_First**(b), which now turns to the second decision at b and calls **Depth_First**(d). Recognizing d as a goal, the procedure publishes it and the program dies in a blaze of glory.

If there were no goals, the procedure would continue, returning control to **Depth_First**(b), which would then return control to **Depth_First**(a), which would call **Depth_First**(c), and so forth. We can visualize the process whereby control is passed as follows. Imagine an ant standing just to the left of r. It starts walking toward a and keeps going, always keeping an edge on its left side without stopping until it either reaches a goal or returns to its starting position. This process is shown for an arbitrary tree in Figure 2.5.

Depth-first search is not always so well behaved. Suppose it's applied to the maze in Figure 2.4 (p. 50) and that elements in $\mathcal{F}$ are examined in alphabetic order. In this case, depth-first search will oscillate forever between r and s. ∎

The following theorem shows that depth-first search always works on a finite tree.

Theorem 2.1   Depth-First Search Sometimes Works

Suppose that the search tree is finite. If a goal exists, Algorithm 2.3 will find a goal. If no goal exists, Algorithm 2.3 will examine all vertices in the search tree.

Proof: The proof follows the recursive definition of trees. The theorem is obviously true for a single vertex. Suppose T is a search tree with root r. Let the possible decisions at r lead to $x_1, \ldots, x_k$ which are roots of the search trees $T_1, \ldots, T_k$. By the induction hypothesis, the theorem is true for the T_i. There are three cases to consider:

- r is a goal, in which case the theorem is true;

- there is no goal;

- there is a goal and r is not a goal.

If there is no goal, when the algorithm reaches x_i, it searches all of T_i by the induction hypothesis; then it moves to the next decision, if any, at r. Since failure at r sends the algorithm to x_1, it searches the entire tree.

Now suppose there is a goal in T_j but in no T_i with $i < j$. A similar argument shows that eventually T_j is reached. By the induction hypothesis, the algorithm will find a goal in T_j. ∎

How much storage does depth-first search require? Except for some overhead, the amount of storage is proportional to the depth of the vertex being considered. This can be seen in various ways.

- One method is to convert the algorithm into a nonrecursive one that simply works with the list of vertices on the path from the root to the current vertex. If you haven't had a course in data structures in which you studied depth-first search, you may find this conversion an instructive exercise.

- Another method involves Figure 2.5. By inspecting it, you can probably convince yourself that there is one layer of **Depth_First** for each vertex on the path from the root to the current vertex.

- One can include depth as a parameter in the algorithm. Change the procedure to **Depth_First**(v, d), change the call within the procedure to **Depth_First**$(x_i, d + 1)$, and begin by calling **Depth_First**$(r, 0)$. It is then easy to see that the second argument will be the depth of the first and that there will be one unfinished call for each d value from 0 through the depth of the current vertex.

Obviously, depth-first search overcomes the storage problem associated with breadth-first search—the amount of storage grows linearly rather than exponentially with depth.

Iterative-Deepening Search

Unfortunately, depth-first search introduces two new problems. First, there is no guarantee that the algorithm will find an existing solution—it may progress downward forever on an infinite branch of the search tree. Second, there is no reason for the goal state found by depth-first search to be the one involving the least number of decisions. *Iterative-deepening search* provides solutions to these problems. There is a price to pay—iterative-deepening search examines more states than breadth-first search. Before tackling iterative-deepening search, let's list the goals, difficulties, and compromises as suggested on page 17.

- **Goals:** We want to find a general search procedure that is (a) as fast and as reliable as breadth-first search and (b) as sparing in storage requirements as depth-first search. Also, it would be nice if the algorithm found a way to reach the goal that requires (nearly) the least possible number of decisions.

- **Difficulties:** We've really already seen these—we have two algorithms that don't achieve the goals. (You should explain why breadth-first search and depth-first search don't achieve the goals.)

- **Compromises:** Except that the algorithm examines more vertices than in breadth-first search, iterative-deepening search achieves the goals. One compromise and difficulty that has been hidden is the fact that we use the search tree rather than the search graph. As a result, some vertices may be examined many times. There appears to be no way to eliminate this repetition without seriously compromising the goals.

The basic idea of iterative-deepening search is simple. Let's modify depth-first search so that it will not go below some given depth D. If no solution is found, we simply increase D and try again. This seems very wasteful—looking at vertices over and over again. We'll see that it's not as bad as it looks, but first we need an explicit statement of the algorithm.

Algorithm 2.4   Iterative-Deepening Search

Let a potential search tree and an increasing sequence of positive integers $D_1, D_2, \ldots$ be given. The following algorithm starts at the root r and attempts to find a goal. To use the algorithm, execute the first line of code.

```
For k = 1, 2, ... call Iterative(r, 0, D_k).


Iterative(v, d, D)
  If v is a goal, publish v and stop all execution.
  If d = D, return.
```

```
    Expand v.
    Set i = 1.
    While there is an ith decision xᵢ at v
        Call Iterative(xᵢ, d + 1, D).
        Set i = i + 1.
    End While.
    Return.
End.
```

Note that, except for testing $d = D$, Iterative is the same as Algorithm 2.3. Thus it is just depth-first search on a tree from which all vertices of depth greater than D have been removed. Before analyzing the algorithm, let's look at it in action.

Example 2.6   Our Trivial Maze Revisited

We observed earlier that depth-first search never terminates when run on the maze in Figure 2.4 (p. 50), provided the elements in $\mathcal{F}$ are examined in alphabetic order. If we use iterative-deepening search, we'll eventually reach a solution: As soon as $D_i \geq 5$, the circled goal in Figure 2.4 will be in the search tree of vertices with depth at most D_i. Since the tree is finite, Theorem 2.1 tells us that depth-first search will find a goal. ∎

The result in the example is typical. If a solution is possible, iterative-deepening search will always find it; furthermore, the solution will have a nearly minimal depth.

Theorem 2.2   Iterative-Deepening Search Works

Let $d_{\min}$ be the minimum depth of all goal states in the search tree rooted at r. Suppose that $D_{\kappa-1} < d_{\min} \leq D_{\kappa}$ (where D_0 is taken to be -1). Algorithm 2.4 will find a goal state whose depth is at most D_{κ}.

Proof: Consider Iterative(D). Since $\mathcal{F}(s)$ is assumed finite for all s and since the tree examined by Iterative(D) has no vertices below depth D, the tree is finite. According to Theorem 2.1, depth-first search will find a goal if one exists.

By the definition of $d_{\min}$, the tree searched by Iterative(D) contains a goal if and only if $D \geq d_{\min}$. Thus, no goal is found until the search with $D = D_{\kappa}$, at which time a goal will be found. Since the goal is in the tree, its depth is at most D_{κ}. ∎

Theorem 2.2 shows that iterative-deepening search overcomes two of the problems of depth-first search—it always finds a solution if any exist and the solution it finds is close to best possible. Like depth-first search, iterative-deepening search overcomes the storage problems of best-first search. It appears, however, that there's a severe running time penalty because some vertices will be expanded many times if κ is large. In fact, some vertices will be expanded κ times! Actually, the running time penalty for iterative-deepening search is not at all severe: If $|\mathcal{F}(s)| > 1$ for all nongoal s and the D_k's are chosen in a reasonable manner, the number of vertices examined (counting multiplicity) is no greater than a constant times the number examined by breadth-first search. Even when the condition on $\mathcal{F}(s)$ is violated, iterative-deepening search is usually fairly efficient.

Theorem 2.3   Iterative-Deepening Search Is Fast

Suppose that $D_k = k$ and that $|\mathcal{F}(s)| > 1$ for all nongoal vertices s. Let I be the number of calls of `Iterative` until a solution is found. Let L be the number of vertices placed on the list Λ by Algorithm 2.1 (p. 49). Then $I < 3(L+1)$.

This result may seem counterintuitive at first since iterative-deepening examines some vertices many times. The key is that as long as a tree keeps branching, the number of vertices at any given depth exceeds the number of vertices above them. Here is a mathematical formulation of the claim.

Claim: Suppose that $|\mathcal{F}(s)| > 1$ for any nongoal vertex s. Let κ be the least depth of any goal. Let $d_=(k)$ be the number of vertices in the search tree at depth k and let $d_<(k)$ be the number at depth less than k. For $k \le \kappa$, it follows that $d_<(k) < d_=(k)$ and $d_<(k-1) < \frac{1}{2}d_<(k)$.   (2.3)

The proof of this is left as an exercise. Let's use it to prove the theorem.

Proof: Note that, as a consequence of (2.3),

$$d_<(k) < \tfrac{1}{2}d_<(k+1) < \left(\tfrac{1}{2}\right)^2 d_<(k+2) < \cdots < \left(\tfrac{1}{2}\right)^{\kappa-k} d_<(\kappa) \qquad (2.4)$$

for $k < \kappa$.

Suppose that the first goal encountered by breadth-first search is the nth vertex at depth κ. Then $L = d_<(\kappa) + n - 1$ because the goal is not placed on the list. The number of calls of `Iterative` for $D_k = k < \kappa$ is $d_<(k) + d_=(k)$. Thus, the total number of calls of `Iterative` is

$$I = \sum_{k=0}^{\kappa-1}\Big(d_<(k) + d_=(k)\Big) + d_<(\kappa) + n$$

$$= \sum_{k=0}^{\kappa-1} d_<(k+1) + d_<(\kappa) + n$$

$$= \sum_{k=1}^{\kappa} d_<(k) + d_<(\kappa) + n$$

$$\leq \sum_{k=1}^{\kappa} \left(\tfrac{1}{2}\right)^{\kappa-k} d_<(\kappa) + d_<(\kappa) + n \qquad \text{by (2.4)}$$

$$= d_<(\kappa) \sum_{i=0}^{\kappa-1} \left(\tfrac{1}{2}\right)^{i} + d_<(\kappa) + n$$

$$< 2d_<(\kappa) + d_<(\kappa) + n \leq 3(L+1).$$

This completes the proof. ∎

Exercises

2.4.B. Give an algorithm for depth-first search and describe the order in which it examines vertices.

2.4.C. Give an algorithm for iterative-deepening search.

2.4.D. What are the major good and bad points of breadth-first search and of depth-first search? How does iterative-deepening search compare with these good and bad points?

2.4.9. The purpose of this exercise is to prove Claim (2.3).

  (a) For $k+1 \leq \kappa$, show that $d_<(k+1) = d_=(k) + d_<(k)$ and that $d_=(k+1) \geq 2d_=(k)$.

  (b) Using (a) and induction, or otherwise, prove that $d_<(k) < d_=(k)$.

  (c) Using (a) and (b), or otherwise, prove that $d_<(k+1) > 2d_<(k)$.

2.4.10. In Exercise 2.4.4 (p. 52), we considered generalizing breadth-first search to allow a cost other than depth.

  (a) Explain why it doesn't make sense to try doing this for depth-first search.

  (b) Describe how to modify Algorithm 2.4 to make use of a more general cost.

2.5   Heuristic Search

> *It is a mark of insincerity of purpose to spend one's time in looking for the sacred Emperor in the low-class tea-shops.*
> —Ernest Bramah (1900)

> *This is a well-known AI lesson: knowledge reduces the need to search.*
> —Kenneth D. Forbus and Johan de Kleer (1993)

In the previous section, the best goals in the search tree were those of least depth—cost was identified with depth. The possibility of a more general cost was briefly discussed in Exercises 2.4.4 (p. 52) and 2.4.10. In this section, we assume that there's a general cost. In addition, we'll look at the possibility of improving algorithms by including a guess about the future in our costs. This idea leads to a cost function that combines some knowledge of the past with a heuristic approximation to the future:

Definition 2.9   (Heuristic) Cost Functions

- $g_u(v)$ is the cost of going from u to v by moving down in the decision tree.

We *assume* that $g_u(v)$ increases as v moves downward in the tree.

- $C^*(v) = \min g_r(z)$, where the minimum is taken over all goals z that are reached by a path through z. This is the cost to reach a goal through g. If there are no such goals, $C^*(v) = \infty$.

- $h^*(v) = C^*(v) - g_r(v)$ is the additional cost to reach a goal through v after paying the cost to reach v.

- $h^e(v)$ is any estimate of $h^*(v)$; this is called a *heuristic cost function*.

- $C^*(v) = g_r(v) + h^*(v)$ is the heuristic cost of reaching a goal through v.

Here's the motivation behind these definitions. The search methods in the previous section use only information about the path from the root to the current vertex v; that is, they use $g_r(v)$. If we knew $C^*(v)$, we could use it instead of $g_r(v)$ and search would be trivial: We'd simply start at the root and, whenever u is the current vertex, move to a $v \in \mathcal{F}(u)$ for which $C^*(v)$ is as small as possible. Unfortunately, we usually know only $g_r(v)$ when we're at v. You should convince yourself that the claims made so far are true.

$\ast$ $\ast$ $\ast$ Stop and think about this! $\ast$ $\ast$ $\ast$

The function $h^*(v) = C^*(v) - g_r(v)$ can be thought of as the additional cost to reach a goal through v. Although it's unlikely that we'd know h^*, we might be able to produce h^e—an estimate of h^*. For example, a good chess player might not *know* how good a position in a game is, but he would be able to make a fairly accurate guess. Since using $C^* = g_r + h^*$ would make search trivial, it seems reasonable that using the estimate $C^e = g_r + h^e$ would improve upon the simple search strategies in the previous section. This is so for some estimates.

Before proceeding, let's take a minute to look at the goals, difficulties, and compromises.

- **Goals:** We want to search a problem space to locate a goal. To do this, we are representing the situation by a search tree.

- **Difficulties:** We cannot examine a sufficient portion of the search tree because of limitations in data, computational resources, or our cleverness in designing an algorithm.

- **Compromises:** We plan to limit our search to what appear to be the most promising portions of the tree. To locate these regions, we will make use of a heuristic cost function h^e as a sort of crystal ball. This function is some sort of approximation to the true cost.

It's quite clear from this that the heuristic function h^e plays a major role in how well our approach will work. Therefore, we need to know more about it, as the following questions indicate.

- **How should we use a heuristic function?** If we are searching until a goal is found, we should probably treat the heuristic as if it were the true cost. In "partial search," other issues may be important.

- **What makes a heuristic function good?** It should reduce our work by indicating promising directions in which to search. Unfortunately, that answer doesn't help us very much.

- **How can we create good heuristic functions?** Solutions are often specific to the problem at hand and knowledge-intensive. Consequently we can't discuss such solutions here.

Aside. Often the cost g is additive in the sense that $g_u(v) + g_v(w) = g_u(w)$ when the path from u to w contains v. Examples of this are the number of edges on a path or the mileage on roads between points. When g is additive, $h^*(v) = \min g_v(z)$. In this case, there is a subtle but important difference in the domains of g and h. We *must* define $g_u(v)$ on the vertices of the search *tree* because it contains information about *how* we reached v from the root. On the other hand, $h^*(v)$ and $h^e(v)$ can be defined on the vertices of the search *graph*.

Admissible Heuristics

> *From a table of irregular verbs:*
> *I heurist. You try and err. He/she flounders.*
>
> —Stan Kelly-Bootle (1981)

We can adapt the simple algorithms of the previous section to use the function $C^e(v) = g_r(v) + h^e(v)$ instead of the depth of v. What properties must $C^e(v)$ have to make breadth-first search and iterative-deepening search reach low-cost goal states? The answer is provided by the following theorem whose proof is left as an exercise.

Theorem 2.4   Guaranteed Heuristic Search
Suppose that $C^e(v) = g_r(v) + h^e(v)$ is such that

(i) for every M only finitely many vertices have $C^e(v) < M$, and
(ii) if z is a goal that can be reached through v, then $C^e(v) \leq C^e(z)$.

It follows that heuristic search using C^e works; that is,

(a) the modified breadth-first search of Exercise 2.4.4 (p. 52) will find the least-cost goal, and
(b) the modified iterative-deepening search of Exercise 2.4.10 will find a nearly least-cost goal as in Theorem 2.2.

Condition (i) simply ensures that the search is finite. Why is a condition like (ii) needed? Suppose that z and z' are goal vertices in the search tree, that z is reached through v, that $C^e(v) > C^e(z') > C^e(z)$, and that z' is reached by a path such that $C^e(w) \leq C^e(z')$ for all w on the path. You should be able to see that if we terminate our search at a cost between $C^e(z')$ and $C^e(v)$, we will reach z', but we won't reach v and hence won't reach z.

While condition (i) is usually easy to verify, condition (ii) may be more troublesome. The next theorem provides one means of verification.

Definition 2.10   Admissible Heuristic
If $h^e(v) \leq h^*(v)$ for all v, then h^e is called an *admissible heuristic*.

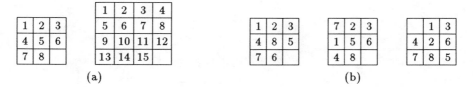

Figure 2.6   (a) The usual goal state for the 8 and 15 puzzles; (b) nongoal states. In the three nongoal states shown here, $h^e = 4$. In the first case, $h^* = 4$, too. In the second case, $h^* = 16$. In the last case, it's impossible to reach the goal from the given state and so $h^* = \infty$.

Theorem 2.5   Admissible Heuristics Are Guaranteed

If h^e is an admissible heuristic, then (ii) in Theorem 2.4 holds.

Proof: Suppose $h^e(v) \leq h^*(v)$ for all v. Then

$$C^e(v) = g_r(v) + h^e(v) \leq g_r(v) + h^*(v),$$

and, by the definition of $h^*(v)$, this is the cost of the least-cost goal reached through v. ■

Example 2.7   Admissible Heuristics Based on Geometric Distance

Suppose we have a road map and are trying to find the shortest route from city R to city Z. We can construct a search space where the states are highway intersections and the cost of going from one state to another is the highway distance between them. We start at the state R and try to reach the goal Z. A simple choice for $h^e(v)$ is the straight line distance from v to Z. Since $h^e \leq h^*$, the theorem applies. Depending on how the roads are laid out, h^e may be fairly close to h^*.

Another geometric measure is the *manhattan distance*:

distance from (x, y) to (x', y') equals $|x - x'| + |y - y'|$.

It receives its name from the way traffic must move in a city like Manhattan. Only east-west and north-south travel is allowed. The manhattan distance is also called the *taxicab distance*. Here's an application

A common puzzle is the $n^2 - 1$ puzzle played on an $n \times n$ board using $n^2 - 1$ tiles numbered 1 through $n^2 - 1$ arranged in a square pattern with one empty square as shown in Figure 2.6. A tile adjacent to the empty square either horizontally or vertically can be slid so as to exchange positions with the empty square. This is a move in the puzzle. The goal is to reach some desired arrangement through a sequence of moves. The cost is the number of moves required. If each tile is one unit on a side, then moving a tile changes the manhattan distance between the tile and any point by one unit. Thus, we must move a tile at least as many times as the manhattan distance between its present location and its desired location. Summing these distances over all

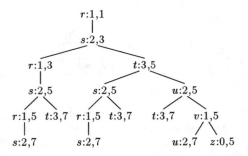

Figure 2.7 The search tree for the maze of Figure 2.4 (p. 50). The goal is z. Numbers at the vertices are taxicab heuristics in the form $h^e(\ ), C^e(\ )$. Only those vertices x with $C^e(x) \leq C^e(z) = 5$ have been expanded.

$n^2 - 1$ tiles, we obtain a lower bound h^e on the number of moves needed to solve the puzzle. Thus, $h^e \leq h^*$ and so is an admissible heuristic. In many positions, h^e seriously underestimates h^*.

The 15 puzzle has been sold as a toy. The 8 puzzle and, to a lesser extent, the 15 and 24 puzzles have all been studied by search theorists. ∎

The next example shows how C^e can be improved during iterative-deepening search. After the example, the method is formalized into an algorithm. You may wish to read the algorithm and example together.

Example 2.8 Our Trivial Maze Yet Again

We can use the taxicab metric for our maze in Figure 2.4 (p. 50). If each edge has unit length, then $g_r(x)$ is simply the depth of x,

$$h^e(r) = h^e(v) = 1, \quad h^e(s) = h^e(u) = 2, \quad \text{and} \quad h^e(t) = 3.$$

Figure 2.7 is the search tree for the maze with those vertices q with $C^e(q) \leq C^e(z) = 5$ expanded. The values of h^e and $C^e = g_r + h^e$ are given for each vertex. Since two occurrences of t haven't been expanded, the taxicab metric heuristic eliminates some vertices from consideration. We can do better if we're willing to change h^e based on information gained in each iteration of the algorithm. Let's perform iterative-deepening with $D_k = k$.

On the first iteration, we expand r, obtaining $\mathcal{F}(r) = \{s\}$. Since $h^e(s) = 2$, we have $C^e(s) = 3$, as shown on the left side of Figure 2.8. This tells us that it must cost at least 3 to reach the goal from r. Thus, we can change $C^e(r)$ to 3 and so $h^e(r) = C^e(r) - g_r(r) = 3 - 0 = 3$. This completes the expansion for $C^e(x) \leq 1$. Since we now have $C^e(r) = 3$, no expansion occurs for $C^e(x) \leq 2$.

The expansion for $C^e(x) \leq 3$ is shown in the middle of Figure 2.8. This tells us that the cost of a solution is at least 5. As a result, we update values: $C^e(r) = C^e(s) = 5$, $h^e(r) = C^e(r) - g_r(r) = 5 - 0 = 4$, and $h^e(s) = C^e(s) - g_r(s) = 5 - 1 = 4$. No expansion occurs for $C^e(x) \leq 4$.

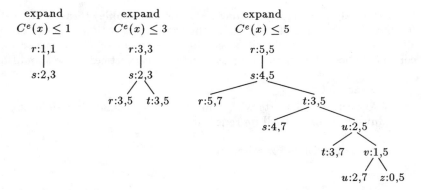

Figure 2.8   The search tree for the maze of Figure 2.4. Information gained in each iteration of iterative-deepening search is used to improve h^e for the next iteration. Above each tree is the description of those vertices that are expanded. Numbers at the vertices are the heuristics at the start of that expansion in the form $h^e(\ ),C^e(\ )$.

The tree for $C^e(x) \leq 5$ is shown on the right side of Figure 2.8. Notice how much smaller it is than the tree in Figure 2.7. ■

Now let's capture the idea behind Example 2.8 in an algorithm.

Algorithm 2.5   Iterative-Deepening Search with Improving Heuristics

Suppose that we are given the situation in Algorithm 2.4 (p. 56), except that now we also have cost function C^e given by $C^e(v) = g_r(v) + h^e(v)$, where h^e is an admissible heuristic whose value depends only on the search graph vertex corresponding to the search tree vertex v. Here is an iterative-deepening algorithm that improves h^e.

```
For k = 1, 2, ... call Iterative(r, D_k)

Iterative(v, D_k)
   If v is a goal, publish v and stop all execution
   If C^e(v) ≥ D_k return C^e(v).
   Set i = 1 and C^e = +∞.
   While there is an ith decision x_i at v
      Set C^e = min(C^e, Iterative(x_i, D_k)).
      Set i = i + 1.
   End While.
   Set h^e(v) = max(h^e(v), C^e - g_r(v)).
   Return g_r(v) + h^e(v).
End.
```

Unfortunately this algorithm may require considerable storage: The values of h^e must be saved for *all* vertices in the search graph that we have examined. We could modify the algorithm so that the values of h^e are saved only for certain vertices, but then we must decide which vertices these should be.

Theorem 2.6
If Algorithm 2.5 starts with a heuristic h^e that is admissible, it remains admissible through all its redefinitions.

The proof is left as an exercise.

Other Heuristics

> *Our motivation is based on the view that heuristics are central to AI, and that we cannot claim to understand them until we have mathematical models which explain experimental results obtained by using them.*
>
> —Stephen V. Chenoweth and Henry W. Davis (1991)

In order to guarantee that the search methods will find (nearly) least-cost solutions, we've required that our heuristic h^e be admissible; that is, $h^e(v) \leq h^*(v)$ for all v. But such a guarantee is of little use if the amount of search required is beyond our computational capabilities. Thus, we may want to abandon admissibility and attempt to construct a better heuristic function.

What makes one choice for h^e better than another? Presumably, less search is better. What if a nonadmissible h^e leads to a goal that is not the least cost solution? How can we analyze a situation when the search method is not guaranteed to find the least-cost solution? Since search time is often much more important than the cost of the solution, perhaps we should ignore the cost and look at just search time.

In general, we can't *guarantee* a search time. Hence, we should probably look at average behavior. This implies some underlying probabilistic model. Do we really need one? What should it be?

Let's look at why we need a probabilistic model. Suppose we have in hand a search tree, h^* and h^e. We could then determine the number of vertices searched by A* search (i.e., iterative-deepening heuristic search) and see how the solution found compares with the least-cost solution. By simply rearranging the order in which the vertices in $\mathcal{F}(v)$ are examined, we can change the number of vertices searched and the cost of the solution found. Thus, search times can be expected to vary considerably, depending on the order in which each $\mathcal{F}(v)$ is expanded. Since there will be a wide spread between best and worst times, some sort of average time is more informative. When we speak of averages, there is, at least implicitly, a probabilistic model present. By making the model explicit and viewing the average as the expectation of a random variable, we gain two things. First, the tools of probability theory make it more likely that we can analyze the situation instead of simply collecting data or giving "reasonable" arguments. Second, by stating our assumptions clearly, we make it easier to criticize and improve them. Since we haven't explored the necessary probabilistic tools yet, we'll postpone this line of discussion until Section 13.3.

Exercises

2.5.A. What is a heuristic and how does it help in search?

2.5.B. What is an admissible heuristic and what does it guarantee about search methods based on breadth-first and iterative-deepening approaches?

2.5.1. In chess, a knight either moves one square horizontally and two vertically or moves two squares horizontally and one vertically. A position can be described by two integers (i, j) indicating horizontal and vertical coordinates on the board. The goal is to reach $(0, 0)$. Design a good heuristic function for which $h \leq h^*$ and prove that it satisfies $h \leq h^*$. Your heuristic function should be something that can be computed quickly in your head from i and j—not some sort of table lookup. Also, it should work for arbitrarily large i and j.

2.5.2. Prove Theorem 2.4.

2.5.3. Apply Algorithm 2.5 to the trivial maze in Figure 2.4 using the trivial admissible heuristic $h^e(s) = 0$ for all s.

2.5.4. Apply Algorithm 2.5 to the mazes shown here and produce pictures like those in Figure 2.8. Use the manhattan distance for the starting heuristic. The starting vertices are labeled r and the goals are labeled z.

    (a) $r - s - t$ (b) $r - s - t$ (c) $r - s - t$

            | | | | | |

            w u $z - w$ u z u

                  | | | |

         $z - v$ v $w - v$

    (d) $r - s - t - u - v - w - z$ Yes, there are two goals:

           $z^* - q$ z^* is unreachable from r.

2.5.5. The purpose of this exercise is to prove Theorem 2.6. Let $C^*(x)$ denote the true cost $g(x) + h^*(x)$ and suppose that s_0 is some particular tree vertex that is not a goal.

   (a) Prove that $C^*(s_0)$ is the minimum of $C^*(t)$ over all $t \in \mathcal{F}(s_0)$.

   (b) Prove that if h^e is admissible, then redefining $h^e(s_0)$ by

$$h^e(s_0) = \left(\min_{t \in \mathcal{F}(s_0)} C^e(t) \right) - g_r(s_0)$$

   leaves h^e admissible.

   *(c) Prove Theorem 2.6.
      Hint. Look at the first time `Iterative(`v, D_k`)` produces an inadmissible value.

2.5.6. We are given a search graph in which it is possible to reach a goal from the starting vertex. Suppose that the cost of reaching a goal is the number of decisions made. Thus $g(t)$ is the depth of t in the search tree. Suppose that $h^e \geq 0$ is a heuristic. We *do not assume* that h^e is admissible.

   (a) Suppose we do iterative-deepening search using $C^e(v)$ instead of the depth of v in Algorithm 2.4. Show that such a search will always find a solution.

   (b) Construct a simple example to show that the algorithm may not find the least-cost solution.

   (c) Prove that iterative-deepening search with improving heuristics (Algorithm 2.5) will find a solution.

2.5.7. Why not search on the search graph instead of the search tree? Suppose that the search space is the simple one described near the end of this section: a graph $G = (V, E)$ in which each nongoal vertex lies on exactly b edges. We will search by always moving from a vertex to the vertex adjacent to it which has the least heuristic cost C^e.

   (a) Describe $\mathcal{F}(v)$ for all $v \in V$ and describe the search graph. Remember, the search graph is a *directed* graph, so it is not quite the same as G.

   (b) Suppose that we are at a vertex v such that $C^e(v) < C^e(w)$ for all vertices w whose distance from v is 1 or 2. (This is not unreasonable in some situations.) Show that if we reach v in our search, we will never move more than one vertex away.

   (c) Explain how tree search avoids the problem in (b).

2.6 Partial Search

I am going in search of a great perhaps.

—François Rabelais (1553)

So far we've discussed search problems as if it were feasible to search until a goal vertex is found. This is often impossible, as the following situations illustrate.

- The search tree may be too large to search within the allotted time. For example, a computer cannot search the entire move tree for chess in order to find the best move.

- Even if we could search the entire tree, it may be best to take some action so as to gain additional information that may significantly reduce the search. For example, by carefully recalling how my program has acted and by consulting manuals, I may be able to determine what is wrong with it. Rather than spend all my time exploring a large search space, I may decide to add some print statements that significantly reduce the search tree.

- There may not be enough information to describe the search space. For example, in "dungeons and dragons" computer games, many aspects of the game are uncovered only through exploration.

All of these examples relate to tradeoffs between time, information, and action. This suggests an important idea for search in AI

$$\boxed{\text{To design a good strategy, we must come to grips with the tradeoffs between time, action, and information.}} \quad (2.5)$$

Many claim that this is the core issue in intelligent behavior. Herbert Simon, for example, made it the focus of his research in both economics and AI.

In view of the breadth of (2.5), we can't attack it head on. In the following discussion, we'll focus on the first example. That is, we'll assume that some sort of time limit has been imposed on making a decision and ignore the influence of potential information gain on choosing an action.

There are some general principles we can use in adapting previous methods to the present situation.

- Use a limited depth search.
- Make a decision and start again.
- Accept the heuristic as reasonably accurate.

We'll elaborate on each of these.

Use a Limited-Depth Search

Whatever search method we choose, it's essential to use a heurisitic function. Otherwise, we'd simply base our decision on g which measures only how far a vertex is from the root, not how close it is to a goal.

We expect that time constraints will cause our search to terminate before reaching a goal. Therefore, a method like depth-first is inappropriate—it's likely to spend its time exploring a few alternatives in depth. Heuristic best-first search and heuristic iterative-deepening search are more appropriate.

There is a problem about how deep to run the search. With best-first, we can simply use the vertex that is being expanded at the time the search is terminated. With iterative-deepening, we can use the best result found with the last k for that an iteration was completed. This leads to algorithms which can be interrupted at any time.

Definition 2.11   Interruptible or Anytime Algorithm

An *interruptible* or *anytime algorithm* is an algorithm that will provide an answer whenever a user stops it, provided some minimum time has elapsed.

From the previous discussion, we see that anytime algorithms must have some ability to assess future prospects, for example, by using a heuristic function h^e. We saw that heuristic depth-first search does not lead to an anytime algorithm, but that heuristic best-first search and heuristic iterative-deepening search do.

The following generalization of iterative-deepening search constructs an anytime algorithm from a large class of algorithms. The basic idea is that we have some algorithm $\Xi(x)$ that provides no useful information unless it is allowed to finish. The parameter x determines running time but not in a way that we can easily control. All we may know is that increasing x increases running time.

Algorithm 2.6   Constructing an Anytime Algorithm

Let $\Xi(x)$ be an algorithm with a parameter x that influences running time in such a way that running time is an increasing function of x, say $t(x)$. We do not assume that $t(x)$ is known. If $x_1 < x_2 < \cdots$ is any sequence increasing without bound, then the following is the interruptible algorithm $\Xi^*(\vec{x})$.

1. **Initialize:** Set $k = 1$ and *out* $= \emptyset$.
2. **Execute:** Begin execution of $\Xi(x_k)$. If interrupted, return *out*; otherwise, go to Step 3 upon completion.
3. **Iterate:** Set *out* to the output of $\Xi(x_k)$, replace k by $k + 1$ and go to Step 2.

The following theorem lists some properties of the algorithm. Its proof is left as an exercise.

Theorem 2.7

We use the notation of Algorithm 2.6.

(a) If $\Xi^*(\vec{x})$ is interrupted at time $T \geq t(x_1)$, it produces the same output as $\Xi(x_k)$ where

$$t(x_1) + \cdots + t(x_k) \leq T < t(x_1) + \cdots + t(x_k) + t(x_{k+1}).$$

(b) Suppose that the quality of the output of Ξ increases with the length of time it runs. Then there exists an $\vec{x}$ with the following property. If $\Xi^*(\vec{x})$ is interrupted at time T, then $\Xi(x)$ must run for time greater than $T/4$ to produce better output.

It may not be possible to determine the vector $\vec{x}$ whose existence is asserted in (b) because it depends on knowing the function $t(x)$. As a result, one may wish to build an anytime algorithm that chooses the value of x_k based on how long $\Xi(x_i)$ ran for $i < k$. Exercise 2.6.3 shows that it is often best to try to choose x_k so that the running time of $\Xi(x_k)$ is double the time for $\Xi(x_{k-1})$.

Make a Decision and Start Again

After searching for some time, we may stop searching without having found the goal. The time limit may be imposed externally—our time has expired and we must do the best we can. This is like a quiz show in which you're given a time in which to decide upon an answer. Alternatively, the time limit may be self-imposed because making a decision is expected to produce benefits; for example, we learn more about our environment, our opponent makes a move, or we gain more time for later search. This happens in a chess match: Although you're given a time limit for the *entire* game, you must decide how much time to spend on each move.

Consider the following situation. After partially exploring a search space using a heuristic function as an aid, we stop and make a single decision. This process is carried out repeatedly, each new search working with the modified search space and/or heuristic that was obtained as a result of our decision.

- **Stopping**: When should we stop searching and make a decision? To make this determination, we would need a way to measure the cost of search time against the expected gain from further searching. In other words, we would need the value-versus-time graph in Figure 1.1 (p. 15).

- **Choosing**: Determining which decision is best can also be a difficult problem. This is particularly true when we expect to gain new data by acting.

For example, suppose I am lost in the wilds. Should I spend time climbing a nearby peak in hopes of obtaining a good view that will help me decide which way to go? Or should I simply follow the stream down because, as streams get larger, they tend to have campgrounds, cabins, villages, and so forth nearby? Or should I do something else? In the first case, I spend a lot of time merely hoping to gain information. In the second case, I start moving in a reasonable direction. To deal with such problems, we would have to incorporate the expected value of information into the problem space.

We won't tackle either of these difficult problems.

There's another, simpler problem that we need to avoid. By making a single decision and then starting anew, we open the door to the possibility of getting stuck in an infinite loop. This is because we're looking, in effect, at the search graph rather than the search tree. Here's an example.

Suppose that the root r and the vertex $s \in \mathcal{F}(r)$ have smaller values of h^e than any vertices that can reached from either of them in d decisions. Also suppose that $r \in \mathcal{F}(s)$ and the cost is simply the number of decisions. If we start at r and search to a depth at most d, we will decide to move to s. Now, starting at s and searching to a depth at most d, we will decide to move back to r.

One way to avoid this is by increasing $h^e(v)$ every time we decide to move to the vertex v. This can be done by keeping a table of previous decisions. Adjusting h^e in this way is the basis of the RTA* algorithm developed by Korf [6]. See also Exercise 2.5.7 (p. 68).

Accept the Heuristic as Reasonably Accurate

If $h = h^*$, we could simply compute $h^e(s)$ for all $s \in \mathcal{F}(r)$ and select that s for which the cost is a minimum. Of course, we expect that $h \neq h^*$; nevertheless, we must expect it to provide some information since we're using it. How much should we trust the information?

Generally, it's reasonable to rely on the heuristic as a guide for searching, but it's unreasonable to expect it to be very accurate when we're making a decision. For instance, if one decision looks almost as good as another, it may be worthwhile to explore those two possibilities further. Deciding when to do so requires probability, so let's discuss it in Chapter 14.

Exercises

2.6.A. What is partial search and why might it be done?

2.6.B. What is an anytime algorithm?

2.6.C. Suppose that $\Xi(x)$ is an algorithm with a time parameter x. How can an anytime algorithm be constructed from $\Xi(x)$?

2.6.1. Suppose that $\Xi(x)$ does depth-first search up to depth at most x in a tree having branching factor b. Assume that running time, in some units, equals the number of vertices that are looked at. We may as well choose $\vec{x}$ to consist of distinct integers since depths are integral and there is no reason to repeat a calculation.

  (a) Using Exercise 2.3.2, show that the running time of $\Xi(n)$ is approximately $\frac{b^{n+1}}{b-1}$ for large n and conclude that the ratio of the running times of $\Xi(m)$ and $\Xi(n)$ is about b^{m-n}.

  (b) Suppose that $x_k = n$ and $x_{k+1} = m$. Further suppose that T is such that $\Xi^*(\vec{x})$ is stopped just short of finishing $\Xi(m)$. Show that, approximately, $T \geq \frac{b^{m+1}+b^{n+1}}{b-1}$. Conclude that Ξ can produce the same output as Ξ^* in $(\frac{1}{b+1})$ the time or less.

  (c) From Theorem 2.7(b) we conclude that there should be a Ξ^* which takes only about four times as long as Ξ. On the other hand we have just shown in (b) that it can take up to about $b+1$ times as long. How can this apparent contradiction be resolved?

2.6.2. The purpose of this exercise is to prove Theorem 2.7.

  (a) Prove Theorem 2.7(a).

  (b) Suppose that $t(x)$ is a strictly increasing function of x so that it makes sense to talk about the inverse function $x(t)$. Choose some t_1 and let $t_k = 2^{k-1}t_1$ and $x_k = x(t_k)$. Prove Theorem 2.7(b) for this choice of $\vec{x}$. *Hint.* Compute the time Ξ must run to produce the output that Ξ^* produces at time T.

  (c) The function $t(x)$ need not be strictly increasing. For example, in depth-first search up to depth x, the time only changes when x passes through an integer value.
     Prove Theorem 2.7(b) in general as follows. Use t_k as in (b) above. In place of the inverse of $t(x)$, let $x(t)$ be such that $\Xi(x(t))$ will finish within time t and such that Ξ cannot produce better output within time t.

2.6.3. Returning to (b) of the previous exercise, we examine how good the procedure of doubling times actually is. Suppose that $x_k = rx_{k-1}$ for some $r > 1$. Show that, instead of $T/4$, we obtain $(r-1)T/r^2$. Also show that this is a maximum at $r = 2$. Thus, $r = 2$ is best in some sense.

2.6.4. In contrast to the previous exercise, $r = 2$ may not be best when the function $t(x)$ is not strictly increasing. In fact, it may lead to $\Xi(x_k)$ and $\Xi(x_{k+1})$ carrying out exactly the same search. By looking at this set of exercises, you can find an example that illustrates this fact. Do so.

*Chess Programs and Search

> *It is quite certain that the operations of the [chess playing] Automaton are regulated by* mind *and by nothing else. Indeed this matter is susceptible of a mathematical demonstration* a priori.

> —Edgar Allan Poe (1836)

Chess players divide the game into three somewhat ill-defined parts: opening, midgame, and endgame. Openings and endgames have been extensively studied and many books have been written on them. Any serious player will have memorized many openings and many endgame techniques. Similarly, openings and endgame techniques are frequently built into chess playing programs. The midgame, however, possesses too much variety for anyone to present such tabular methods for playing it. As a result, players rely on partial heuristic search. Here are some things we need to build such a program for the middle game.

- An internal representation of the game: We'll ignore this.
- A method for generating possible moves: This is a method for expanding the vertices in an AND/OR search tree; in other words, creating $\mathcal{F}(s)$ given s. $\mathcal{F}(s)$ may contain just "reasonable" moves—not all possible moves. We'll ignore move generation, too.
- A heuristic for unexpanded positions: This is an evaluation of the "goodness" of a position, based on such factors as strength of material (a weighted count of pieces) and a crude, rapid assessment of position (using pawn formation, pieces attacked, etc.). We'll ignore how such a heuristic is produced, but will look at what it means and how it might be used.
- A combining rule: This tells how to obtain $h^e(s)$ given the values of $h^e(t)$ for all $t \in \mathcal{F}(s)$. It moves heuristic information back toward the root so that a move can be selected.
- A method for deciding when and where to search: Which vertices should be expanded? When is further search not warranted due to a low expected rate of return per unit time?

Before studying h^e, we need to know what h^* should measure. It should measure how my position compares with my opponent's. At first glance, this evaluation seems to be based just on the board position. That's not entirely true. Since chess is played with time limits, complexity can be important. For example, if I'm short on time, I might try to simplify the position. Another factor that can enter in is knowledge of my opponent's abilities. One position may invite an inexperienced player into a trap, whereas a stronger player might turn the position against me. Programs have difficulty incorporating time into h^e, and they don't attempt to incorporate knowledge of the opponent's abilities.

Once we have $h^e(s)$ for all reasonable moves from the root, we can select the best move. To get this information, we need a "combining rule" that allows us to move information up the tree. What should it be? Chess playing progams use combining rules that are adaptations of the the max-min method for AND/OR trees, which we'll study in the next section.

All players use some sort of search procedure for choosing moves during the midgame. Current chess playing programs rely on doing extensive search—more than humans are capable of by many orders of magnitude. This compensates for the fact that their heuristics are much poorer than those of expert players. Some systems rely mainly on search, depending on a quickly evaluated h^e and special hardware. Others take more time on h^e, operating on the assumption that a better h^e will reduce the need to search. Some search tactics used by humans have been incorporated into chess programs. Here are two examples:

- Suppose that, for one line of play A, the opponent has many apparently reasonable responses, while for another line B, all her responses seem poor except for one apparently very good response B'. Based on this assessment, A is preferable to B; however, a player may explore the response B' further to see if it is really as good as it appears. If it is not, B may be preferable to A. In other words, the tree will be explored to a greater depth at B than at A. B' is an example of a *singular move*—one that is much better than the alternatives. The general heuristic is this: If a move is dependent on knowing the value of a singular move accurately, that move should be explored in greater depth.

- Suppose search reveals that the opponent has a very good response R to one line of play. In other words, R "refutes" that line of play. Now suppose we are in another part of the tree at the same depth where R was used. It makes sense to see if R refutes the present line before trying the many other possible moves for the opponent. This is called the *killer heuristic* because R kills the line of play.

See Section 13.3 for a discussion of some probabilistic aspects of partial search.

*2.7   AND/OR Trees and Related Species

Imagine a problem space in which each state is a game position as well as
an indication of whose move it is. Suppose our opponent is a good player. To
reach our goal (a win for us), we must now find a path where the nature of the
decision alternates from vertex to vertex—best for us on our move and worst
for us on our opponent's move. ("Worst" because we assume our opponent
will choose her best move.) This situation can be handled by modifying our
search tree idea as follows.

Definition 2.12   AND/OR Trees

An AND/OR *tree* looks like a decision tree. Each vertex is labeled either
AND or OR so that the labels alternate on each path from the root to each
vertex. Every vertex s for which $\mathcal{F}(s) = \emptyset$ is assigned a value $p(s)$ in some
manner. The value of every other vertex is defined inductively:

$$p(s) = \begin{cases} \min_{t \in \mathcal{F}(s)} p(t), & \text{if } s \text{ is labeled AND,} \\ \max_{t \in \mathcal{F}(s)} p(t), & \text{if } s \text{ is labeled OR.} \end{cases} \quad (2.6)$$

This definition allows us to propagate values of p from the leaves upward
to the root. An AND/OR tree is also called a *max-min tree*.

If $p(s) = 1$ indicates true and $p(s) = 0$ indicates false, min and max compute
logical "and" and "or," respectively—hence the name AND/OR. More generally,
if $0 \le p(s) \le 1$, the tree computes fuzzy "and" and "or."

For game playing, the vertex tells us the position and whose move it is,
and we use the following:

$$\begin{aligned} \mathcal{F}(s): &\quad \text{moves that can be made from } s, \\ \max: &\quad \text{computed at our turn,} \\ \min: &\quad \text{computed at opponent's turn,} \\ p(s) = 1: &\quad \text{position is a win for us,} \\ p(s) = 0: &\quad \text{position is a loss for us.} \end{aligned}$$

This makes sense if we imagine players choosing their best possible moves:
Our goal is to maximize and our opponent's is to minimize.

AND/OR graphs arise in planning. Starting at an OR vertex, we choose an
initial plan for reaching our goal. If the plan s contains several separate goals
$\mathcal{F}(s)$, which much each be attained to fulfill the plan, the vertex for the plan
is an AND vertex. Each of the separate goal vertices now functions like our
original goal vertex—an OR. And so on.

Here are two procedures, **p_max_at** and **p_min_at** for computing the value
of the root of an AND/OR tree. If the root r is an OR, we compute **p_max_at**(r);
otherwise, we compute **p_min_at**(r).

```
Procedure p_max_at(s)   /* Return p(s) = max p(t),  t ∈ F(s). */
   If F(s) = ∅, then return p(s).
   Set M = −∞.
   For t ∈ F(s), do
      Set m = p_min_at(t).
      If m > M, then set M = m.
   End for.
   Return M.
End Procedure.
```

```
Procedure p_min_at(s)   /* Return p(s) = min p(t),  t ∈ F(s). */
   If F(s) = ∅, then return p(s).
   Set M = ∞.
   For t ∈ F(s), do
      Set m = p_max_at(t).
      If m < M, then set M = m.
   End for.
   Return M.
End Procedure.
```

For convenience, we adopt the convention that

p_*_at stands for either p_max_at or p_min_at, as appropriate.

The algorithms interact recursively. At a given nonleaf vertex s, the value of $p(s)$ is computed by first computing $p(t)$ for all $t \in \mathcal{F}(s)$. This involves recursion: The procedures move deeper into the tree until a leaf is reached. You should convince yourself that the code does what is claimed. (See Exercise 2.7.2.)

The functions max and min can be replaced by others. For example, if we are operating in an uncertain environment, our opponent becomes "Nature," which selects a vertex in $\mathcal{F}(s)$ in some apparently random manner. Thus, "average" ("expectation") may be a more appropriate function than "minimum."

In studying an AND/OR tree you can make use of a heuristic function and apply the ideas in Section 2.5. Thus, instead of calling p_*_at to evaluate a node, we can call a heuristic function, say p_*_heurist, to evaluate the situation. Since we are not concerned with distance to a goal in games, what should p_*_heurist return? A reasonable idea is to let it return a measure of how promising the vertex appears. Thus a high value indicates that we think a win is likely while a low value indicates that we think it unlikely. This idea is used in partial search. See the discussion of chess in the previous section.

Exercises

2.7.A. Define an AND/OR tree.

2.7.1. Define the concept of an AND/OR graph so that it will bear the same relation to an AND/OR tree as a search graph bears to a search tree.

2.7.2. The purpose of this exercise is to prove that p_*_at returns correct values for *finite* AND/OR trees. The form of the proof is a type of induction, with (a) playing the role of the starting value 1 for simple induction and (b) playing the role of the inductive step. In (c), you prove that induction works in this situation.

(a) Prove that p_*_at(s) returns the correct value when s is a leaf.

(b) Prove that p_*_at returns the correct value if all its calls to p_*_at return correct values.

(c) Prove that p_*_at returns correct values.
Hint. Suppose the proposition is false and let s be the deepest vertex such that p_*_at(s) returns an incorrect value.

The next exercise assumes some knowledge of expectation from probability theory, which we don't discuss until Chapter 12.

*2.7.3. Many games involve an element of chance, often using cards or dice. These games do not fit the "complete information" framework we construct for games like chess. Suppose we can assign a probability to each of the chance events that might occur in such games and suppose that what happens on one move is independent of what happens on another, as is true, for example, when rolling fair dice. By breaking a move into pieces to separate out the chance and the action of the player, explain how to modify (2.6) in Definition 2.12.

Alpha-Beta Pruning

An elegant method for reducing the search in AND/OR trees is *alpha-beta pruning*, also written *α-β pruning*. This somewhat subtle method is easy to misunderstand—the term is sometimes used, incorrectly, to refer to methods less powerful than true α-β pruning. In terms of using given information, α-β pruning is best possible.

The key to understanding α-β pruning is to realize that some of the calculations done in p_*_at are unnecessary. Suppose that we want to compute

$$\max\Big(4, \min(3, \ldots)\Big).$$

After seeing the 4, we know that the value of the max will be at least 4. After seeing the 3, we know that the value of the min will be at most 3. Thus, the values indicated by ... have no effect on the value of the max. Of course, they may affect the value of the min, but that is of no interest to us—we are only interested in the value of the max.

In terms of AND/OR trees, suppose we want to know $p(s)$ for an OR vertex s. After computing $p(t)$ for some $t \in \mathcal{F}(s)$, we know that $p(s) \geq p(t)$. If we encounter some other vertex v after this and learn that $p(v) \leq p(t)$, the exact value of $p(v)$ is irrelevant. All that matters is that it does not exceed $p(t)$. As a result, we could falsely report that $p(v) = p(t)$ without affecting the value that would be computed for $p(s)$.

The idea in the previous paragraph can be incorporated into p_*_at: At all times, we keep track of a lower bound such that values below this bound will not affect computations back toward the root. Thus, if a p_max_at call has seen a value M, nothing is of interest among its descendents unless it exceeds M. As a result, any value not exceeding M can be reported simply as M. Here's the modified code for p_*_at.

```
Procedure p_max_at(s, α)
/* Return p(s) if it exceeds α; else, return at most α. */
   If F(s) = ∅, then return p(s).
   Set M = α.
   /* M is the current lower bound on the values */
   /* of p that must be reported correctly. */
   For t ∈ F(s), do
      Set m = p_min_at(t, M).
      If m > M, then set M = m.
   End for.
   Return M.
End Procedure.


Procedure p_min_at(s, α)
/* Return p(s) if it exceeds α; else return at most α. */
   If F(s) = ∅, then return p(s).
   Set M = ∞.
   For t ∈ F(s), do
      Set m = p_max_at(t, α).
      If m < M, then
         If m ≤ α, then return α.
         Set M = m.
      End if.
   End for.
   Return M.
End Procedure.
```

If the root r is an OR, we compute $\textbf{p\_max\_at}(r, -\infty)$; otherwise, we compute $\textbf{p\_min\_at}(r, -\infty)$. Before proceeding, convince yourself that the code for $\textbf{p\_*\_at}(r, \alpha)$ is correct.

*   *   *   Stop and think about this!   *   *   *

What we have done for lower bounds and OR can also be done for upper bounds and AND. To do this, we introduce β. The result is the complete α-β pruning algorithm:

```
/* Each procedure returns p(s) if it lies in (α, β). */
/* If p(s) < α, the procedures return at most α. */
/* If p(s) > β, the procedures return at least β. */

Procedure p_max_at(s, α, β)
  If F(s) = ∅, then return p(s).
  Set M = α.
  /* M is the current lower bound on the values */
  /* of p that must be reported correctly. */
  For t ∈ F(s), do
    Set m = p_min_at(t, M, β).
    If m > M, then
      If m ≥ β, then return β.
      Set M = m.
    End if.
  End for.
  Return M.
End Procedure.


Procedure p_min_at(s, α, β)
/* Return p(s) if it exceeds α; else return at most α. */
  If F(s) = ∅, then return p(s).
  Set M = β.
  /* M is the current upper bound on the values */
  /* of p that must be reported correctly. */
  For t ∈ F(s), do
    Set m = p_max_at(t, α, M).
    If m < M, then
      If m ≤ α, then return α.
      Set M = m.
    End if.
  End for.
  Return M.
End Procedure.
```

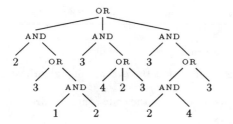 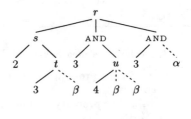

Figure 2.9 The left-hand figure is an AND/OR tree with the values of $p(s)$ specified at the leaves. In the right-hand figure, those vertices which were not consulted are removed. An α or β has been inserted to indicate whether it was the result of alpha pruning or beta pruning, respectively.

Example 2.9   An Illustration of α-β Pruning

Figure 2.9 illustrates the use of the α-β pruning algorithm. The vertices in $\mathcal{F}(s)$ are listed immediately below s and are examined from left to right. If $\mathcal{F}(s) = \emptyset$, the value of $p(s)$ is given; otherwise, the nature of the vertex is given. For simplicity, we speak of one vertex calling another.

We start computation with a call to the root r by **p_max_at**$(r, -\infty, \infty)$.

The root calls the leftmost AND by **p_min_at**$(s, -\infty, \infty)$. After s calls its left child, it has $M = 2$ and so calls t by **p_max_at**$(t, -\infty, 2)$. After obtaining 3 from its left child, t returns 2 to s. This amounts to (beta) pruning the right child of t, which is indicated in the figure by β.

After some calls, u is called from the middle AND by **p_maxat**$(u, 2, 3)$. The value 4, which exceeds 3, causes the pruning shown by the two β's below u. The value 3 is returned by u and the middle AND.

The root then calls the rightmost AND. When it sees 3, alpha pruning takes place as indicated. You should be able to fill in the details. Do so. ∎

If a tree has branching factor b, what is the effective branching factor for α-β search? That is, what would the branching factor be for a tree that examined the same number of vertices at depth d? Suppose that $p(s)$ is assigned randomly to the leaves. If the vertices in each $\mathcal{F}(s)$ are examined in the best possible order, which normally happens only by accident, the effective branching factor is $\sqrt{b}$. If the vertices in each $\mathcal{F}(s)$ are examined in a random order, the effective branching factor is $\frac{x}{1-x}$ where $x \in (0, 1)$ is a root of $x^b + x - 1$. It can be shown that $1 - x \approx \frac{\log b}{b}$ and so the effective branching factor is about $b/\log b$—not a great improvement.

Exercises

2.7.B. Explain the idea behind alpha-beta pruning.

2.7.4. Adapt the proof in Exercise 2.7.2 to show that the α-β pruning algorithm is correct.

Hint. The property that needs to be proved is in the code comment preceding the algorithm.

2.7.5. The 4×4 game of fox and hounds was described in Exercise 2.3.3 (p. 47). Assume that you are the fox.

   (a) Draw the AND/OR tree and the AND/OR graph for the first three moves—but don't assign values.

   (b) Determine the values $p(s)$ for all vertices s of the search tree in (a), using 1 to indicate a fox win and 0 to indicate a fox loss.

2.7.6. Let $x \in (0, 1)$ be a root of $x^b + x - 1 = 0$. In this exercise, you estimate y when b is large.

   (a) Write $x = 1 - y$. Using $\log(1 - y) \approx -y$, show that $e^{-by} \approx y$ and so $y \approx |\log y|/b$.

   (b) Write $y \approx b^{-t}$. Use the previous part to show that $b^{1-t} \approx t \log b$. Suppose that we want to choose a constant value for t so that this approximation is as good as possible as b becomes very large. Show that the best constant approximation for t is $t = 1$.

   (c) Write $y \approx z/b$, deduce that $z \approx \log b - \log z$, and thence conclude that $z \approx \log b$.

*2.7.7. In this exercise you will prove that α-β pruning is best possible in the following sense:

> Let Ξ be an algorithm that is correct and visits leaves in the same order that α-β pruning does. In any AND/OR tree, $\mathcal{A}$ must examine at least those leaves examined by α-β pruning.

Suppose there is some AND/OR tree $\mathcal{T}$ such that Ξ omits some leaves that α-β pruning examines. Let l be the first leaf examined by α-β pruning and not by Ξ. The idea of the proof is to modify $p(l)$ and all leaves that could be examined after $p(l)$ so that $p(r) = p(l)$. It will be convenient to use **p_*_at** to denote either **p_max_at** or **p_min_at**.

   (a) Suppose that t is the last vertex visited by α-β pruning in $\mathcal{F}(s)$ and that it is called via **p_*_at**(t, α, β). Show that, if $\alpha < p(t) < \beta$, then $p(s) = p(t)$ and $\alpha' < p(s) < \beta'$, where s was called via **p_*_at**(s, α', β').

   (b) Let $\mathcal{T}^*$ be the tree obtained by removing all vertices from $\mathcal{T}$ that could be visited after l. Suppose that l was visited by **p_*_at**(l, α, β). Prove that, in $\mathcal{T}^*$, $p(r) = p(l)$ whenever $\alpha < p(l) < \beta$.

   (c) Let $\mathcal{T}^*$, α and β be as above. Define a new tree $\mathcal{T}'$ as follows. It has the same shape as $\mathcal{T}$. For all leaves $v \neq l$ in $\mathcal{T}^*$, $p(v)$ is unchanged. If v is a leaf not in $\mathcal{T}^*$, define $p(v)$ to be α if the last vertex on the path from r to v an OR vertex and define $p(v)$ to be β otherwise. The value of $p(l)$ satisfies $\alpha < p(l) < \beta$. This completes the description of $\mathcal{T}'$. Prove that $p(r) = p(l)$.

(d) Conclude that for some value of $p(l)$, the algorithm Ξ does not compute $p(r)$ correctly for $\mathcal{T}'$.

Notes

Additional introductory material on recursion can be found in books on some areas of computer science and in books on combinatorics. (The latter often have "finite mathematics" in their titles.) The best reference is probably the book by Roberts [11], which is devoted to a practical exposition of basic recursive methods for computer science. Material on trees can also be found in some computer science and combinatorics texts. For material on both trees and recursion, I have an obvious bias toward [1].

Few books are devoted entirely to search. I'm aware of the text by Bolc and Cytowski [2] and the volume edited by Kanal and Kumar [3]. Two similar, readable surveys of search are the contributions by Korf [5] and Pearl and Korf [10].

Russell and Wefald [12] explore questions related to partial search. In particular, they discuss anytime algorithms extensively on pp. 178–182. (See also [13].) For more information on chess playing programs and search in chess, see the book [7] by Levy and Newborn or Newborn's article [8].

For more results on α-β pruning as well has historical information, see the article [4] by Knuth and Moore and the search survey [10] by Pearl and Korf.

References

1. E. A. Bender and S. G. Williamson, *Foundations of Applied Combinatorics*, Addison-Wesley, Reading, MA (1991). This book refers to ordered trees as RP-trees.

2. L. Bolc and J. Cytowski, *Search Methods for Artificial Intelligence*, Academic Press, London (1992).

3. L. Kanal and V. Kumar (eds.), *Search in Artificial Intelligence*, Springer-Verlag, Berlin (1988).

4. D. E. Knuth and R. W. Moore, An analysis of alpha-beta pruning, *Artificial Intelligence* **6** (1975) 293–326.

5. R. E. Korf, Search: A survey of recent results. In H. E. Shrobe (ed.) *Exploring Artificial Intelligence: Survey Talks from the National Conferences on Artificial Intelligence*, Morgan Kaufmann, San Mateo, CA (1986) 197–237.

6. R. E. Korf, Real-time heuristic search, *Artificial Intelligence* **42** (1990) 189–211.

7. D. Levy and M. Newborn, *How Computers Play Chess*, Computer Science Press and W. H. Freeman, New York (1991).

8. M. Newborn, Computer chess: Ten years of significant progress. In M. C. Yovits (ed.), *Advances in Computers*, Vol. 29, Academic Press, San Diego (1989) 197–250.

9. J. Pearl, *Heuristics: Intelligent Search Strategies for Computer Problem Solving*, Addison-Wesley, Reading, MA (1984).

10. J. Pearl and R. E. Korf, Search techniques. In J. F. Traub (ed.), *Annual Review of Computer Science*, Vol. 2, Annual Reviews, Palo Alto, CA (1987) 451–467.

11. E. S. Roberts, *Thinking Recursively*, John Wiley and Sons, New York (1986).

12. S. Russell and E. Wefald, *Do the Right Thing: Studies in Limited Rationality*, MIT Press, Cambridge, MA (1991).

13. S. Russell and S. Zilberstein, Composing real-time systems. In *Proceedings of the Twelfth International Joint Conference on Artificial Intelligence* (IJCAI-90), Morgan Kaufmann, San Mateo, CA (1991) 212–216.

3

The Concepts
of
Predicate Logic

It is reasonable to hope that the relationship between computation and mathematical logic will be as fruitful in the next century as that between analysis and physics in the last. The development of this relationship demands a concern for both applications and mathematical elegance.

—John McCarthy (1967)

Introduction

Reasoning, which is the central problem of AI, is an everyday action for human beings. To develop methods for computer reasoning, we might ask "How do people do it?" Unfortunately, many aspects of everyday reasoning are, at best, poorly understood. There is one exception: logical deduction—the method *supposedly* used by Sherlock Holmes and by mathematicians. Since deductive logic has a firm foundation and sound algorithms, it's a reasonable place to start the quest for reasoning tools.

Mathematical logic formalizes the structures and procedures used in the deductive manipulation of information. Since such manipulations do not require any "understanding," algorithms for logic are ideally suited for use in computer programs. As a result, mathematical logic plays an important role in the quintessential information processing discipline, computer science. Perhaps it's reasonable to embrace mathematical logic as *the* tool for AI. This approach has been taken by some AI researchers, especially in the early years of AI. How does this fit the goals–difficulties–compromises pattern?

- Goal: To develop a method for reasoning about the world.
- Difficulty: A lack of tools, perhaps due to a poor understanding of how people reason.
- Compromise: To use the *deductive* reasoning tools of formal logic, which, it turns out, offer only one possible approach to reasoning.

As we develop these tools, it will become apparent that our compromise creates severe limitations, but provides a basis for overcoming these limits by using other systems of logic. Whether these other logics will be adequate is debated by AI researchers. Currently, most feel that more is needed.

What Is Mathematical Logic?

Mathematical logic distinguishes *syntax* (structure) from *semantics* (meaning). Syntax describes how certain formal "structures" can be built. Semantics tells us how to interpret our structures. In the case of logic, our interpretation is in terms of "true" and "false." You may have encountered syntax and semantics in compiler design. The syntax tells you how to parse the code and the semantics tells you how to translate it into machine instructions.

Given the syntax and semantics, we want methods for manipulating syntactic structures so we can decide if a given structure is true, false, or neither. Such methods of manipulation are called *proof methods*. This agrees with ordinary mathematical usage where a proof of a theorem is a sequence of manipulations that establishes the truth of the theorem.

The simplest level of logic is called *propositional logic*. It formalizes what is meant when we apply the connectives "and," "or," "not," "if...then," and "if and only if" to statements (which are called formulas). Is this enough for our purposes? No. A deeper analysis of syntax and semantics is needed for AI. This is provided by *first-order predicate logic* (FOPL)—often called simply *predicate logic* or *first-order logic* (FOL). In addition to the connectives just mentioned, predicate logic introduces objects, properties and quantifiers. An object may or may not have a particular property. Quantifiers formalize the notions of "for all" and "for some." What does this gain us? For one thing, predicate logic allows us to examine the internal workings of a statement such as "Every person has a mother." (For all P, there is some M such that M is the mother of P.) In contrast, propositional logic treats the statement as an undigestible lump since it contains no connectives that allow us to split it up.

These two logics are also called *caculi*: specifically, *propositional calculus* and *predicate calculus*. A *calculus* is simply a method of calculation. (The mathematics course referred to simply as "calculus" is more properly called "differential and integral calculus.") Although the term "predicate logic" emphasizes structural aspects while "predicate calculus" emphasizes computational aspects, they are used interchangeably.

Propositional logic and first-order predicate logic are covered in this chapter and the next one. Although useful, they are too limited for many AI needs; however, they provide the starting point for other logics, some of which we'll explore in Chapter 6.

Logic and AI

A proof technique is a method for establishing true statements. In AI, we start with a knowledge base and attempt to derive true statements from it. It's possible to mechanize the manipulations required in the steps of a proof. Unfortunately, being able to carry out a step mechanically does not tell us *which* step to carry out. We want efficient algorithms for selecting steps that will allow us to decide if something is true or false. Although this wish cannot be completely fulfilled, useful algorithms have been found.

A particularly efficient algorithm for predicate logic is called resolution of Horn clauses. Since the programming language Prolog is based on this algorithm, we'll use Prolog to provide direction for our study of logic. Prolog is briefly introduced in the next section, and additional syntax is discussed with the relevant aspects of logic. This will not make you a Prolog programmer, but it may give you a better appreciation of predicate logic.

This chapter emphasizes the concepts of predicate logic from an AI perspective, culminating with a partial description of Horn clause resolution. The next chapter provides the algorithmic details and theoretical underpinnings. In Chapter 6, we'll look at extensions.

Prerequisites: The material on ordered trees (Section 2.1) and that on depth-first search (Section 2.2) are needed to understand Prolog.

Used in: This chapter is essential for Chapters 4 and 6. The propositional logic concepts in Section 3.2 are referred to briefly in some nonessential examples in Chapter 7.

3.1   What Is Prolog?

> *Prolog is a programmer's and software engineer's dream.*
> *It is compact, highly readable, and arguably the*
> *"most structured" language of them all.*
>
> —Peter H. Schnupp (1989)

The workhorse languages of computer programming—Pascal, C, Fortran and the like—are called *procedural languages* because we write code to describe the procedures that are to be carried out. In contrast, Prolog is primarily a *declarative language.* Declarative code states information instead of describing manipulations as procedural code does. Put another way, procedural knowledge (and hence code) answers "How?" while declarative knowledge answers "What?" The collection of declarative facts and rules that encapsulate knowledge about a particular subject is referred to as a *knowledge base.*

Prolog manipulates its code by using depth-first search to attempt Horn clause resolution. Purely declarative Prolog code often fits the framework of predicate logic. Procedural statements take us outside that domain. In this book we'll limit ourselves to the predicate logic aspects of Prolog.

Example 3.1   Family Relationships

Looking at a simple example based on family relationships, we can examine some statements about how relationships interact:

1. If X is a parent of A
    and X is a parent of B
    and A and B differ,   then A and B are siblings.
2. If X and Y are siblings
    and X is a parent of A
    and Y is a parent of B, then A and B are cousins.

And we can make some statements about specific relationships:

3. Mary is a parent of Jane.   4. Mary is a parent of John.
5. Jane is a parent of Karen.   6. Jane is a parent of Bill.
7. John is a parent of Jim.

Given the set of rules for interaction and the specific data, we can deduce various facts. For example, "Karen and Bill are siblings."

Prolog provides a language for writing such rules and facts and provides a mechanism for deducing other facts. Here are some rules for translating from

English to Prolog. Upper- and lowercase refer only to the *initial letter*:

English	Prolog
Constant	Lowercase string
Variable	Uppercase string
Relationship	Functional notation
And	Comma
If α then β.	$\beta := \alpha$.

(3.1)

Using this, we can rewrite the previous statements in Prolog:

```
siblings(A,B) :- parent(X,A), parent(X,B), A\=B.    % 1

cousins(A,B) :- siblings(X,Y),                      % 2

    parent(X,A), parent(Y,B).                       % 2

parent(mary,jane).      parent(mary,john).          % 3, 4

parent(jane,karen).     parent(jane,bill).          % 5, 6

parent(john,jim).                                   % 7
```

Suppose we want to locate a sibling for Karen. We would give Prolog the question :- siblings(karen,W). [For the present, don't ask why Prolog considers this to be a question.] As a result of this question, Prolog would attempt to use the first statement with the variable A set to karen and the variable B set to W. Remember that this statement says that the left side of :- is true provided the right side is. The first condition is parent(X,karen), which is true when X is jane because of the fifth statement, parent(jane,karen). The next part of statement 1 is now parent(jane,W). Prolog will try the statements that "define" parent in order. Statements 3 and 4 are quickly discarded because parent(mary,...) cannot agree with parent(jane,...). Statements 5 and 6 are okay; however, 5 fails because it leads to karen\=karen, which is false. Finally, Prolog tells us that W=bill is a solution.

A Prolog question is also referred to as a *query* or *goal clause*. ∎

Here's another simple toy example based on crossword puzzles.

abalone
abandon
anagram
connect
elegant
enhance

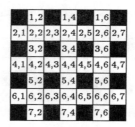

Figure 3.1   The six words on the left are to be placed in the crossword grid. The squares in the grid have been numbered for use in the Prolog program. One solution is shown on the right.

Example 3.2   A Crossword

Suppose that we want to place the words given in Figure 3.1 into the crossword grid shown there. Let `Lij` denote the entry in position (i,j) of the array, numbering from top to bottom and left to right. The following Prolog code solves the problem.

```
word(a,b,a,l,o,n,e).
word(a,b,a,n,d,o,n).
word(a,n,a,g,r,a,m).
word(c,o,n,n,e,c,t).
word(e,l,e,g,a,n,t).
word(e,n,h,a,n,c,e).
:- word(L21,L22,L23,L24,L25,L26,L27),
   word(L41,L42,L43,L44,L45,L46,L47),
   word(L61,L62,L63,L64,L65,L66,L67),
   word(L12,L22,L32,L42,L52,L62,L72),
   word(L14,L24,L34,L44,L54,L64,L74),
   word(L16,L26,L36,L46,L56,L66,L76).
```

One solution is shown in Figure 3.1. Prolog will also find a second solution which is the transpose of this one, that is, the result of interchanging rows with columns. ∎

These examples lead naturally to a variety of questions about Prolog. Among these are

- In more detail, how does Prolog work?
- Why does Prolog work?
- Can Prolog make a mistake?
- What are the limitations of Prolog?

To fully understand and answer these questions, we must explore logic, which we'll do in the next section.

In the meantime, you need to understand a bit more about Prolog to do the exercises.

- Prolog works only with the form of the statements, not the meaning. For example, we know that if *a* and *b* are siblings, then so are *b* and *a*; however, Prolog will not be able to deduce this unless we give it a statement like `siblings(X,Y):-siblings(Y,X)`.

- Variables that appear in statements should be regarded as local variables; that is, the same name may appear elsewhere. You can think of this in terms of procedural languages. Each statement corresponds to a separate procedure; there are no global variables, and Prolog works like call-by-reference (not call-by-value) programming languages.

- The procedural analog extends even further. In attempting to establish the left side of a statement, Prolog tries to verify each of the clauses on the right *in the order in which they are written*. While attempting to verify a clause, Prolog may try to apply a rule. To do this, it attempts to verify, in order, the clauses in the rule. In other words, Prolog uses depth-first search.

- If Prolog's attempt to apply a statement fails, Prolog "forgets" the identifications it may have made. Consider, for example, our previous application of statement 3 in answering `:- siblings(karen,W)`. When Prolog encounters `parent(X,B)`, where X=jane and B=W, it would first try B=karen because it finds the statement "`parent(jane,karen).`" This fails when Prolog attempts to verify that A and B are not equal. As a result, Prolog forgets its B=karen attempt at `parent(jane,B)` and looks for another choice. It then finds "`parent(jane,bill).`" which works because `karen\=bill`.

Exercises

3.1.A. What is the difference between declarative and procedural code?

3.1.B. What is a proof method?

3.1.C. What is the form (syntax) of a Prolog fact? a Prolog rule? a Prolog query?

The following exercises refer to the Prolog code appearing in this section. Since we have not fully explored how Prolog works, your answers might be somewhat vague at some points.

3.1.1. (*Answer follows*) Explain how the question "`:- siblings(W,karen).`" is dealt with.

3.1.2. (*Answer follows*) Assume Prolog can be told to find all possible solutions—which it can. What does it do with the question ":- siblings(A,B)."?

3.1.3. (*Answer follows*) Explain how the question ":- cousins(karen,T)." is dealt with.

3.1.4. (*Answer follows*) Explain how the question ":- cousins(jane,X)." is dealt with.

We'll expand our Prolog example by introducing the predicate sex(X,Y) where X is the name of a person and Y is the person's sex, either male or female. Thus ":- sex(X,male)." will find X whose sex is male.

3.1.5. (*Answer follows*) Using the available predicates, define the predicate mother(X,Y) so that Y is the mother of X.

3.1.6. Using the available predicates, define the predicate brother(X,Y) so that Y is the brother of X.

3.1.7. Using the available predicates, define predicates for aunt and nephew.

*3.1.8. Using the available predicates, define ancestor(X,Y) so that Y is an ancestor of X.

Answers

3.1.1. Prolog would again use the third statement. This time A=W and B=karen. Prolog now tries to satisfy parent(X,W). All parent statements work here. The first two give X=mary. Prolog then fails when it attempts to satisfy parent(X,karen) with X=mary. The third and fourth parent statements give X=jane. Prolog is then able to satisfy parent(X,karen). The third parent statement leads to failure at A\=B because A=karen and B=karen. The fourth parent statement works because A=bill. Thus, Prolog finally tells us that W=bill is a solution.

3.1.2. Prolog would begin by simply identifying A and B with A and B in the first statement. Then the work starts. Prolog tries all possibilities for parent(X,A). Each of these gives a value for X which is then used to try the possibilities for parent(X,B). Finally, the test A\=B is applied. In this way, Prolog will find the following four solutions in the order given.

> A=jane, B=john;   A=john, B=jane;
> A=karen, B=bill;   A=bill, B=karen.

3.1.3. Statement 2 would be applied. Proceeding as in the previous answer, Prolog would first try to satisfy siblings(X,Y). All of these would fail at the next part, parent(X,karen), except for the solution X=jane, Y=john. With this, Prolog attempts to satisfy parent(john,T). This has only one solution, namely T=jim. Thus, Prolog produces the answer T=jim.

3.1.4. Prolog would fail in attempting to satisfy this. The first requirement of statement 2, `siblings(X,Y)` has four solutions—see Exercise 3.1.2. None of these solutions satisfies `parent(X,jane)`, the second requirement in statement 2.

3.1.5. This is somewhat confusing because the order of X and Y is reversed from what might be expected given the order in `parent(X,Y)`. If you do this sort of thing in your Prolog code, you're likely to get confused. Where there is a natural ordering as in ancestry, you should set up all predicates relating to it so that they have the same ordering.

$$\text{mother}(X,Y):-\text{parent}(Y,X),\text{sex}(X,\text{female}).$$

3.2 Propositional Logic

Two different aspects of Prolog statements appeared in the previous section: They have connectives such as "and" and "if...then," and they have properties (called predicates) such as "parent" and "siblings." The formalization and study of the connectives belongs to propositional logic, while that of predicates belongs to predicate logic.

We'll now look at the syntax (form) and semantics (meaning) of propositional logic. In the next section, we'll do the same for predicate logic.

Syntax

A *proposition*, or *formula*, is simply a statement such as "this book is boring" or "if this book is boring, then I'll fall asleep." Lowercase Greek letters will be used to denote formulas. Connectives produce new formulas from old, as in

$$\alpha \text{ and } \beta \qquad \text{and} \qquad \text{if } \alpha \text{ then } \beta.$$

This leads to a recursive definition of formulas. To start the recursive definition of formulas, we'll need some basic formulas, which are often called *propositional letters*. *Connectives* such as "and" and "if...then" are needed to build new formulas from old. Here's a list of connectives and the everyday concept to which they (nearly) correspond—their "meanings."

connective	meaning
$\vee$	or
$\wedge$	and
$\neg$	not
$\rightarrow$	implies (if...then)
$\equiv$	if and only if (iff)

(3.2)

Finally, to avoid ambiguity, let's use parentheses.

Definition 3.1   Syntax of Propositional Logic

Let S be a set whose elements will be called *propositional letters*. We'll denote the propositional letters by p, q, and so on. A *propositional logic language* $\mathcal{L}$ with propositional letters S is the collection of formulas determined by the following four conditions:

(a) All propositional letters are formulas.

(b) If α is a formula, so is $(\neg\alpha)$.

(c) If α and β are formulas, so are $(\alpha \wedge \beta)$, $(\alpha \vee \beta)$, $(\alpha \rightarrow \beta)$, and $(\alpha \equiv \beta)$.

(d) All formulas are obtained in this manner.

Some people call formulas *well formed formulas*, or simply *WFFs.*

Example 3.3   Some Formulas and Nonformulas

Suppose that α, β, and γ are formulas. Then so are

$$(\alpha \rightarrow ((\alpha \rightarrow \beta) \rightarrow \beta)) \qquad ((\neg(\neg\alpha)) \equiv \alpha) \qquad ((\neg(\alpha \vee \beta)) \equiv ((\neg\alpha) \wedge (\neg\beta))).$$

According to the definition, all these parentheses are necessary; however, it's not uncommon to be sloppy and omit pairs of parentheses when the resulting string of symbols corresponds unambiguously to a formula. Don't, for example, be surprised if you see $\alpha \rightarrow \beta$ instead of $(\alpha \rightarrow \beta)$.

To show that the above expressions are formulas, we must show how Definition 3.1 applies. Let's take the second one.

By (b) of the definition, $(\neg\alpha)$ is a formula.

By (b) of the definition, $(\neg(\neg\alpha))$ is a formula.

By (c) of the definition, $((\neg(\neg\alpha)) \equiv \alpha$ is a formula.

We can also use the definition to show that some expressions (or strings) are not formulas. For example, $(\rightarrow \alpha)$ is not a formula: (a) does not apply since it is not a propositional letter; (b) does not apply since there is no "$\neg$"; and (c) does not apply because it requires two formulas separated by a connective. Looking at this in another way, we could say that the definition does not apply because it requires a formula before the connective "$\rightarrow$." ∎

Since formulas are defined recursively, proofs about formulas are usually done by induction. We can induct on the length of the string of symbols or on the number of connectives (since each application of the definition increases both of them)—or we can induct on the number of applications of the definition itself. The following example is based on the idea enunciated in (2.2) (p. 42).

Example 3.4   An Inductive Proof about Formulas

As an illustration of (2.2), let's prove the following almost trivial result.

> Claim: In a formula, the number of left parentheses, the number of
> right parentheses, and the number of connectives are all equal.

Proof: Let $l(\alpha)$, $r(\alpha)$, and $c(\alpha)$ be the number of each in the formula α.
We'll induct on the number of applications of Definition 3.1. The induction
hypothesis, then, is the claim that the claim is true for all formulas obtained
by fewer applications of Definition 3.1 than are needed to produce the formula
γ being considered.

Suppose γ was obtained by applying Definition 3.1(a). Then α is simply
a propositional letter, so $l(\alpha) = 0$, $r(\alpha) = 0$, and $c(\alpha) = 0$.

Suppose γ was obtained by applying Definition 3.1(b). By the induction
hypothesis α is a formula with l, r, and c all equal. In applying (b), we increase
each of l, r, and c by 1 and so they remain equal.

Suppose γ was obtained by applying Definition 3.1(c). By the induction
hypothesis α is a formula with l, r, and c all equal as is β. The application of
Definition 3.1(c) gives one more left parenthesis, one more right parenthesis,
and one more connective. Thus, the numbers remain equal. We can express
this algebraically as

$$l(\alpha * \beta) = l(\alpha) + l(\beta) + 1,$$
$$r(\alpha * \beta) = r(\alpha) + r(\beta) + 1,$$
$$c(\alpha * \beta) = c(\alpha) + c(\beta) + 1,$$

where $*$ is any of the connectives in (c).

Instead of inducting on applications of the definition, we could have in-
ducted on some measure of the formula's complexity, as suggested just before
the example. In this case the number of connectives would be a natural measure of complexity. ■

Exercises

3.2.A. What is the difference between syntax and semantics?

3.2.B. Define the syntax of propositions (also called the formulas of propositional
   logic). What are connectives?

3.2.1. (*Answer follows*) Suppose that α, β, and γ are formulas. In each case, either

   (i) explain why the expression is a formula by showing how Definition 3.1 applies or

   (ii) explain why Definition 3.1 does not apply.

   (a) $\neg\alpha$

   (b) $(\alpha \equiv (\neg\beta))$

   (c) $((\alpha \lor \beta))$

   (d) $(\alpha\neg\beta)$

   (e) $(\alpha \lor \beta \lor \gamma)$

3.2.2. Let $l(\alpha)$ be the number of propositional letters in the proposition α and let $k(\alpha)$ be the number of connectives other than $\neg$. In both cases, repetitions are counted. For example,

$$l\big((p \to (q \land (\neg p)))\big) = 3 \quad \text{and} \quad k\big((p \to (q \land (\neg p)))\big) = 2.$$

Prove that $l = k + 1$.

Answers

3.2.1. Only (b) is a formula. It's obtained by using (b) then (c) in the definition. Parentheses are missing in (a) and (e) while (c) has extra parentheses—(b) and (c) in the definition state when parentheses are used. Negation is used incorrectly in (d)—see (b) in the definition.

Semantics

> *We have to renounce a description of phenomena based on the concept of cause and effect.*
>
> —Niels Bohr (1933)

The meaning of a formula is given in terms of the notions of truth and falsity. We assume that the truth and falsity of propositional letters is known and then recursively compute the truth and falsity of all formulas by paralleling the recursive construction of formulas. For example, "(it is raining) and (the barometer is falling)" is true if "(it is raining)" and "(the barometer is falling)" are both true. If either of the two building blocks is false, so is the compound statement. We can describe this idea by using a *truth table*:

α	β	$\alpha \land \beta$
F	F	F
F	T	F
T	F	F
T	T	T

The first line in the table says that if formula α is false and formula β is false, then formula $\alpha \wedge \beta$ is false. Here's the general definition.

Definition 3.2   Semantics of Propositional Logic

For each propositional letter, we are told whether it is true or false. In view of Definition 3.1, the following truth table describes how to determine the truth or falsity of any formula recursively. The first two columns list all possible combinations of true and false assignments to α and β. The remaining columns list the values of formulas built from α and β. True and false are indicated by T and F, respectively.

α	β	not $\neg\alpha$	or $\alpha \vee \beta$	and $\alpha \wedge \beta$	implies $\alpha \to \beta$	iff $\alpha \equiv \beta$
F	F	T	F	F	T	T
F	T	T	T	F	T	F
T	F	F	T	F	F	F
T	T	F	T	T	T	T

$$(3.3)$$

You may be inclined to disagree with some entries in (3.3). For example, the everyday usage of "α or β" is ambiguous. When we say "α or β" is true we may mean either

(a) at least one of α and β is true or

(b) exactly one of α and β is true.

Logicians have adopted (a) as the meaning for "or" and refer to (b) as "exclusive or." That was simple. Implication—the phrase "if α then β"—is more troublesome, as the next example shows.

Example 3.5   Varieties of Implication

Implication is the most troublesome entry in table (3.3) because necessity and causality are intertwined with our notion of implication. In the everyday usage of "if P then Q," we consider the statement to be true only when there is some connection between the meaning of P and Q. In contrast, FOL (first-order logic) is concerned only about truth values: "If P then Q" is true provided Q is true whenever P is true. This is called *material implication*.

Since necessity and causality are outside the scope of FOL, how should FOL deal with implication? FOL must base the truth and falsity of $(\alpha \to \beta)$ solely on the truth and falsity of α and β, not on any other information about them. This leaves open the question of why $(\alpha \to \beta)$ should be given the interpretation in (3.3). Based on everyday usage, we insist that $\alpha \to \beta$ means that β must be true whenever α is true. This explains why $(\alpha \to \beta)$ is true when α and β are both true, and it explains why $(\alpha \to \beta)$ is false when α is true and β is false. In logic we must consider *all* possibilities. What is the truth of $(\alpha \to \beta)$ when α is false?

Some people argue that we can say nothing—when α is false, it gives us no information about β. That argument is beside the point—we are trying to determine the truth or falsity of $(\alpha \rightarrow \beta)$, not the truth or falsity of β. One way to reach a decision is to take a somewhat different approach. Ask yourself, "How can the claim that α implies β be wrong?"

$*$ $*$ $*$ Stop and think about this! $*$ $*$ $*$

It will be wrong when it happens that α is true but β is not. But this is equivalent to saying that $(\alpha \rightarrow \beta)$ is true otherwise—just what (3.3) states.

Let's digress to look briefly at some other approaches to implication.

One approach hinges on the notion of necessity. "If P then Q" is considered true when Q would necessarily follow if P were true, without regard to whether P is true or not. Consider the two statements,

"If Newtonian mechanics is correct, it is possible to exceed light speed."
"If Newtonian mechanics is correct, it is impossible to exceed light speed."

Both are logically true because we know that Newtonian mechanics is incorrect. However, a physicist would probably tell you that the first is true and the second is false because, *if* Newtonian mechanics *were* true, it could be shown that arbitrarily high speeds are possible. This is an example of *strict implication*, which we'll discuss a bit more on page 228.

The slippery notion of causality suggests another approach to implication. The nature of causality in the everyday use of implication varies:

"If it rained, then the ground is wet."	α caused β;
"If the ground is wet, then it rained."	α was caused by β;
"If roads are wet, then the ground is wet."	a common cause for α and β.

Causality issues may also arise when interpreting other connectives. Thayse [20, p. 6] gives the following example with "and":

"He became afraid and killed the intruder."

versus

"He killed the intruder and became afraid."

Do you see how implied causality leads to very two different meanings? We'll discuss causality further in Chapter 8.

People also use implication procedurally. For example, I might say

"If it's hotter than 78°F, then turn on the air conditioning."

This is far outside the domain of logic because logic deals with truth, not action.

The fact that $\alpha \rightarrow \beta$ leads to so much discussion is a warning sign:

Be cautious! Take extra care in translating between $\alpha \rightarrow \beta$ in mathematical logic and "if α then β" in everyday discourse.

(3.4)

Remember that the first-order logic meaning of implication is *completely* described by the $\alpha \to \beta$ column of (3.3). ∎

We can think of a truth table as a tabular representation of a function. For example, if $\mathcal{V}$ is the function that maps formulas to the truth values $\{\text{T}, \text{F}\}$ according to FOL rules, the table says

$$\mathcal{V}(\alpha \wedge \beta) = \begin{cases} \text{T}, & \text{if } \mathcal{V}(\alpha) = \mathcal{V}(\beta) = \text{T}, \\ \text{F}, & \text{otherwise.} \end{cases}$$

When working with the functional viewpoint, we can replace T and F with 1 and 0, respectively. This enables us to write the truth value of statements built with connectives algebraically. For example,

$$\mathcal{V}(\neg\alpha) = 1 - \mathcal{V}(\alpha),$$
$$\mathcal{V}(\alpha \vee \beta) = \max\big(\mathcal{V}(\alpha), \mathcal{V}(\beta)\big) = \mathcal{V}(\alpha) + \mathcal{V}(\beta) - \mathcal{V}(\alpha)\mathcal{V}(\beta),$$
$$\mathcal{V}(\alpha \wedge \beta) = \min\big(\mathcal{V}(\alpha), \mathcal{V}(\beta)\big) = \mathcal{V}(\alpha)\mathcal{V}(\beta), \tag{3.5}$$
$$\mathcal{V}(\alpha \to \beta) = \max\big(1 - \mathcal{V}(\alpha), \mathcal{V}(\beta)\big),$$
$$\mathcal{V}(\alpha \equiv \beta) = \mathcal{V}(\alpha)\mathcal{V}(\beta) + \big(1 - \mathcal{V}(\alpha)\big)\big(1 - \mathcal{V}(\beta)\big).$$

You should check that these formulas are correct by verifying that they compute the same values given in the tabular definitions (3.3).

Using the truth table idea, we can compute the truth and falsity of more complicated statements. The following table shows that $((\neg(\neg\alpha)) \equiv \alpha)$ is always true. A statement that is always true is called a *tautology*. We also say that the formula is *valid*.

α	$(\neg\alpha)$	$(\neg(\neg\alpha))$	$((\neg(\neg\alpha)) \equiv \alpha)$
F	T	F	T
T	F	T	T

The first column in this table gives all possible truth values for α. The second column is obtained from the third column of (3.3). The third column is obtained by using the second column of this table and the semantics for $\neg$ in (3.3). Finally, the last column is obtained from the first and third column and the semantics for $\equiv$. We can also carry out such calculations using the algebraic form of $\mathcal{V}$, as is done in the proof of the next theorem. Which method is better depends on the situation and your personal taste.

In algebra, we have the fundamental rule "Equals may be substituted for equals." This means that if we know $A = B$, then we may replace any occurrence of A with B without changing the truth of an algebraic statement. In logic, formulas with the same truth values play the role of equals. The connective $\equiv$ plays the role that $=$ plays in algebra: If we know that $\alpha \equiv \beta$, then we may replace α by β without changing the truth of a formula.

Theorem 3.1

Let $\alpha_1, \ldots, \alpha_n$ be formulas. Let $\mathcal{V}$ be 1/0-valued rather than T/F-valued.

(a) Let β be a proposition formed by connecting all of α_i in any order using only the connective $\vee$. Then $\mathcal{V}(\beta) = \max(\mathcal{V}(\alpha_1), \ldots, \mathcal{V}(\alpha_n))$.

(b) Let γ be a proposition formed by connecting all of $\neg\alpha_i$ in any order using only the connective $\wedge$. Then $(\neg\beta) \equiv \gamma$ is a tautology.

(c) Let δ be a proposition formed by connecting all of α_i in any order using only the connective $\wedge$. Then $\mathcal{V}(\delta) = \min(\mathcal{V}(\alpha_1), \ldots, \mathcal{V}(\alpha_n))$.

(d) Let ζ be a proposition formed by connecting all of $\neg\alpha_i$ in any order using only the connective $\vee$. Then $(\neg\delta) \equiv \zeta$ is a tautology.

Parts (a) and (c) imply the *associative* and *commutative* laws for $\vee$ and $\wedge$, respectively.   Recall from elementary algebra:

> associativity: grouping doesn't matter;
> commutativity: order doesn't matter.

Associativity tells us that, although there are many ways to parenthesize $\alpha_1 \vee \alpha_2 \vee \cdots \vee \alpha_n$, they all have the same truth value. Commutativity tells us that the order of the α_i's is also irrelevant. Parts (b) and (d) yield *de Morgan's laws*:

$$\neg(\alpha_1 \vee \cdots \vee \alpha_n) \equiv ((\neg\alpha_1) \wedge \cdots \wedge (\neg\alpha_n))$$
$$\neg(\alpha_1 \wedge \cdots \wedge \alpha_n) \equiv ((\neg\alpha_1) \vee \cdots \vee (\neg\alpha_n)). \tag{3.6}$$

Proof: First note that $\mathcal{V}(\neg\phi) = 1 - \mathcal{V}(\phi)$ for any formula ϕ. You should be able to show that this observation together with (a) and (c) can be used to prove (b) and (d).

Since the proofs of (a) and (c) are similar, let's prove only (a). The proof involves induction on n. When $n = 1$, β is simply α_1, and so there is nothing to prove.

Suppose $n > 1$. Since β is a formula and the only connective used was $\vee$, β must be $\beta_1 \vee \beta_2$ for some formulas β_1 and β_2. Let $S_1 = \{\, j \mid \alpha_j \text{ appears in } \beta_1 \,\}$ and define S_2 similarly. Since $|S_i| < n$ and β_i is built using only $\vee$, the induction hypothesis tells us that

$$\mathcal{V}(\beta_1) = \max_{i \in S_1} \mathcal{V}(\alpha_i) \quad \text{and} \quad \mathcal{V}(\beta_2) = \max_{i \in S_2} \mathcal{V}(\alpha_i).$$

By (3.5), $\mathcal{V}(\beta_1 \vee \beta_2) = \max(\mathcal{V}(\beta_1), \mathcal{V}(\beta_2))$. The theorem now follows. ∎

The following result allows us to eliminate occurrences of "implies."

Theorem 3.2

Let α_1, ..., α_n and β be formulas. Then the following is a tautology. (Parentheses have been omitted to avoid clutter.)

$$\left((\alpha_1 \wedge \cdots \wedge \alpha_n) \to \beta\right) \equiv \left((\neg\alpha_1) \vee \cdots \vee (\neg\alpha_n) \vee \beta\right).$$

Proof: Use a truth table to show that

$$(\delta \to \beta) \equiv ((\neg\delta) \vee \beta). \tag{3.7}$$

is a tautology. Let δ be $\alpha_1 \wedge \cdots \wedge \alpha_n$ and apply the second of de Morgan's laws. ∎

Example 3.6   More Complicated Connectives

The connectives discussed so far involve only one or two propositions. What about more complicated connectives? Consider

$$\text{if } \alpha \text{ then } \beta, \text{ else } \gamma, \tag{3.8}$$

which is a popular construct in programming. In natural language, "else" is usually replaced by "otherwise" as in

> If it's sunny, I'll cut AI class; otherwise I'll attend.

Logically, (3.8) is equivalent to $((\alpha \to \beta) \wedge ((\neg\alpha) \to \gamma))$. Consequently we don't need a new connective to express this idea. We saw earlier that $(\alpha \to \beta)$ is equivalent to $((\neg\alpha) \vee \beta)$. Thus the connective "$\to$" is also unnecessary. This suggests the general question:

> What connectives do we need to be able to express everything?   (3.9)

To answer this, we need to clarify "express everything."

Since the focus of logic is truth, all that matters about a statement is its truth table. In other words, a statement in propositional logic can be viewed as a function $f(\alpha_1, \ldots, \alpha_n)$ whose domain is $\{\text{T}, \text{F}\}^n$ and whose range is $\{\text{T}, \text{F}\}$. From this viewpoint, building new formulas from old using Definition 3.1 (p. 94) is simply a matter of functional composition. For example, $((\neg\alpha) \vee \beta)$ is the "or" function applied to two arguments, the first being the result of applying the "not" function to α and the second being simply β.

Now, rephrasing our vague question (3.9), we can ask precisely: What functions are needed to obtain, via functional composition, all possible functions from $\{\text{T}, \text{F}\}^n$ to $\{\text{T}, \text{F}\}$? One answer is found in the next theorem.

Theorem 3.3   NAND Suffices

Define the function $\text{NAND} : \{\text{T}, \text{F}\}^2 \to \{\text{T}, \text{F}\}$ by

$$\text{NAND}(\alpha, \beta) = (\neg(\alpha \wedge \beta)).$$

It's also written $(\alpha \uparrow \beta)$. For every $n > 0$, every function from $\{\text{T}, \text{F}\}^n$ to $\{\text{T}, \text{F}\}$ can be obtained from NAND by functional composition.

Since this theorem is particularly important in certain technologies for building logic circuits, you may have seen it in another computer science course. (It's important because NAND is easily built and the theorem tells us that it's enough.)

Proof: First, let's see how to build up some of the common functions. You can easily verify that

$$\mathcal{V}(\neg\alpha) = \mathcal{V}(\alpha \uparrow \alpha)$$

and so

$$\mathcal{V}(\alpha \wedge \beta) = \mathcal{V}\Big((\alpha \uparrow \beta) \uparrow (\alpha \uparrow \beta)\Big).$$

By de Morgan's law, $(\alpha \vee \beta)$ is equivalent to $\big(\neg((\neg\alpha) \wedge (\neg\beta))\big)$, which is equivalent to

$$\Big((\alpha \uparrow \alpha) \uparrow (\beta \uparrow \beta)\Big).$$

We've created $\neg$, $\wedge$, and $\vee$ from NAND.

Now let's prove by induction on n, the number of variables, that any function can be built from $\neg$, $\wedge$, and $\vee$, and hence also from NAND.

When $n = 1$, there are only four possible functions $f(\alpha)$. You should be able to list them and show how to construct them using NAND. It helps to note that $(\neg(\alpha \wedge (\neg\alpha)))$ is always true.

If $n > 1$, you should be able to show that $f(\alpha_1, \ldots, \alpha_n)$ is equivalent to

$$\Big(\alpha_n \wedge f(\alpha_1, \ldots, \alpha_{n-1}, \mathrm{T})\Big) \vee \Big((\neg\alpha_n) \wedge f(\alpha_1, \ldots, \alpha_{n-1}, \mathrm{F})\Big).$$

This completes the proof, because $f(\alpha_1, \ldots, \alpha_{n-1}, \mathrm{T})$ and $f(\alpha_1, \ldots, \alpha_{n-1}, \mathrm{F})$ are both just functions of $n-1$ variables. (They are, in general, not the *same* function because the first is obtained by setting the nth variable to T and the second by setting it to F.) ■

Exercises

3.2.C. Define the semantics of formulas in propositional logic.

3.2.D. How does causality enter into interpretations of implication and why does logic ignore it?

3.2.E. What is a truth table?

3.2.3. Suppose we have a formula that contains k propositional letters. Show that the truth table for this formula contains 2^k rows.

3.2.4. Verify that each of the algebraic expressions for $\mathcal{V}$ of connectives given in (3.5) agrees with those (3.3).

3.2.5. Construct truth tables to determine which of the following formulas are tautologies.

   (a) $(\alpha \vee (\neg \alpha))$

   (b) $(\alpha \wedge (\neg \alpha))$

   (c) $(\alpha \rightarrow (\alpha \vee \beta))$

   (d) $(\alpha \rightarrow (\alpha \wedge \beta))$

   (e) $((\alpha \wedge \beta)) \rightarrow \alpha)$

   (f) $((\alpha \vee \beta)) \rightarrow \alpha)$

   (g) $((\alpha \rightarrow \beta) \equiv ((\neg \alpha) \vee \beta))$

   (h) $((\alpha \rightarrow \gamma) \rightarrow (\alpha \rightarrow (\beta \rightarrow \gamma)))$

   (i) $((\alpha \rightarrow \beta) \rightarrow (\alpha \rightarrow (\beta \rightarrow \gamma)))$

3.2.6. Complete the proof of Theorem 3.1.

3.2.7. Prove that (3.7) is a tautology.

3.2.8. Here are four rearrangements of implication

$$\alpha \rightarrow \beta \qquad \text{(original formula)}$$
$$\beta \rightarrow \alpha \qquad \text{(converse)}$$
$$(\neg \alpha) \rightarrow (\neg \beta) \qquad \text{(inverse)}$$
$$(\neg \beta) \rightarrow (\neg \alpha) \qquad \text{(contrapositive).}$$

(a) Show that the original formula and the contrapositive are equivalent; that is,

$$(\alpha \rightarrow \beta) \equiv ((\neg \beta) \rightarrow (\neg \alpha))$$

is a tautology.

(b) For the other five possible equivalences, which are equivalent and which are not? (In particular, the original formula and its converse are *not* equivalent; however, people sometimes assume that they are.)

(c) Can you recall an example from real life where someone assumed that an implication and its converse were equivalent?

3.2.9. Complete the proof of Theorem 3.3:

(a) List all four functions from $\{T, F\}$ to $\{T, F\}$ and express them using NAND.

(b) Fill in the details for $n > 1$.

3.3 Predicate Logic

If your thesis is utterly vacuous
Use first-order predicate calculus.
With sufficient formality
The sheerest banality
Will be hailed by the critics: "Miraculous!"

—Henry A. Kautz (1986)

Predicate logic allows us to look into the structure of phrases that propositional logic treats as "black boxes" denoted by propositional letters. Consider the statement

"If John is human, then John has a human mother."

In propositional logic, we could write this statement as

$$(p \rightarrow q) \text{ where } p = \text{"John is human" and}$$
$$q = \text{"John has a human mother."} \tag{3.10}$$

To explore the structure of p and q, we use predicate logic as follows:

$$p = \text{human(john)} \text{ and } q = \text{has\_human\_mother(john)}. \tag{3.11}$$

Here, "human" and "has_human_mother" are *predicates*. When a predicate is applied to its argument(s), the result is either true or false. The arguments of predicates may be constants, like "john," or variables. The arguments of predicates are called terms. To mimic Prolog, we'll use lowercase for predicates and constants and uppercase for variables.

What purpose do variables serve? The statement in (3.10) and (3.11) actually applies to all things, not just John. In other words,

$$\Big(\text{human}(X) \rightarrow \text{has\_human\_mother}(X)\Big)$$

for all choices of X. We express this by saying

$$\forall X \Big(\text{human}(X) \rightarrow \text{has\_human\_mother}(X)\Big). \tag{3.12}$$

The string $\forall X$ is read "for all X." Various expressions in English are equivalent to "for all"; for example, "for every" and "for each." "For all" is often tacitly assumed in an implication; for example, "if $x > 0$, then ..." means $\forall x ((x > 0) \rightarrow \ldots)$. (Note that this is *not the same* as $(\forall x (x > 0)) \rightarrow \ldots$.)

Although (3.12) captures much more than does $p \rightarrow q$ and even more than (3.10) and (3.11) combined, it still lacks something. We may not think of "has_human_mother" as a predicate with a single argument; instead,

"John has a human mother"

may be thought of as

> "There is someone who is human and who is John's mother."

We can capture this with predicate logic; in fact, we can state it for all people instead of just for John:

$$\forall X\left(\text{human}(X) \rightarrow \Big(\exists Y\big(\text{human}(Y) \wedge \text{mother}(Y, X)\big)\Big)\right), \qquad (3.13)$$

where $\exists Y$ is read "for some Y" and mother(Y, X) means "Y is the mother of X." Other English versions of $\exists$ are "there exists" and "there is."

We can express the idea in (3.13) by using functions as follows: When the predicate human(X) is true, the function mother(X) will have as its value the mother of X. We don't care how mother(X) is defined when human(X) is false. Using the function "mother," we could restate the information in (3.13) as

$$\forall X\left(\text{human}(X) \rightarrow \text{human}\big(\text{mother}(X)\big)\right).$$

In this, human( ) is a predicate and mother( ) is a function.

We use the same notation for functions and predicates. Is this because they're the same? No!

> A function produces a value that is a term.
> but
> A predicate produces only "true" and "false."

Predicates are truth-valued functions of *terms* and are defined *when interpretations are given*. Connectives are truth-valued functions of *formulas* and are defined *in propositional logic by* (3.3) (p. 97). For example, we insist that $\neg \alpha$ is true if and only if α is false.

Syntax

The previous discussion serves as the foundation for the following series of rather lengthy definitions.

Definition 3.3   The Elements of Predicate Logic Language

A predicate logic language $\mathcal{L}$ consists of the following symbols:

- an infinite set of *variables*, denoted by uppercase letters;
- a set of *constants*, denoted by lowercase letters, usually a, b, etc.;
- a set of *predicates*, denoted by lowercase letters, usually p, q, etc.;
- a set of *functions*, denoted by lowercase letters, usually f, g, etc.;
- the *connectives* $\neg$, $\vee$, $\wedge$, $\rightarrow$, and $\equiv$;
- the *quantifiers* $\forall$ and $\exists$; and
- the parentheses) and (.

Predicates and functions take "arguments" in a manner to be specified later. (We'll usually assume that all the sets are *countable*; that is, they are either finite or can be put into one-to-one correspondence with the positive integers. Since we're concerned with ideas that can be implemented on a computer, this isn't a severe restriction.)

The notation in the definition is consistent with that for propositional logic: The connectives play the same role in both logics and the predicates are a generalization of propositional letters. In fact, predicate logic using only

- connectives,
- parentheses, and
- predicates that take zero arguments

is propositional logic.

Definition 3.4   The Syntax of Predicate Logic Terms

The terms in $\mathcal{L}$ are defined recursively as follows:

(a) Every variable and every constant is a term.

(b) If $t_1, \ldots, t_n$ are terms and f is a function that takes n arguments, then $f(t_1, \ldots, t_n)$ is a term.

(c) Every term is obtained in this manner.

Terms with no variables are called *variable-free terms*.

In the syntax of predicate logic, functions are merely symbols. In the semantics (interpretation) of predicate logic, functions are functions in the ordinary sense and their ranges and domains are the constants of Definition 3.3. To distinguish between the symbol and its interpretation, some authors speak of "function symbols," rather than functions in the preceding definitions.

Definition 3.5   The Syntax of Predicate Logic Formulas

The formulas in $\mathcal{L}$ are defined recursively as follows.

(a) If $t_1, \ldots, t_n$ are terms and p is a predicate that takes n arguments, then $p(t_1, \ldots, t_n)$ is a formula, called an *atomic formula*.

(b) If α is a formula, so is $(\neg \alpha)$.

(c) If α and β are formulas, so are $(\alpha \vee \beta)$, $(\alpha \wedge \beta)$, $(\alpha \to \beta)$, and $(\alpha \equiv \beta)$.

(d) If V is a variable and α is a formula, then $(\forall V \, \alpha)$ and $(\exists V \, \alpha)$ are formulas.

(e) Every formula is obtained in this manner.

In $(\forall V \, \alpha)$ and $(\exists V \, \alpha)$, we say that α is the *scope* of the quantifier and that all occurrences of V in α are *bound* (by the quantifier). If a formula contains a variable that is not bound, we say that the variable has a *free occurrence* in the formula.

Bound variables are sometimes called "dummy variables." For example, t is a dummy (or bound) variable in $\int_0^x f(t)\, dt$. In this case, the fact that t is bound is indicated by the dt. The t could be replaced by any other variable without changing the meaning of the formula. In fact, we could rewrite it as $\int_0^x f(x)\, dx$. This can lead to confusion. For example,

$$\int_0^1 \left(\int_0^x f(x)\, dx \right) dx \quad \text{and} \quad \int_0^1 \left(\int_0^x f(t)\, dt \right) dx$$

are exactly the same since in either case we first integrate f as its argument ranges from 0 to x to obtain a new function of x which is then integrated from 0 to 1. In integral calculus, people normally use different names for different bound variables to avoid confusion. They also usually use different names for bound and unbound variables.

The same sort of confusion can arise in predicate calculus. Consider

$$((\exists X \, p(X)) \vee q(X)) \quad \text{and} \quad \Big(\exists X \, ((\forall X \, p(X)) \to q(X)) \Big).$$

In the first formula, the X in $p(X)$ is bound and the X in $q(X)$ is free. In the second formula, both occurrences of X are bound, but they are bound by different quantifiers. We can avoid such problems by insisting that every quantifier in a formula refer to a different variable. Specifically, in Definition 3.5(c), we could insist that any bound variable in α be different from all variables in β and every bound variable in β be different from all variables in α. To conform with this, we can rewrite our preceding formulas as

$$((\exists V \, p(V)) \vee q(X)) \quad \text{and} \quad \Big(\exists V \, ((\forall X \, p(X)) \to q(V)) \Big).$$

Exercises

3.3.A. What is the symbol for a universal quantifier? an existential? What do they correspond to in everyday discourse?

3.3.B. What is the definition of formulas in predicate logic?

3.3.C. What are atomic formulas? free variables? bound variables?

3.3.D. Explain how to translate a statement such as "for all x such that $x > 0$ ..." into predicate logic. Do the same with "for some x such that $x > 0$ ..."

3.3.1. (*Answer follows*) For each of the following, indicate whether or not it is a formula in predicate logic and, if not, why not. Do not be concerned about parentheses.

(a) $X \vee p(X)$

(b) $p(X) \vee p(X)$

(c) $\exists X \, q(Y)$

(d) $\forall X (\exists X \, p(X))$

(e) $p(q(X), Y)$

(f) $\forall X (\exists p \, p(X))$

(g) $p(X \vee Y)$

3.3.2. (*Answer follows*) For each of the following formulas, identify the bound variables and rewrite the formulas so that the bound variables have unique names. Do not change the names of any free variables.

(a) $\forall X \Big(\big(\exists Y (p(X) \equiv q(Y)) \big) \rightarrow r(X, Y) \Big)$

(b) $q(X, Y) \rightarrow \big((\exists X \, p(X)) \vee \exists X \, r(X) \big)$

(c) $\forall X \Big(\big(\forall X \, p(X) \big) \rightarrow p(X) \Big)$

3.3.3. How would you translate "there is no X such that ..." into predicate logic?

Answers

3.3.1. It may be unclear *why* (c) and (d) are formulas, so let's explain that, too.

(a) This is not a formula since $X \vee p(X)$ can be a formula only if both X and $p(X)$ are formulas; however, X is a term, not a formula.

(b) This is a formula.

(c) This is a formula; there's no requirement that the variable being bound actually appear in the formula.

(d) This is a formula, but is somewhat confusing because of the repeated name for bound variables. The inner $\exists X$ bounds the X, leaving no variables for the outer $\forall X$ to bind. In a formula like $\forall X\big((\exists X\, p(X)) \vee q(X)\big)$, the X in $p(X)$ is bound by the $\exists X$ while the X in $q(X)$ is bound by $\forall X$. Thus, the formula is equivalent to $\forall Y\big((\exists X\, p(X)) \vee q(Y)\big)$.

(e) This is not a formula; the arguments of a predicate must be terms and $q(X)$ is not a term.

(f) This is not a formula; we don't have quantifiers for predicates. (That would be *second*-order predicate calculus.)

(g) This is not a formula; the arguments of predicates must be terms and $X \vee Y$ is not a term.

3.3.2. The bound variables are underlined. Subscripts are used to provide unique names.

(a) $\forall \underline{X}\Big((\exists \underline{Y}(p(\underline{X}) \equiv q(\underline{Y}))) \to r(\underline{X}, Y)\Big)$ becomes

$\forall X_1\Big((\exists Y_1(p(X_1) \equiv q(Y_1))) \to r(X_1, Y)\Big)$

(b) $q(X, Y) \to \big((\exists \underline{X}\, p(\underline{X})) \vee \exists \underline{X}\, r(\underline{X})\big)$ becomes

$q(X, Y) \to \big((\exists X_1\, p(X_1)) \vee \exists X_2 r(X_2)\big)$

(c) $\forall \underline{X}\big((\forall \underline{X} p(\underline{X})) \to p(\underline{X})\big)$ becomes $\forall X_1\big((\forall X_2\, p(X_2)) \to p(X_1)\big)$

Semantics

> *He who would distinguish the true from the false*
> *must have an adequate idea of what is true and false.*
>
> —Benedict Spinoza (1677)

The syntactic definitions tell us how to construct everything in the language of first-order predicate calculus. As in propositional logic, the syntax tells us nothing about what our formulas "mean." To associate meaning with predicate logic formulas, we must know how to interpret them. Unfortunately, this is more complicated than the simple true/false of propositional logic. The approach logicians use to define semantics involves the discussion of models. We'll take a more informal approach now; but we'll have to be more careful in the next chapter.

Definition 3.6   Informal Semantics for Predicate Logic

Our semantics will specify when a formula is true in a recursive manner that parallels the syntactic definition of the formula. The range of a function is the set of constants. The domains of functions and predicates are n-tuples of constants. Our definition depends on knowing the functions and knowing when predicates are true—an "interpretation."

(a) If p is a predicate and none of the terms $t_1, \ldots, t_n$ contains variables, then $p(t_1, \ldots, t_n)$ is either true or not according to the interpretation.

(b) and (c) If the truths of α and β are known, then the truth of connectives is determined by (3.3) (p. 97).

(d) Let V be a variable and α a formula. If there is some constant c such that replacing every free occurrence of V in α with c gives a true formula, then $(\exists V \, \alpha)$ is true. (The restriction to free V is needed because a quantifier in α might bind some occurrences of a variable that is also called V.) $\exists$ is called an *existential quantifier*.

(d') Let V be a variable and α a formula. If, for every constant c, replacing every free occurrence of V in α with c gives a true formula, then $(\forall V \, \alpha)$ is true. $\forall$ is called a *universal quantifier*.

A formula is called *valid* or a *tautology* if and only if it is true for all possible interpretations.

There are a couple of important facts to note about Definition 3.6. First, it defines the truth and falsity of formulas only when there are no free occurrences of variables; so don't try to apply it to a formula where a variable occurs freely. Second, as in propositional logic, the definition is often applied in the reverse direction of the definition of syntax—while syntax builds up, semantics tears down. For example, consider

$$((\forall X \, p(X)) \rightarrow (\exists X \, p(X))). \tag{3.14}$$

To determine the truth of (3.14) we must first determine the truth of $(\forall X \, p(X))$ and $(\exists X \, p(X))$. There are three relevant possibilities for the truth of $p(\;)$. Here they are, along with their consequences.

- $p(c)$ is true for all c. In this case, both $(\forall X \; p(X))$ and $(\exists X \; p(X))$ are true. Thus (3.14) is true.

- $p(c)$ is false for all c. In this case, both $(\forall X \; p(X))$ and $(\exists X \; p(X))$ are false. Thus (3.14) is true.

- $p(c)$ is true for some c and false for some c. In this case, $(\forall X \; p(X))$ is false and $(\exists X \; p(X))$ is true. Thus (3.14) is true.

Example 3.7   Positive Integers

For this example, let our constants be the positive integers $\{1, 2, 3, \ldots, \}$. Our functions are addition and multiplication, which we'll write in infix notation. We'll also omit some pairs of parentheses. The following show how additional concepts can be defined in terms of the predicate "equal":

$$\forall X \Big(\mathrm{odd}(X) \equiv \big(\exists Y \; \mathrm{equal}(2 \times Y, X + 1) \big) \Big)$$
$$\forall X \Big(\mathrm{odd}(X) \equiv \big(\neg \exists Y \; \mathrm{equal}(2 \times Y, X) \big) \Big) \tag{3.15}$$
$$\forall Z \Big(\mathrm{prime}(Z) \equiv \big(\neg (\exists X \; \exists Y \; \mathrm{equal}((X + 1) \times (Y + 1), Z)) \big) \Big).$$

These can be regarded as definitions of the concepts on the left-hand side of the $\equiv$. For example, the first line says that X is odd if and only if there is an integer whose double equals $X + 1$. These formulas are true in one interpretation—the positive integers in which all the functions and predicates have their usual meanings. But they are not tautologies because there are interpretations in which they are not true. For example, if we make no changes in the interpretation except to interchange the meanings of the functions $\times$ and $+$, the formulas will not be true in this interpretation. ∎

The concluding sentences of Example 3.7 appear to suggest an insurmountable problem: If we look just for valid formulas (i.e., every interpretation is true), we're unlikely to make much progress. Actually, the reverse is true! Because every interpretation must make a formula true before we call it valid, it turns out we need not be concerned with the real world meaning of interpretations when carrying out computer manipulations. This is fortunate because computers are not aware of the real world either.

The real world must enter *somehow*. It enters through our knowledge base. Suppose that $\alpha_1, \alpha_2, \ldots$ are statements in our knowledge base that express what we know to be true for our *intended* interpretation of the predicates, functions and constants. Suppose that

$$(\alpha_1 \wedge \alpha_2 \wedge \ldots) \rightarrow \beta \tag{3.16}$$

is a tautology. In particular, it is true with the real world interpretation we have in mind, so β tells us something about the real world. On the other hand, since (3.16) is a tautology, it is possible to establish it without appealing to its meaning—a perfect job for a computer program. This is just the sort of thing Prolog does.

Let's look at another example of how things are expressed in predicate logic terms. This one deals with the definition of a limit, which is notorious for causing problems for calculus students. These problems often arise because it's not easy to understand

- what the quantifiers imply,
- what order they appear in, and
- what manipulations are allowed when formulas contain quantifiers.

The next example addresses the first two issues.

Example 3.8   The Definition of Limit

We say that $\lim_{x \to a} f(x) = L$ if

$$\text{for all } \epsilon > 0, \text{ there is a } \delta > 0 \text{ such that} \atop |f(x) - L| < \epsilon \text{ whenever } |x - a| < \delta \text{ and } x \neq a. \tag{3.17}$$

We want to translate (3.17) into predicate calculus notation. Our constants will be the real numbers $\mathbb{R}$. Again, we'll write functions such as absolute value in the usual form. Note that a statement like $\epsilon > 0$ involves a predicate. We could write it as greater$(\epsilon, 0)$, but we'll stick with the more standard mathematical notation.

As written, the definition contains an implicit quantifier. "For all" and "there is" are clearly quantifiers for ϵ and δ, respectively. After "such that" the variable x begins to appear, but no quantifier is mentioned. How is it quantified? A more accurate definition would have said "such that for all x" instead of merely "such that."

There are two main problems in carrying out this formalization of the definition in FOL: (i) What does "whenever" mean and (ii) how dow we include the condition $\epsilon > 0$ when we introduce a quantifier for ϵ?

First, "α whenever β" means "if β then α." This changes (3.17) to

for all $\epsilon > 0$ there is a $\delta > 0$ such that for all x

$$\left(\left((|x - a| < \delta) \wedge (x \neq a) \right) \to (|f(x) - L| < \epsilon) \right). \tag{3.18}$$

What does "for all $\epsilon > 0$ γ" mean, where γ is a predicate logic formula? If you think about it a bit, you should be able to convince yourself that it means $\forall \epsilon ((\epsilon > 0)) \to \gamma)$. What about the phrase "there exists $\delta > 0$ such that γ"? Again, if you think about it, you should realize that it means $\exists \delta ((\delta > 0) \wedge \gamma)$. Notice the important difference in translation of universal (for all) versus existential (there is) quantifiers with conditions. The discussion shows that

$$\boxed{\begin{array}{ll} \text{"for all } X \text{ with } P(X) \text{ we have } \gamma\text{"} & \text{means} \quad \forall X \left(P(X) \to \gamma \right) \\ \text{"there is } X \text{ with } P(X) \text{ such that } \gamma\text{"} & \text{means} \quad \exists X \left(P(X) \wedge \gamma \right) \end{array}} \tag{3.19}$$

If this is unclear, try to think of additional examples. Incorporating this into (3.18) leads to the FOL translation of the definition. Here it is, with braces used in place of some parentheses to improve readability,

$$\forall \epsilon \left\{ (\epsilon > 0) \to \left(\exists \delta (\delta > 0) \wedge \phi \right) \right\}$$

where ϕ stands for $\hspace{8cm} (3.20)$

$$\forall x \left(\{ (|x - a| < \delta) \wedge (x \neq a) \} \to (|f(x) - L| < \epsilon) \right). \quad \blacksquare$$

As we've seen, the process of translation can be tricky. In particular, anyone setting up a rule-based expert system must be careful when doing translations. On the one hand, this example is more convoluted than we usually encounter in practice. On the other hand, the imprecision of everyday English leads to other problems. The next example and some of the exercises provide practice in translating from English to predicate logic.

Example 3.9   A Lewis Carroll Example

The Oxford geometer Charles Dodgson is famous for writing *Alice in Wonderland* and *Through the Looking Glass* under the pen name Lewis Carroll. Two years before his death, he published a symbolic logic text containing delightful problems. We'll look at one of his problems in this example and at some others in the exercises.

"(1) Coloured flowers are always scented;

(2) I dislike flowers that are not grown in the open air;

(3) No flowers grown in the open air are colourless."   [3, p. 115]

How can we recast these in terms of predicate logic?

Our constants will be flowers and our predicates will be as follows:

$c(X)$: indicates X is coloured,
$d(X)$: indicates I dislike X,
$g(X)$: indicates X is grown in the open air,
$s(X)$: indicates X is scented.

You should be able to verify that the following are translations of the statements:

1. $\forall X \Big(c(X) \to s(X) \Big)$

2. $\forall X \Big((\neg g(X)) \to d(X) \Big)$

3. $\neg \Big(\exists X \Big(g(X) \wedge (\neg c(X)) \Big) \Big)$

Other translations are possible, but I believe these are the most direct and also reflect the meaning of the English.

Later we'll learn a proof method for predicate logic that let's us draw conclusions from these formulas. In this case, one possible conclusion is

$$\forall X \Big((\neg s(X)) \to d(X) \Big);\tag{3.21}$$

that is, I dislike all flowers that are not scented.

This example illustrates some of the problems of translating natural language to predicate calculus.

1. Although the first two statements did not contain overt implications, both were translated that way. A statement of the form, "X [that are] p are q" usually translates as $\forall X \big(p(X) \to q(X) \big)$.

2. In a statement like the second, "that are" is the same as "whenever," so we have a statement of the form "$p(X)$ whenever $q(X)$." This translates as $\forall X \big(q(X) \to p(X) \big)$.

3. In the third statement, the predicate "colourless" appears. This is simply the negation of the predicate "coloured." In fact, English employs various devices to negate predicates; for example, "known" negates to "unknown" and "like" roughly negates to "dislike." When a predicate and its negation appear as words, one should be replaced by the negative of the other. Sometimes negation is more subtle as in "dead" and "alive." What about "long" and "short"? ■

The next theorem is useful for moving quantifiers past the connectives $\neg$, $\vee$, and $\wedge$. In our discussion of propositional logic, we saw that these connectives are more than sufficient.

Theorem 3.4 Quantifiers

Let α be a formula in which any occurrences of X and Y are free and let β be a formula with no free X. Let $*$ be either $\vee$ or $\wedge$. Then the following formulas are tautologies:

$$(\forall X(\forall Y\ \alpha)) \equiv (\forall Y(\forall X\ \alpha)) \quad \text{and} \quad (\exists X(\exists Y\ \alpha)) \equiv (\exists Y(\exists X\ \alpha)), \ (3.22)$$
$$(\neg(\forall X\ \alpha)) \equiv (\exists X\ \neg\alpha) \quad \text{and} \quad (\neg(\exists X\ \alpha)) \equiv (\forall X\ \neg\alpha), \qquad (3.23)$$
$$((\forall X\ \alpha)*\beta) \equiv (\forall X(\alpha*\beta)) \quad \text{and} \quad ((\exists X\ \alpha)*\beta) \equiv (\exists X(\alpha*\beta)). \ (3.24)$$

Proof: Abuse notation and write $\alpha(X)$ to indicate the free occurrences of X in α. Use $\Longleftrightarrow$ to stand for the phrase "if and only if." Let's prove (3.23) here and leave the rest as an exercise. By the definition of the semantics,

$$\neg(\forall X\ \alpha(X)) \text{ is true} \iff (\forall X\ \alpha(X)) \text{ is false}$$
$$\iff \text{there is some constant } c \text{ such that } \alpha(c) \text{ is false}$$
$$\iff (\exists X\ \neg\alpha(X)) \text{ is true}$$

The right side of (3.23) follows similarly. ∎

We'll use the following corollary to convert Prolog rules to *Horn clauses*.

Corollary 3.4.1

Let $Y_1,\ldots,Y_k$ be variables that do not appear in β and let $X_1,\ldots,X_j$ be distinct from the Y_i's. The following three formulas are either all true or all false:

$$\forall X_1\ldots\forall X_j\left(\left(\exists Y_1\ldots\exists Y_k(\alpha_1\wedge\cdots\wedge\alpha_n)\right)\to\beta\right)$$
$$\forall X_1\ldots\forall X_j\forall Y_1\ldots\forall Y_k\left((\alpha_1\wedge\cdots\wedge\alpha_n)\to\beta\right) \qquad (3.25)$$
$$\forall X_1\ldots\forall X_j\ \forall Y_1\ldots\forall Y_k\left(\beta\vee(\neg\alpha_1)\vee\cdots\vee(\neg\alpha_n)\right).$$

Proof: Start with the first formula and convert the implication $(\gamma\to\beta)$ to $(\beta\vee(\neg\gamma))$. Next, move the negation $\neg$ through the existential quantifiers $\exists Y_i$ using (3.23), then use (3.24) to move the resulting universal quantifiers $\forall Y_i$ outside. Now, either convert the "$\vee$" back to an implication to obtain the second formula or use de Morgan's law to move $\neg$ through the $\wedge$'s to obtain the third. ∎

c	d	g	s	1	2	3	all
T	T	T	T	T	T	T	*
T	T	T	F	F	T	T	
T	T	F	T	T	T	T	*
T	T	F	F	F	T	T	
T	F	T	T	T	T	T	*
T	F	T	F	F	T	T	
T	F	F	T	T	F	T	
T	F	F	F	F	F	T	

c	d	g	s	1	2	3	all
F	T	T	T	T	T	F	*
F	T	T	F	T	T	F	*
F	T	F	T	T	T	T	*
F	T	F	F	T	T	T	*
F	F	T	T	T	T	F	
F	F	T	F	T	T	F	
F	F	F	T	T	F	T	
F	F	F	F	T	F	T	

Figure 3.2   A truth table for the formulas in (3.26). The argument "(a)" has been omitted from the predicates to save space. The rows in which all three formulas are true are indicated by an asterisk.

Example 3.10   The Lewis Carroll Example Revisited

Let's take the third statement in Example 3.9 and move the negation through the existential quantifier using (3.23). We then have three universally quantified statements, namely

$$\forall X \Big(c(X) \to s(X) \Big), \quad \forall Y \Big((\neg g(Y)) \to d(Y) \Big), \quad \text{and} \quad \forall Z \Big((\neg g(Z)) \vee c(Z) \Big),$$

where some variable names have been changed to emphasize that there is no relationship between them.

Each of these formulas corresponds to an infinite set of formulas which are obtained by replacing the variable by all choices of constants. For example, we have

$$\Big(c(a) \to s(a) \Big), \quad \Big((\neg g(a)) \to d(a) \Big), \quad \text{and} \quad \Big((\neg g(a)) \vee c(a) \Big). \tag{3.26}$$

Since $c(a)$, $s(a)$ and so on are either true or false, we can treat them just like we treated propositional letters in the propositional calculus. In particular, we could construct a truth table, as shown in Figure 3.2.

Those rows in the table for which all three formulas in (3.26) are true are marked with an asterisk. Since $\big((\neg s(a)) \to d(a) \big)$ is true for all these rows, it follows from (3.26). Since a is just an arbitrary constant, we've shown that (3.21) is true. ∎

Example 3.10 shows that deriving results from universally quantified statements can sometimes be done purely in propositional logic. It's tempting to try to extend this idea; however, it doesn't cover enough of predicate logic for the needs of AI. In the next section, we'll examine how Prolog reasons. In Chapter 4 we'll put this reasoning, called *resolution* and *unification*, on a solid theoretical foundation.

Aside. Here we've allowed variables to replace only constants, but we might want variable predicates as well. For example, we might claim that the relationship between Al and Jo is the same as the relationship between Ken and Barbie. That means, we have some predicate P such that both $P(\text{Al}, \text{Jo})$ and $P(\text{Ken}, \text{Jo})$ are true. In other words,

$$\exists P\Big(P(\text{Al}, \text{Jo}) \wedge P(\text{Ken}, \text{Jo})\Big)$$

is true. We can't say this in *first*-order predicate logic because variable predicates are not allowed. *Second*-order predicate logic allows variable predicates. Thus second-order logic is more powerful than first-order. Unfortunately we pay a high price—there is no algorithmic procedure comparable to resolution.

Exercises

3.3.4. Write formulas as in Example 3.7 to define each of the following concepts:

   (a) One number is greater than another.

   (b) One number divides another.

   (c) A number has all of its prime factors distinct.

   (d) A number is a power of 2.

   *(e) A number is a power of 6.

3.3.5. What role does "only" play? Consider the statement "Only AI students do this problem."

   (a) Using predicates **AI_student** and **do_this_problem**, write a Prolog rule corresponding to the English one.

   (b) Write an FOL formula for the statement.

   (c) Describe the general method for translating statements containing "only."

3.3.6. Complete the proof of Theorem 3.4.

3.3.7. A popular TV ad says "Nobody doesn't like Sara Lee®." With $p(X)$ meaning "X likes Sara Lee," translate the statement to FOL and rearrange to obtain a simpler statement.

3.3.8. A Texas bumper sticker reads "If you ain't a cowboy, you ain't ****" (expletive deleted). Introduce predicates, write in FOL and simplify. Let the universe of constants be bumper sticker readers.

3.3.9. In contrast to (3.22), give an example that shows $\big(\exists X(\forall Y\ \alpha)\big)$ is not equivalent to $\big(\forall Y(\exists X\ \alpha)\big)$; however, prove that the former implies the latter.

3.3.10. Let α and β be formulas in which X is the only free variable. Precisely two of the following are true for all such α and β; that is, they are tautologies.

$$\Big((\forall X(\alpha \vee \beta)) \equiv ((\forall X\, \alpha) \vee (\forall X\, \beta))\Big)$$
$$\Big((\forall X(\alpha \wedge \beta)) \equiv ((\forall X\, \alpha) \wedge (\forall X\, \beta))\Big)$$
$$\Big((\exists X(\alpha \vee \beta)) \equiv ((\exists X\, \alpha) \vee (\exists X\, \beta))\Big)$$
$$\Big((\exists X(\alpha \wedge \beta)) \equiv ((\exists X\, \alpha) \wedge (\exists X\, \beta))\Big)$$

(a) Identify those that are not tautologies and give an example to show that they are not.

(b) Identify the tautologies and prove that they are, in fact, tautologies.

3.3.11. Negating quantifiers with qualifications, as in the definition of a derivative, is nontrivial. Nevertheless, it's an essential tool for constructing proofs by contradiction and is a frequent source of student errors in such proofs. We'll look at it a bit. For this exercise, let $\forall[X:\alpha]$ mean "for all X with α" and let $\exists[X:\alpha]$ mean "for some X with α," as in (3.19).

(a) Prove the following:

$$\Big(\neg\forall[X:\alpha]\,\beta\Big) \equiv \Big(\exists[X:\alpha]\,(\neg\beta)\Big)$$
$$\Big(\neg\exists[X:\alpha]\,\beta\Big) \equiv \Big(\forall[X:\alpha]\,(\neg\beta)\Big).$$

In order to translate $[X:\alpha]$ into predicate logic, use (3.19). You may also find some previous theorems useful.

(b) Do these agree with your usual understanding of the expressions written out in words? Explain.

(c) Using the above and the predicate logic definition of a limit, write out a predicate logic definition of its negation, that is, a definition of $\lim_{x \to a} f(x) \neq L$.

(d) Convert the previous result to ordinary calculus prose. Does this agree with what you think it should say? If not, you should (i) correct your solution to the previous part, (ii) correct your translation to prose, or (iii) reexamine and correct your thoughts about what it should say.

3.3.12. Here are some definitions. Translate them into predicate logic notation as we did for the definition of a limit.

  (a) We have two types of objects, real numbers and vectors. Use the predicate $\mathbb{R}(b)$ to determine if b is a real number. We say that the vectors $\vec{u}$, $\vec{v}$, and $\vec{w}$ are independent if, whenever $a\vec{u} + b\vec{v} + c\vec{w} = \vec{0}$, it follows that $a = b = c = 0$.

  (b) Using the previous exercise, convert the definition in part (a) into a definition for $\vec{u}$, $\vec{v}$, and $\vec{w}$ being dependent, that is, not independent.

  (c) We say that the points in S form a circle centered at the origin if and only if there is some radius $r > 0$ such that $x^2 + y^2 = r^2$ for just those points $(x, y) \in S$.
  Hint. You can represent any set by a predicate. Since $(x, y) \in S$ is either true or false, we can think of it as a predicate, say $S(x, y)$.

Each of the following exercises contains a list of statements taken from Carroll's text on symbolic logic. Write down predicate logic equivalents for them as in Example 3.9 and then deduce the indicated conclusion as done in Example 3.10. Remember to specify the set of constants. Indicate what your predicates stand for.

3.3.13. Translate the following statements into predicate logic:
  "(1) Babies are illogical;
  (2) Nobody is despised who can manage a crocodile;
  (3) Illogical persons are despised."   [3, p. 112]
  Show that they imply "Babies cannot manage crocodiles."

3.3.14. Translate the following statements into predicate logic:
  "(1) All my sons are slim;
  (2) Nobody is healthy who takes no exercise;
  (3) Gluttons are always fat;
  (4) No daughter of mine takes any exercise."   [3, p. 116]
  Let the universe of constants be my children. Using the facts that (a) children are either sons or daughters and (b) a person is either slim or fat, show that the statements imply "All gluttons, who are children of mine, are unhealthy."

3.3.15. Translate the following statements into predicate logic:
  "(1) No kitten, that loves fish, is unteachable;
  (2) No kitten without a tail will play with a gorilla;
  (3) Kittens with whiskers always love fish;
  (4) No teachable kitten has green eyes;
  (5) No kittens have tails unless they have whiskers."   [3, p. 118]
  Show that they imply "No kitten with green eyes will play with a gorilla."

3.4   An Algorithm for Prolog

In this section we'll describe the Prolog algorithm intuitively as well as in terms that are amenable to the proof techniques of predicate logic. The intuitive approach is straightforward—as long as we ignore the details.

We'll start by looking at the Prolog algorithm intuitively. Then we'll translate Prolog to FOL. Finally, we'll explore the Prolog algorithm as an example of Horn clause resolution. Some of the details and all of the proofs will be postponed until the next chapter.

The Prolog Algorithm

In brief, Prolog uses depth-first search, Algorithm 2.3 (p. 53). To describe the search tree, we need some terminology.

Definition 3.7   Head and Body of a Prolog Clause

Prolog statements are also called *clauses*, and we'll use that term here. The part of a Prolog clause to the left of :- is called the *head of the clause* and the part to the right is called the *body of the clause*. We follow the convention that a fact consists of a head with no body and a query consists of a body with no head.

Prolog treats the body of each clause as a list, with the entries in the order they appear in the clause. For the purposes of depth-first search, Prolog treats the knowledge base as a list of clauses and goes through the list in order when making a depth-first search decision.

Prolog's decision tree (also called a search tree) for depth-first search is constructed as follows. At each vertex there is a list, the list at the root being the query. Essentially, the list at a vertex is what needs to be "proved" to establish the query. Thus, if Prolog ever obtains an empty list, a positive answer to the query has been found.

Suppose Prolog is working in propositional logic and we are at some vertex v that has the list p,q,.... If there is a clause in the knowledge base with head p and body r,s,..., then Prolog can select that clause. Thus, there is a vertex in the decision tree where the possible decisions correspond to all knowledge-base clauses with head p. According to the specification of the Prolog language, the order of these decisions is the same as the order of the corresponding clauses in the knowledge base. Suppose Prolog selects the clause p :- r,s,.... This decision leads to a vertex whose list is r,s,... ,q,..., in other words, the body of the knowledge base clause followed by the remainder of the list at v.

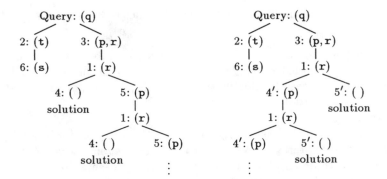

Figure 3.3 Prolog search trees for the query :- **q**. See Example 3.11 for the knowledge base. Each vertex is labeled with the knowledge-base statement that led to it and the list of what remains to be proved. The parts of the trees indicated by vertical ellipses go on forever alternating between **p** and **r**, with an empty clause after each **r**. The right-hand search tree arises when statements 4 and 5 are interchanged in the knowledge base.

Example 3.11   A Prolog Algorithm Search Tree for Propositions

Let's use lowercase roman letters to indicate propositional letters in this example. Suppose that our Prolog knowledge base is as follows, where the translation to propositional logic on the right is based on (3.1) (p. 89) and (3.2) (p. 93):

p :- r.	% 1	$r \to p$
q :- t.	% 2	$t \to q$
q :- p,r.	% 3	$(p \wedge r) \to q$
r.	% 4	r
r :- p.	% 5	$p \to r$
t :- s.	% 6	$s \to t$

Let our query be :- **q**. It asks Prolog if **q** is true given the knowledge base. The associated decision tree, which is infinite, is shown (in part) in Figure 3.3.

Of course, Prolog doesn't actually construct this tree beforehand. It constructs (and destroys) the search tree as it attempts to establish :- **q** by reaching the empty list. Since Prolog uses depth-first search, it will succeed by reaching the leftmost empty list in the left-hand tree of Figure 3.3. Let's be more explicit. We'll refer to the vertices by their numeric labels.

- Prolog starts at the query and goes down to 2.
- From 2 it goes down to 6.
- At 6, no decisions are possible, so it returns to 2.
- At 2, no further decisions are possible, so it returns to the query.

- At the query, it finds a second choice and goes to 3, then to 1, and finally to 4.

In constrast to what we've just looked at, if statements 4 and 5 were reversed in the knowledge base, we'd obtain the tree on the right-hand side of Figure 3.3. In this case, Prolog would run until it ran out of time or storage, never finding a solution. ■

We've just seen that depth-first search can lead to disaster in Prolog. But in the previous chapter, we saw that both breadth-first and iterative-deepening search—Algorithms 2.1 and 2.4—never fail. Why doesn't Prolog use them?

In the first place, breadth-first search requires considerable storage. Although there are other objections, this may be sufficient to dismiss it from consideration. The reasons for not using iterative deepening are more subtle. Three interrelated major problems are side effects, clarity, and speed.

- **Side Effects**: Prolog is more than logic—it includes procedural code. For example, it interacts with the user through the terminal and it can alter the content of the knowledge base. To use procedural code correctly, the programmer must understand when Prolog accesses knowledge-base statements. This brings us to

- **Clarity**: It is easier to visualize the process of depth-first search than that of iterative-deepening search, primarily because iterative deepening traverses the same statements many times. Even so, many programmers initially have difficulty with Prolog because they do not understand the somewhat subtle ways in which recursion interacts with depth-first search.

- **Speed**: Suppose the other objections are swept away. It is frequently possible to order the statements in a Prolog knowledge base so that depth-first search will usually be much faster than iterative deepening: Less likely and/or more costly branches are placed further to the right in the list of decisions at a vertex. Recall that Theorem 2.3 says iterative-deepening search is fast. Don't these comments contradict that? No. Look at Theorem 2.3 (p. 58) and try to see why before continuing.

    *   *   *   Stop and think about this!   *   *   *

In Theorem 2.3, iterative-deepening search is compared with breadth-first search; not with depth-first search. Depth-first search can be much faster if the clauses with a given head are listed in an order that begins with those most likely to lead to a quick solution.

All these objections notwithstanding, there are times when iterative-deepening search may be preferable, particularly when procedural code is absent. If you are a Prolog programmer, you might implement iterative-deepening search within Prolog. This can be done with varying degrees of sophistication.

Let's look at how Prolog handles predicates. Suppose the first entry in the list at some vertex involves the predicate **p**. The possible decisions at this vertex in the decision tree are those knowledge-base clauses with head **p** whose arguments can be made to agree with the arguments of **p** at the vertex. For example, if we have **p(X,a,Y)** at the vertex v and **p(b,U,V)** in the knowlege base, we obtain agreement by setting **X** equal to **b**, setting **U** equal to **a**, and identifying the variables **Y** and **U**. Setting variables to constants and variables in this manner is called *unifying* the arguments of the predicates. This decision to use a particular knowledge-base clause C at the vertex v leads to a child vertex whose list is the body of C followed by the list at the parent vertex v with the first predicate removed. In this new list, all these unifications are made. Of course, when Prolog moves back toward the root by rejecting a decision, it must forget the unifications the decision caused. Except for these added complications, the search is the same as for propositional logic.

Example 3.12   The Prolog Algorithm with Predicates

In the following knowledge base, **sib**, **sis**, and **p** stand for the predicates sibling, sister, and person, respectively. The formula **sis(X,Y)** is true when **Y** is a sister of **X**. The formula **p(X,S,M,F)** is true when **X** is a person whose sex is **S**, whose mother is **M**, and whose father is **F**.

```
sib(X,Y) :- p(X,SX,M,FX),p(Y,SY,M,FY).          % 1
sib(X,Y) :- p(X,SX,MX,F),p(Y,SY,MY,F).          % 2
sis(X,Y) :- sib(X,Y), p(Y,f,M,F).               % 3
p(a1,m,b1,b2).  p(a2,m,b1,b3).  p(a3,f,b4,b2).  % 4, 5, 6
```

The definition of sibling is not quite correct since it does not include the check **X\=Y**. We'll omit this check to simplify things a bit.

Consider the query **:- sis(a1,X)**. It asks Prolog to find a constant such that setting **X** equal to the constant makes **X** a sister of **a1**. Prolog begins with the third rule, unifying the **X** of the rule with **a1** and the **Y** of the rule with the **X** of the query. This is shown at the top of Figure 3.4. Using the information in the figure caption, you should be able to see why the rest of the Prolog search tree is as shown. Remember that Prolog always replaces the first predicate in a list by using a clause from the knowledge base whose head can be unified with that predicate.

    *   *   *   Stop and think about this!   *   *   *

Because the clauses in the knowledge base are in a poor order for this query, Prolog must traverse the entire search tree to find a solution. ■

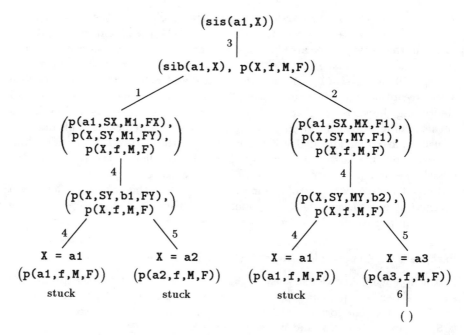

Figure 3.4 The entire Prolog search tree for the query :- sis(a1,X). The knowledge-base clause numbers appear on edges because of limited space. Whenever a knowledge-base clause is resolved with a clause that leads to a conflict of variable names, a subscript is added to the name of the knowledge-base clause variable. When variables are unified, the knowledge-base clause variable names are replaced with the new ones in the unification. When the X of the query is unified with a constant, that is indicated in the tree. "Stuck" indicates that this branch of the search tree terminates because there is no knowledge-base clause that can be unified with the predicate.

Exercises

3.4.A. What is a Prolog clause? What are its head and body? Which of the following lack a head (or body): fact, rule, query?

3.4.B. Why does Prolog use depth-first search rather than breadth-first search or iterative-deepening search?

3.4.1. (*Answer follows*) Return to the left-hand tree of Figure 3.3. Give a description in ordinary English, free of Prolog and propositional logic terminology, of what Prolog is trying to do at each step and what happens. For example, you might start with "Prolog wants to see if q is true. It knows that if t is true, then q is true, so it decides to try to prove t."

3.4.2. Suppose that `sis(X,Y)` is changed to mean that X is a sister of Y in Example 3.12.

(a) Rewrite Statement 3 so that it is correct.

(b) Draw the search tree for the query `:- sis(X,a1)`. For your own information, compare it to the left-hand tree in Figure 3.4.

3.4.3. Consider the query `:- q.` for the following knowledge base:

```
p :- t.              % 1
q :- p,r.            % 2
q :- t.              % 3
t :-r.               % 4
t.                   % 5
```

(a) Draw the search tree for this query and knowledge base.

(b) Indicate explicitly how Prolog traverses the tree.

Answers

3.4.1. There are many ways to do this; here's one.

- Prolog wants to see if **q** is true. It knows that if **t** is true, then **q** is true, so it decides to try to prove **t**.

- It knows that if **s** is true, then **t** is true, so it decides to try to prove **s**.

- Since it has no way to prove **s**, it looks for another way to prove **t**, but finds none.

- As a result, Prolog is back to looking for another way to verify the original query **q**.

- It knows that if **p** and **r** are true, then **q** is true so it turns its attention to them.

- It knows that if **r** is true, then **p** is true, so it only needs to verify **r**.

- Prolog knows that **r** is true—it's a fact in its knowledge base.

*Prolog Lists and Recursion

We've mentioned recursion in connection with Prolog. The next example is a classic Prolog example of recursion. To understand it, you need to know something about how Prolog represents lists. A list is enclosed in brackets and the entries in the list are separated by commas as in [a,b,c,d] and [], where the first list has four items and the latter is the empty list. A particularly useful notation allows us to separate off the first element in a list: In the syntax [X|Y], X is the first entry in the list and Y is the remainder of the list—which could be the empty list. For example, [a,b,c,d] can be unified with [X|Y] by setting X equal to a and Y equal to the list [b,c,d]. In other words, [a|[b,c,d]] and [a,b,c,d] are the same list in Prolog. In still other words, [X|Y] is a list whose first element is X and whose remaining elements are the elements in the *list* Y. We cannot unify [] with [X|Y] because X must be the first element of the list and there is none.

Example 3.13   A Recursive Prolog Example: append

We want to write Prolog code for a predicate append(L,M,LM) so that the list LM will consist of the elements of the list L followed by those of the list LM. Here's a solution.

```
append([],L,L).                              % 1
append([X|L],M,[X|LM]) :- append(L,M,LM).    % 2
```

The first line of code obviously describes what to do when the first list is empty. But what does the second line do? It simply *asserts*—makes a declarative statement—that, if the variable LM is the result of appending M to L, then [X|LM] is the result of appending M to [X|L]. Phrased that way, it's obviously the correct thing to *say*.

This explanation of the second line of code usually engenders howls of outrage (at least subvocally) from procedural programmers. After all, there's no code to *do* anything! That's the beauty of Prolog. If we can say in Prolog's logic what it takes for something to be correct, Prolog will usually be able to do the rest. The above code does just that:

- The first line states the fact that the result of appending a list L to an empty list is just L.

- When the first list is not empty, the second statement tells how to decide if **append** is correct by looking at a portion of the first list.

You can think of what we have as a recursive definition.

It may be helpful to follow a simple example. It is important to remember that the variables in the Prolog statement are bound—that is, for procedural programmers, regarded as local. To keep repeated usage straight, let's employ subscripts. Consider the query :- `append([1,2],[a,b],X)`. Statement 1 cannot be unified with the query because `[1,2]` is not an empty list. Thus Prolog tries the second statement, obtaining

$$\text{append}([1|[2]],[a,b],[1|LM_1]) : -\text{append}([2],[a,b],LM_1).$$

In order to satisfy the right side, Prolog must use Statement 2 again:

$$\text{append}([2|[]],[a,b],[2|LM_2]) : -\text{append}([],[a,b],LM_2).$$

(Note that the third argument on the left is also called LM_1, so that LM_1 equals $[2|LM_2]$.) Now the first statement applies to the right side, giving `[a,b]` for LM_2. Going backward in the sequence of equalities, LM_1 is $[2|[a,b]]$, which is `[2,a,b]`, and $[1|LM_1]$ is thus `[1,2,a,b]`.

It's tempting to think of the code as the core of a recursive algorithm, but this can be misleading. We tend to think of an algorithm as running in one direction. For example, you might set the query

$$: -\text{append}([1,2],[a,b],X). \quad \text{and get the answer} \quad X = [1,2,a,b]$$

as we did above. However, you're less likely to consider

$$: -\text{append}(X,Y,[a,b,c,d]).$$

Nevertheless, this is a perfectly acceptable query. Using the first knowledge-base clause, Prolog finds the solution

$$X = [], \ Y = [a,b,c,d].$$

If asked to find another solution, Prolog will try the second statement, in which `X` is `a` and `LM` is `[b,c,d]`. This leads Prolog to seek a solution to `append(L,M,[b,c,d])`, which can be found by the first clause. This leads to the solution

$$X = [a], \ Y = [b,c,d].$$

Again, Prolog can be asked to find another solution and will do so. Altogether, it will find all five possible solutions. Details are left as an exercise. ■

The Prolog list structure has no direct analog in FOL; however, we can implement lists in FOL by using a function. Here's a way to do it. Let the constant e be the empty list and let $l(X,Y)$ be the function that produces a list whose first entry is X and whose remaining entries are the elements of Y. Some accommodation must be made for the situation in which y is not a list; for example, the value of l could be a special constant indicating that the arguments are bad. We may also want functions that extract the first element and remainder of a list.

Exercises

3.4.C. Explain Prolog's list notation.

3.4.4. Draw the entire search tree for the query `:- append(X,Y,[a,b,c,d])`.

3.4.5. Suppose you want to reverse the order of elements in a Prolog list. Explain why this is described declaratively by

```
reverse([],[]).
reverse([X|L],Y) :- reverse(L,M), append(M,[X],Y).
```

3.4.6. Using the idea sketched after Example 3.13, express the definition of **append** in FOL.

3.4.7. An ordered tree is defined in Definition 2.3 (p. 38). The definition can be modified to define unlabeled ordered trees, such as those shown in (a) below. Instead, we'll take a Prolog approach. An unlabeled ordered tree is either the empty list (a single vertex) or a list containing one or more unlabeled ordered trees.

   (a) How are the following trees represented as Prolog lists?

   (b) We call an ordered tree *binary* if each nonleaf vertex has exactly two children. On page 253 of *Foundations of Applied Combinatorics* by Bender and Williamson, a function f that establishes a 1:1 correspondence between n-vertex ordered trees an n-leaf binary ordered trees was given. It was defined recursively to map from all ordered trees to the binary ordered trees. With the trees in Prolog notation, the function is given by $f([\,]) = [\,]$ and $f([T_1,\ldots,T_k]) = \left[f(T_1), f([T_2,\ldots,T_k]) \right]$. If $k = 1$, the rightmost f is $f([\,])$. Write Prolog code for a predicate **f(Any,Binary)** where **Any** is an arbitrary unlabeled ordered tree and **Binary** is the corresponding binary one.

   (c) What tree is produced when the previous algorithm is applied to the leftmost tree in (a)? Show your work. Give the answer as a picture and as a Prolog list.

Translating Prolog to Logic

Here we'll examine a translation mechanism $\mathbb{T}$ for converting Prolog to predicate logic.

Prolog has three types of statements—facts, rules, and queries. The facts ϕ_i and the rules ρ_i constitute the knowledge base $\mathcal{K}$ with which we are working:

$$\text{knowledge base } \mathcal{K} = \{\phi_1, \ldots, \phi_f, \rho_1, \ldots, \rho_r\}.$$

All statements in the knowledge base are true, that is, the "and" of these statements is true. Thus we have

$$\mathbb{T}(\mathcal{K}) = \mathbb{T}(\phi_1) \wedge \cdots \wedge \mathbb{T}(\phi_f) \wedge \mathbb{T}(\rho_1) \wedge \cdots \wedge \mathbb{T}(\rho_r).$$

Since a fact consists of a predicate with constant arguments, $\mathbb{T}(\phi_i)$ is simply ϕ_i read as a formula in predicate logic.

Rules are a bit more complex. A typical Prolog rule has the form

$$\texttt{p(X,Y) :- q(X,Z), r(Z,c,Y).}$$

The prose version of this Prolog, namely "If $\texttt{q(X,Z)}$ and $\texttt{r(Z,c,Y)}$ for some $\texttt{Z}$, then $\texttt{p(X,Y)}$," is ambiguous because $\texttt{X}$ and $\texttt{Y}$ have not been quantified. The Prolog statement is supposed to hold for all $\texttt{X}$ and $\texttt{Z}$. We can be precise by translating to predicate logic:

$$\mathbb{T}\Big(\texttt{p(X,Y) :- q(X,Z), r(Z,c,Y)}\Big)$$
$$= \forall X \, \forall Y \Big(\big(\exists Z (q(X,Z) \wedge r(Z,c,Y))\big) \rightarrow p(X,Y)\Big).$$

By Corollary 3.4.1 (p. 115), we can rewrite this translation as

$$\mathbb{T}\Big(\texttt{p(X,Y) :- q(X,Z), r(Z,c,Y)}\Big)$$
$$= \forall X \, \forall Y \, \forall Z \Big(p(X,Y) \vee (\neg q(X,Z)) \vee (\neg r(Z,c,Y))\Big).$$

This entire process can be extended to any Prolog rule: Let $X_1, \ldots, X_k$ be the variables in the rule ρ. Let C be the "atomic formula" (Definition 3.5) on the left of $\texttt{:-}$ and let $H_1, \ldots, H_n$ be those to its right. Then

$$\mathbb{T}(\rho) = \forall X_1 \ldots \forall X_k \big(C \vee (\neg H_1) \vee \cdots \vee (\neg H_n)\big).$$

This translation is an example of a Horn clause.

Definition 3.8   Horn Clause

A *Horn clause* is a formula of the form $\forall X_1 \ldots \forall X_n (\alpha_1 \vee \cdots \vee \alpha_k)$ containing no free variables and such that each α_i is an atomic formula or the negation of one. Furthermore, at most one of the atomic formulas is not negated. The notation $\{\alpha_1, \ldots, \alpha_k\}$ is used for a Horn clause.

Warning: Do not confuse the comma in a Horn clause—where it means "or"—with a comma separating Prolog predicates—where it means "and."

According to the translation rules, we have the following:

- A Prolog fact translates to a Horn clause with one atomic formula, and it is not negated.

- A Prolog rule translates to a Horn clause with all but one of the atomic formulas negated.

- A Prolog query translates to a Horn clause in which all atomic formulas are negated, since a query is written like a rule with no head.

The translation of a query looks strange. Since we want to know if it can be satisfied, why in the world would we translate it into a Horn clause with *negated* atomic formulas? For example, if the query is :- p(X), it means, "Does there exist an X such that p(X) is true?" In other words, the query means "Is the statement $\mathbb{T}(\mathcal{K}) \to (\exists X\, p(X))$ true?" With a little work, you should be able to show that

$$\Big(\alpha \to (\exists X\, p(X)) \Big) \equiv \Big((\neg\alpha) \vee (\exists X\, p(X)) \Big) \equiv \neg\Big(\alpha \wedge (\forall X(\neg p(X))) \Big)$$

Thus, proving that $\mathbb{T}(\mathcal{K}) \to (\exists X\, p(X))$ is true is equivalent to proving that

$$\mathbb{T}(\mathcal{K}) \wedge \Big(\forall X \big(\neg p(X) \big) \Big) \tag{3.27}$$

is false. In other words, we can verify that the query can be satisfied by proving that (3.27) is false. Thus the translation scheme converts the Prolog problem into a collection of Horn clauses that must be proved contradictory.

Is showing that something is false better than using the decision tree description of Prolog? Actually, you'll see in the next chapter that both methods carry out the same manipulations! In that case, why use contradiction? The reason for using the contradiction approach is that it's sometimes easier to prove theorems about contradictions. As a result, logicians often think about Prolog inference as a process of obtaining contradictions in the way we've just described. Does this mean you should think about Prolog's algorithm as a proof by contradiction? Sometimes. From a practical viewpoint, it's simpler to think of Prolog as trying to find solutions to the query. However, when theoretical insight is needed to study the power and limitations of Prolog, the "solution by inconsistency" viewpoint is often better.

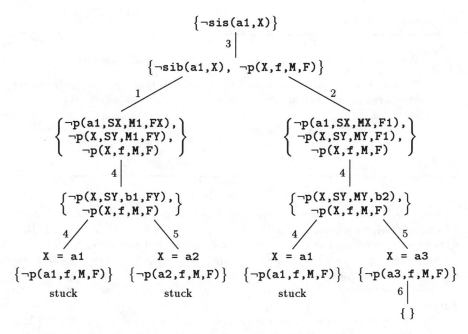

Figure 3.5 The Prolog search tree of Figure 3.4 (p. 124) written in Horn clause resolution form.

Our translation of Prolog, including the query, consists of universally quantified Horn clauses joined by "and." In predicate logic terms, the Prolog search process involves resolving Horn clauses. Suppose we have two Horn clauses with no variable names in common. Let one clause contain $p(\ldots)$ and the other contain $\neg p(\ldots)$. Their *resolution* is another Horn clause that contains all the atomic formulas from both clauses except the $p(\ldots)$ and $\neg p(\ldots)$ just mentioned. In addition, the variables and constants have been unified to make these two predicates agree except for the $\neg$. To obtain the Horn clause resolution search tree for Prolog, simply take the search trees drawn earlier, negate each predicate, and convert lists to clauses. Figure 3.5 illustrates this procedure. Horn clause resolution will be taken up more fully in the next chapter.

Exercises

3.4.D. What are Horn clauses and how are two Horn clauses resolved?

3.4.E. Explain how a goal clause (query) is translated into logic for the purposes of resolution and explain why this *a priori* counterintuitive translation is correct.

Notes

The discussion of FOL in this chapter is far from complete: It lacks theorems and proofs (which is remedied in the next chapter), and its definition of FOL is not the most general. The semantics defined here do not specify interpretations for variables, so we're limited to interpreting formulas with no free occurrences of variables. The most general formulation of FOL avoids these limitations; however, the limited version is more easily explained and adequate for our purposes.

In the United States, early experiments with high-level, logic-based languages "proved" that such languages were hopelessly inefficient. But in the early 1970s, the foundations of Prolog were developed by Robert Kowalski and demonstrated by Maarten van Emden, both at the University of Edinburgh. Alain Colmerauer's group at the University of Marseille-Aix developed a Fortran implementation of Prolog. Then, in the mid-1970s, David Warren of Edinburgh developed an efficient implementation that made Prolog practical. In light of their own earlier experience, Americans largely ignored these European developments. Although Prolog gained considerable popularity in Europe, Lisp was essentially the only AI language in the United States. When the Japanese selected Prolog as the language for their Fifth-Generation computer project in 1981, Americans first criticized the choice and then looked at Prolog more seriously. Most current researchers regard both Prolog and Lisp as major AI programming languages, each with its strengths and weaknesses.

In addition to the usual manuals, various texts on Prolog are available. They usually contain examples of interest in AI and some discussion of Prolog's logical foundations. Among these books are the texts by Bratko [2], Covington et al. [4], Maier and Warren [7], Shoham [14], and Sterling and Shapiro [19]. Some familiarity with Prolog is advisable (but not necessary) when reading [19]. Merritt's text [8] is devoted to expert systems and assumes familiarity with Prolog.

Lisp is the other major programming language for AI. My neglect of it should not be regarded as taking sides in the Lisp versus Prolog debate. I've introduced Prolog simply because its close connection with predicate logic makes it easy to illustrate aspects of logic without delving deeply into a language.

Anyone doing serious symbolic programming in AI should be familiar with both Lisp and Prolog. Among the introductory Lisp texts, Touretzky's [21] may be the gentlest. Winston and Horn's text [22] discusses applications to AI, but the introduction to Lisp is fast-paced. Revesz's [11] and Stark's [18] books provide the mathematical background for Lisp. Some familiarity with Lisp is assumed in Norvig's extensive text on Lisp programming for AI [9].

"Logic" puzzles of various sorts have been popular for ages. The logician Raymond Smullyan has written books, such as [15, 16], that teach some logic through the medium of puzzles.

All general AI texts discuss logic to some extent, but usually not to the depth I'm doing. On the other hand, texts on logic for computer scientists usually have deeper and broader coverage than that found here. Among these are the texts by Thayse [20], Genesereth and Nilsson [5] and Sperschneider and Antoniou [17]. However, the text by Schöning [13] covers just propositional and predicate calculus and Prolog—but in greater depth than here.

Example 3.5 dealt briefly with some problems concerning the meaning (or, rather, meanings) of "If P, then Q." Implication has been debated in philosophy and logic for more than two millenia. The first part of [12] gives the history. Reprints of some research appear in [6]. Nute [10] provides an introduction which you'll probably find easier to read after Chapter 6. I briefly compare implication and conditional probability in Example 7.8 (p. 273). Various philosophical attempts to deal with causality are discussed in [1].

References

1. M. Brand (ed.), *The Nature of Causation*, University of Illinois Press, Urbana, IL (1976).

2. I. Bratko, *Prolog Programming for Artificial Intelligence*, 2d ed., Addison-Wesley, Reading, MA (1990).

3. L. Carroll, *Symbolic Logic. Part I: Elementary*, MacMillan and Co., London (1896).

4. M. A. Covington, D. Nute, and A. Vellino, *Prolog Programming in Depth*, 2d ed., Prentice Hall, Englewood Cliffs, NJ (1995).

5. M. R. Genesereth and N. J. Nilsson, *Logical Foundations of Artificial Intelligence*, Morgan Kaufmann, San Mateo, CA (1987).

6. W. L. Harper, R. Stalnaker, and G. Pearce (eds.), *Ifs: Conditionals, Belief, Decision, Chance and Time*, D. Reidel, Dordrecht, The Netherlands (1980). Some of the material requires more background than I've provided.

7. D. Maier and D. S. Warren, *Computing with Logic: Logic Programming with Prolog*, Benjamin/Cummings, Menlo Park, CA (1988).

8. D. Merritt, *Building Expert Systems in Prolog*, Springer-Verlag, Berlin (1989).

9. P. Norvig, *Paradigms of Artificial Intelligence Programming: Case Studies in Common Lisp*, Morgan Kaufmann, San Mateo, CA (1992).

10. D. Nute, *Topics in Conditional Logic*, Dordrecht, The Netherlands (1980).

11. C. Revesz, *Lambda-Calculus, Combinators, and Functional Programming*, Cambridge University Press, Cambridge, Great Britain (1988).

12. D. H. Sanford, *If P, Then Q: Conditionals and the Foundations of Reasoning*, Routledge, London (1989).

13. U. Schöning, *Logic for Computer Scientists*, Birkhäuser, Boston (1989).

14. Y. Shoham, *Artificial Intelligence Techniques in Prolog*, Morgan Kaufmann, San Mateo, CA (1993).

15. R. Smullyan, *To Mock a Mockingbird and Other Logic Puzzles Including an Amazing Adventure in Combinatory Logic*, Alfred A. Knopf, New York (1985).

16. R. Smullyan, *What Is the Name of This Book? The Riddle of Dracula and Other Logical Puzzles*, Prentice Hall, Englewood Cliffs, NJ (1978).

17. V. Sperschneider and G. Antoniou, *Logic. A Foundation for Computer Science*, Addison-Wesley, Reading, MA (1991).

18. W. R. Stark, *LISP, Lore, and Logic: An Algebraic View of LISP Programming, Foundations, and Applications*, Springer-Verlag, Berlin (1990).

19. L. Sterling and E. Shapiro, *The Art of Prolog: Advanced Programming Techniques*, 2d ed., MIT Press, Cambridge, MA (1994).

20. A. Thayse (ed.), *From Standard Logic to Logic Programming*, John Wiley and Sons, New York (1988).

21. D. Touretzky, *Common Lisp: A Gentle Introduction to Symbolic Computation*, Benjamin/Cummings, Redwood City, CA (1989).

22. P. H. Winston and B. K. P. Horn, *Lisp*, 3d ed., Addison-Wesley, Reading, MA (1989).

4

The Theory
of
Resolution

A mathematical work that attempted to glide over all the
difficulties of the subject matter would be completely unfit for
training a reader in mathematical thinking and giving insight into
this special field

—Heinrich Tietze (1965)

[Niels Bohr] never trusted a purely formal or mathematical
argument. "No, no," he would say, "You are not thinking; you
are just being logical."

—Otto R. Frisch (1979)

Introduction

The focus of the previous chapter was conceptual: What are the ideas of
predicate logic and how do they relate to Prolog? The focus of this chapter
is more theoretical and more limited: It concentrates on the theoretical basis
for the resolution method of proof and its specialization to Prolog.

Some practical people may dispute the need for this chapter, arguing

Many people have used Prolog. It has produced correct answers. In
the face of such evidence any theoretical work is an unneeded ivory
tower pursuit.

To this, we can come up with several objections:

- There are important issues about Prolog itself that cannot be answered empirically: Will Prolog always find all the answers? If not, when and why? Will it find wrong answers? If so, when?

- There may be difficulties in the Prolog interpreter: How can we identify and understand them? What can we use besides guessing and testing to overcome them?

- Predicate logic is an inadequate foundation for AI: How can it be extended to include some kind of "default" and/or "nonmonotonic" capabilities? There have been a variety of attempts to do this, some of which we'll explore in Chapter 6. These attempts provide overwhelming evidence that theory is needed.

At the same time, we don't need all the details of the theory, particularly in a first text. As a result, we won't fully explore the theoretical foundations of predicate logic.

In the next section, we'll expand on the previous chapter's discussion of the distinction between truth and proof. Next, the theories of propositional and predicate caculi are studied, with an emphasis on resolution proofs. In the following section, we return to Prolog for a discussion of its features in light of the previous material. Finally, we briefly discuss FOL in general and Prolog in particular as a tool for AI.

Prerequisites: Chapter 3.

Used in: The material in this chapter is useful in Chapter 6; however, only Section 4.1 is required.

4.1   Truth versus Proof

What is truth?
—Pontius Pilate (ca 30 AD)

Mathematics takes us into the region of absolute necessity, to which not only the actual world, but every possible world, must conform.

—Bertrand Russell (1902)

Since truth is defined in the semantics of a language, there is only one notion of truth in a given language. In contrast, there may be many methods of proof because a method of proof is simply a method for manipulating the syntax to obtain valid results. You've already seen two distinct methods of proof for propositional calculus—truth tables and resolution. What is the connection between proof and truth? A proof method exists solely as a means of extracting truth. Truth is the more fundamental notion.

Truth

The notion of truth for propositional logic is given in Definition 3.2 (p. 97) and that for predicate logic in Definition 3.6 (p. 110). These ideas are based on truth assignments to the propositional letters and predicates, respectively. There are three interrelated truth notions for a formula:

- **valid**: All possible interpretations make the formula true.

- **satisfiable**: Some possible interpretation makes the formula true.

- **unsatisfiable**: No possible interpretations make the formula true.

In particular, it follows that α is valid if and only if $(\neg\alpha)$ is unsatisfiable. The latter two notions are extended to a set S of formulas, but the terminology differs:

- **consistent**: Some interpretation makes all the formulas in S simultaneously true.

- **inconsistent**: No interpretation makes all the formulas in S simultaneously true.

In propositional logic, "all possible interpretations" are easy to list: Simply consider all possible assignments of T and F to the propositional letters in the formula. This shows that validity, satisfiability, and unsatisfiability of a formula can all be determined in a finite amount of time.

In predicate logic, the situation is more complicated: The number of interpretations is infinite. As a result, the semantics of predicate logic does not automatically provide an algorithm for validity the way the semantics of propositional logic does. In Definition 3.6 (p. 110), it's unclear how many constants, functions, and predicates our interpretation must allow. Our interpretation must allow a number of constants at least equal to the total number of constants and variables in the formula we're looking at. As far as functions are concerned, we should have all possible functions. Even this is not quite enough because, when we see $f(c)$, we may want its value to be different from other constants we're considering.

As you can see, the "universe" for an interpretation can be quite large. Too large? What does that mean? Mathematicians have become quite adept at conjuring up and manipulating *very* large sets, but even they are interested in keeping infinities "small" when possible. Of course, any infinity is too large for an algorithm. How large a universe *must* we allow? The answer is provided by what are known as *Herbrand universes*, which are infinite, but not *too* large. We won't discuss these here, but a discussion can be found in almost any text on logic for computer scientists or mathematicians. Now, laying aside the notion of semantics, let's look at the notions of validity, satisfiability, and unsatisfiability, which can be extended to a situation where we have a set S of hypotheses. Here's a formulation for validity.

Definition 4.1   Consequences

Let S be a set, possibly empty, of formulas in the syntax of some logic and let α be another formula in that logic. We say that α is a *consequence* of S if, whenever all the formulas in S are true, α is also true. This is written as $S \models \alpha$. If $S = \emptyset$, we usually write $\models \alpha$, call α a *tautology*, and say that α is *valid*. Note that if S is inconsistent, then $S \models \alpha$ for *all* formulas α in the language because there is no interpretation in which S is true and α is not.

If S is a finite set consisting of the formulas $\sigma_1, \ldots, \sigma_n$, then

$$S \models \alpha \quad \text{if and only if} \quad \models ((\neg\sigma) \vee \alpha), \tag{4.1}$$

where σ is the formula $\sigma_1 \wedge \cdots \wedge \sigma_n$. This is easily shown: By definition, $S \models \alpha$ if and only if $\sigma \models \alpha$. By definition, the latter is true if and only if, whenever σ is true, α is also true. In other words, either α must be true or σ must be false for every interpretation. This is simply the definition of $\models (\alpha \vee (\neg\sigma))$.

Example 4.1   Simple Examples of Consequences

In propositional logic, we have

$$\{\beta, (\beta \rightarrow \gamma)\} \models \gamma.$$

You can see this by constructing a truth table:

β	γ	$\beta \rightarrow \gamma$
T	T	T
T	F	F
F	T	T
F	F	T

Since β is true in the first and second rows and $\beta \rightarrow \gamma$ is true in the first, third, and fourth rows, all the formulas in S are true only in the first row. In the first row, γ is also true.

When dealing with predicate logic, we cannot simply appeal to truth tables. For example, consider the following

$$\{\exists X \, \forall Y \, p(X,Y)\} \models \forall Y \, \exists X \, p(X,Y).$$

This is correct, but it's not so evident how to establish it. ■

Proof

> *Our difficulty is not in the proofs, but in learning what to prove.*
>
> —Emil Artin (ca 1950)

Consider the statement

$$\text{This statement cannot be proved.} \qquad (4.2)$$

Let's try to prove by contradiction that it is true. Suppose it is false. In that case, what it claims is not true. Hence (4.2) can be proved; but if it can be proved, that means it's true. We've obtained a contradiction from the assumption that (4.2) is false. Hence we have proved that (4.2) is true, which means it cannot be proved. But we just proved it!

This discussion shows that we must have a precise definition of what is meant by a proof if we hope to avoid paradoxes.

Definition 4.2 Proof Method, Soundness, Completeness

A *proof method* is a procedure for manipulating the syntax to deduce conclusions from assumptions. If the set of assumptions is the formulas in $\mathcal{S}$ and if the formula α is a conclusion that the proof method deduces for $\mathcal{S}$, we write $\mathcal{S} \vdash \alpha$. If $\mathcal{S} = \emptyset$, we usually write $\vdash \alpha$. *Note that $\vdash$ depends on the proof method.*

For a proof method to be useful, it should not prove false formulas, that is, whenever $\mathcal{S} \vdash \alpha$, we also have $\mathcal{S} \models \alpha$. Such a proof method is called *sound*.

It's also helpful if all consequences are provable; that is, whenever $\mathcal{S} \models \alpha$, we also have $\mathcal{S} \vdash \alpha$. Such a proof method is called *complete*.

When a method of proof is sound and complete, precisely those formulas that are consequences of $\mathcal{S}$ are provable. Except for the simplest logics, such proof methods do not exist. Yes, they do not exist—which is much stronger than saying they are not known. More on this later.

In addition to soundness and completeness, certain aspects of proof methods are important from a practical point of view. In computer science, we want to use such a method as the basis of an algorithm for carrying out proofs. Thus, we want the method to lead to a reasonably efficient algorithm. Satisfying this requirement is far from trivial.

Thus, given a method of proof, there are at least three questions we should address:

- Is the method sound? (If not, it is probably useless.)
- Is the method complete? (If not, it might still be useful.)
- Can the method be used as the basis for a reasonable computer algorithm?

We'll take up these issues in the next couple of sections, focusing on SLD-resolution—Prolog's proof method. (The acronym SLD derives from the fact that the resolution has a **S**election function, is **L**inear, and deals with **D**efinite clauses. Now you can forget what SLD means.)

There is another practical aspect of proof methods. A proof method is an algorithm for proving the truth of a *given* conclusion. It does not address the issue of how to *discover* interesting things to prove. Discovery is an aspect of learning and is in a more primitive state of development than proof methods.

Aside. The syntax and semantics of a logic seldom give an operational method for establishing $S \models \alpha$, whereas proof methods give techniques for establishing $S \vdash \alpha$. Thus it seems resonable to reserve the term *logic* for the syntax and semantics and the term *calculus* for the proof methods. This is seldom done—*predicate logic* and *predicate calculus* are used interchangeably.

Truth Tables

The manipulation of syntax via truth tables is a method of proof in propositional calculus. This method is the same as the definition of semantics. Hence the truth table method of proof is complete and sound when the formulas in S contain only a finite number of propositional letters. Unfortunately, the amount of work required to construct a truth table is exponential in the number of propositional letters since a truth table for n propositional letters has 2^n rows. As a result, the truth table method is impractical when we have a large number of propositional letters. A more serious drawback is the fact that the method cannot be extended to predicate logic, and we need predicate logic for AI.

Axiomatics

You're already familiar with one style of proof method for FOL from Euclidean geometry. This method is used throughout mathematics. The subject being considered is described by a series of *axioms* and the methods for manipulating the axioms are described by *rules of inference*. Let S be the set of axioms together with hypotheses of a theorem we wish to prove and let α be the theorem's conclusion. The rules of inference comprise a proof method for establishing $S \vdash \alpha$.

This approach to reasoning goes back, at least, to Aristotle (384–322 BC) and has been taught as the method of logical reasoning for centuries. In the 1890s it was popularized by David Hilbert (1862–1943) as a method for providing a firm foundation for mathematics. Typically in a mathematical area, the axioms specific to that area are spelled out. However, the more general axioms of mathematics and the rules of inference are rarely stated explicitly.

The specific axioms and rules of inference vary from system to system, but are all equivalent. It is standard to include at least *modus ponens*. This says that, if we are given α and $\alpha \rightarrow \beta$, we may infer β. You might say, "Why do we need this—it's obvious from truth tables." But truth tables are *another* proof method, so they are not included in the axiomatic method.

Unfortunately, axiomatics suffers from a severe drawback that everyone who has attempted a proof has encountered: "What should the next step be?" In order to implement an axiomatic system on computer, guidance concerning the next step is needed. It's not necessary to determine the step completely, because we can always use a search strategy as long as the amount of search needed is not excessive. Some success has been achieved using heuristics; however, finding other proof methods has proved more fruitful for automating mathematical logic. For this reason, we'll say no more about axiomatics.

Resolution

Resolution was introduced in the previous chapter as the basis for Prolog. Before resolution can begin, the formulas must be in clausal form. Resolution then constructs a proof by contradiction; that is, it proves that the formulas are inconsistent.

It turns out that the resolution method is sound and complete for FOL. Unfortunately, it has some drawbacks. First, it can be time-consuming to rewrite formulas in clausal form. Second, there is still the problem of "What should the next step be?" To some extent, we have to expect such problems because establishing validity in propositional logic is NP-complete. (NP-completeness is discussed *very* briefly on page 15.)

Prolog overcomes the first problem by limiting its syntax to statements that are easily converted to clausal form. It partially overcomes the second problem by further limiting syntax so that Horn clauses are obtained. This allows us to use a limited form of resolution called SLD-resolution without sacrificing completeness. It is easier to decide "what next" in SLD-resolution than it is in general resolution. But we pay a price for these improvements:

Prolog does not include all of FOL. For example, we cannot say $p \rightarrow (q \vee r)$ or $(\neg p) \rightarrow q$ in Prolog.   (4.3)

The remainder of this chapter is devoted to resolution and Prolog.

Exercises

4.1.A. What does it mean for a formula to be valid? satisfiable? unsatisfiable?

4.1.B. What is the distinction between truth and proof?

4.1.C. What are the meanings and usage of $\models$ and $\vdash$?

4.1.D. What do satisfiable and unsatisfiable mean? What can you say about $\mathcal{S} \models \alpha$ if $\mathcal{S}$ is unsatisfiable?

4.1.E. Why didn't we discuss truth tables further?

4.1.F. What is axiomatics and why are we not discussing it further?

4.1.G. If our goal is a proof method for FOL, what are some problems and why is Prolog a compromise?

4.1.1. Suppose we have two proof methods, Method A and Method B, for a given system of logic such that anything provable by Method A is provable by Method B.

   (a) Show that if Method A is complete, then Method B is complete. Why is the converse not necessarily true?

   (b) Show that if Method B is sound, then Method A is sound. Why is the converse not necessarily true?

4.1.2. Write both of the formulas in (4.3) in clausal form and explain why neither is a Horn clause.

4.2   Resolution and Propositional Calculus

In the first part of this section, we discuss the resolution method for propositional calculus and prove that it is sound and complete. In the second part, we do the same for SLD-resolution of Horn clauses.

The Resolution Method

There are two features of the resolution method you need to keep in mind to avoid confusion. First, it is a method of proof by contradiction. Second, it requires that formulas be in clausal form.

Clausal Form

Definition 4.3   Literals and Clausal Form

A *literal* is either a propositional letter or its negation. The former is called a *positive literal* and the latter, a *negative literal*. If l is a literal, $\bar{l}$ denotes its negation with $\neg\neg$ canceled; that is, $\bar{p}$ is $\neg p$ and $\overline{\neg p}$ is p. Recall that a formula α is in *clausal form* if it has the form

$$\alpha_1 \wedge \cdots \wedge \alpha_n \quad \text{where} \quad \alpha_i \text{ is } \beta_{i,1} \vee \cdots \vee \beta_{i,k_i}$$

and each $\beta_{i,j}$ is a literal. Each of the α_i's is called a clause and is usually written in the form $\{\beta_{i,1}, \ldots, \beta_{i,k_i}\}$. One often thinks of α as a set of sets, each of which corresponds to a clause. Clausal form is also called *conjunctive normal form*, or simply *CNF*.

The important result about clausal form is given by the following theorem.

Theorem 4.1   Clausal Form for Propositional Calculus

If α is a formula in propositional calculus, then there exists a formula γ such that (a) γ is in clausal form and (b) $\models (\alpha \equiv \gamma)$. Furthermore, there exists an algorithm for obtaining γ from α. (Warning: There may be more than one such γ for a given α.)

Let's recall what $\models (\alpha \equiv \gamma)$ means. Given any assignment whatsoever of truth values to the propositional letters, $(\alpha \equiv \gamma)$ will be true. In other words, $(\alpha \equiv \gamma)$ is valid; that is, it's a tautology.

Proof: By exhibiting the algorithm and showing that it works, we'll automatically prove the existence of γ. We'll need the following tautologies in the proof:

$$((\alpha \to \beta) \equiv ((\neg\alpha) \vee \beta)), \qquad \text{for step 1;}$$

$$((\alpha \equiv \beta) \equiv ((\alpha \wedge \beta) \vee ((\neg\alpha) \wedge (\neg\beta)))), \qquad \text{for step 2;}$$

$$((\neg(\alpha \vee \beta)) \equiv ((\neg\alpha) \wedge (\neg\beta))), \qquad \text{for step 3;}$$

$$((\neg(\alpha \wedge \beta)) \equiv ((\neg\alpha) \vee (\neg\beta))), \qquad \text{for step 3;}$$

$$((\neg(\neg p)) \equiv p), \qquad \text{for step 4;}$$

$$(((\alpha \wedge \beta) \vee \gamma) \equiv ((\alpha \vee \gamma) \wedge (\beta \vee \gamma))), \qquad \text{for step 5.}$$

You can prove these tautologies by using truth tables. (Two of them are special cases of de Morgan's laws.)

Imagine applying the tautologies by replacing occurrences of a formula to the left of $\equiv$ with the formula to the right, repeating this procedure until such replacements can no longer be made. This will transform the α given in the theorem to a new formula. Note that when we substitute one side of $\equiv$ in a tautology for the other, we obtain a statement with the same truth value for every assignment of truth values to the propositional letters. Since we'll obtain γ from α by just such a series of substitutions, $(\alpha \equiv \gamma)$ will be valid.

The first step involves using the first formula to eliminate $\rightarrow$. Suppose $\rightarrow$ occurs in the formula we are transforming. Move leftward from the $\rightarrow$ until the number of left parentheses from your position to the connective exceeds the number of right parentheses. Find the right parenthesis that is paired with this left parenthesis. You will now have found a formula contained in your original formula that has the form $\alpha \rightarrow \beta$. Replace it using the first tautology on the list. For a complete proof, we would have to show that such parentheses can always be found and that they do in fact delimit a formula. We won't do that. The process must eventually stop because each application eliminates an $\rightarrow$.

The second step involves essentially the same process to eliminate $\equiv$.

The third step involves moving negations inside parentheses using the third and fourth tautologies. The left parenthesis now immmediately precedes the $\neg$ and the right parenthesis is found as in the first step. This process must eventually stop because each application moves a $\neg$ "deeper" inside parentheses. (A formal proof requires a more precise formulation of the last statement.)

The fourth step involves using the fifth tautology to eliminate multiple negations on a propositional letter. When this is done, we are left with literals connected by $\vee$ and $\wedge$.

The fifth step involves moving $\vee$ inward and $\wedge$ outward simultaneously. In this case, we use the last tautology and the one that results from the commutativity of $\vee$. If you think of $\vee$ as addition and $\wedge$ as multiplication, this process is like expanding a complicated combination of sums and products to obtain an expression that is a sum of products. It can be done in a manner similar to that used in the first and second steps. It will eventually stop because each application moves a $\vee$ deeper inside parentheses. ∎

An alternative approach to proving Theorem 4.1 is given in Exercise 4.2.1.

Given a clause, we can sometimes simplify it: First, if a propositional letter appears more than once in the clause, it can be eliminated. Second, the same is true for the negation of a propositional letter. This elimination of duplicates agrees with the notation of writing a clause as a set since the elements of a set are all distinct. Third, if one clause is a subset of another, the smaller clause may be dropped because

$$((\alpha \wedge (\alpha \vee \beta)) \equiv (\alpha \vee \beta))$$

is valid. Finally, if both p and $\neg p$ appear in a clause, we can eliminate the clause because it is a tautology. (For any truth assignment, either p or $\neg p$ will be true.)

Example 4.2 Converting Formulas to Clausal Form

Let's convert $((p \to q) \to (p \land q))$ to clausal form. Eliminating $\to$ one at a time gives

$$(((\neg p) \lor q) \to (p \land q))$$

and then

$$((\neg((\neg p) \lor q)) \lor (p \land q)).$$

Moving $\neg$ inward gives

$$(((\neg(\neg p)) \land (\neg q)) \lor (p \land q)).$$

Eliminating $\neg\neg$ gives

$$((p \land (\neg q)) \lor (p \land q)).$$

Distributivity now gives, with some parentheses omitted,

$$(p \lor p) \land (p \lor q) \land ((\neg q) \lor p) \land ((\neg q) \lor q).$$

Finally, the simplifications give

$$(p \lor q) \land ((\neg q) \lor p).$$

Of course, you needn't slavishly follow the steps in the proof—you can use any manipulations that don't change truth values. ■

A Resolution Algorithm

Let's recall how two clauses are resolved. Suppose that we have two clauses C_1 and C_2, one containing the propositional letter p and the other containing $\neg p$. The *resolvent* of C_1 and C_2 on p is a clause consisting of all elements of C_1 and C_2 except p and $\neg p$. We speak of resolving C_1 and C_2 on p.

If C_1 and C_2 consist of only p and $\neg p$, their resolvent is empty and is denoted by { } or □. What does it mean? Since the elements of a clause are joined by "or," at least one element of a clause must be true for the clause to be true. Since our formula is a conjunction ("and") of clauses, a formula with an empty clause will be false.

The following is a nondeterministic algorithm for resolution proof in propositional logic. (The nondeterministic part is due to the use of "choose.") At the end of the previous chapter, we used an algorithm similar to this one for Prolog.

Algorithm 4.1   Resolution Proof in Propositional Logic

Let $\mathcal{S}$ be a set of formulas and let α be a formula.

- Imagine converting $\neg\alpha$ and all of the formulas in $\mathcal{S}$ to clausal form. (We can only "imagine" because $\mathcal{S}$ may be infinite.) This gives a set $\mathcal{C}$ of clauses.
- Repeatedly:
  - Choose a propositional letter p and two clauses C_1 and C_2 in $\mathcal{C}$ that can be resolved on p.
  - Create the resolvent, simplifying by eliminating duplicate elements.
  - Discard the clause if it contains both a propositional letter and its negation.
  - If the resolvent is not in $\mathcal{C}$, add it to $\mathcal{C}$.

  If it's possible to obtain the empty clause in this manner, we say that $\mathcal{S} \vdash \alpha$ (by resolution).

The following theorem answers two basic questions about the algorithm.

Theorem 4.2   Propositional Logic Resolution

The resolution method of proof is sound and complete. That is $\mathcal{S} \vdash \alpha$ if and only if $\mathcal{S} \models \alpha$.

Proof: Suppose that $\mathcal{S}$ is finite. By (4.1) (p. 138), we can move $\mathcal{S}$ to the right side of $\models$. Thus, we're reduced to proving $\models \alpha$ if and only if $\vdash \alpha$ for all α.

Recall that a collection of formulas is *satisfiable* if some possible assignment of truth values to the propositional letters makes all the formulas true. Also recall that a collection of formulas which is not satisfiable is called *unsatisfiable*. These terms are often abbreviated SAT and UNSAT.

Soundness:

Proof by resolution is a proof by contradiction; it converts $\neg\alpha$ to clausal form and resolves clauses to obtain $\square$. We must show that if this happens, then $\models \alpha$. Let's give a proof by contradiction. Thus, we assume $\vdash \alpha$ and $\not\models \alpha$ and want to deduce a contradiction.

The statement $\not\models \alpha$ means that $\neg\alpha$ is SAT. To see this, note that $\not\models \alpha$ means there is some assignment of T and F to the propositional letters so that α is F and so $\neg\alpha$ is T. Fix this assignment.

We now claim that, for this assignment, the resolution of two clauses will also be T. This claim is equivalent to "If $\mathcal{V}(p \vee \alpha) = \text{T}$ and $\mathcal{V}((\neg p) \vee \beta) = \text{T}$, then $\mathcal{V}(\alpha \vee \beta) = \text{T}$." It can be proved directly or, instead, the equivalent tautology

$$\Big(\big((p \vee \alpha) \wedge ((\neg p) \vee \beta) \big) \to (\alpha \vee \beta) \Big) \qquad (4.4)$$

can be proved by using a truth table. The claim shows that all clauses produced by resolution will be true for our fixed assignment of truth to the propositional letters. Since $\square$ is always false, it cannot be obtained by resolution. Thus $\not\vdash \alpha$, a contradiction.

Completeness:

Suppose that $\mathcal{C}$ is the clausal form of $\neg\alpha$. Completeness means that if $\mathcal{C}$ is UNSAT, then resolution can deduce $\square$ from $\mathcal{C}$.

Since α is a formula, it, and hence $\mathcal{C}$, contains only a finite number of propositional letters. We'll prove completeness by induction on the number of propositional letters. Let's start the induction with a single letter, say q. The only UNSAT collection of clauses is $\{\{q\},\{\neg q\}\}$, from which it's easy to deduce $\square$.

The idea behind the induction proof is to construct from $\mathcal{C}$ some clauses $\mathcal{C}^p$ that contain neither p nor $\neg p$ and then prove

(a) if $\mathcal{C}$ is UNSAT, then $\mathcal{C}^p$ is UNSAT;

(b) if resolution can deduce $\square$ from $\mathcal{C}^p$, then it can do so from $\mathcal{C}$.

Before showing how to construct $\mathcal{C}^p$, let's prove that (a) and (b) are enough to complete the induction proof. Suppose that $\mathcal{C}$ is UNSAT. Apply (a). Apply the induction hypothesis: completeness for fewer letters tells us that, since $\mathcal{C}^p$ is UNSAT, resolution can deduce $\square$ from $\mathcal{C}^p$. Finally apply (b).

It remains to define $\mathcal{C}^p$ and to prove (a) and (b).

The set of clauses $\mathcal{C}^p$ is defined as follows:

- If $C \in \mathcal{C}$ contains neither p nor $\neg p$, then it is in $\mathcal{C}^p$.

- If $C_1, C_2 \in \mathcal{C}$, $p \in C_1$, and $(\neg p) \in C_2$, then the resolvent of C_1 and C_2 on p is in $\mathcal{C}^p$.

Result (b) follows from the definition: The clauses in $\mathcal{C}^p$ are either in $\mathcal{C}$ or obtained by resolving two clauses in $\mathcal{C}$. Thus, anything obtainable from $\mathcal{C}^p$ by resolution is also obtainable from $\mathcal{C}$.

Result (a) takes more work. We'll prove it by establishing the contrapositive; that is, we'll assume that $\mathcal{C}^p$ is SAT and prove that then $\mathcal{C}$ is SAT. Since $\mathcal{C}^p$ is SAT, we may fix an assignment of values to all the propositional letters (except p) so that all clauses in $\mathcal{C}^p$ are true. Let $\mathcal{C}(p)$ be those clauses in $\mathcal{C}$ that contain p and let $\mathcal{C}(\neg p)$ be those that contain $\neg p$.

If every clause in $\mathcal{C}(p)$ is such that at least one of its elements (other than p) is true, then defining $\mathcal{V}(p) = \text{F}$ makes all clauses in $\mathcal{C}(p)$ and $\mathcal{C}(\neg p)$ true. Since all other clauses in $\mathcal{C}$ lie in $\mathcal{C}^p$, this would make $\mathcal{C}$ SAT and we would be done. In other words, either we are done or there is a clause $C_1 \in \mathcal{C}(p)$ all of whose elements except p are false. Interchanging the roles of p and $\neg p$, we conclude that either we are done or there is a clause $C_2 \in \mathcal{C}(\neg p)$ all of whose elements except $\neg p$ are false.

It suffices to show that at least one of C_1 and C_2 doesn't exist. If both exist, resolve them on p to obtain a clause in $\mathcal{C}^p$ all of whose elements are false. This contradicts the assignment of truth values, which was based on C^p being SAT.

Compactness:

We began by assuming that S is finite. In actual applications, our knowledge base can contain only a finite number of statements, so S will be finite. In general, this need not be true. Logicians usually assume that there can be an infinite but "countable" number of propositional letters. They then prove *compactness*: α is a consequence of S if and only if α is a consequence of some finite subset of S. You should readily see that the "if" direction is trivial. The proof of the "only if" direction requires the Axiom of Choice. Rather than develop this tool, let's leave the proof incomplete for infinite S. ∎

Example 4.3   Double Resolution—A Common Mistake

How can the clauses $\{p, \neg q, r\}$ and $\{\neg p, q, s\}$ be resolved? A common mistake is to "resolve" on both p and q to obtain $\{r, s\}$. This is WRONG. It may be instructive to look at the reason from different viewpoints.

Before doing so, let's look at the correct way to resolve these clauses. We can resolve on p to obtain $\{\neg q, r, q, s\}$. Unfortunately, this clause is useless because it is always true—either q is true or $\neg q$ is true. Similarly, resolving on q gives the useless clause $\{p, r, \neg p, s\}$.

One way to see that double resolution is wrong is by looking at a simpler case: $\{p, \neg q\}$ and $\{\neg p, q\}$. We'd obtain the empty clause, indicating inconsistency; however, making both predicates true satisfies the clauses. This shows that the method is wrong, but it does not explain why this is wrong while resolution is not.

To explore the why, let's look at the idea behind resolution. It's based on the tautology (4.4). A double resolution form of this tautology would be

$$\Big(\big((p \vee l \vee \alpha) \wedge ((\neg p) \vee (\neg l) \vee \beta)\big) \rightarrow (\alpha \vee \beta)\Big), \tag{4.5}$$

where l is q or $\neg q$. I'll leave it to you to prove that this is not valid. ∎

Example 4.4   Two Resolution Proofs

Let's begin with a resolution proof for

$$\{((\neg p) \rightarrow r), (p \rightarrow s), (r \rightarrow q), (s \rightarrow (\neg t)), t\} \vdash q. \tag{4.6}$$

We'll go through the details of all the various conversions that set the stage for resolution and then turn to the resolution.

Calling the formulas on the left $\sigma_1, \ldots, \sigma_5$, rewrite (4.6) as described in (3.25) (p. 115):

$$\vdash \big(\neg(\sigma_1 \wedge \cdots \wedge \sigma_5) \vee q\big).$$

Resolution attempts to derive a contradiction to the negation of the right side of this, that is, a contradiction to $\sigma_1 \wedge \cdots \wedge \sigma_5 \wedge (\neg q)$. Let C_i be the clausal form of σ_i and let C_6 be $\{\neg q\}$. Replacing $\alpha \to \beta$ with $(\neg\alpha) \vee \beta$ converts the σ_i's to clausal form. We get

$$C_1 = \{p, r\} \qquad C_2 = \{\neg p, s\} \quad C_3 = \{q, \neg r\}$$
$$C_4 = \{\neg s, \neg t\} \quad C_5 = \{t\} \qquad C_6 = \{\neg q\} \tag{4.7}$$

for the given clauses.

Here's our procedure for doing resolution: Set $i = 1$, resolve C_i with all other clauses, increment i and repeat resolution and incrementation until the empty clause is reached. (Stop if no new clauses are generated for a given i. This happens if there is no proof.) To keep things straight, let $C_i * C_j$ denote the resolution of C_i and C_j. It will always be evident what propositional letter we resolved on. This procedure leads to

$$C_7 = C_1 * C_2 = \{r, s\}, \qquad C_8 = C_1 * C_3 = \{p, q\},$$
$$C_9 = C_2 * C_4 = \{\neg p, \neg t\}, \qquad C_{10} = C_2 * C_8 = \{q, s\},$$
$$C_{11} = C_3 * C_6 = \{\neg r\}, \qquad C_{12} = C_4 * C_5 = \{\neg s\},$$
$$C_{13} = C_4 * C_7 = \{r, \neg t\}, \qquad C_{14} = C_4 * C_{10} = \{q, \neg t\},$$
$$C_{15} = C_5 * C_9 = \{\neg p\}, \qquad C_{16} = C_5 * C_{13} = \{r\},$$
$$C_{17} = C_5 * C_{14} = \{q\}, \qquad C_{18} = C_6 * C_8 = \{p\},$$
$$C_{19} = C_6 * C_{10} = \{s\}, \qquad C_{20} = C_6 * C_{14} = \{\neg t\},$$

and $C_6 * C_{17} = \square$. This completes the proof.

There are shorter proofs. One is

$$((((C_3 * C_6) * C_1) * C_2) * C_4) * C_5 = \square.$$

This proof has another, more important feature than its length. It starts with the negation of the goal and then repeatedly resolves it with "given" information—the σ_i's.

This approach is not always possible. Consider the problem of resolving the four clauses

$$\{p, q\}, \ \{p, \neg q\}, \ \{\neg p, q\}, \ \{\neg p, \neg q\} \tag{4.8}$$

to obtain the empty clause. This is easily done: Resolve the first two to obtain $\{p\}$, the last two to obtain $\{\neg p\}$, and these two results to obtain $\square$. The proof that linear resolution is impossible is left as an exercise. ∎

Example 4.5   The Lewis Carroll Example Revisited

In Example 3.10 (p. 116) we used a truth table argument to prove a result in one of Lewis Carroll's problems. Now we'll use propositional calculus resolution. In that example we argued that, although the statements are in predicate logic, it suffices to use propositional logic. In propositional logic notation, we're given

$$(c \to s), \quad ((\neg g) \to d), \quad ((\neg g) \vee c),$$

and want to prove $((\neg s) \to d)$.

In clausal form, the given conditions are

$$((\neg c) \vee s), \quad (g \vee d), \quad ((\neg g) \vee c).$$

Since the negation of the conclusion is $((\neg s) \wedge (\neg d))$, it gives rise to two clauses. Our list of clauses is

$$C_1 = \{(\neg c), s\}, \quad C_2 = \{g, d\}, \quad C_3 = \{(\neg g), c\},$$
$$C_4 = \{\neg s\}, \qquad C_5 = \{\neg d\}. \tag{4.9}$$

Rather than go through the systematic procedure of the algorithm, let's simply state the result using the notation of the previous example:

$$(((C_2 * C_5) * C_3) * C_1) * C_4 = \square. \quad \blacksquare$$

It should be obvious after the last two examples that we need to understand resolution better. As it stands, Algorithm 4.1 has problems.

- First, there are typically many choices of clauses to resolve. How should we choose among the possibilities?

- Second, it may be that proof is impossible so we should give up. How do we decide when we've tried enough; that is, when we can conclude that the result is not true?

As noted earlier, SLD-resolution of Horn clauses will partially solve the first problem.

The second problem has a solution in propositional logic when $\mathcal{S}$ is finite: Since the number of propositional letters present is finite, there are only a finite number of possible clauses. At some point, we will either obtain the empty clause or discover that all of the (finite number) of possible resolutions give clauses already present in $\mathcal{C}$. In the latter case, no further resolutions are possible, so we are done.

Unfortunately, the second problem has no solution in FOL. In fact, even a weaker problem—the "decision problem"—has no solution:

> It is impossible to produce an algorithm that takes as input an arbitrary formula α in FOL and produces as output the answer to the question "Is α valid?"   (4.10)

(In 1928, Hilbert and Ackermann, who called this the *Entscheidungsproblem*, had declared it to be the principal problem of mathematical logic.) You may be familiar with the halting problem—design a Turing machine $\mathcal{H}$ such that:

- The input of $\mathcal{H}$ consists of (a) the description of any Turing machine $\mathcal{T}$ and (b) the input on which $\mathcal{T}$ is to run.

- The output of $\mathcal{H}$ is the answer to the question "Will $\mathcal{T}$ stop?"

Turing proved that it's impossible to design such a machine $\mathcal{H}$. The decision problem for FOL is equivalent to the halting problem.

Exercises

4.2.A. What is a positive literal? a negative literal?

4.2.B. What is clausal form?

4.2.C. What is the resolution method of proof? How does it use contradiction?

4.2.D. The resolution method of proof is sound and complete (Theorem 4.2). What does this mean?

4.2.1. The purpose of this exercise is to use truth tables to prove Theorem 4.1. (The exercise uses truth tables because each $\mathcal{V}$ in (c) corresponds to a row of the truth table for $\neg\alpha$.)

(a) A formula is in *disjunctive normal form*, or *DNF*, if it is the "or" of one or more formulas, each of which is the "and" of one or more literals. Suppose that $(\neg\alpha) \equiv \gamma$ is a tautology, where γ is in DNF. Let δ be the result of interchanging the symbols $\vee$ and $\wedge$ and replacing every literal l with $\bar{l}$ in γ. Prove that δ is in CNF and that $\alpha \equiv \delta$ is a tautology.

(b) In view of the above, it suffices to construct a DNF for $\neg\alpha$. Let the propositional letters appearing in α be $p_1, \ldots, p_n$. Prove that, if $\neg\alpha$ is always false, then $(p_1 \wedge (\neg p_1))$ is a DNF for $\neg\alpha$.

(c) Suppose that $\neg\alpha$ is satisfiable. If $\mathcal{V}$ is an assignment of T and F, define

$$\gamma(\mathcal{V}) = l_1 \wedge \cdots \wedge l_n, \quad \text{where} \quad l_i = \begin{cases} p_i, & \text{if } \mathcal{V}(p_i) = \text{T}, \\ \neg p_i, & \text{if } \mathcal{V}(p_i) = \text{F}. \end{cases}$$

Let γ be the "or" of those $\gamma(\mathcal{V})$ for which $\mathcal{V}(\neg\alpha) = \text{T}$. Now prove that $(\neg\alpha) \equiv \gamma$ and conclude that Theorem 4.1 is true.

4.2.2. Prove that formula (4.5) is not a tautology; that is, prove that it is not valid.

4.2.3. Suppose that clauses (4.8) are used for resolution in Algorithm 4.1.

(a) Show that after the first resolution you will have one of the four clauses $\{p\}$, $\{q\}$, $\{\neg p\}$ and $\{\neg q\}$. (Ignore clauses that the algorithm discards.)

(b) Show that resolving any of the four clauses in (a) with a clause in (4.8) gives another clause in (a).

(c) Conclude that it's impossible to deduce the empty clause from (4.8) by repeated resolution starting with one of those clauses and repeatedly resolving with any of the other three clauses in any order.

4.2.4. Redo Exercises 3.3.13–3.3.15 (p. 119) using resolution as in Example 4.5.

4.2.5. Let $\mathcal{R}(\mathcal{C})$ be the set of all clauses that can be obtained from the clauses $\mathcal{C}$ by repeated resolution—including $\mathcal{C}$. In other words, this is just the augmented $\mathcal{C}$ produced by the resolution algorithm, without stopping if $\square$ is reached. For each of the following sets of clauses, compute $\mathcal{R}(\mathcal{C})$.

(a) Clauses C_1, C_2 and C_3 of (4.7).

(b) All the clauses in (4.7).

(c) $\Big\{\{\neg p, \neg q, \neg r\},\ \{p, \neg q, \neg s\},\ \{\neg p, q, \neg r\},\ \{\neg q, \neg r, s\}\Big\}$

SLD-Resolution of Horn Clauses

Recall that a Horn clause is a clause containing at most one positive literal. We saw in Section 3.4 (p. 120) that, in simple Prolog, each knowledge-base entry is a clause containing exactly one unnegated propositional letter and the query corresponds to a clause with all propositional letters negated; that is, they are Horn clauses. A Horn clause with no positive literals is called a *goal clause*.

Algorithm 4.2   SLD-Resolution for Horn Clauses

Suppose that we have some rule for ordering the literals in any goal clause and assume all goal clauses are ordered using this algorithm. We are given a set of Horn clauses containing exactly one (ordered) goal clause G_0.

Here is the SLD-resolution algorithm. For $i = 0, 1, \ldots$, select a clause C_i to resolve with the first literal of G_i, thereby producing a new (ordered) goal clause G_{i+1}. In contrast to Algorithm 4.1 (p. 146), duplicate literals need not be removed from the G_i's.

If there is some choice of C_i's that leads to $\square$, we say that $\neg G_0$ has been proved by SLD-resolution.

Note that SLD-resolution is a special case of general resolution because it specifies that the resolution proceeds by repeatedly resolving one goal clause to obtain another. Furthermore, the literal to resolve on is dictated by the ordering algorithm. This ordering may depend on anything, including all the previous steps in the resolution. The basic result is stated in the next theorem.

Theorem 4.3   SLD-Resolution for Horn Clauses

Regardless of what ordering algorithm is used for the G_i, SLD-resolution is sound and complete for the propositional calculus of Horn clauses.

Proof: Since removing duplicate literals only makes it easier to obtain □ and since Algorithm 4.1 is consistent, this algorithm is also consistent. Thus, we only need to prove completeness. That is, if S contains nongoal Horn clauses and G is a goal clause, then $S \models (\neg G)$ implies that there is a Horn clause resolution. The proof will be in three stages:

1. We show that if there is a fact $\{p\}$ in S, we can assume that no other clause contains the positive literal p

2. We show that if we can *select* the ordering rule, then an SLD-resolution proof exists.

3. We show how to convert such a proof into one for which the ordering rule is arbitrary.

From now on, we'll assume that $S \models (\neg G)$. As in Theorem 4.2 (p. 146), we'll assume that S is finite, thus avoiding a compactness argument.

First Stage:
For every propositional letter p such that $\{p\} \in S$, remove from S all clauses containing both the positive literal p and at least one negative literal. Call the resulting set of clauses S'.

Suppose that $S' \not\models (\neg G)$. This means that there is some assignment of truth values $\mathcal{V}$ such that $\mathcal{V}(C) = \text{T}$ for all $C \in S'$ and for G. Now suppose that $D \in S$ and $D \notin S'$. Then D contains a positive literal p such that $\{p\} \in S'$. By the definition of $\mathcal{V}$, we have $\mathcal{V}(\{p\}) = \text{T}$. Hence $\mathcal{V}(D) = \text{T}$. This shows that $\mathcal{V} = \text{T}$ for all clauses in S. Since we also have $\mathcal{V}(G) = \text{T}$, this contradicts the fact that $S \models (\neg G)$. It follows that the assumption $S' \not\models (\neg G)$ is incorrect. Hence, $S' \models (\neg G)$. This completes the first stage.

Second Stage:
Use induction on the total number of distinct propositional letters. If there is only one letter p, then $G = \{\neg p\}$ and the only possible clause in S' is $\{p\}$. Hence the proof is trivial.

By Theorem 4.2, a resolution proof $S' \models (\neg G)$ exists. The resolution of two clauses containing a negative literal produces another clause with a negative literal. Thus the resolution proof must use some clause that has no negative literals, that is, a clause $\{p\}$. Let S'' be S' with $\{p\}$ removed. Let G' be G with $\neg p$ removed if it is present.

We claim that $S'' \models (\neg G')$. The proof is similar to that in the first stage for $S' \models (\neg G)$: Suppose $\mathcal{V}$ is a truth assignment that proves $S'' \not\models (\neg G')$. Note that $\mathcal{V}(p)$ need not be defined since p does not appear in S'' or G'. Extend $\mathcal{V}$ to p by defining $\mathcal{V}(p) = \text{T}$. You should be able to show that $\mathcal{V}$ gives the contradiction $S' \not\models (\neg G)$. Thus the assumption $S'' \not\models (\neg G')$ is false.

By induction, there is an SLD-resolution proof of $S'' \vdash (\neg G')$ where the ordering of goal clauses is under our control. Take this proof and insert $\neg p$ in those clauses from which it was removed in creating S'' and G'. The resolution now leads to a clause containing some number of copies of $\neg p$ rather than to □. Repeated resolution with $\{p\}$ leads to □. This completes the induction.

Third Stage:

We now have an SLD-resolution with

$$G = G_0, G_1, \ldots, G_{n+1} = \square \quad \text{and} \quad C_0, C_1, \ldots, C_n,$$

but it may not involve resolving in the order dictated by the ordering rule we are given. It remains to prove that the resolution can be rearranged to satisfy that requirement.

There are various ways to establish that resolution can be so rearranged. One simple method is by counting. Since duplicate literals are not eliminated after resolution, each resolution eliminates some p and its negation. Since we end up at $\square$, it follows that the number of times p appears in G_0 and $C_0, C_1, \ldots$ must equal the number of times $\neg p$ appears.

Put all the C_i's in a collection and set $G_0' = G$. Now let's use the order rule. It tells us to resolve G_i' on some p. Remove a clause C_i' containing p from the collection and use it. The counting argument in the last paragraph guarantees that we'll always be able to find such a clause. ∎

Exercises

4.2.E. What are Horn clauses?

4.2.F. Describe an algorithm for SLD-resolution. How does it differ from general resolution?

4.2.6. Use SLD-resolution of the Horn clauses

$$\mathcal{S} = \Big\{ \{q\}, \; \{p, \neg q, \neg s\}, \; \{\neg p, q, \neg r\}, \; \{\neg q, \neg r, s\} \{\neg q, r\} \Big\}$$

and the negation of $\gamma = (p \wedge q)$ to show that $\mathcal{S} \models \gamma$.

4.3   First-Order Predicate Calculus

As we'll soon see, the presence of constants makes the truth table proof method untenable for predicate logic. Axiomatic methods can be extended. In fact, that's the way theorems are normally proved in mathematics. Of course, we don't actually reformulate the statements in terms of predicate calculus the way we did for the definition of a limit in Example 3.8 (p. 112). As we saw in the last chapter, resolution can also be used. After discussing the semantics of predicate calculus, we'll explore resolution theoretically and algorithmically in this section.

To carry out a resolution proof we must rewrite a formula so that all the quantifiers are universal and on the outside while the formula inside the quantifiers is in clausal form. This can be done as follows:

- Adapt the algorithm in the proof of Theorem 4.1 (p. 143) to produce a clausal form containing quantifiers. Use (3.23) (p. 115) to move negation inward through quantifiers.

- Assign unique names to all quantified (=bound) variables and then use (3.24) to move the quantifiers to the left side of the formula. The result is said to be in *prenex form*.

- Use "Skolemization" to eliminate existential quantifiers.

This process isn't needed for Prolog. For a simple Prolog statement, we can use (3.25) (p. 115) to put it in normal form. To combine Prolog statements, all we need do is make sure that no two statements use the same name for a variable. The fact that this procedure is so easy for Prolog is one of the features that make Prolog practical.

Once this normal form has been achieved, resolution can begin; however, it's complicated by the need for "unification." The unification must be done so as not to impose any equalities that are not absolutely required—the "most general unification." Unification is no simpler in Prolog than it is in general. In fact, you'll see that Prolog interpreters usually cheat when doing unification.

Skolemization

Suppose we have a formula $\beta = \forall X_1 \ldots \forall X_n \exists Y \ \alpha$, where α may involve quantifiers. It's possible to create another formula β' not containing Y such that β' is valid if and only if β is valid. The idea is simple enough if you think about what validity of β means:

> For every choice of values for $X_1, \ldots, X_n$ from the set of constants, we can find a value for Y such that α is valid when $X_1, \ldots, X_n, Y$ are replaced by these values.

Thus, Y depends on the formula α and the values assigned to $X_1, \ldots, X_n$. In other words, Y is a function of $X_1, \ldots, X_n$, the actual nature of the function depending on α. We can express this by creating a new function name, say f, and saying that Y is $f(X_1, \ldots, X_n)$. This leads to a simple algorithm for Skolemization.

Algorithm 4.3   Skolemization

Let $\mathbf{Q}$ stand for a quantifier and let α be a formula without quantifiers. Denote the formula $\mathbf{Q}X_1 \ldots \mathbf{Q}X_n \, \alpha$ by β_0. In general, let β_{i-1} be

$$\forall X_{a_1} \ldots \forall X_{a_k} \mathbf{Q} X_i \, \gamma.$$

The Skolemization of β_0 is β_n where β_i is computed as follows.

- If X_i is universally quantified, β_i is β_{i-1}.

- If X_i is existentially quantified, β_i is $\forall X_{a_1} \ldots \forall X_{a_k} \delta$, where δ is obtained from γ by replacing all occurrences of X_i with $f(X_{a_1}, \ldots, X_{a_k})$ and where f is a new function name that is added to the list of functions in the language.

It's possible that there are no universal quantifiers preceding X_i. In this case, $k = 0$ and the new function f would have no arguments; that is, it is a new constant.

This is a departure from what we've done previously. All previous manipulations of FOL formulas have involved working within the given language. In this manipulation, we've created new, Skolem, functions and added them to the language. If we could choose the interpretations we wanted for these functions, β_0 would be valid if and only if β_n was. Unfortunately, in the notion of validity, we must allow *all possible* interpretations of these functions, and not all of these preserve validity. Thus

> Skolemization of a formula does not, in general, preserve validity.

In order to understand what Skolemization does give us, we need the concept of *satisfiable*, which was defined earlier for propositional logic. Here it is for FOL.

Definition 4.4   Satisfiability

Let α be a formula with no free variables. We say that α is *satisfiable* if there is some interpretation in which α is true.

Since a rigorous proof of the following theorem requires more precise attention to semantics, we won't give one. The discussion preceding the definition makes the theorem plausible, and even indicates the way a proof would proceed.

Theorem 4.4 Skolemization and Satisfiability

Let α be a formula with no free variables, let β be its Skolemization, and let $f_1, \ldots, f_n$ be the functions introduced in the process of Skolemization. Let $\mathcal{I}$ be an interpretation that does not mention $f_1, \ldots, f_n$ and let $\mathcal{I}^*$ be an extension of $\mathcal{I}$ to include $f_1, \ldots, f_n$.

- If β is valid in $\mathcal{I}^*$, then α is valid in $\mathcal{I}$.
- If α is valid in $\mathcal{I}$, then there is an extension $\mathcal{I}^*$ in which β is valid.

We say that α and β are *equisatisfiable*.

Early in the previous chapter, we claimed that validity is a good idea since it does not depend on interpretation. Since satisfiability does depend on interpretation, can it possibly be any good? Yes, sometimes. The most important situation is when something is not satisfiable:

$$\text{A formula } \alpha \text{ is not satisfiable if and only if } (\neg\alpha) \text{ is valid.} \qquad (4.11)$$

You should be able to prove this. (If not, see p. 137.) Unsatisfiability is the idea behind the resolution method and some other proof methods. By proving $(\neg\beta)$ is unsatisfiable, we prove that β is valid.

Example 4.6 Clausal Form and Skolemization

To illustrate how the previous ideas apply, let's convert the formula defining a limit to the form we've been discussing. We can rewrite (3.20) (p. 113) in more standard FOL form as

$$\forall Z \Big\{ p(Z) \rightarrow \Big(\exists Y \big\{ p(Y) \wedge \forall X \big((q(X,a,Y) \wedge r(X,a)) \rightarrow q(f(X),l,Z) \big) \big\} \Big) \Big\}.$$

Now let's work on this. First, getting rid of the $\rightarrow$ gives

$$\forall Z \Big\{ (\neg p(Z)) \vee \Big(\exists Y \big\{ p(Y) \wedge \forall X \big(\{\neg(q(X,a,Y) \wedge r(X,a))\} \vee q(f(X),l,Z) \big) \big\} \Big) \Big\}.$$

Moving negations in and quantifiers out, we get

$$\forall Z \, \exists Y \forall X \Big\{ (\neg p(Z)) \vee \Big(p(Y) \wedge \big((\neg q(X,a,Y)) \vee (\neg r(X,a)) \vee q(f(X),l,Z) \big) \Big) \Big\}.$$

Finally, replacing Y with $g(Z)$ and using the distributive law to obtain a clausal form, we have

$$\forall Z \, \forall X \Big\{ \big((\neg p(Z)) \vee p(g(Z)) \big)$$

$$\big((\neg p(Z)) \vee (\neg q(X,a,g(Z))) \vee (\neg r(X,a)) \vee q(f(X),l,Z) \big) \Big\}.$$

The function $g(Z)$—or $g(\epsilon)$ in the original form of the definition—is explored in calculus. To prove that a limit exists, we must give a rule for determining δ as a function of ϵ. The rule we obtain is simply the function $g(Z)$ that replaces Y. You may find it interesting to translate the two clauses back to something more like English. The first says that either $\epsilon \leq 0$ or $g(\epsilon) > 0$. This simply guarantees that $\delta = g(\epsilon)$ is greater than zero whenever ϵ is. The second clause is more interesting. It states that either $|f(x) - L| < \epsilon$ or one of the three conditions

$$\epsilon > 0, \quad |x - a| < \delta, \quad x \neq a$$

fails to hold. Perhaps the process of Skolemization and finding clausal form has helped you understand the notion of limits better—or maybe it made limits more confusing! ∎

Unification

The unification problem is as follows. Suppose we have the Skolemized clausal form of a formula. Let $p(t_1, \ldots, t_n)$ and $\neg p(t'_1, \ldots, t'_n)$ be predicates from two different clauses, where t_i and t'_i are terms. Unification is the process of substituting terms for some (or all) variables so that, after the substitution, t_i and t'_i are the same term for $1 \leq i \leq n$. This is *unification,* and the substitution is called *a unifier* of $p(t_1, \ldots, t_n)$ and $p(t'_1, \ldots, t'_n)$.

I said "a unifier" because many unifications are possible. For example, a unifier of $p(X)$ and $p(Y)$ is obtained by substituting $f(a, Z)$ for both X and Y. Another unifier is obtained by replacing X and Y with c. Still another is obtained by substituting X for Y. The last unification is "more general" because the first two can be obtained by a further substitution for the variable X in the last substitution—either $f(a, Z)$ or c, respectively. The goal of unification is to obtain a "most general" unifier; that is, a unifier from which all others can be obtained by further substitutions. *A priori,* it's not clear that a *most* general unifier exists.

There are various ways of describing an algorithm for finding the most general unifier. The most easily understood involves depth-first traversal of ordered trees associated with predicates. The trees are called *formation trees.* The description of the tree construction parallels Definition 3.4 (p. 106) and part of Definition 3.5. Here's the definition and the tree construction in parallel.

The terms in $\mathcal{L}$ and the associated formation trees are defined recursively as follows.:

D1. Every variable as well as every constant is a term.

T1. A vertex labeled with any variable or constant is a formation tree.

D2. If $t_1, \ldots, t_n$ are terms and f is a function that takes n arguments, then $f(t_1, \ldots, t_n)$ is a term.

T2. If $t_1, \ldots, t_n$ are terms and f is a function that takes n arguments, then the ordered tree whose root is labeled f and whose ith child is the formation tree for t_i is a formation tree.

Finally, we have the atomic formulas:

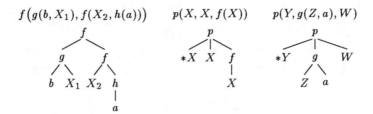

Figure 4.1 Some formation trees. Above each tree is the term or atomic formula to which it corresponds. The asterisks are explained in Example 4.7.

D3. If $t_1, \ldots, t_n$ are terms and p is a predicate that takes n arguments, then $p(t_1, \ldots, t_n)$ is an atomic formula.

T3. If $t_1, \ldots, t_n$ are terms and p is a predicate that takes n arguments, then the ordered tree whose root is labeled p and whose ith child is the formation tree for t_i is a formation tree.

Figure 4.1 shows some formation trees.

Notice that, since a variable takes no arguments, it can occur only as the leaf of a formation tree. Here's the algorithm for finding a most general unifier.

Algorithm 4.4   Most General Unifier

Let T_0 and T_1 be two formation trees. The algorithm terminates with both trees displaying the result of the most general unifier, or it reports failure. Start at the root of each tree and carry out a depth-first search and substitution as described below until both trees have been traversed.

1. **Compare:** Let L_i be the label at the present vertex in T_i. If $L_0 = L_1$, go to Step 3.

2. **Substitute:** Let U_i be the formation tree rooted at the current vertex in T_i. Since the roots L_0 and L_1 are not equal, there are three possibilities:

   (a) Neither L_0 nor L_1 is a variable: In this case, stop with failure.

   (b) One of L_i is a variable V and V appears in U_{1-i}. In this case, stop with failure.

   (c) One of L_i is a variable V and V does not appear in U_{1-i}. In this case, replace all leaves labeled V with a copy of U_{1-i}. (This substitutes U_{1-i} for V.)

3. **Advance:** If there are no more vertices in the depth-first search, terminate with success. Otherwise, move to the next depth-first vertex in both T_0 and T_1 and go to Step 1.

The subscript $1 - i$ is used to refer to the tree other than T_i. This works since $(i = 0) \rightarrow (1 - i = 1)$ and $(i = 1) \rightarrow (1 - i = 0)$.

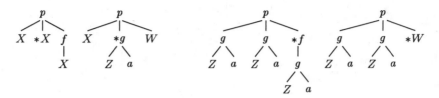

Figure 4.2 Application of the unification algorithm. See Example 4.7 for details.

Example 4.7   Unification

Let's see how the algorithm works on the rightmost trees in Figure 4.1. The asterisks indicate where the first disagreement occurs in the depth-first search. In this case, L_0 and L_1 are both variables so we can replace X with Y or Y with X. Choosing the former leads to the two trees on the left side of Figure 4.2.

Again, the vertices where disagreement occurs are marked by an asterisk. Now $L_0 = X$ and $L_1 = g$, so we must substitute U_1 for X everywhere. The results are shown on the right side of Figure 4.2. Finally, the last substitution that the algorithm requires is replacing W with $f(g(Z, a))$. The final result is $p\big(g(Z, a), g(Z, x), f(g(Z, a))\big)$.

Now let's make a slight change in the original T_1. Replace the variable W with X. In this case, the algorithm reaches its third disagreement at f in T_0 and g in T_1 (g instead of X because an earlier substitution for X). As a result, it reports failure because no substitution for a variable will change the function symbols.

Let's make a different change. Replace W with Z in T_1. The third disagreement is at the same place as before, but now there is a Z in T_1 and U_0 contains Z because U_0 corresponds to $f(g(Z, A))$. The algorithm ends in failure. Why can't we simply substitute $f(g(Z, A))$ for Z everywhere? If we were to do so, we would have to make the substitution in $f(g(Z, a))$ as well, and this would lead to an infinite repetition: $f(g(Z, a))$ becomes $f(g(f(g(Z, a)), a))$, which becomes $f(g(f(g(f(g(Z, a)), a)), a))$, and so on. It's a bit easier to see what's happening by looking at X and $h(X)$. Repeated substitution leads to $h(h(h(\ldots)))$, where the ellipsis indicates an infinite sequence of nested h's. ∎

Theorem 4.5   Most General Unifier

Algorithm 4.4 always terminates. If it terminates in failure, no unification is possible. If it produces a unification, it is a most general one; that is, every other unification is obtainable from that produced by the algorithm by substituting for the variables it contains.

Since the algorithm starts with two trees, which are finite, it may seem that termination is obvious. This is not the case. Conceivably, the process of substitution could result in larger and larger trees such that the depth-first traversal is never finished. In fact, if we forget to include (b) in the algorithm, the algorithm can go on forever as described at the end of the previous example. This is what most (perhaps all) implementations of Prolog actually do.

Proof: How is termination proved? Very simply. At the start, the number of variables in the two formation trees is finite since the entire trees are finite. Because of (b) in the algorithm, each substitution eliminates all occurrences of a variable. Hence the number of substitutions done by the algorithm is finite. Consequently, the trees are changed only a finite number of times and so do not grow without bound.

Any disagreement that is found by the algorithm must be eliminated if a unification is to be found. The only substitutions that are allowed are those replacing a variable with a term. This explains why there must be a failure when (a) occurs in the algorithm. Case (b) must lead to failure because it leads to an infinite chain of substitutions as discussed near the end of the previous example.

This leaves (c). The substitution made there is the least possible to create unification. Mightn't a more restrictive substitution somehow allow more freedom elsewhere in the trees? Perhaps you intuitively see that this is not the case—and perhaps you don't, in which case some experimentation will probably convince you. To give a proof up to present-day standards of mathematical rigor, we'd have to spend time looking carefully at properties of substitutions. We won't do that, so the proof is not quite complete. ■

Aside. The entire discussion of unification could have been formulated with several atomic formulas in the predicate p rather than just two. You should find it relatively easy to adapt the algorithm.

Resolution

To prepare for resolution, we first put formulas in clausal form, ignoring the location of quantifiers. Then we use Skolemization to eliminate existential quantifiers and place universal quantifiers on the leftmost side of the formula. At this point, we're ready to obtain a contradiction by means of resolution. The process is the same as that for propositional calculus, except that we must make use of unification. We select two clauses, one containing an atomic formula $p(t_1, \ldots, t_n)$ and another containing the negation of the predicate, say $\neg p(t'_1, \ldots, t'_n)$. A most general unifier is found. The literals in the two clauses, with the exception of $p(t_1, \ldots, t_n)$ and $\neg p(t'_1, \ldots, t'_n)$, are placed in a new clause and the substitutions of the most general unifier are applied.

The main differences between resolution proofs for the propositional and predicate calculi lie in the need for Skolemization and unification in the latter. Examples of predicate calculus proofs with unification appear in the Prolog discussion of Section 3.4 (p. 120). Because of the nature of Prolog, Skolemization is not required. The following example involves all the various features.

Example 4.8   Lewis Carroll Yet Again

We've been pursuing an example from Lewis Carroll in a series of examples—3.9, 3.10, and 4.5—primarily from a propositional calculus viewpoint. This has required some adjustments because the original statements are more naturally stated in the predicate calculus. In fact, according to Example 3.9 (p. 113), the given formulas are

$$\forall X \Big(c(X) \to s(X) \Big)$$
$$\forall X \Big((\neg g(X)) \to d(X) \Big)$$
$$\neg \Big(\exists X \Big(g(X) \wedge (\neg c(X)) \Big) \Big)$$

and we wish to deduce

$$\forall X \Big((\neg s(X)) \to d(X) \Big).$$

The propositional forms of the three hypotheses were put in clausal form in Example 4.5 (p. 149). Those manipulations are easily adapted to give the FOL clausal forms. The negation of the conclusion is a bit trickier since it requires Skolemization. To begin with,

$$\Big(\neg \forall X \Big((\neg s(X)) \to d(X) \Big) \Big) \equiv \Big(\exists X \, \neg \Big(s(X) \vee d(X) \Big) \Big)$$
$$\equiv \Big(\exists X \Big((\neg s(X)) \wedge (\neg d(X)) \Big) \Big).$$

Inside the quantifier we have a clausal consisting of two clauses with one literal each. To Skolemize, we replace X with a function f. Since there are no variables besides X, the function has no arguments. The result is

$$(\neg s(f())) \wedge (\neg d(f())).$$

Combining this with the clausal forms of the three hypotheses, we have the following set of clauses:

$$C_1 = \{\neg c(X)\, s(X)\}, \quad C_2 = \{g(Y), d(Y)\}, \quad C_3 = \{\neg g(Z), c(Z)\},$$
$$C_4 = \{\neg s(f())\}, \qquad C_5 = \{\neg d(f())\},$$

similar to (4.9). Resolution follows the same order as that with the clauses (4.9):

- Resolving C_2 and C_5 using the unification of Y with $f()$ gives $\{g(f())\}$.
- Resolving this with C_3 using the unification of Z with $f()$ gives $\{c(f())\}$.

- Resolving this with C_1 using the unification of X with $f(\ )$ gives $\{s(f(\ ))\}$.
- Resolving this with C_4 gives $\square$.

Since a function with no arguments is simply a constant, the effect of C_5 is to to unify the variables X, Y, and Z with this constant. Since the same constant appears in C_4, a contradiction is reached in the final step. ∎

Soundness and Completeness

So far we've been concerned only with methods. What about a proof that resolution is sound and complete and that SLD-resolution is sound and complete for Horn clauses?

First, there is the matter of Skolemization. As noted in the discussion concerning Theorem 4.4 (p. 157), this is not a problem for resolution proofs. Let's review. Suppose we want to prove $\mathcal{S} \models \gamma$. The approach is to show that $\alpha = (\mathcal{S} \wedge (\neg\gamma))$ is unsatisfiable. Let β be the Skolemization of α. With a bit of thought, you should be able to see that the contrapositive of Theorem 4.4 asserts that α is unsatisfiable if and only if β is. (Recall that the contrapositive of "if A, then B" is "if $\neg B$, then $\neg A$.") As a result, we can simply work with the Skolemized formulas.

Soundness can be proved by a reduction to propositional calculus as follows. Imagine a resolution proof leading to $\square$. In the process of the proof, various substitutions have been made. Propagate the substitutions backward. For example, if we resolve $\{p(X), q(X)\}$ and $\{\neg p(Y), r(f(Y))\}$, the substitution of X for Y is made. If we then resolve the result with $\{\neg q(b)\}$, the substitution of b for X is made. Propagating these substitutions backward, our three clauses become $\{p(b), q(b)\}$, $\{\neg p(b), r(f(b))\}$, and $\{\neg q(b)\}$, respectively. The result of such replacements is a resolution derivation of $\square$ that involves no unifications. Finally, replace any variables that appear in this derivation with arbitrary constants to obtain a resolution derivation of $\square$ without variables. As discussed in Example 3.10 (p. 116), this derivation can be viewed as a resolution proof in propositional logic, where we know consistency holds.

Can this idea of reduction to propositional logic be exploited to prove completeness? Yes, but more argument is required. What needs to be shown is the following. If a set $\mathcal{S}$ of universally quantified clauses is inconsistent, then the same is true for a finite set $\mathcal{T}$ of clauses containing no variables, where each clause in $\mathcal{T}$ is obtained from a clause in $\mathcal{S}$ by substituting terms for all bound variables and deleting the universal quantifiers. One clause in $\mathcal{S}$ may give rise to several in $\mathcal{T}$ through various substitutions. Except for the following example, we won't discuss the proof of completeness further here. You'll have to look in a text on FOL logic for the details.

*Example 4.9   Reduction to Propositional Calculus

Recall the Prolog definition of **append** from Example 3.13 (p. 126):

```
append([],Y,Y).                                              % 1
append([X|L],M,[X|LM]) :- append(L,M,LM).                    % 2
```

Let's look at the resolution solution to the query :- append([a,b],[y,z]).
We'll mix Prolog and FOL notation, but it shouldn't be confusing.
    The clauses are

$$C_1 = \{\texttt{append}([\,], \texttt{Y}, \texttt{Y})\},$$
$$C_2 = \{\texttt{append}([\texttt{X}|\texttt{L}], \texttt{M}, [\texttt{X}|\texttt{LM}]), \neg\texttt{append}(\texttt{L}, \texttt{M}, \texttt{LM})\},$$
$$C_3 = \{\neg\texttt{append}([\texttt{a}, \texttt{b}], [\texttt{c}, \texttt{d}], \texttt{Z})\}.$$

The resolution proof that Prolog constructs is $((C_3 * C_2) * C_2) * C_1$. The first
resolution requires that **a** be substituted for **X**, that **[b]** be substituted for **L**,
that **[c,d]** be substituted for **M**, and that **[X|LM]** be substituted for **Z**. (The
expression **[X|LM]** is simply a function of **X** and **LM**.) To distinguish the **LM**
occurring in this result from the **LM** in C_2, let's prime it. Putting all these
substitutions together, Prolog uses

$$C_3' = \{\neg\texttt{append}([\texttt{a}, \texttt{b}], [\texttt{c}, \texttt{d}], [\texttt{a}|\texttt{LM}'])\}$$
$$C_2' = \{\texttt{append}([\texttt{a}|[\texttt{b}]], [\texttt{c}, \texttt{d}], [\texttt{a}|\texttt{LM}']), \neg\texttt{append}([\texttt{b}], [\texttt{c}, \texttt{d}], \texttt{LM}')\}.$$

to obtain $\{\neg\texttt{append}([\texttt{b}], [\texttt{c}, \texttt{d}], \texttt{LM}')\}$. This is resolved again with C_2. The result
of the substitution is that **LM'** becomes **[b|LM]** and various other substitutions
are made so that Prolog uses

$$C_2'' = \{\texttt{append}([\texttt{b}|[\,]], [\texttt{c}, \texttt{d}], [\texttt{b}|\texttt{LM}]), \neg\texttt{append}([\,], [\texttt{c}, \texttt{d}], \texttt{LM})\}$$

to obtain $\{\neg\texttt{append}([\,], [\texttt{c}, \texttt{d}], \texttt{LM})\}$. After unification by replacing **Y** and **LM** with
[c,d], this is resolved with C_1 to obtain $\square$. Making the backward substitution
for **LM** and **LM'**, we see that Prolog has used the clauses

$$C_3' = \{\neg\texttt{append}([\texttt{a}, \texttt{b}], [\texttt{c}, \texttt{d}], [\texttt{a}, \texttt{b}, \texttt{c}, \texttt{d}])\},$$
$$C_2' = \{\texttt{append}([\texttt{a}|\texttt{b}], [\texttt{c}, \texttt{d}], [\texttt{a}|[\texttt{b}|[\texttt{c}, \texttt{d}]]]), \neg\texttt{append}([\texttt{b}], [\texttt{c}, \texttt{d}], [\texttt{b}|[\texttt{c}, \texttt{d}]])\},$$
$$C_2'' = \{\texttt{append}([\texttt{b}|[\,]], [\texttt{c}, \texttt{d}], [\texttt{b}|[\texttt{c}, \texttt{d}]]), \neg\texttt{append}([\,], [\texttt{c}, \texttt{d}], [\texttt{c}, \texttt{d}])\},$$
$$C_1' = \{\texttt{append}([\,], [\texttt{c}, \texttt{d}], [\texttt{c}, \texttt{d}])\}$$

to obtain $\square$. (I've taken the liberty of using the fact that, in Prolog, a term
like **[a,b]** is just different notation for the term **[a|[b]]**.) Since none of
these involves variables, we can simply treat them as being propositional logic
statements:

$$C_3' = \{\neg p\}, \quad C_2' = \{p, \neg q\}, \quad C_2'' = \{q, \neg r\}, \quad C_1' = \{r\},$$

where p is the proposition **append([a, b], [c, d], [a|LM'])**, and so forth.

Don't be misled by this example. It doesn't show you how to carry out a reduction to propositional calculus that will lead to a resolution proof. What we did was start with a resolution proof and derive the propositional calculus reduction that lies behind it. The difficulty is in showing there is such a reduction *without having a resolution proof to work from*. To prove completeness, we must first do the reduction and then use completeness of propositional logic resolution to obtain a resolution proof. ∎

Decidability

A logic is *decidable* if there exists a method such that, for each formula α, the method will tell us whether or not α is satisfiable. Since validity of α is equivalent to nonsatisfiability of $\neg\alpha$ (p. 137), decidability is equivalent to being able to determine if a formula is valid.

If a proof method is sound and complete, it will be able to prove the validity of any valid formula and will never "prove" the validity of an invalid formula. However, there is no guarantee that the proof method will be able to tell that a formula is not valid. How can this be? Certainly, if the proof method terminates, we have an answer one way or the other. But it's possible that, for some invalid formula, the proof method will not terminate.

The discussion in the previous paragraph shows that the notion of decidability is stronger than the notion of soundness and completeness. Does it matter? That is, are there logics of interest that are not decidable and do we care? A resounding "yes" on both counts. For the first, we have

Theorem 4.6   Semidecidability of FOL

FOL is not decidable; that is, it's impossible to construct an algorithm that will determine whether or not a given formula in FOL is valid. (Or, equivalently, it's impossible to construct an algorithm to determine satisfiability.) However, there exist proof methods for FOL that are sound and complete. This situation is described by saying that FOL is *semidecidable*.

We omit the proof of the first statement in the theorem. The equivalence of the parenthetic statement to the first statement was already proved in the preceding discussion. The soundness and completeness of resolution proves the final claim.

Why do we care about decidability? When using FOL, it might be useful to know that a formula is not valid. In some forms of "nonmonotonic" reasoning, it's essential. (See Chapter 6.) As a result, the nonexistence of such algorithms in FOL implies the nonexistence of sound and complete proof methods for some nonmonotonic logics.

Exercises

4.3.A. What does a formula in prenex form look like? Give an example of one in prenex form and one not in prenex form.

4.3.B. What is Skolemization? Give an example of a formula and its Skolemization.

4.3.C. Explain what the statement "Skolemization preserves satisfiability but not validity" means.

4.3.D. What is unification?

4.3.E. For FOL, describe the method of using resolution to obtain a proof by contradiction.

4.3.F. Since a proof method should establish validity and not merely satisfiability, why is Skolemization okay when giving a proof by the resolution method?

4.3.G. Explain the statement "FOL is semidecidable."

4.3.1. Earlier you were asked to give propositional calculus resolution proofs for Exercises 3.3.13–3.3.15 (p. 119). Now give FOL resolution proofs for them.

The remaining exercises have asterisks because they require familiarity with Prolog's list notation.

*4.3.2. What is the FOL resolution proof of :- `append([a,b],Z,[a,b,c])` in Prolog? Reduce it to propositional logic.

*4.3.3. Prolog finds three answers to :- `append(V,W,[a,b])`. What are the FOL resolution proofs? Reduce them to propositional logic.

*4.3.4. Consider the Prolog clauses

`sub([],L).`	rule 1
`sub([X|L],[X|M]) :- sub(L,M).`	rule 2
`sub(L,[X|M]) :- sub(L,M).`	rule 3

(a) Give a simple non-Prolog explanation of when a query `sub(L,M)` will succeed, where L is the list $a_1, \ldots, a_k$ and M is the list $b_1, \ldots, b_m$.

(b) Draw the complete Prolog search tree for query `sub([a,X],[a,b,a])`, indicating which rule is used at each edge. Indicate those terminal vertices (leaves) where Prolog finds a solution.

(c) Translate the rules to clausal form and give an FOL resolution proof for the query that leads to X=a.

4.4 Prolog

Let's review basic Prolog in light of the information in this chapter. The main goal in doing this is to obtain a better understanding of what happens when a clean mathematical theory collides with the real world of programming.

Let $\mathcal{K}$ denote the knowledge base and $\mathbb{T}(\mathcal{K})$ the translation of $\mathcal{K}$ into FOL. Each knowledge-base fact or rule leads to a Horn clause in which all the variables are universally quantified. Since each knowledge-base clause contains exactly one predicate that is not negated, it is easily shown that $\mathbb{T}(\mathcal{K})$ is consistent: Simply interpret all predicates to always be true. The interpretation of functions and constants is arbitrary.

When we give Prolog a goal, we are asking it to exhibit some values of the variables which make the goal true. Thus, the quantifiers on a goal are existential. However, Prolog negates the goal and attempts to show that it is inconsistent with the knowledge base. Since negation converts existential to universal quantifiers, the translation of the goal can be written as a clause γ in which all variables are universally quantified and all predicates are negated. We now have a collection of Horn clauses in Skolem normal form, just the form needed for resolution.

Resolution proceeds by resolving clauses, using unification, until the empty clause is found. When this state is reached, the unifications have made substitutions that establish the inconsistency of $\mathcal{K} \wedge (\neg\gamma)$. This is just what we wanted: First, the inconsistency of $\mathbb{T}(\mathcal{K}) \wedge \gamma$ is equivalent to $\mathbb{T}(\mathcal{K}) \models (\neg\gamma)$. Second, the resolution proof has determined values for the variables in the goal in the process of establishing inconsistency.

Prolog uses SLD-resolution. As discussed earlier, its use of depth-first search rather than breadth-first or iterative-deepening leaves open the possibility that Prolog may fail to terminate even when an SLD-resolution proof exists.

When Prolog uses unification, it ignores the possibility that the term being substituted for a variable may contain the variable. As discussed earlier, this lack of an "occurs check" can lead to a nonterminating algorithm. It will probably become standard for Prolog interpreters to implement the occurs check part of unification.

Now let's turn to some features of Prolog that are beyond standard FOL. From the logician's viewpoint, these are (necessary?) evils. From the programmer's viewpoint, these are features that are needed to make Prolog a viable language.

The Cut Operator

Perhaps you've been exposed to the debate over the use of **goto** in procedural programming. At its height, the argument ranged from ardent demands for abolition of this habit-forming construct to impassioned defenses of its moderate use. Among the arguments against **goto** are:

- It is most often used because the concept being programmed is poorly formulated.

- It makes it harder to verify that code is correct.

- It encourages poor thought and programming practices.

Virtually the same arguments have been raised in the debate over the Prolog "cut" operator.

What does the cut operator do and why does it cause problems? Recall that, at any given stage of its SLD-resolution, Prolog is looking at knowledge-base clauses whose heads involve some predicate p. If the clause being tried contains a cut, Prolog will not try any other clauses with head p (after this one) at this step of the SLD-resolution. How does this fit with the objections to **goto**?

Poor formulation: People often use the cut operator because they don't see how to structure their code to avoid infinite depth-first search. However, with a better understanding of the problem and what Prolog does, such structuring is often possible.

Correctness: By eliminating some parts of the search tree, the cut operator stops Prolog from fully implementing SLD-resolution. Consequently, we can no longer appeal to the fact that SLD-resolution is complete.

Poor thinking: The cut operator encourages procedural thinking. To write good Prolog code, we must think declaratively rather than procedurally as much as we can. Thus, the cut operator encourages a mode of thinking (procedural) that leads to poor code. For example, imagine programming the predicate **append(X,Y,XY)**, which is intended to ensure that the list **XY** is the result of appending the list **Y** to the list **X**. When writing the code procedurally, you may think of **X** and **Y** as being instantiated. When writing the code declaratively, you are less likely to do so. If your code is well written, someone will be able to use it when **XY** is instantiated and both **X** and **Y** are not.

Negation

Suppose we were to allow the Prolog statement "**p :- q, ¬r.**", where ¬ is FOL negation. The corresponding clause is $\{p, \neg q, r\}$, which is not a Horn clause because it contains two positive literals. Thus, Prolog cannot allow negation. Since negation is important, how does Prolog get around this problem?

The Prolog **not** connective means failure, not negation. It is often called *negation as failure*.

Suppose Prolog encounters **not p(X,Y)** and it has unified **Y** with **c**. In this case, Prolog temporarily takes **p(X,c)** as a new goal. If the goal fails, then **not p(X,Y)** succeeds, and conversely.

What does this mean in FOL terms? Let $\sim$ denote the Prolog **not**. Prolog attempts to show that $\neg p(X,c)$ is inconsistent with the $\mathbb{T}(\mathcal{K})$. Suppose it succeeds. Since SLD-resolution is a complete proof method, this means that

$$\mathbb{T}(\mathcal{K}) \models (\exists X \, p(X,c)) \text{ and so } \mathbb{T}(\mathcal{K}) \not\models (\neg\exists X \, p(X,c)).$$

Now suppose that Prolog fails. Similar reasoning leads to

$$\mathbb{T}(\mathcal{K}) \not\models (\exists X \, p(X,c)) \text{ and, we hope, } \mathbb{T}(\mathcal{K}) \models (\neg\exists X \, p(X,c)).$$

Unfortunately, the hope is not justified, as a simple propositional calculus example shows: We have both $p \not\models q$ and $q \not\models p$—simply take the propositional letter the the left of $\not\models$ to be true and the other to be false.

Removing some of the negations from the previous paragraph, it is simply the statement that

The fact "α is not consequence of $\mathbb{T}(\mathcal{K})$" does not imply the fact "$\neg\alpha$ is a consequence of $\mathbb{T}(\mathcal{K})$."

Since this is the case, how can we justify using negation as failure in Prolog? Various justifications are offered.

The most common justification is the *closed world assumption*, abbreviated *CWA*. The closed world assumption says that all facts and rules relevant to the problem at hand are contained in the knowledge base $\mathcal{K}$. What does this mean? From a Prolog viewpoint, it means that anything not deducible from $\mathcal{K}$ may be assumed false.

What does the CWA mean from a logical viewpoint? Let α be a formula. In FOL, the CWA means that the truth value of α is the same in every interpretation where $\mathbb{T}(\mathcal{K})$ is true. In other words, every formula is either valid or unsatisfiable. In practice, this is unlikely to be the case, unless we assume that there are some "hidden facts" in $\mathcal{K}$, namely all formulas of the form $\neg p(a, b, \ldots)$ that are consistent with the knowledge base.

Another justification of negation as failure is the *completed data base assumption*. This means that the Prolog knowledge base gives all the ways in

which a predicate can be true. (Data base and knowledge base are used interchangeably.) Here's what this means. Suppose we attempt to unify the heads
of clauses with $p(a, b)$. Those unifications that succeed produce a collection of
bodies $\alpha_1, \ldots, \alpha_n$. The Prolog data base asserts that

$$(\alpha_1 \vee \cdots \vee \alpha_n) \rightarrow p(a, b).$$

The completed data base assumption strengthens this to

$$(\alpha_1 \vee \cdots \vee \alpha_n) \equiv p(a, b).$$

Using this approach, researchers have established a framework in which Prolog's negation as failure is based on a proof method that is sound and complete.
Thus, Prolog's negation as failure is an elegant work-around for the problem
of introducing negation in the context of Horn clauses. (As noted earlier in the
discussion, simply allowing logical negation takes us out of the Horn clause
domain.)

Equality, Arithmetic, and Procedural Code

Prolog contains various features not contained in FOL. These include equality,
arithmetic, and procedural code. How do these features fit into the scheme of
things?

Equality is the least disruptive. It can be included in FOL with some
effort. People speak of "first-order logic with equality." The idea is twofold.

First some axioms must be added to the knowledge base concerning the
equality predicate. Writing $e(a, b)$ to indicate $a = b$, the axioms logicians
usually add are

- $e(X, X)$.
- $\Big(e(X_1, Y_1) \wedge \cdots \wedge e(X_n, Y_n)\Big) \rightarrow e\Big(f(X_1, \ldots, X_n), f(Y_1, \ldots, Y_n)\Big)$ for all
  n-ary functions f and all n.
- $\Big(e(X_1, Y_1) \wedge \cdots \wedge e(X_n, Y_n)\Big) \rightarrow e\Big(p(X_1, \ldots, X_n) \rightarrow p(Y_1, \ldots, Y_n)\Big)$ for all
  n-ary predicates p and all n.

This list contains infinitely many axioms because there are infinitely many
choices for f, p and n.

Second it must be proved that proof methods for FOL remain sound and
complete when we insist that $e(a, b)$ be interpreted the way we understand
equality. Don't the axioms guarantee that equality has this property? No. For
example, we might take $e(X, Y)$ to be true for all X and Y.

This does not entirely solve the problem of allowing Prolog's equality
because Prolog has another meaning for it: In Prolog $X = Y$ means that X
and Y can be unified.

Prolog allows arithmetic. This leads to complications that aren't found in FOL. Turing and Church showed that it is impossible to solve the validity problem; that is, there can never be an algorithm whose input is a general formula and whose output is the validity/nonvalidity of the formula.

Worst of all, Prolog has procedural code. Procedural code causes theoretical problems when it interacts with the declarative code. For example, **assert** or **retract** may be used to alter the knowledge base in the process of a depth-first search attempt to find a resolution proof. These alterations are not undone by backtracking. More than the soundness and completeness of SLD-resolution are at stake. We can't attach *any* FOL meaning to $\mathbb{T}(\mathcal{K}) \models \alpha$ when our "proof method" changes $\mathbb{T}(\mathcal{K})$!

Exercises

4.4.A. What is the CWA (closed world assumption)?

4.4.B. How does Prolog handle negation? When might this be reasonable? unreasonable?

4.4.C. How does Prolog procedural code present problems from a theoretical viewpoint?

4.5 FOL and Prolog as AI Tools

> *Expressing information in declarative sentences is far more modular than expressing it in segments of a computer program or tables. ... The same fact can be used for many purposes, because the logical consequences of collections of facts can be available.*
> —John McCarthy (1987)

Let's begin this section by reviewing the goals, problems, and compromises that led to the logical aspects of Prolog. Then we'll look at the much fuzzier issue of the suitability of FOL and Prolog for AI.

Goals and Compromises

The goal was to provide a system that could be used as the basis of intelligent reasoning—the sort human beings supposedly do. Unfortunately, the lack of understanding of human reasoning, and commonsense reasoning in particular, makes it impossible to decide what should be done. As a result, deductive logic was proposed as a compromise.

Deductive logic is not one thing—you've already seen propositional and predicate logics. Thus, the question becomes "What deductive logic?" There are two conflicting measures of desirability. On the one hand, a limited form of logic is more likely to be tractable both algorithmically and theoretically. On the other hand, a powerful form of logic is more likely to provide an adequate framework of AI.

In such a situation, the best route is usually to start at the simple end and introduce more complexity only when the present level is well understood and a serious limitation has been found. The understanding and the serious limitation provide powerful guides for adding complexity. Furthermore, understanding the simpler level helps in developing and understanding the more complicated level.

Propositional logic is probably the simplest level. Theoretically, it has some very nice features. The semantics is defined by truth tables which are finite in nature. Thus, truth tables lead to a proof method that is consistent and complete. Furthermore, truth tables show that, given any formula, we can decide whether or not it is valid—the decision problem. (This is different from completeness, which only ensures that a valid formula has a proof.) Finally, truth tables can provide the basis of a computer algorithm.

As discussed earlier, propositional logic is inadequate for our needs. FOL is much more promising. There exist sound, complete proof methods for FOL, which is good. One of the nice features of propositional logic is lost, however: It's impossible to design an algorithm that will take as input an arbitrary formula in FOL and produce as output the answer to "Is it valid?" (This is the decision problem for FOL.)

The goal of constructing an adequate reasoning method has led from propositional logic to the compromise of FOL. Even FOL is inadequate as a general language for reasoning in AI. Various modifications and extensions have been explored by philosophers and logicians. An explosion of interest and study began in the 1970s when AI researchers became interested in such logics and their algorithms. Before considering extensions of FOL, we should look at another goal that is implicit in all of AI: The existence of adequate algorithms.

What can we say about algorithms for FOL? Resolution is perhaps the best known proof method. It requires that the formulas be rewritten in clausal form, a task which is NP-hard. The resolution algorithm itself can be quite slow. If such behavior is typical and no better algorithms can be found, we'll have to compromise. Horn clauses provide a reasonable choice for a compromise because

- Horn clauses are already in clausal form.

- We can use SLD-resolution, which involves much less search than general resolution.

- When we make statements in FOL, they are often in Horn clause form.

What have we lost in restricting ourselves to Horn clauses? Horn clauses restrict our use of negation. For example, we can't say

$$(p \wedge (\neg q)) \to r \text{ or, equivalently, } p \to (q \vee r).$$

Prolog is based on Horn clauses and SLD-resolution, but it goes beyond them in order to provide a useful programming language. (See the discussion in the previous section.) Prolog's addition of procedural features make it practically impossible to define, much less prove, the soundness and completeness of its algorithm.

Prolog also compromises by using depth-first search rather than breadth-first or iterative-deepening search in its resolution. Even limited to Horn clause logic, such an algorithm is not complete—it can fail to halt when a proof exists. This compromise forces a further compromise on Prolog, the cut operator. Why use depth-first search? It's usually significantly faster than the alternatives—a compromise for speed. Also, the effect of procedural code is more easily visualized and controlled by the programmer with depth-first search. Thus, adding procedural statements almost forces this compromise. As computers become faster and more parallel, the possibility of using a different search method for nonprocedural Prolog code becomes more attractive.

We've discussed the limitations of FOL and Horn clauses mainly from a programming perspective. Limitations of FOL from an AI perspective are discussed briefly in the next chapter as a prelude to chapters on other logics and on quantitative reasoning methods.

Exercises

The following two exercises explore two simple idealized situations in Prolog search. In a way, the exercises lie at opposite extremes. The second is probably more realistic than the first.

4.5.1. Suppose that a Prolog knowledge base is such that each predicate appears as the head of exactly k statements. Suppose that the search tree for SLD-resolution has all its leaves at a depth of d.

(a) Show that the tree has k^d leaves.

(b) Suppose that the goal is reached by taking, on the average, the lth choice at each vertex where l is much smaller than k. (This is not unrealistic in many situations because the programmer controls the order of statements in the knowledge base.) Since we lack the tools to deal with averages, assume it is always *exactly* the lth choice. Show that the number of nodes examined by DFS (depth-first search) is

$$\frac{l-1}{k-1}\frac{k^{d+1}-1}{k-1} + \frac{(k-l)(d+1)}{k-1}.$$

There's some ambiguity here: Just when is a node examined? Consider it to be examined when it is expanded, rejected as a nongoal leaf or accepted as a goal.
Hint. Let D_d be the number and express D_{i+1} in terms of D_i.

(c) Under the previous assumptions, show that the number of nodes examined by BFS (breadth-first search) is $\frac{l(k^d-1)}{k-1} + 1$.
Hint. Using the approach for D_i, compute a formula for the position of the goal leaf among the k^d leaves.

(d) Show that when k^d is large, the ratio of the number of vertices examined by DFS to those examined by BFS is approximately $\frac{(l-1)k}{l(k-1)}$. What does this say about the comparative running times of DFS and BFS?

4.5.2. Suppose that a Prolog knowledge base is such that each predicate appears as the head of exactly k statements. Suppose that failure leaves on the search tree for SLD-resolution all occur after f steps off the correct path and that the goal lies at a depth of d. Suppose that the goal is reached by taking the lth choice at each vertex. Show that the number of nodes examined by depth-first search is

$$\frac{d(l-1)(k^f-1)}{k-1} + 1.$$

Are FOL and Prolog Good Choices for AI?

Asking about the appropriateness of FOL and Prolog for AI can stir up traditional AI debates. We'll avoid these by taking a narrower view.

People have proposed various criteria for evaluating knowledge representation and manipulation methods. Those that have been proposed have much in common. Here's one possible list of interrelated criteria.

- **Completeness**: Does the representation method allow us to represent all relevant knowledge? Does the manipulation method allow us to do what we wish with the knowledge?

- **Flexibility**: Can the representation of a given set of knowledge be used in many ways? Can the method be used in a variety of situations?

- **Efficiency**: Is it easy to enter knowledge? (This usually requires modularity and a natural representation.) Do the algorithms for manipulation produce results in a reasonable amount of time?

- **Understandability**: Is knowledge entered and stored in a clear and natural manner? How easily can a skeptical user understand the manipulation method so as to be able to accept or reject it?

- **Modularity**: To what extent can modifications of the knowledge base be made locally with little concern for global effects? (Besides simplifying initial construction, modularity makes extending and updating easier.)

- **Debugging**: Does the representation method tend to enforce correct usage? How easy is it to debug the entered knowledge?

For any representation and manipulation method, the answers will depend to some extent on the use to which it is being put and on the person answering the questions. Nevertheless, some general observations are possible. The following assessments are based on my own opinions and not just on facts. Thus, assessments by others may differ because they are based on different opinions.

Aside. Evaluations of AI methodologies later in the text will be *much* briefer. I strongly encourage you to carry out such evaluations on your own or in class discussion. It will help you appreciate the strengths and weaknesses of the methodologies.

Completeness

Completeness depends heavily on the problem at hand. Some limitations of FOL and Prolog have been pointed out in this chapter and others will be discussed in the next chapter. The limitation of Prolog to Horn clauses and negation as failure is often not important. The use of DFS resolution in Prolog can have more severe consequences. First, it makes representation of some information impossible or unnatural, even with the use of a cut. For example, suppose we have an equivalence relation such as "sibling." (To make it an equivalence relation, we'll allow someone to be his own sibling.) Call the relation **p**. We enter certain facts in the form **p(a,b)** and would like to be able to deduce others. The natural way to represent the fact that **p** is an equivalence relation is via the definition of an equivalence relation:

$$p(X, X). \qquad p(X, Y) : -p(Y, X). \qquad p(X, Y) : -p(X, Z), p(Z, Y).$$

You should be able to explain why this will not work with DFS and you should have considerable difficulty attempting to find an adequate remedy.

A potential difficulty with FOL and Prolog is the size of the knowledge base. For a well defined limited problem, this doesn't arise. On the other hand, vaguer and broader situations may raise severe difficulties.

Score: It depends on the problem.

Flexibility

Both FOL and Prolog score high marks for flexibility. Declarative representation of knowledge is usually quite flexible. It's more difficult to achieve flexibility when we think about knowledge procedurally. However, this distinction is not as sharp as it may at first appear. For example, Kowalski observed that a Horn clause such as **p :- q,r** can be interpreted procedurally: to solve **p**, do **q** and **r**.

Score: Good.

Efficiency

There can be difficulty in translating ordinary English into FOL; however, it is often fairly straightforward to translate more technical information. The algorithms for FOL can be quite time-consuming.

Prolog has a much more efficient algorithm than that for general FOL, but it imposes some problems on entering knowledge. The speed of the DFS can depend heavily on the order in which clauses with the same head are entered. In extreme cases, changing the order can convert a reasonably efficient knowledge base into one that leads to infinite searches. Efficiency also relates to the cut operator and the previous discussion of flexibility.

Score: Good.

Understandability

Often, the knowledge bases in FOL and Prolog are rather easy to understand; however, some recursively defined predicates can be rather tricky. Naturally, texts have a tendency to overemphasize such definitions, this one being no exception. On the positive side, it is usually possible to define and use complex data structures in Prolog in an understandable way. (It's also usually possible to do so in a confusing manner.)

There are two levels at which a user might be skeptical. The first deals with the soundness and completeness of Prolog's proof method. We've discussed that in this chapter. The second deals with the particular application: Does the knowledge base accurately reflect the skeptic's view of reality? Rather than looking at the knowledge base directly, users need a mechanism that explains the steps that led to a particular conclusion and/or the reason certain information is being requested. There is no explanatory mode built into SLD-resolution. On the other hand, the sequence of steps in the resolution proof can easily be adapted to provide an explanation. Expert system shells based on SLD-resolution frequently include this and other automatic explanations as options. They also include options for programmer supplied explanations.

Score: Good.

Modularity

Like any language, Prolog can be used to create or subvert modularity. The form of Prolog statements and the fact that the inference process is controlled by the interpreter both support the production of modular code.

At another level, Prolog supports and encourages the separation of knowledge statements from reasoning statements (how knowledge should be used). This promotes clearer thinking, modularity, and reusability. We haven't discussed the implementation of new reasoning statements in Prolog; however, texts relating Prolog to AI often contain such examples.

Score: Very Good.

Debugging

As with any other language, debugging Prolog can be a frustrating experience. This is made worse because of the tension between declarative and procedural code. Programmers are trained to think procedurally and an SLD-algorithm (or any other) is by nature procedural. On the other hand, much of Prolog is declarative in nature and it is probably best to approach Prolog programming with a declarative mindset. When, and if, this adjustment has been made, Prolog seems much more natural.

An additional complication is provided by the SLD-algorithm. Programmers are used to having rather limited system code lurking in the background (e.g., converting floating point to ASCII for output). The hidden, all embracing DFS version of SLD-resolution is a new experience. Fortunately, Prolog provides a trace facility that lets the programmer step through the resolutions.

Another, subtle point goes deeper than declarative versus procedural and the hidden Prolog resolution engine. This is the different usages for the Prolog implication. Purely procedural usage, as in obtaining user input, is fairly benign. A definitional use can cause problems if it involves recursion or is bidirectional, as in

$$married(X, Y) : -married(Y, X).$$

(which is likely to lead to an infinite loop). More subtle are "real-world" usages. They frequently involve a time flow, as the following show:

- Diagnostic time flow is from head to body, as in

$$rain(Locale) : -wet(Street, Locale).$$

- Causal time flow is from body to head, as in

$$wet(Street) : -within(Locale, Street), rain(Locale).$$

- Action-producing time flow is from body to head, as in

$$heat(Room, on) : -temp(Room, low), occupied(Room).$$

Why is time flow important? The fact is,

> Unless we're very careful, including statements with different time flow directions in a knowledge base will probably cause difficulties.

Why is this so? If we can move both forward and backward in time, we have the potential for creating loops. Unfortunately, Prolog is incapable of detecting time flow; moreover, the structure of the language doesn't make considering time flow a natural part of programming.

Score: Mixed.

Exercises

4.5.A. Give at least three criteria for evaluating knowledge representation and manipulation methods.

Notes

Aristotelian logic, which predates Aristotle, is a primitive form of FOL; however, its use of quantifiers is rudimentary and no solid foundation is given. The publication of George Boole's *The Mathematical Analysis of Logic* in 1847 can be regarded as the birth of mathematical logic and the beginnings of propositional logic as a mathematical subject. A variety of richer logics—richer even than FOL—were studied in the years that followed. (Logicians use "stronger" rather than "richer.") First-order logic was recognized as a distinct entity in lectures given by David Hilbert in 1917; however, the ideas had been developed several years earlier by others. Hilbert advocated basing mathematics on a system of formal logic. Gödel made an important contribution to this program when he demonstrated the completeness of FOL in his doctoral thesis. Hopes for the program were soon dashed, as Gödel began proving incompleteness theorems. The first of these is roughly as follows.

> Let a consistent logic containing FOL and the ability to add and multiply integers be given. There exists a valid formula in that logic which cannot be proved by a sound proof method.

The second incompleteness theorem showed that it is impossible to prove the consistency of any system this rich without recourse to a still richer system. Thus, any attempt to prove consistency would lead to an infinite regression of richer and richer systems of logic. At about the same time, Church and Turing showed that it is possible to write a program to decide the validity of formulas. For a discussion of the interactions between logic and the foundations of mathematics, see Moore [4]. Smullyan [11] discusses and proves the incompleteness theorems.

The search for practical computer algorithms for FOL began in the late 1950s. Robinson's 1963 discovery of resolution and his proof of its completeness was the turning point in achieving practicality. The restriction to SLD-resolution of Horn clauses made logic programming practical and resulted in the birth of Prolog. Robinson has written a nice brief history of logic programming and the discovery of resolution [9] as have Lobo et al. [3]. The pre-Prolog classical papers of computational logic are reproduced in [10]. For references on Prolog and on logic for computer science, see the previous chapter. For a discussion of the computational complexity of logic, see Part II of [7].

There are a variety of texts on mathematical logic oriented toward mathematics students. One such is by Nerode and Shore [6]. It contains an extensive annotated bibliography arranged by subject and so can be used as a source for further references in the history, theory, and applications of logic.

Unification is an example of a situation in which worst-case analysis of an algorithm can be misleading. There exists an algorithm whose worst-case running time is linear in the input size. However, the best average-case time appears to be provided by an algorithm whose worst-case time is exponential in the input size. See [1] for details.

Prolog texts usually go deeper into the issues that I touched on in the last two sections. See the Notes of Chapter 3 for some references. In addition Naish [5] goes much deeper into the problem of negation. Researchers have explored ways of overcoming Prolog's shortcomings. For example, see the monograph [8] and its bibliography.

Biographical Sketches

Kurt Gödel (1906–1978)

Born in Brünn in the Austro-Hungarian empire, he received his mathematical education from the University of Vienna. In 1929, he began a decade of pioneering work:

- In his 1929 thesis he proved the completeness of FOL.

- In 1931 he proved the incompleteness of any system of logic that was powerful enough to allow elementary arithmetic. That is, in any logical system allowing mathematics, one can construct statements that can neither be proved nor disproved within that system. In this regard, John Barrow quipped that, if one defines a religion to be any system of thought espousing unprovable truths, then only mathematics has been *proved* to be a religion.

- In 1937, he proved that the *axiom of choice* and the *generalized continuum hypothesis* are both consistent with standard set theory. "β is consistent with α" means that, if α contains no contradictions, then adding the assumption that β is true will not introduce any contradictions. Essentially, the axiom of choice says that, given an infinite collection S of sets, one can assume the existence of a function f whose domain is the collection of sets such that, for all $S \in \mathcal{S}$, $f(S) \in S$. That is, f selects one element from each set. The generalized continuum hypothesis talks about sizes of infinite sets. The continuum hypothesis asserts that there are as many real numbers as there are sets of integers. In 1963 Paul Cohen proved that assuming both the axiom of choice and the generalized continuum hypothesis are false is also consistent with standard set theory.

In 1940 Gödel moved from Vienna to the Institute for Advanced Study in Princeton, New Jersey. From then on, he worked mostly in philosophy. However, in 1949, he found a solution to Einstein's equations of general relativity that described a rotating universe in which time travel is possible. (Such research is important in exploring the limits of general relativity and in attempting to determine the nature of our universe.)

This material is based primarily on the biography in [2].

John Alan Robinson (1930–)

Born in Halifax, England, he received his B.A. degree in classics from Corpus Christi College, Cambridge, in 1952. In 1956 he received his Ph.D. in philosophy from Princeton University. His interest in automatic theorem proving and computational logic began in 1960. Working at Rice University, he developed resolution and refined unification. In 1967, Robinson moved to Syracuse University where he continues his research in computational logic and automatic theorem proving. For more details on the history of resolution, see [9].

References

1. L. Albert, R. Casas, F. Fages, and P. Zimmerman, Average case analysis of unification algorithms, *STACS* 91, Springer-Verlag, Berlin (1991) 196–213. (*STACS* is *Symposium on Theoretical Aspects of Computer Science*.)

2. S. Feferman (ed.), *Kurt Gödel: Collected Works*, Vol. 1, Oxford University Press, New York (1986).

3. J. Lobo, J. Minker, and A. Rajasekar, *Foundations of Disjunctive Logic Programming*, MIT Press, Cambridge, MA (1992).

4. G. H. Moore, A house divided against itself: The emergence of first-order logic as the basis for mathematics. In E. R. Phillips (ed.), *Studies in the History of Mathematics*, *MAA Studies in Mathematics*, Vol. 26, The Mathematical Association of America (1987), 98–136.

5. L. Naish, *Negation and Control in Prolog*, Springer-Verlag, Berlin (1986).

6. A. Nerode and R. A. Shore, *Logic for Applications*, Springer-Verlag, Berlin (1993).

7. C. H. Papadimitriou, *Computational Complexity*, Addison-Wesley, Reading, MA (1994).

8. S. Raatz, *Graph-Based Proof Procedures for Horn Clauses*, Birkhäuser, Boston (1990).

9. J. A. Robinson, Logic and logic programming, *Communications of the ACM* **35**, no. 2 (March 1992) 40–65.

10. J. Siekmann and G. Wrightson, *Automation of Reasoning. Classical Papers on Computational Logic*, 2 Vols., Springer-Verlag, Berlin (1983).

11. R. M. Smullyan, *Gödel's Incompleteness Theorems*, Oxford University Press, New York (1992).

5

Let's Get Real

Nor is there requir'd such profound knowledge to discover the present imperfect condition of the sciences, but even the rabble without doors may judge from the noise and clamour, which they hear, that all goes not well within.

—David Hume (1740)

The great danger in computer implementation of approximate reasoning is the use of inappropriate, unjustified, ad hoc models. Newcomers in the domain of common-sense reasoning could be overwhelmed by the multitude of models.

—Philippe Smets (1991)

Introduction

Predicate calculus is designed for determining eternal truths in tidy worlds. Since the "real world" is seldom so tidy, FOL must be extended, modified, or abandoned to meet the needs of AI.

After exploring some of the problems the real world causes for FOL, we'll look at some partial solutions. You should be wary whenever many partial solutions or explanations exist. Such a multiplicity may indicate that we've confused the issue by lumping together different phenomena. Or it may indicate that we need a new way to look at the phenomenona. Or, worst of all, it may indicate both. You probably think, "Now he'll tell us which of these applies to reasoning in AI." Sorry. Cogent arguments exist for almost any opinion.

The chapter concludes with a brief discussion of missing and conflicting information, including the Arrow Impossibility Theorem, which asserts that there is no ideal method for resolving conflicts.

Prerequisites: Nothing.

Used in: This chapter sets the stage for Chapters 6–9, but isn't needed in them.

5.1   Real-World Issues and FOL

> *There are more things in heaven and earth, Horatio,*
> *Than are dreamt of in your philosophy.*
>
> —William Shakespeare (*Hamlet*)

When we reason about the real world, we usually lack the certainty provided by FOL—often due to information problems:

- Relevant information may be missing.

- Like weather predictions, the information we have may not be in a simple yes/no form.

- Perhaps worst of all, some information may simply be wrong.

The "obvious" solution of gathering all possible information won't work: The sheer volume of information available in the real world is too great ever to be collected and stored. Worse yet, the very piece of information that's needed may be unobtainable or may require too much effort to obtain. In summary, we must often reason with an uncertain, incomplete information base—just the opposite of what FOL requires.

As an added complication, the world changes. In some situations, this can be ignored or can be taken care of by changing our knowledge base. In other situations we may have to change the reasoning process itself. For example, in the statement "If I do this, then that will happen and then ...," time and causality are integral parts of the reasoning process.

To develop new tools, we have to be clear on what the difficulties are. The following list classifies the problems discussed above and adds some new ones. Let's begin with the organization of knowledge.

- **Time:** Some things must be done in a certain order; for example, you can't bake the ingredients for a cake and mix them afterwards. FOL has no time sense.

- **Relevance:** What information (some of which may not be available) is relevant in the situation being considered? There is simply too much information to collect it all, so an AI system must decide what it needs. This is essentially the "frame problem," which is discussed a bit more on page 190.

- **Causality:** We often use causal relationships in reasoning. Sometimes we reason from cause to effects (deduction) and at other times, from effects to cause (abduction). Causality is closely related to time and probably plays a role in the commonsense solution to the problem of relevance. Unfortunately, it's unclear how to incorporate it into a reasoning method.

Now let's look at some attributes of the pieces of knowledge we have. One possibility is that it is wrong. Although detecting and correcting erroneous "facts" is an important problem for reasoning systems, let's ignore it. Here are other aspects of knowledge.

- **Exceptions:** General rules usually have exceptions. For example, the statement "Birds can fly" has exceptions related to species, health, age, and so on. Exceptions may, in turn, have exceptions.

- **Uncertainty:** Betting odds and weather predictions are day-to-day events that reflect uncertain knowledge about the world. Day-to-day use of information about uncertainty is often confused and inconsistent.

- **Ignorance:** Many researchers believe there's a subtle distinction between ignorance and uncertainty, but some disagree. I am *uncertain* about what to include in this text, but I'm *ignorant* about whether you're enjoying it or not.

- **Vagueness:** Some concepts are vague. Suppose you ask Joe's friends if he is tall. Some may say "yes," others "no," and still others something in between. This is because "tall" is a vague or "fuzzy" concept.

Exercises

5.1.A. List some difficulties that FOL encounters in the real world.

5.1.B. Why is causality important?

5.2   Some Alternative Reasoning Tools

> Alice: *"Would you tell me, please, which way I ought to go from here?"*
> Cheshire Cat: *"That depends a good deal on where you want to get to."*
>
> —Lewis Carroll (1865)

> *We're really in desperate need of some kind of overall architecture that would enable us to integrate something like a probabilistic model and something like the classification and inheritance models into a coherent framework where the right pieces play the right roles with respect to each other.*
>
> —Bill Woods (1991)

A variety of tools have been proposed for dealing with the limitations of the predicate calculus. There will undoubtedly be more in the future, as well as improvements of existing ones. Most tools are either primarily qualitative (nonnumeric) or primarily quantitative. Here's a partial list, with references to further discussion.

Qualitative Methods

Qualitative methods avoid the use of numeric information. Sometimes numeric information is not relevant. At other times it may be unavailable or unreliable. Furthermore, human reasoning processes do not normally employ numeric information. Researchers have proposed a variety of *nonstandard logics*—any logics other than FOL. Here are three overlapping categories of nonstandard logics.

- **Nonmonotonic Logics:** The logic we've studied so far is *monotonic*, which means that once a result has been established, it remains true regardless of additional information. In nonmonotonic logic, something that was presumably true may be found to be false in light of additional information.

- **Temporal Logics:** Temporal logics incorporate time explicitly in the syntax of the logic. Although time is numeric in nature, temporal logics are usually qualitative because they are concerned with the *order* in which events occur, not their durations.

- **Autoepistemic Logics:** FOL focuses on validity and unsatisfiability, largely ignoring the vast middle ground where something is neither certain nor impossible. Autoepistemic logic reasons about belief. This approach is based on *modal logics*, which introduce other truth values "between" T and F to reflect concepts like possibility and belief.

Representational methods other than logic are used for qualitative methods. They include the following:

- **Rule Systems:** Superficially, rule systems resemble logic—"If A then B." In logic such a statement says that the truth of A implies the truth of B. Furthermore, the contrapositive asserts that the falsity of B implies the falsity of A. In rule systems, the truth of A implies that the action B should be carried out. (This action might be assigning a truth value to a statement as in FOL.) The contrapositive plays no role. Issues of temporal order arise when "If A then C" is also present and the actions B and C interact.

- **Graphical Systems:** Several types of knowledge systems are based on graphical representations, usually with directed graphs. (See Definition 2.1 (p. 36).) The vertices of the graph contain information about conceptual units and the edges indicate relationships between units.

C_1 = "It rained."
C_2 = "The automatic sprinkler was on."
R_1 = "The grass looks wet."
R_2 = "The sidewalk looks wet."

Figure 5.1 Multiple possible causes C_i and results, R_i. Suppose the sprinkler system is activated only if a built-in moisture sensor decides the ground is too dry. A simple rule-based system could equally well have the four probable rules "If C_i, then R_j" or the four probable rules "If R_j, then C_i." They must be dealt with in different ways: if R_1 is true, then it is much *more* likely that R_2 is true; if C_1 is true, then it is much *less* likely that C_2 is true.

Most of the subjects listed above are discussed in the next chapter. Those discussions are intended to give you a brief introduction to the ideas, accomplishments, and problems of some of the methods.

Quantitative Methods

Commonsense human reasoning tends to break down when information is far from certain. Qualitative AI methods are useless then, too. The canonical mathematical method for dealing with uncertainty is probability theory. AI researchers have used it and have also developed alternative numeric methods for dealing with uncertainty.

- **Bayesian Networks:** This method uses probability theory to deal with uncertainty in causality problems. Roughly, "If A then B" has an associated probability measuring the chance that A will cause B. Figure 5.1 shows a simple example.

- **Belief Theory:** This theory incorporates both ignorance and uncertainty and includes Bayesian networks as a special case.

- **Fuzzy Sets:** The notion of set membership is "crisp"; that is, either $s \in S$ or not—there is no intermediate situation. Fuzzy sets are based on the idea that not all membership is crisp. For example, consider the set of good students. Some people certainly belong and some certainly do not, but others are somewhat in and somewhat out—so the concept of "good student" is fuzzy. Logic can be based on set theory and, similarly, "fuzzy logic" can be based on fuzzy sets.

Bayesian nets are discussed in some detail in Chapter 8. Chapter 7 contains the probability theory on which Bayesian nets are based. An introduction to belief theory and fuzzy logic is provided in Chapter 9.

The notion of fuzzy logic contains a mixture of the qualitative aspects of logic with quantitative methods. Other such marriages have been attempted using probability theory. In attempting such a fusion, the meaning of the probabilities is an issue. Suppose you're given a statement S and associated probability p. How should it be interpreted? One possibility is a statistical statement about the world—p measures the frequency with which S is true. Another is probabilistic logic, which considers a collection of possible worlds such that p is the probability of choosing one in which S is true. Another is based on the idea the probability reflects belief. See Sections 7.1 and 9.1 for further discussion on the meaning and appropriateness of probability.

Why Such Diversity?

The reasons for the variety of methods fall into three main categories related to knowledge and its uses:

- **Knowledge Content**: *What* we know influences how it is described. In this text we use the numeric (quantitative) versus purely qualitatitive knowledge distinction to divide up the methods listed above as well as the presentations in the following chapters. In the preceding brief descriptions, you should be able to find less obvious examples of how content influences method.

- **Understandability and Modularity**: If a large system does not have its knowledge organized in an understandable and modular way, it will collapse. First, a lack of understandability almost guarantees that someone trying to develop a knowledge base will enter some information incorrectly, even though it will *appear to be correct*. This makes debugging difficult. Second, a lack of modularity almost guarantees that unexpected undesirable interactions between parts of the knowledge base will occur. Attempts to correct these problems will probably be ad hoc and create new problems.

- **Knowledge Manipulation**: *How* we intend to manipulate the knowledge can strongly influence our choice of method. To overcome the need for such a choice, we might try to design a very general system that could reason in many ways. It would probably be faced with severe computational problems that could be overcome only by restricting the system. Various methods for restricting reasoning lead to different types of nonmonotonic logics and to rule systems.

Exercises

5.2.A. What do nonmonotonic logics allow that FOL does not? modal logics? temporal logics?

5.2.B. What is the basic idea behind fuzzy sets?

5.2.C. What numeric method focuses on causality?

5.2.D. Why is there such a large variety of reasoning systems?

5.3 Incomplete Information

> *A problem faced in most reasoning situations is that all the information that may be relevant is not available and that which is available is confusing and not necessarily relevant.*
>
> —Raj Bhatnagar and Laveen N. Kanal (1992)

> *Life is the art of drawing sufficient conclusions from insufficient premises.*
>
> —Samuel Butler (1912)

Providing complete information on a subject in a knowledge base may be impractical or even impossible. One approach to the problem is to be as thorough as possible and then make the *closed-world assumption,* or CWA: *All* relevant information is contained in the data and rules of the system. (We've discussed the CWA in connection with Prolog negation on page 169.)

While the CWA is reasonable in certain circumscribed domains such as commercial expert systems, it is usually unrealistic in daily life. The amount of information potentially available even in simple situations is so vast that collecting it and reasoning about it are both impractical. This problem is related to *bounded rationality*—reasoning when data and computational power are limited. People's brains deal with this superabundance of information as a matter of course. We somehow use what knowledge we have plus "common sense." At times, common sense will tell us to make certain "reasonable" assumptions about missing information; at other times, it will tell us to seek more information. Designing a system with these abilities is a core problem for AI.

Two key problems have been identified in connection with deciding what information is (likely to be) relevant in a given situation:

- The *qualification problem* is concerned with missing information that may invalidate a general rule. General rules about the day-to-day world are incomplete. That is, a general rule requires so many preconditions that it is infeasible and perhaps impossible to list all of them. Needless to say, it is therefore impossible to check them. How can we deal with this? A classic example is

  general rule: Birds can fly.
  application: Tweety is a bird; therefore Tweety can fly.

The general rule should have preconditions that eliminate penguins, nestlings, birds with broken wings, and so forth.

- The *frame problem* is concerned with the fact that, potentially, any change in the environment may influence any other part of the environment. How can we reasonably determine what conclusions persist as time passes and actions are taken? As Hayes [13, p. 125] put it,

  One feels that there should be some economical and principled way of succinctly saying what changes an action makes, without having to explicitly list the things it doesn't change as well; yet there doesn't seem to be any way to do it. That is the frame problem.

The classic example of the frame problem is the *Yale shooting problem*, named after the university where it was proposed:

  Rule: If a loaded gun is fired at someone, the person dies.
  Fact: Fred is alive at time T_0.
  Fact: A gun is loaded at time T_1.
  Fact: The gun is fired at Fred at time $T_1 > T_0$.

If $T_1 - T_0$ isn't large, it's reasonable to assume Fred will be dead because it's reasonable to assume that no intervening event unloaded the gun. However, it's also "reasonable" to assume that Fred is still alive (and so the gun was somehow unloaded), because people who are alive at T_0 are unlikely to be dead at T_1. Thus one abnormal event or the other must occur (Fred dies or the gun becomes unloaded). How can a reasoning system be built with the common sense to decide correctly? Crockett [5] believes that a solution to the frame problem is the key to designing intelligent systems.

People continually deal with the qualification and frame problems in everyday life by using common sense. They do not always reach the correct conclusion, but that is to be expected in any system where bounded rationality is important. Similarly, any AI system will also sometimes fail to reach the correct conclusion. Designing a system that exhibits such common sense will require coming to grips with the qualification and frame problems or finding some unsuspected way of sidestepping them. These problems are the subject of research and debate in AI, cognitive science, and philosophy.

Exercises

5.3.A. What is the closed-world assumption?

5.3.B. What are the qualification and frame problems and why are they important in AI?

5.4 Inaccurate Information and Combining Data

> *It is better to know nothing than to know what ain't so.*
> —Josh Billings (1874)

Data may be based on expert opinion or on observations. Variations in data affect results. The media often report data problems related to global issues such as nuclear winter or global warming where obtaining accurate data is difficult. The classic example of this is Forrester's *World Dynamics* [8], which presents a model of what will happen in the next few decades assuming there is no nuclear war. The model requires enormous amounts of data, most of which are not available and some of which can never be obtained. As a result, most of the "data" presented in the book are simply the guesses of researchers, who found that the model made dire predictions that could be averted only by rapid, major, counterintuitive global action. As the debate developed, other researchers used guesses they thought were better and obtained other results. It may never be known whose guesses were best.

Inaccuracies can be of various types, ranging from totally incorrect information to slightly inaccurate numerical estimates. *Sensitivity analysis* studies how slight variations in input or computational accuracy affect output. These variations may be either quantitative or qualitative. (An example of qualitative is the ranking of alternatives from best to worst.) Problems where slight input and/or computational variations produce major output variations are called *ill-conditioned* and should be avoided if at all possible. Unfortunately, sensitivity analysis is a difficult subject in all but the simplest cases.

Example 5.1   Simple Examples of Ill Conditioning

Suppose you want to find the roots of the quadratic equation $x^2 + bx + c = 0$, where b is a large positive number and c is small. By the quadratic formula, one of the roots is given by

$$x = \frac{-b + \sqrt{b^2 - 4c}}{2} \quad \text{which is the same as} \quad x = \frac{2c}{-b - \sqrt{b^2 - 4c}}.$$

Mathematically, either formula is an equally good choice. In actual computations, the last formula is preferable because of roundoff error. To see why this is so, imagine an error δ in computing the square root. In the left-hand formula, the leads to an error of $\delta/2$ in the estimate for x, which is a large *relative* error because x is small, about $-c/b$. (The relative error is about $b\delta/2c$.) In contrast, setting $D = b + \sqrt{b^2 - 4c}$, the second formula leads to an error that is about

$$\frac{2c}{-D} - \frac{2c}{\delta - D} = \frac{2c\delta}{D^2 - D\delta} \approx \frac{2c\delta}{D^2}.$$

Since $D \approx 2b$, the relative error is about $\delta/2b$, which is much smaller than $b\delta/2c$.

Estimating parameters from data may be ill-conditioned. For example, suppose you believe a sample consists of two unknown radioactive elements in unknown proportions. From the theory of radioactive decay, the amount of radioactivity produced by the first element at time t should be about Ae^{-at}; by the second, Be^{-bt}. (We use the term "about" because the theory only predicts average behavior.) The total radioactivity is then about

$$r(t) = Ae^{-at} + Be^{-bt} \quad \text{where } a, b, A, \text{ and } B \text{ are unknown.}$$

Since you can measure $r(t)$ at various times by using a geiger counter or other instrument, you can collect several data points and estimate the unknown parameters. Unfortunately, this problem can be ill-conditioned, a situation that has trapped the unwary neophyte. ∎

In the design of expert systems, data is often based on the estimates provided by an expert. One approach to inaccurate estimates is to consult several experts and then create a reasonable compromise based on their estimates. Such reconciliation is an important problem in expert system implementation [2].

Example 5.2   Sex Discrimination

Here's a simple example of problems with numerical data. The numbers are fictional, but the example is based on a situation that occurred at the University of California, Berkeley.

Suppose that a school is divided into an Engineering Division and an Arts and Sciences Division. In a particular year, 2,000 men and 500 women applied to the Engineering Division. Of these 70% of the men and 80% of the women were admitted. For Arts and Sciences, there were 1,000 male applicants and 2,500 female. The admissions rates were 45% and 50%, respectively. It appears that women were favored over men. But when we combine the figures for both divisions (do it!), we find that of the 6,000 applicants (3,000 of each sex), we find that the university admitted 1,850 (62%) of the men and 1,650 (55%) of the women. It now appears that men were favored over women! ∎

The Arrow Impossibility Theorem (proved by K. J. Arrow) shows that reconciling qualitative data can also be a problem. What does the theorem tell us? Suppose several experts individually rank some alternatives $A, B, \ldots, Z$ from most likely to least likely. Thus, each expert gives us a list that is simply a rearrangement of $A, B, \ldots, Z$. We want to use these lists to prepare a master list from most likely to least likely, without interjecting any opinions except our prior assessment of each expert's accuracy. Imagine a computer program into which we can feed a list of alternatives, an assessment of how reliable each expert is, and the experts' lists. The program will then produce our master list, resolving conflicts in some reasonable, consistent manner using some method. What method? The *Arrow Impossibility Theorem* asserts that there is no "reasonable" method for the general problem. Thus, we must force our experts to alter their opinions or we must make some compromises about what is "reasonable."

*The Arrow Impossibility Theorem

> *[The Arrow Impossibility Theorem] has deeply influenced theoretical welfare economics, moral and political philosophy, and mathematical approaches to microeconomic theory.*
> —Jerry S. Kelly (1978)

> *The axiomatic method has many advantages over honest work.*
> —Bertrand Russell (1872–1970)

Let's proceed in three steps:

- define an "election procedure," simply called an "election";
- list the axioms a "fair" election must satisfy; and
- prove that the axioms are inconsistent.

Since the axioms for a fair election are inconsistent, it follows that fair elections are impossible. An election turns out to be simply the resolving of conflicts to produce a consensus among experts. Thus, Arrow's theorem asserts that there is no "fair" way to do that in general.

Uppercase letters will denote candidates and lowercase letters will denote voters. In an *election*, each voter provides some ordering of the candidates. We'll refer to the ordering provided by voter i as a ranking by voter i and denote it $\geq_i$. For example, with candidates A, B, C and voters i, j, k, m, we might have the rankings

$$A >_i B =_i C, \quad A =_j C >_j B, \quad C >_k A >_k B, \quad B =_m C =_m A,$$

where $A >_i B$ is read as "voter i prefers A to B" and $A =_j C$ as "voter j prefers A and C equally." Now let's say what we mean by an ordering. An *ordering* satisfies

(a) $A \geq_i A$,

(b) $A \geq_i B \geq_i C$ implies $A \geq_i C$,

(c) exactly one of $A >_i B$, $B >_i A$, and $A =_i B$ holds.

Suppose we have some fixed set of at least three voters. An *election program* is a computer program that incorporates our assessment of the voters' reliability, takes the rankings by the voters as input, and produces an ordering "$\geq$" of the candidates as output.

Arrow stated certain axioms that the ranking $\geq$ should reasonably satisfy for the election program to be fair. He proved that the axioms were inconsistent; that is, it is impossible to create a fair election program. Here are the axioms:

Axiom 1: All conceivable rankings by the voters are allowed as input.

Axiom 2: A unanimous desire is obeyed; that is, if $A \geq_i B$ for all i, then $A \geq B$ with equality if and only if $A =_i B$ for all i.

Axiom 3: Let primes denote rankings in a second election. Suppose that $A >_i' B$ whenever $A >_i B$. If the program concludes that $A > B$, it must also conclude that $A >' B$.

Axiom 4: There is no dictator; that is, given i and $\geq_i$, there is some ranking by the other voters such that $\geq_i$ and $\geq$ are different rankings.

Theorem 5.1   Arrow Impossibility Theorem

The axioms for fair elections are inconsistent; that is, fair elections are impossible.

Proof: In the course of the proof, various special cases arise because the consensus ranking may have ties. Dealing with ties complicates and obscures the proof, without introducing any new ideas. Therefore, let's consider only the case without ties.

The proof will be by contradiction. We assume Axioms 1, 2, and 3 and then prove that there must be a dictator. This contradicts Axiom 4.

A set $\mathcal{D}$ of voters is *decisive* for $A > B$ if, whenever $A >_i B$ for all $i \in \mathcal{D}$, we have $A > B$. In other words, if the voters in $\mathcal{D}$ agree that A is better than B, then the fair election program must agree, too. Note that the set of all voters is decisive for every A and B by Axiom 2.

The proof that there is a dictator consists of two steps:

- First, show that if $\mathcal{D}$ is decisive for $A > B$, then it is decisive for $C > D$ for all candidates C and D.

- Second, show that if $\mathcal{D}$ is decisive for $A > B$ and has more than one element, then a proper subset of $\mathcal{D}$ is decisive for some pair of voters.

Putting these two results together, we can repeatedly reduce the size of the decisive set for A and B until it contains only one element. By the first step, the voter in this set is a dictator.

Before carrying out these two steps, we need a simple observation. Suppose we have an election in which $A > B$. Then the set $\mathcal{D} = \{\, i \mid A >_i B \,\}$ is decisive for $A > B$. This follows from Axiom 3: The role of the first election is played by the election that was used to construct $\mathcal{D}$. Any election for which $A >'_i B$ for all $i \in \mathcal{D}$ would play the role of the second election in the axiom.

We now take the first step. Let $\mathcal{D}$ be decisive for $A > B$. Consider an election in which $C >_i A >_i B$ for $i \in \mathcal{D}$ and $B >_j C >_j A$ for $j \notin \mathcal{D}$. By decisiveness, $A > B$. By unanimity, $C > A$. Putting these together, $C > B$. Looking at the election we just created, it follows that $\mathcal{D}$ is decisive for $C > B$. In a similar manner, using $C >_i B >_i D$ and $B >_j D >_j C$, we get that $\mathcal{D}$ is decisive for $C > D$.

Now let's take the second step. Let $\mathcal{D}_1$ and $\mathcal{D}_2$ be nonempty disjoint sets whose union is $\mathcal{D}$. Consider the election in which

$$C >_i A >_i B, \quad \text{for } i \in \mathcal{D}_1,$$
$$A >_i B >_i C, \quad \text{for } i \in \mathcal{D}_2,$$
$$B >_i C >_i A, \quad \text{for } i \notin \mathcal{D}.$$

Since we must have $A > B$, only three possibilities exist:

- $C > A > B$ and so $\mathcal{D}_1$ is decisive for $C > B$;
- $A > B > C$ and so $\mathcal{D}_2$ is decisive for $A > C$;
- $A > C > B$ and so $\mathcal{D}_1$ is decisive for $C > B$ and $\mathcal{D}_2$ for $A > C$.

This completes the second step and hence the proof. ■

Exercises

5.4.A. Without recourse to mathematical symbolism or terminology, explain the concept of fair elections. How do fair elections relate to the problem of resolving conflicting expert opinions?

5.4.1. Over a period of 35 years, birdwatchers observed the following numbers per year of a rare species of bird stopping at a swamp along the migratory route.

$$
\begin{array}{cccccccccc}
9 & 7 & 7 & 8 & 11 & 12 & 11 & 8 & 6 & 6 \\
11 & 15 & 10 & 8 & 0 & 3 & 9 & 11 & 4 & 18 \\
21 & 9 & 6 & 6 & 7 & 11 & 14 & 9 & 7 & 6 \\
7 & 9 & 11 & 11 & 10 & & & & &
\end{array}
$$

A conservation group, realizing that the data fluctuates too much to examine it year by year, decides to divide it into seven-year blocks and examine the total for each block. To their horror, they find that the totals are decreasing steadily from 65 at the start to 61 at the end. (Do the calculations.) You want to drain the swamp for a housing development. Explain why grouping the data into five-year blocks might help your cause.

*5.4.2. Majority voting on ranking seems an easy way to satisfy the axioms. Suppose for simplicity that the number of voters is odd and that equality is not allowed. Define $A > B$ if and only if $A >_i B$ for more than half of the voters.

(a) Verify that the axioms are satisfied.

(b) Show that $>$ may not be an ordering—this is a "hidden assumption" that was stated in our definition of the computer program.
Hint. Look for an example involving three voters and three alternatives.

(c) Where does the proof of the Arrow Impossiblity Theorem break down for the so-called ordering produced by using majority vote? In other words, why does the proof fail for the example you constructed in (b)?

Notes

Shafer and Pearl [16] have collected a variety of papers on uncertain reasoning, arranged them by topic, and provided introductions to the various topics. These papers are primarily quantitative. Ginsberg [12] has done the same for the qualitative case. Nilsson gives a discussion of his probabilistic logic in [14]. A good reference on the problems of combining logic and probability is Bacchus [4]. Although a monograph, it's like an introductory text. I'll give more references when I discuss relevant topics in Chapters 6 and 9.

Bounded rationality is discussed by Russell and Wefald [15]. Some recent discussions of the frame problem are found in papers in [7]. Most of the papers are rather technical, but the introduction by Ford and Hayes is not.

The situation in Example 5.2, where combining information reverses an ordering, is known as *Simpson's paradox*. See [9] for two examples drawn from baseball.

Arrow published his theorem in 1950. The proof here is based on [3], which contains the full proof and a discussion of related topics. Feldman [6] proves the theorem and discusses other methods of evaluating social alternatives. For further discussion of the problems encountered in combining quantitative data see [10] and [11]. They require some background in probability theory, so you may need to read Chapters 7 and 12 first. Abidi and Gonzalez [1] discuss some methods for combining input from several robotic sensors.

References

1. M. A. Abidi and R. C. Gonzalez, *Data Fusion in Robotics and Machine Intelligence*, Academic Press, San Diego (1992).

2. S. M. Alexander and G. W. Evans, The integration of multiple experts: A review of methodologies. In E. Turban and P. R. Watkins (eds.), *Applied Expert Systems*, Elsevier, Amsterdam (1988) 47–53.

3. K. J. Arrow and H. Raynaud, *Social Choice and Multicriterion Decision-Making*, MIT Press, Cambridge, MA (1986).

4. F. Bacchus, *Representing and Reasoning with Probabilistic Knowledge: A Logical Approach to Probabilities*, MIT Press, Cambridge, MA (1990).

5. L. J. Crockett, *The Turing Test and The Frame Problem: AI's Mistaken Understanding of Intelligence*, Ablex, Norwood, NJ (1994).

6. A. M. Feldman, *Welfare Economics and Social Choice Theory*, Martinus Nijhoff, Boston (1980).

7. K. M. Ford and P. J. Hayes (eds.), *Reasoning Agents in a Dynamic World: The Frame Problem*, JAI Press, Greenwich, CT (1991).

8. J. W. Forrester, *World Dynamics*, Wright-Allen Press, Cambridge, MA (1971).

9. R. J. Friedlander, Ol' Abner has done it again, *American Mathematical Monthly* **99** (1992) 845.

10. D. P. Gaver, Jr., et al. (a National Research Council panel), *Combining Information. Statistical Issues and Opportunities for Research*, National Academy Press, Washington, D.C. (1992).

11. C. Genest and J.V. Zidek, Combining probability distributions: A critique and an annotated bibliography, *Statistical Science* **1** (1986) 114–135. See also the comments by experts and the rejoinder on pp. 135–148.

12. M. L. Ginsberg, *Readings in Nonmonotonic Reasoning*, Morgan Kaufmann, San Mateo, CA (1987).

13. P. Hayes, What the frame problem is and isn't. In Z. W. Pylyshyn (ed.), *The Robot's Dilemma: The Frame Problem in Artificial Intelligence*, Ablex, Norwood, NJ (1987).

14. N. J. Nilsson, Probabilistic logic, *Artificial Intelligence* **28** (1986) 71–87. Reprinted in [16], 680–688.

15. S. Russell and E. Wefald, *Do the Right Thing: Studies in Limited Rationality*, MIT Press, Cambridge, MA (1991).

16. G. Shafer and J. Pearl (eds.), *Readings in Uncertain Reasoning*, Morgan Kaufmann, San Mateo, CA (1990).

6

Nonmonotonic Reasoning

It is logically impossible to reason successfully about the world around us using only deductive reasoning. All interesting reasoning outside of mathematics involves defeasible steps.

—John L. Pollock (1989)

First ponder, then dare.

—Helmuth von Moltke (1800–1891)

Introduction

Inquiring AI researchers want to know "What is common sense?" because they want to mimic it. Since most people have common sense, discovering the answer should be easy. But it isn't—duplicating common sense has proved surprisingly difficult.

- Some researchers believe we understand common sense well enough and that it involves knowing a considerable amount about the way the world is. They believe that, once we've captured this information in a knowledge base, we'll be able to implement commonsense reasoning. Proposed implementations are based on some form of nonmonotonic reasoning.

- Most researchers believe that the problem is more fundamental—we don't really know how our commonsense reasoning works. Much of the research on the problem has involved nonmonotonic reasoning methods. Many people believe that further research in this area will eventually lead to a solution.

What is nonmonotonic reasoning and why is it so important?

Ordinary first-order predicate logic (FOL) is timid—if something is uncertain, FOL makes no assumptions. Human beings would get nowhere doing this in the real world. We all must make (tentative) assumptions all the time—I assume the building I'm in is structurally sound, you assume the other drivers will obey the traffic signals, and so forth. If an assumption seems wrong at the moment, we discard it for that particular situation, a process called nonmonotonic reasoning. Some methods of nonmontonic reasoning are based on extensions of FOL; others on tools outside mathematical logic. Which is best? Researchers disagree on this, but they generally agree that it's important to find approaches that can provide a solid theoretical foundation—the problem seems too complex and our ignorance too great to rely solely on intuition and experiment. Since classical mathematical logic has a solid foundation, some people look for ways of extending it to deal with uncertainty. Others look for solutions outside mathematical logic, arguing that commonsense reasoning employs methods that don't comfortably fit into the framework of mathematical logic.

Where does the name *nonmonotonic reasoning* come from? As a result of FOL's lack of assumptions, results stay proved. That is, if you prove the truth of a statement using only part of a knowledge base, you won't discover that your proof is invalid when you look at the entire knowledge base. Such a theory is called *monotonic* because adding to the knowledge base never invalidates previously discovered truths. In *non*-monotonic reasoning conclusions reached by using part of the knowledge base may become invalid when the entire knowledge base is used. For example, suppose I tell you that Prolog is a programming language and that good programmers usually learn new programming languages easily. From this you'd conclude that a good programmer would probably learn Prolog easily. Now add the information that good programmers frequently have difficulty learning Prolog. Your previous conclusion is invalid. This problem is attributable to the presence of words like "usually" and "frequently," which occur in commonsense reasoning but are not a part of FOL.

In the next section, we'll examine some issues in commonsense reasoning and sketch some methods developed to deal with them. In the remaining sections, we'll discuss these nonmonotonic approaches in more detail.

Prerequisites: You'll need some of the graph theory concepts from Section 2.1. Knowledge of first-order logic is required—you should be familiar with the concepts and results in Chapter 3 and in Section 4.1. Other parts of Chapter 4 are helpful but not necessary.

Used in: No other chapters require this material.

6.1   Coming Attractions

> *What's most depressing is the realization that everything we*
> *believe will be disproved in a few years.*
>
> —Sidney Harris (ca 1975)

Suppose that $\mathcal{K}$ and $\mathcal{K}' \supseteq \mathcal{K}$ are knowledge bases and that we use FOL to show that $\mathcal{K} \vdash \alpha$. Then $\mathcal{K}' \vdash \alpha$, too. What does this mean in ordinary English? It simply says that if we're able to conclude α from $\mathcal{K}$, then the additional knowledge in $\mathcal{K}'$ won't invalidate α. This property of FOL is referred to as *monotonicity* because the set of conclusions that can be reached increases as the knowledge base increases. It has an extremely important practical consequence: Suppose we can use a small part $\mathcal{K}$ of a large knowledge base $\mathcal{K}'$ to reach a conclusion. Since information in the rest of the knowledge base can't invalidate the conclusion, we can ignore the rest of the knowledge base.

By definition, nonmonotonic reasoning methods allow us to reach conclusions that may become invalid as we gain further knowledge. As a result:

> It is *a priori* necessary to examine the entire knowledge base before reaching a conclusion.

This leads to a core problem for nonmonotonic methods whose solution is crucial for designing an algorithm that runs in a reasonable time:

$$\boxed{\text{How can we limit our attention to a (small) part of the knowledge base and still guarantee that the conclusion reached will be correct?}} \quad (6.1)$$

In FOL, Prolog avoids examining the entire knowledge base by using Horn clauses. As a result, Prolog needs to look only at clauses whose heads can be unified with the predicate currently being examined. Perhaps one should look for something similar in nonmonotonic reasoning.

In commonsense reasoning we *somehow* make "reasonable" assumptions about missing data. This leads to another core problem for nonmonotonic reasoning methods:

$$\boxed{\text{During the reasoning process, how do we detect missing knowledge and make reasonable assumptions about it?}} \quad (6.2)$$

The detection problem is closely connected with problem (6.1). The extent to which the assumption problem is dealt with varies from method to method.

- At the most basic level, an approach would tell us what choices of assumptions could be justified as being "rational." However, it would not give any reason for favoring one possibility over another. For example, it's rational to assume that a car will come speeding around the corner and it's rational to assume that no car will speed around the corner; however, it's not rational to make both assumptions.

- At the next level, some procedure for selecting among assumptions is part of the reasoning mechanism. Because there is a red light and drivers usually obey traffic signals, it's more reasonable to assume that no car will speed around the corner.

- Finally, the decision problem can be brought to the fore: After determining the gains and losses that can result from various assumptions or actions, a procedure is used to decide which is best. This requires quantitative methods, which are outside this chapter's scope.

Exercises

6.1.A. What distinguishes monotonic reasoning from nonmonotonic reasoning?

6.1.B. What are two major problems that nonmonotonic reasoning must face?

Types of Qualitative Nonmonotonic Reasoning

> *The first step in analyzing nonmonotonic logic is to determine what sort of nonmonotonic reasoning it is meant to model. After all, nonmonotonicity is a rather abstract* syntactic *property of an inference system, and there is no* a priori *reason to believe that all forms of nonmonotonic reasoning should have the same logical basis.*
>
> —Robert C. Moore (1983)

As already mentioned, we can distinguish two approaches to qualitative nonmonotonicity—those motivated by a desire to expand FOL beyond its limitations and those motivated by a desire to replace mathematical logic with more appropriate methods. This distinction has blurred as researchers have found ways to translate between various reasoning methods. Nevertheless, it's still a useful distinction.

Nonmonotonic Logics

We want to adapt FOL by adding statements of the form "typically α," meaning that α is true except in "unusual situations." For example, "typically birds fly," which could be written in an FOL-like form as

$$\forall X \Big(\text{typically}\big(\text{bird}(X) \to \text{flies}(X)\big)\Big). \tag{6.3}$$

Unfortunately, this doesn't tell us how to detect the unusual situations in which α is not true. We could specify unusual situations in FOL by ANDing the negation of a variety of conditions with bird(X). This would give us something like

$$\forall X \Big(\big(\text{bird}(X) \wedge \neg\big(\text{penguin}(X) \vee \text{injured}(X) \vee \cdots\big)\big) \to \text{flies}(X)\big)\Big), \tag{6.4}$$

where the ellipsis indicates other possible conditions. This is unacceptable for various reasons. Here are two.

- **Objection 1**: Every time we discover a new condition, we must modify (6.4), eventually obtaining an extremely long statement.

- **Objection 2**: More important, (6.4) doesn't accomplish what we want. Unless we can actually *prove* that X is neither a penguin, nor injured, nor ... , we can't use FOL and (6.4) to conclude that X can fly. In practice, when you know Tweety is a bird, you *assume* that Tweety can fly unless you have some information that makes Tweety abnormal (a penguin, injured, etc.).

Such objections show that FOL is inadequate for formulating nonmonotonic ideas such as (6.3).

Various nonmonotonic logics have been proposed. Perhaps there are different aspects of reality for which different logics are needed. Perhaps researchers have not yet discovered the right logic. Or perhaps both are true. Only time will tell. Here are the four main approaches to nonmonotonic logic.

- **Default Logic**: Default logic works with statements like (6.4), but it treats the negation appearing there in a different manner. Instead of proving that penguin(X) and so on are each false, default logic insists only that it not be possible to prove them true. In other words, default logic meets Objection 2 head on. With this modified interpretation of negation, statements like (6.4) are called *defaults*. Default statements are rules of inference— after we check the left side of (6.4), we can conclude that X flies. Such statements cannot be manipulated as in FOL. In particular, we can't replace $(\alpha \to \beta)$ with $(\beta \vee (\neg\alpha))$. (More on this later.)

  You may have noticed that the preceding discussion sounds like Prolog's negation as failure: Instead of proving $\neg$penguin(X), it determines that it cannot prove penguin(X). We discussed Prolog's negation as failure on page 169 and will explore its connection with nonmonotonic reasoning in this chapter.

At first sight, default logic seems to be a complete resolution of the nonmonotonic reasoning problem. Unfortunately it's not. First, default logic doesn't tell us *how* to reason; that is, how to derive results. Second, default logic says that, if we can't prove that X is a penguin, then we *may* assume that X is not a penguin—not that we *must* assume it. Which defaults should we assume? What if defaults contradict one another? Default logic lets us determine if a certain set of defaults can be assumed by a "rational" person, but it doesn't tell us which defaults should be assumed.

- **Defeasible Reasoning**: Superficially, defeasible reasoning is like default reasoning in that it augments FOL with rules of inference for dealing with rules that have exceptions. There are important differences. Default logic emphasizes the logical concepts whereas defeasible reasoning emphasizes the reasoning methods.

- **Autoepistemic Logic and Modal Logics**: Epistemology is the branch of philosophy that investigates the nature and limits of human knowing. Autoepistemic reasoning involves reasoning about one's own knowledge. For example, if I have no knowledge that I am an ax murderer, I may conclude that I'm not an ax murderer. Not all lack of knowledge can be used this way. For example, although I do not know that all students will do exceptionally good work in this course, I should not conclude that they will—although they might like me to do so. Somehow, I must distinguish between knowledge I would have if it were true (ax murderer) and knowledge that there is no reason to believe I would have (future student performance).

    Autoepistemic logic is an example of *modal logic*, which is an extension of FOL that introduces one or more operators to indicate states of belief or (partial validity).

- **Circumscription**: Circumscription extends the knowledge base rather than the language of FOL. For example, suppose we have

$$\forall X \Big(\big(\mathrm{bird}(X) \wedge \mathrm{normal}(X)\big) \rightarrow \mathrm{flies}(X)\Big).$$

A circumscriptive approach might attempt to make the predicate "normal" apply to as many terms as possible. This sounds very much like default logic and, indeed, there are connections between the two approaches.

Temporal reasoning—reasoning about situations in which time plays a central role—is often based on adaptations of these methods.

All the above methods modify the interaction between FOL and a knowledge base in some manner in order to deal with beliefs. Here's a quick description of some major differences:

- *Default logic isolates beliefs* in a new type of knowledge-base statement, the *default rule*, which is used to produce additional FOL statements (in other words, assumptions). These assumptions supply information that is incomplete in the FOL part of the knowledge base. Thus, beliefs are contained in a part of the knowledge base that has limited interaction with the remainder and all reasoning is done in FOL. The theory describes how to decide if a set of assumptions is acceptable, but it doesn't tell us how to find such sets.

- *Defeasible reasoning introduces new reasoning rules* as well as new types of knowledge-base statements on which these rules act. The amount of FOL that is allowed depends on the default logic. Some limit the FOL formulas to Prolog-like statements.

- *Autoepistemic logic extends the language of FOL* by adding the ability to indicate that something is believed—a modal operator. Just as it is necessary to develop a new theory when we move from propositional logic to FOL, it's necessary to develop a new theory when we move from FOL to autoepistemic logic.

- *Circumscription introduces predicates for abnormality*, but makes no changes to FOL and introduces no new types of statements. Typically, we introduce one or more predicates whose truth indicates an abnormal situation. Beliefs are expressed as predicate logic rules in the knowledge base by insisting that the abnormal predicate be false—the situation is not abnormal. If the abnormal predicates were always false, the knowledge base would be inconsistent. To maintain consistency, circumscriptive reasoning adds new FOL statements to the knowledge base that assert that abnormality is true in certain situations. The number of additional statements (abnormalities) should be kept small.

Other Approaches

Another way of overcoming the problems with FOL is to look for approaches that are not based on mathematical logic. This strategy leads to alternative methods for both monotonic and nonmonotonic reasoning. Here are three important ones:

- **Rule Systems**: Although we've viewed Prolog clauses as a part of FOL, we can instead view them as rules: Whenever the conditions in the body of a clause are satisfied, the rule (i.e., clause) gives us the head of the clause as a new fact. Obviously, the FOL and rule system approaches to Prolog are quite similar.

- **Graphical Methods (Semantic Nets)**: Some people—including me—use the term *semantic nets* to designate graphical reasoning methods in general;

others use it to designate a particular type of graphical reasoning structure. Semantic nets use directed graphs with knowledge stored at both the vertices and the edges. Roughly speaking, vertices contain facts and edges describe relationships. For example, "a cat is a feline" could be represented by "cat" and "feline" vertices joined by an "IS-A" edge. The rules for reasoning with a semantic net are described in terms of the net's local structure; for example, if there are IS-A edges from u to v to w, we can deduce the edge u IS-A w.

Rule systems and semantic nets were developed in the search for methods where the representation and manipulation of knowledge seemed more natural than FOL and nonmonotonic logics. While the inventors of nonmonotonic logics emphasized provable results over representation and algorithms, the inventors of alternative systems emphasized representation and manipulation, often ignoring theoretical foundations. Unfortunately, intuition alone doesn't provide answers to questions like "Are the algorithms guaranteed to produce consistent results?" For that, mathematical rigor is required.

In their search for secure theoretical foundations, researchers have often translated alternative approaches into nonmonotonic logics. Thus, graphical methods can usually be viewed as alternative representations of certain nonmonotonic logics. Should we therefore abandon such systems in favor of mathematical logic? No. In the first place, the interpretations are sometimes incomplete. In the second place, the form in which something is represented is important. In other words, choose the tool to fit the task. While mathematical logic might be the best tool for *proving* something about a reasoning method, it may not be the best tool for *representing* knowledge or *using* the method. You've already seen this in connection with Prolog: The resolution algorithm is based on finding a contradiction, but the Prolog interpretation of it is more direct—it traverses a tree looking for solutions. (See p. 129.)

Exercises

6.1.C. What are the four main approaches to nonmonotonic logic? In FOL we have a reasoning method and a knowledge base. How do these three methods adjust the reasoning method and/or knowledge base to allow for nonmonotonic reasoning?

6.1.D. What are two approaches to (nonmonotonic) reasoning that aren't based on logic?

6.1.E. Why is a semantic net useful even if it can be described by a nonmonotonic logic?

How Well Do Nonmonotonic Methods Work?

Nonmonotonic reasoning has turned out to be far more difficult than any of us expected. The fundamental advantage of working with formal methods—making falsifiable claims—is that we are forced to address these difficulties, to confront the nasty surprises that lie between speculation and practice.

—Matthew L. Ginsberg (1991)

How much will nonmonotonic methods contribute to AI in general and to commonsense reasoning in particular? What will be the relative importance of qualitative methods versus quantitative methods? It's too soon to tell; however, it's possible to make some general observations. Let's do this against the backdrop of goals, difficulties, and compromises.

Goal

The holy grail of nonmonotonic researchers is commonsense reasoning. It's generally accepted that an important facet of such reasoning is the ability to reach sensible conclusions when faced with incomplete and/or uncertain information. Hence, designing methods for doing so is a primary goal of AI research on nonmonotonic reasoning.

Difficulties

Nonmonotonic methods have been quite successful in some limited environments, but seem far from being able to duplicate commonsense reasoning. Some believe the main problem is inadequate knowledge bases; others believe it is the reasoning tools. Still others believe that both the tools and knowledge bases need major improvements. Another possibility is that the problem is too complex to attack directly; instead, a reasoning system and a basic knowledge base should be combined with a learning system that augments the knowledge base. A few believe that commonsense reasoning is an impossible goal. Only time will tell who is right.

In an earlier discussion we identified some core problems for nonmonotonic reasoning: limiting search of the knowledge base, detecting missing knowledge, and making reasonable assumptions. (See (6.1) and (6.2).) Extending FOL makes these problems essentially intractable:

- **Problem 1. FOL is NP-hard:** Since no efficient algorithm is known or likely to exist for FOL, the same is true for any extension of it. (As we noted on p. 172, preparing a general FOL knowledge base for Horn clause resolution is NP-hard.)

- **Problem 2. FOL is semidecidable:** Missing information is detected by being unable to prove something: If the truth of ($\neg\alpha$) is not implied by the data, we may be allowed to assume that α is true. Unfortunately, FOL is semidecidable. This means that although we can create algorithms (such as resolution) that allow us to prove any statement which is true in an FOL language, it is impossible to design an algorithm that will allow us to decide if statements are true or false. (See Theorem 4.6 (p. 165) for more details.) What this means in practice is that there cannot be an algorithm to decide if statements are undecidable; that is, we can't create an algorithm to decide if additional information is needed to determine the truth or falsity of a statement.

Regardless of whether or not the method uses FOL, making assumptions is a source of difficulties. Here are some:

- **Problem 3. Consistency is hard:** Whatever assumptions are made should be consistent with one another and with the data base. It is often difficult to prove consistency.

- **Problem 4. Selecting assumptions is unclear:** If the consistency problem is overcome, how should the reasoning system choose between competing consistent assumptions? The problem here is both conceptual (what *should* the decision criteria be) and procedural (how can they be implemented in a reasonable manner).

Compromises

The first two problems with FOL can be dealt with in at least three ways:

1. The expressive capability of the language can be limited to the point where both problems disappear.
2. The difficulty can be dumped onto the shoulders of the knowledge-base designer.
3. They can be ignored. In Problem 2, for example, if a proof of β is not obtained in a reasonable period of time, it could be assumed that β cannot be proved.

The usual compromise is a blend of the first two methods. Prolog is an example of such a compromise in FOL. While the third way is probably akin to the human approach, its lack of precision poses theoretical difficulties and researchers have generally avoided it. I believe major breakthroughs in large-scale nonmonotonic reasoning await the development of good methods for trading time and information that are implicit in the third approach.

Limiting the expressive capability of the language can make dealing with assumptions easier, too—particularly the problem of determining their consistency.

One "solution" to the problem of choosing between competing assumptions is to declare Problem 4 to be a separate issue outside the reasoning process. That is, the reasoning process determines what assumptions are reasonable and the user must employ some other method for choosing between them. Splitting a problem up this way is called *divide and conquer*, which is an effective tool for attacking complicated problems.

If the reasoning method must actually decide what to assume, an obvious approach is to assume as little as possible. This is not as easy as it sounds because sets of assumptions frequently conflict with one another. A conservative compromise is to make only those assumptions that, in some sense, *must* be true. Unfortunately, this approach may result in too few assumptions. For example, if two completely conflicting assumptions are possible, the system would assume nothing. Another solution is to have some method for choosing one set of assumptions over another. For example, given two assumptions ("Birds fly" and "Penguins don't fly"), we choose the less general one ("Penguins don't fly"). Compromises are needed to avoid slow algorithms. The following example illustrates some problems facing anyone who must decide among conflicting assumptions.

Example 6.1   Which Defaults Should Be Assumed?

Nixon-Republican-Quaker: Nixon is a Quaker and a Republican. Without contrary evidence, Quakers are doves and Republicans are not. (Although Nixon is deceased, we've retained the present-tense phrasing in this classic 1960s example.)

Consider the query "Is Nixon likely to be a dove?" Using the "Quaker," information, we could assume that Nixon is a dove. Unfortunately, this is opposed by the "Republican" information, which leads to the belief that Nixon is not a dove. What should we do? If no other knowledge is available, the most reasonable response to the query is "Insufficient information."

Mollusca: Species in the phylum of Mollusca, which includes clams and snails, normally have external shells. Mollusca contains the class Cephalopoda, which includes squid and octopi. Cephalopods normally lack external shells; however, they include the genus *Nautilus*, and nautiloids normally have external shells.

Consider the query "Is the chambered nautilus likely to have an external shell?" Since nautiloids normally have external shells, we might answer "Yes." On the other hand, since nautiloids are cephalopods and since cephalopods generally lack external shells, perhaps we should say "No." On the third hand, since cephalopods are mollusks, which generally have external shells, perhaps "Yes" is correct after all.

What should we do? Researchers generally agree that the most restrictive condition should be used. Since the three conditions are nautiloid, cephalopod, and mollusk from most restrictive to least, we should use the information that nautiloids normally have external shells. To carry this out in general, we need an operational definition of "more restrictive."   ■

Exercises

6.1.F. What are some problems encountered in designing a method for nonmonotonic reasoning?

6.1.G. What are some methods for dealing with these problems?

6.2   Default Reasoning

> *If I knew what was* true, *I'd probably be willing to sweat and strive for it.*
> —H. L. Mencken (1918)

Introduced by R. Reiter in 1980, default logic is still an active research area. We'll begin with an informal discussion of the meaning and usage of defaults and then explore the formal ideas. The definitions are phrased in terms of FOL; however—

> All references to FOL in the following definitions can be replaced by references to propositional logic with no changes.

Let's assume such replacements whenever we use propositional logic.

Recall the semantic and syntactic definitions of *consistent*: Semantically, we say that β is consistent with $\mathcal{S}$ if there is an interpretation in which both β and the formulas in $\mathcal{S}$ are true. Syntactically, β is consistent with $\mathcal{S}$ if $\mathcal{S} \nvdash (\neg\beta)$. These definitions are equivalent in FOL because of the soundness and completeness of the proof method for FOL. We'll extend the syntactic definition to default logic.

Definition 6.1   What Defaults Mean

If α, β, and γ are formulas in FOL with no free variables, then

$$\frac{\alpha : \beta}{\gamma} \quad \text{which we can also write as} \quad (\alpha : \beta) \rightharpoonup \gamma$$

is a *default rule*, also called a *default*. (The notation $(\alpha : \beta) \rightharpoonup \gamma$ is *not* standard.) This rule means "if α is true and β is consistent with what is true, then we may assume that γ is true." There are two special cases:

- The rule $(\alpha :) \rightharpoonup \gamma$ means "if α is true, then we *may* assume that γ is true." This is not the same as $\alpha \rightarrow \gamma$, which says that if α is true, then we *must* assume that γ is true.

- The rule $(: \beta) \rightharpoonup \gamma$ means "if β is consistent with our beliefs, then we may assume that γ is true."

To indicate that $(\alpha(X) : \beta(X)) \rightarrow \gamma(X)$ is a default rule when any substitution is made for X, we write

$$\forall X \left(\frac{\alpha(X) : \beta(X)}{\gamma(X)} \right) \quad \text{or, more simply,} \quad \frac{\alpha(X) : \beta(X)}{\gamma(X)}. \tag{6.5}$$

The extension of this definition to more than one variable should be obvious.

Often, $\mathcal{D}$ and $\mathcal{W}$ denote the sets of default and FOL formulas, respectively.

Definition 6.1 says that we *may* assume γ is true—not that we *must* assume it is true. Because of the may/must distinction, it's important to regard (6.5) as representing a collection of default rules instead of one single rule. If the substitution of t for X gives an inconsistent $\beta(t)$, then we cannot assume $\gamma(t)$. If the substitution of t for X gives a consistent $\beta(t)$ and a true $\alpha(t)$, then we may assume $\gamma(t)$ but need not do so. Consequently, we might assume some $\gamma(t)$'s and not others.

Let's look at Example 6.1 in terms of default logic.

Example 6.2   Nixon and the Nautiloids Revisited

Nixon-Republican-Quaker: Let's use predicates for dove, Quaker, and Republican and denote them by initials. Each takes one argument—a person. With this notation, the Nixon information translates into the FOL and default formulas

$$\mathcal{W} = \big\{ q(n), \, r(n) \big\} \quad \text{and} \quad \mathcal{D} = \left\{ \frac{q(X) : d(X)}{d(X)}, \, \frac{r(X) : \neg d(X)}{\neg d(X)} \right\}, \tag{6.6}$$

respectively. The only constant in the language $\mathcal{L}$ is n and the only predicates are q, r, and d.

Since $q(n)$ is true and $d(n)$ is consistent with $\mathcal{W}$, the first default tells us that we may assume $d(n)$. If we do so, we have the "extension" $\mathcal{S} = \{q(n), r(n), d(n)\}$. Since $\neg d(n)$ is not consistent with $\mathcal{S}$, we may not apply the second default rule.

On the other hand, if we'd started with the second default rule instead of the first, we could have obtained $\mathcal{S}' = \{q(n), r(n), \neg d(n)\}$.

Mollusca: In this case, the translation is

$$\mathcal{W} = \big\{ \forall X(n(X) \rightarrow c(X)), \, \forall X(c(X) \rightarrow m(X)) \big\}$$

$$\mathcal{D} = \left\{ \frac{m(X) : e(X)}{e(X)}, \, \frac{c(X) : \neg e(X)}{\neg e(X)}, \, \frac{n(X) : e(X)}{e(X)} \right\}. \tag{6.7}$$

The only predicates in the language $\mathcal{L}$ are m, c, n, and e. The only constant will be a. By adding $n(a)$ to $\mathcal{W}$, we'd be in a position to answer the query posed in Example 6.1. Let's consider two other cases first.

If we add $m(a)$ to $\mathcal{W}$ instead, the first default rule tells us that we may assume $e(a)$. There is no way we could use any other default rule, so it's reasonable to assume $e(a)$.

Suppose we add $c(a)$ to $\mathcal{W}$ instead. Now $\mathcal{W}$ tells us that $m(a)$ is also true. Consequently, we may assume either $e(a)$ using the first default or $\neg e(a)$ using the second default.

Finally, suppose we add $n(a)$ to $\mathcal{W}$. Now $\mathcal{W}$ tells us that $c(a)$ and $m(a)$ are also true. Consequently, we may assume either $e(a)$ using the first default, $\neg e(a)$ using the second default, or $e(a)$ using the third default. ∎

Default logic extends FOL by allowing default rules in the knowledge base. A central concept is the idea of an extension, which consists of everything that can be deduced given the FOL part of the knowledge base and some additional formulas. Such formulas γ are given by default rules $(\alpha : \beta) \longrightarrow \gamma$. A recursive definition is provided for deciding when something is an extension—that is, deciding whether something is based on reasonable assumptions concerning the default rules. The theory provides neither an algorithm for producing extensions nor an algorithm for choosing which of several possible extensions is best. This limits its usefulness.

The difficulty with producing extensions is essentially a consistency issue. Normal default rules are an important special case of default rules because they overcome this problem. As a result, an algorithm exists for producing extensions when all the default rules are normal, provided the semidecidability problem of FOL is dealt with—we must be able to decide that a formula cannot be proved. This can be dealt with by restricting the allowable formulas to some subset of FOL as discussed on page 208. Fortunately, many default knowledge bases can be written in normal form.

Extensions

The idea of assuming defaults can be described explicitly by the notion of an extension. Underlying the idea of an extension is the concept of a rational set of beliefs. Suppose we have $\Delta = (\mathcal{D}, \mathcal{W})$. Our beliefs should certainly include $\mathcal{W}$ and its consequences in FOL; however, we may believe additional formulas. Suppose that $(\alpha : \beta) \longrightarrow \gamma$ is in $\mathcal{D}$, that α follows from our beliefs, and that β is consistent with them. Then it's reasonable to believe γ as well. Is more required for rationality? Yes, what we believe should be what is deducible from applying FOL to $\mathcal{W}$ and those γ's obtained from $\mathcal{D}$ by the above procedure. Writing this out in an explicit manner requires a somewhat convoluted and cryptic definition. We'll need to exercise a bit of care in stating the definition in order to get it correct. The rest of this subsection is devoted to the definition of an extension and a discussion of what it means.

Definition 6.2 Some Terminology

Suppose we are given a predicate logic language $\mathcal{L}$ as in Definition 3.3 (p. 106). For any collection $\mathcal{S}$ of FOL formulas, let $\text{Th}_{\mathcal{L}}(\mathcal{S})$ be the set of all FOL formulas α such that $\mathcal{S} \vdash \alpha$; that is, $\text{Th}_{\mathcal{L}}(\mathcal{S})$ is the collection of all results that can be derived from $\mathcal{S}$ in $\mathcal{L}$. Usually, we'll simply write $\text{Th}(\mathcal{S})$, with $\mathcal{L}$ being understood.

Again, assuming a FOL language $\mathcal{L}$, let $\mathcal{W}$ be a collection of FOL formulas and let $\mathcal{D}$ be a collection of default rules. We call $\Delta = (\mathcal{D}, \mathcal{W})$ a *default theory*.

In order to define an extension of a default theory, we need to make use of what is known as a *nonconstructive definition*; that is, a definition that doesn't tell you how to construct the object being defined. Nonconstructive definitions don't guarantee that the thing being defined exists or that it is unique. You've already encountered such problems in calculus, as the following example illustrates.

Example 6.3 Nonconstructive Definitions in Calculus

The derivative of $f(x)$ is defined by

$$ f'(x) = \lim_{h \to 0} \frac{f(x+h) - f(x)}{h}. $$

Although this definition appears to be constructive, it relies on the notion of limit, which is nonconstructive:

We say $\lim_{x \to a} g(x) = L$ if, for every $\epsilon > 0$, there is a $\delta > 0$ such that $|f(x) - L| < \epsilon$ whenever $0 < |x - a| < \delta$.

Because this definition is nonconstructive, we must prove that if a limit exists, it has only one value; that is, we can't have $\lim_{x \to a} g(x) = L$ and $\lim_{x \to a} g(x) = M$ with $L \neq M$. To avoid the problems of nonconstructive definitions, we prove theorems that allow us to compute derivatives constructively; e.g., $(x^n)' = nx^{n-1}$ and $(f(g(x))' = f'(g(x))g'(x)$.

We could define the indefinite integral of a function $f(x)$ to be a function $F(x)$ such that $F'(x) = f(x)$. This definition is also nonconstructive. One of the basic results of calculus states that if $F(x)$ is *any* function which is the indefinite integral of $f(x)$, then *every* indefinite integral of $f(x)$ is of the form $F(x) + C$. This result is satisfactory in one way because it tells us what all indefinite integrals of $f(x)$ look like in terms of one of them. In other ways, however, it is unsatisfactory. For example, it doesn't tell us whether $f(x)$ has an indefinite integral. (In fact, it may not.) ∎

In attempting to define an extension, let's begin with an appealing, but incorrect, definition and then explore how to correct it. We're doing this because the complexity of the correct definition should be justified.

Definition 6.3   Extension of a Default Theory (Incorrect)

Let a default theory $\Delta = (\mathcal{D}, \mathcal{W})$ be given. For any collection $\mathcal{S}$ of FOL formulas, define $\Delta_{\mathcal{S}}$ to consist of $\mathcal{W}$ plus all formulas $\alpha \to \gamma$ for which

$$\neg\beta \notin \mathcal{S} \quad \text{and} \quad \frac{\alpha : \beta}{\gamma} \in \mathcal{D}.$$

If $\text{Th}(\Delta_{\mathcal{S}}) = \mathcal{S}$, we (incorrectly) call $\mathcal{S}$ an extension of Δ.

In this definition, we can think of $\mathcal{S}$ as a set of beliefs and we can think of $\Delta_{\mathcal{S}}$ as $\mathcal{W}$ plus those FOL formulas our beliefs tell us we can obtain from $\mathcal{D}$. The equation $\text{Th}(\Delta_{\mathcal{S}}) = \mathcal{S}$ then says that what we can deduce from our beliefs and Δ is precisely our beliefs.

Where does the definition go wrong? The heart of the problem is as follows. If we are given $a \to b$ and $\neg b$ in FOL, then we can conclude $\neg a$. If we are given $(a :) \to b$ and $\neg b$, we find nothing in the definition of defaults about concluding $\neg a$. In other words, defaults work only from left to right, but ordinary implication works in both directions because of the contrapositive, $(\neg b) \to (\neg a)$. This causes problems in Definition 6.3 because $\neg b \in \mathcal{W}$ and $a \to b \in \mathcal{D}$ lead to

$$\neg b \in \Delta_{\mathcal{S}} \quad \text{and} \quad (a \to b) \in \Delta_{\mathcal{S}}$$

for all $\mathcal{S}$. In other words, $a \to b$ allows more reasoning than $(a :) \to b$ because of "contrapositive" reasoning based on $(\neg b) \to (\neg a)$. How can this problem be resolved?

- An obvious solution is to allow such "contrapositive" reasoning with default rules. There's a good argument against allowing it: Sometimes it's not wanted; and if it ever is wanted, we can simply include $(\neg\gamma : \beta) \to (\neg\alpha)$ as another default rule. As a result, such "contrapositive" reasoning is not allowed with defaults and another resolution of the problem with Definition 6.3 must be found.

- Another approach is to try to define a new connective, say $\to$, that prevents such reasoning. Unfortunately, this can't be done within FOL—see Exercise 6.2.1.

- Yet another approach is to check that α is true and then include γ. This leads to a better definition of $\Delta_{\mathcal{S}}$—in fact, to the one that is used.

A careful refinement of the last idea could lead to the following definition. In keeping with standard notation, we'll use $\Gamma(\mathcal{S})$ in place of $\text{Th}(\Delta_{\mathcal{S}})$ and add the condition $\text{Th}(\Gamma(\mathcal{S})) = \Gamma(\mathcal{S})$ to ensure that $\Gamma(\mathcal{S})$ does in fact include all the FOL formulas that can be deduced from something.

> To simplify future discussion, we'll assume that there are no rules of the form $\frac{\alpha :}{\beta}$ in $\mathcal{D}$. With some care, this constraint can be removed.

Definition 6.4   Extension of a Default Theory

Let a default theory $\Delta = (\mathcal{D}, \mathcal{W})$ be given and let $\mathcal{S}$ be any collection of FOL formulas. Define $\Gamma(\mathcal{S})$ to be the set of FOL formulas such that

(a) $\text{Th}(\Gamma(\mathcal{S})) = \Gamma(\mathcal{S}) \supseteq \mathcal{W}$.

(b) If $\frac{\alpha:\beta}{\gamma} \in \mathcal{D}$, $\alpha \in \Gamma(\mathcal{S})$, and $\neg\beta \notin \mathcal{S}$, then $\gamma \in \Gamma(\mathcal{S})$.

(c) $\Gamma(\mathcal{S})$ is a minimum with respect to (a) and (b); that is, if Γ' satisfies (a) and (b), then $\Gamma' \supseteq \Gamma(\mathcal{S})$.

Finally, if $\Gamma(\mathcal{S}) = \mathcal{S}$, we call $\mathcal{S}$ an *extension* of Δ.

Definition 6.4 is one of the more convoluted definitions in this text. Think about it this way. $\mathcal{S}$ is what someone says is a reasonable set of beliefs given Δ. We then construct $\Gamma(\mathcal{S})$ to be those things we ought to believe based on Δ and the beliefs $\mathcal{S}$. By (a), what we ought to believe should be closed under logical deduction $(\text{Th}(\Gamma(\mathcal{S})) = \Gamma(\mathcal{S}))$, and we ought to believe what is absolutely true $(\Gamma(\mathcal{S}) \supseteq \mathcal{W})$. By (b), we ought to believe results of default rules if we believe α and if β doesn't contradict the reasonable set of beliefs. By (c), we don't include anything in what we ought to believe unless it's forced upon us by Δ and $\mathcal{S}$. Finally, if the original set of beliefs was reasonable, this procedure should merely have reconstructed it; that is, $\Gamma(\mathcal{S}) = \mathcal{S}$.

The definition contains two rather nasty twists. First, in (b) we use $\Gamma(\mathcal{S})$ to help define itself. Is such a procedure valid? Yes, when used *carefully*. In this case, we can simply use (b) iteratively to build up $\Gamma(\mathcal{S})$—every time we add more to $\Gamma(\mathcal{S})$, we may find more rules with $\alpha \in \Gamma(\mathcal{S})$. In fact, this iterative application is the basis of Theorem 6.1 (p. 217). Second, in (c) we insist that $\Gamma(\mathcal{S})$ be a minimum. How do we know there is a minimum? Example: There is no smallest rational number greater than 0. Example: There is no smallest set $X \subseteq A = \{1, 2, 3, 5\}$ such that every element of A is either in X or a sum of two numbers from X—both $X = \{1, 2, 3\}$ and $X = \{1, 2, 5\}$ are *minimal* (nothing smaller works) but neither is a *minimum* (everything that works is larger) because neither contains the other. The proof that the minimum in (c) exists is left as Exercise 6.2.3.

Some examples of finding extensions should clarify how the definition works.

Example 6.4   Some Extensions

Nixon-Republican-Quaker: Consider the default theory (6.6). It has two default extensions, namely

$$\mathrm{Th}(\{q(n), r(n), d(n)\}) \text{ and } \mathrm{Th}(\{q(n), r(n), \neg d(n)\}). \qquad (6.8)$$

Why are these extensions and why are there no other extensions?

Suppose $\mathcal{E}$ is an extension. By definition, it must contain $\mathcal{W}$. There are four cases to consider regarding $d(n)$—whether or not it and/or its negation are in $\mathcal{E}$. Here are the cases and the results.

- $d(n) \in \mathcal{E}$ and $\neg d(n) \in \mathcal{E}$:   We cannot apply the first default in (6.6) since $\neg d(n) \in \mathcal{E}$. We cannot apply the second default since $d(n) \in \mathcal{E}$. Hence $\Gamma(\mathcal{E}) = \Gamma(\mathcal{W})$ and so does not contain $d(n)$ or $\neg d(n)$. Hence we cannot have $\Gamma(\mathcal{E}) = \mathcal{E}$. Since $\mathcal{E}$ was assumed to be an extension, this case doesn't occur.

- $d(n) \in \mathcal{E}$ and $\neg d(n) \notin \mathcal{E}$:   We can apply the first default to conclude that $d(n) \in \Gamma(\mathcal{E})$; however, we cannot apply the second since $d(n) \in \mathcal{E}$. Thus $\Gamma(\mathcal{E}) = \mathrm{Th}(\mathcal{W} \cup \{d(n)\})$.

- $d(n) \notin \mathcal{E}$ and $\neg d(n) \in \mathcal{E}$:   Arguing as in the previous case, we get $\Gamma(\mathcal{E}) = \mathrm{Th}(\mathcal{W} \cup \{\neg d(n)\})$.

- $d(n) \notin \mathcal{E}$ and $\neg d(n) \notin \mathcal{E}$:   Now both defaults apply and so $d(n)$ and $\neg d(n)$ are both in $\Gamma(\mathcal{E})$, which is therefore larger than $\mathcal{E}$. Since $\mathcal{E}$ is an extension, this case can't occur.

What decision should we reach based on the two extensions in (6.8)? A very conservative approach is to make only those assumptions that lie in all extensions. Doing this means no assumptions would be made concerning the truth or falsity of $d(n)$.

Mollusca: In a manner similar to the above, you should be able to determine the extensions of $(\mathcal{D}, \mathcal{W}')$ where $\mathcal{D}$ and $\mathcal{W}$ are given by (6.7) and either

$$\mathcal{W}' = \mathcal{W} \cup \{m(a)\} \text{ or } \mathcal{W}' = \mathcal{W} \cup \{c(a)\} \text{ or } \mathcal{W}' = \mathcal{W} \cup \{n(a)\}.$$

For the first $\mathcal{W}'$, there is one extension, $\mathrm{Th}(m(a), e(a))$. Each of the others has two extensions. Unfortunately, default logic does not tell us how to choose among the possible extensions. The conservative approach suggested for Nixon would lead to the conclusion that nothing should be concluded about a chambered nautilus having a shell.

A Theory with No Extensions: Extensions need not exist. To see this, consider the following simple default theory, which is written in propositional logic:

$$\mathcal{W} = \emptyset \text{ and } \mathcal{D} = \left\{ \frac{: a}{\neg a}, \frac{: \neg a}{a} \right\}$$

Let $\mathcal{E}$ be an extension. We claim that $\neg a \in \mathcal{E}$. If not, the first default can be applied to obtain $\neg a \in \Gamma(\mathcal{E}) = \mathcal{E}$. Likewise, the second default tells us that

$a \in \mathcal{E}$. Unfortunately, if $a, \neg a \in \mathcal{E}$, then neither default can be applied and so neither a nor $\neg a$ is in $\Gamma(\mathcal{E})$, a contradiction. ∎

You may have concluded from the previous example that determining possible extensions and verifying that they are extensions is nontrivial. This is indeed the case. The following theorem makes verification easier. However, the determination of extensions is, in general, much more difficult than solving an NP-complete problem. As a result, researchers have looked for special situations in which a reasonable algorithm for finding extensions can be constructed. Soon we'll discuss a classic case—normal default theories.

Theorem 6.1   Reiter's Test for Extensions
Let $\Delta = (\mathcal{D}, \mathcal{W})$ be a default theory and let $\mathcal{E}$ be a set of formulas. Define $\mathcal{E}_0 = \mathcal{W}$ and, for $i \geq 0$,

$$\mathcal{E}_{i+1} = \text{Th}(\mathcal{E}_i) \cup \left\{ \gamma \;\middle|\; \exists \alpha\, \exists \beta \left((\alpha \in \mathcal{E}_i) \wedge (\neg\beta \notin \mathcal{E}) \wedge \frac{\alpha : \beta}{\gamma} \in \mathcal{D} \right) \right\}.$$

Then $\mathcal{E}$ is an extension of Δ if and only if $\mathcal{E} = \cup_{i=0}^{\infty} \mathcal{E}_i$.

Computer programs deal with finite situations. Since $\mathcal{E}_i \subseteq \mathcal{E}_{i+1}$, they eventually have $\mathcal{E}_n = \mathcal{E}_{n+1}$ for some n. You should be able to easily show that $\mathcal{E}_i = \mathcal{E}_n$ for all $i \geq n$ and so $\cup_{i=0}^{\infty}\mathcal{E}_i = \mathcal{E}_n$. The theorem therefore leads to an algorithm for verifying that a set $\mathcal{E}$ is an extension. It does not provide an algorithm for generating extensions since we must already have $\mathcal{E}$ in order to define $\mathcal{E}_{i+1}$—the definition includes $\neg\beta \notin \mathcal{E}$.

The following proof of the theorem is closely reasoned and a bit tricky, so it will probably require some study on your part.

Proof: Let $\mathcal{E}' = \cup_{i=0}^{\infty}\mathcal{E}_i$. In the proof, the obvious relations

$$\mathcal{E}' \supseteq \mathcal{E}_{i+1} \supseteq \text{Th}(\mathcal{E}_i) \supseteq \mathcal{E}_i \supseteq \mathcal{E}_0 = \mathcal{W}$$

will be useful. We'll begin by proving

(a) $\text{Th}(\mathcal{E}') = \mathcal{E}' \supseteq \mathcal{W}$.
(b) If $\frac{\alpha : \beta}{\gamma} \in \mathcal{D}$, $\alpha \in \mathcal{E}'$, and $\beta \notin \mathcal{E}$, then $\gamma \in \mathcal{E}'$.

We have $\mathcal{E}' \supseteq \mathcal{E}_0 = \mathcal{W}$. Since $\text{Th}(\mathcal{T}) \supseteq \mathcal{T}$ for any $\mathcal{T}$, (a) will be proved when we show that $\text{Th}(\mathcal{E}') \subseteq \mathcal{E}'$. Suppose $\alpha \in \text{Th}(\mathcal{E}')$. It is deduced from a finite set of formulas in $\mathcal{E}'$ (possibly just one, α itself). Any formula in $\mathcal{E}'$ must be in one of the $\mathcal{E}_i$'s whose union constitutes $\mathcal{E}'$. Since the $\mathcal{E}_i$'s increase with i, the *finite* set of formulas needed to deduce α must be contained in some $\mathcal{E}_k$. Since $\mathcal{E}' \supseteq \text{Th}(\mathcal{E}_k)$, we have $\mathcal{E}' \supseteq \text{Th}(\mathcal{E}')$, and so (a) is proved. Claim (b) follows because α must be in some $\mathcal{E}_i$ and so $\gamma \in \mathcal{E}_{i+1}$.

Comparing (a) and (b) with the corresponding parts of Definition 6.4 and using the minimality of Γ, it follows that

$$\Gamma(\mathcal{E}) \subseteq \mathcal{E}' \qquad\qquad (6.9)$$

for any set $\mathcal{E}$.

Since the theorem is an if and only if, we'll consider two separate implications.

First, assume that $\mathcal{E} = \Gamma(\mathcal{E})$ and prove that $\mathcal{E} = \mathcal{E}'$. By (6.9) and $\mathcal{E} = \Gamma(\mathcal{E})$, it suffices to prove $\mathcal{E}' \subseteq \mathcal{E}$. We claim that $\mathcal{E}_i \subseteq \mathcal{E}$. Here's a proof by induction. The case $i = 0$ is simple. Now assume that $\mathcal{E}_{i-1} \subseteq \mathcal{E}$. Then $\mathrm{Th}(\mathcal{E}_{i-1}) \subseteq \mathrm{Th}(\mathcal{E}) = \mathcal{E}$, the last because $\mathcal{E}$ is an extension. Let $\frac{\alpha:\beta}{\gamma} \in \mathcal{D}$ be such that $\alpha \in \mathcal{E}_{i-1}$ and $\neg\beta \notin \mathcal{E}$. Then $\alpha \in \mathcal{E}$ and so $\gamma \in \Gamma(\mathcal{E}) = \mathcal{E}$. This proves the claim. From the claim, $\mathcal{E}' \subseteq \mathcal{E}$.

Second, assume that $\mathcal{E} = \mathcal{E}'$ and prove that $\mathcal{E} = \Gamma(\mathcal{E})$. By (6.9), it suffices to show that $\mathcal{E}' \subseteq \Gamma(\mathcal{E})$, for then $\Gamma(\mathcal{E}) = \mathcal{E}' = \mathcal{E}$. To prove $\mathcal{E}' \subseteq \Gamma(\mathcal{E})$, it suffices to prove by induction that $\mathcal{E}_i \subseteq \Gamma(\mathcal{E})$. The case $i = 0$ is simple: $\mathcal{E}_0 = \mathcal{W} \subseteq \Gamma(\mathcal{E})$. For purposes of induction, assume $\mathcal{E}_{i-1} \subseteq \Gamma(\mathcal{E})$. We need to check two things. First, we have $\mathrm{Th}(\mathcal{E}_{i-1}) \subseteq \mathrm{Th}(\Gamma(\mathcal{E})) = \Gamma(\mathcal{E})$. Second, if $\alpha \in \mathcal{E}_{i-1}$, $\neg\beta \notin \mathcal{E}$, and $\frac{\alpha:\beta}{\gamma} \in \mathcal{D}$, then $\alpha \in \Gamma(\mathcal{E})$ and so $\gamma \in \Gamma(\mathcal{E})$. This proves that $\mathcal{E}_i \subseteq \Gamma(\mathcal{E})$, completing the proof. ∎

Example 6.5   Verifying Extensions

Let's use the theorem to verify the extensions in the previous example. First, consider (6.6) with

$$\mathcal{E} = \mathrm{Th}\big(\{q(n), r(n), d(n)\}\big).$$

By the theorem,

$$\mathcal{E}_0 = \mathcal{W} = \{q(n), r(n)\},$$
$$\mathcal{E}_1 = \mathrm{Th}(\{q(n), r(n)\}) \cup \{d(n)\},$$
$$\mathcal{E}_2 = \mathrm{Th}(\{q(n), r(n), d(n)\}) \cup \{d(n)\},$$
$$\mathcal{E}_3 = \mathrm{Th}(\{q(n), r(n), d(n)\}) \cup \{d(n)\} = \mathcal{E}_2.$$

Since $\mathcal{E} = \mathcal{E}_2$, it is an extension.

Next consider the Mollusca example with $\mathcal{E} = \mathrm{Th}(\mathcal{W} \cup \{e(a)\})$ and $n(a) \in \mathcal{W}$. Since the only constant is a, we can simplify $\mathcal{W}$ accordingly. By the theorem,

$$\mathcal{E}_0 = \mathcal{W} = \{(n(a) \to c(a)), (c(a) \to m(a))\, n(a)\},$$
$$\mathcal{E}_1 = \mathrm{Th}(\mathcal{W}) \cup \{e(a)\},$$
$$\mathcal{E}_2 = \mathrm{Th}(\mathcal{W} \cup \{e(a)\}) \cup \{e(a)\},$$
$$\mathcal{E}_3 = \mathrm{Th}(\mathcal{W} \cup \{e(a)\}) \cup \{e(a)\} = \mathcal{E}_2.$$

Finally consider the Mollusca example with $\mathcal{E} = \mathrm{Th}(\mathcal{W} \cup \{\neg e(a)\})$ and $n(a) \in \mathcal{W}$. By the theorem,

$$\mathcal{E}_0 = \mathcal{W} = \{(n(a) \to c(a)), (c(a) \to m(a))\, n(a)\},$$
$$\mathcal{E}_1 = \mathrm{Th}(\mathcal{W}),$$
$$\mathcal{E}_2 = \mathrm{Th}(\mathcal{W}) \cup \{\neg e(a)\},$$

$$\mathcal{E}_3 = \text{Th}(\mathcal{W} \cup \{\neg e(a)\}) \cup \{\neg e(a)\},$$
$$\mathcal{E}_4 = \text{Th}(\mathcal{W} \cup \{\neg e(a)\}) \cup \{\neg e(a)\} = \mathcal{E}_3.$$

In the first Mollusca case, we picked up $e(a)$ immediately because $n(a) \in \mathcal{E}_0$. In the second case, we needed $c(a)$ to get $\neg e(a)$ and this did not appear until $\mathcal{E}_1$. ■

You've just seen how Theorem 6.1 provides an algorithm for checking proposed extensions. It can also be used in place of Definition 6.4 to simplify proofs.

Here are two facts about the nature of extensions. Their proofs are left as exercises.

- **Incomparability of extensions**: If $\mathcal{E} \subseteq \mathcal{E}'$ are extensions of Δ, then $\mathcal{E} = \mathcal{E}'$. In other words, given any two extensions, each must contain formulas not contained in the other.   (6.10)

- **Consistency of extensions**: If Δ has an inconsistent extension, then that is its only extension. If $\mathcal{W}$ is consistent, then all extensions are consistent.   (6.11)

Aside. I didn't give the general definition of defaults. A general default has the form $(\alpha : \beta_1, \ldots, \beta_m) \rightarrow \gamma$, where the commas are like ANDs. The requirement $\neg\beta \notin \mathcal{E}$ is then replaced by the set of m requirements $\neg\beta_i \notin \mathcal{E}$ for $1 \leq i \leq m$. Let $\beta = \beta_1 \wedge \cdots \wedge \beta_m$. The default $(\alpha : \beta) \rightarrow \gamma$ is not quite equivalent to the original default. See Exercise 6.2.6 for more details.

Exercises

6.2.A. Various desiderata for an extension were listed. What is the correspondence between each of them and the parts of Definition 6.3? Definition 6.4?

6.2.B. Let $\mathcal{E}$ be a set of FOL formulas. Prove that $\text{Th}(\mathcal{E}) = \mathcal{E}$ if and only if $\mathcal{E}$ consists of all the FOL formulas that can be deduced from $\mathcal{E}$.
Hint. This requires nothing more than an understanding of the problem and the definition of Th.

6.2.1. We want to define a connective $\rightarrow$ in FOL such that

   (i) $(\alpha \wedge (\alpha \rightarrow \beta)) \models \beta$ and

   (ii) it is possible to have $(\neg\beta) \wedge (\alpha \rightarrow \beta)$ true, but this provides no information about α.

   (a) Show that the first condition implies that whenever α and $\alpha \rightarrow \beta$ are true, β is true.

   (b) Show that the second condition implies that whenever β is false, we must have $\alpha \rightarrow \beta$ true, regardless of the value of α.
   Hint. What happens if $\alpha \rightarrow \beta$ is both true and false when β is false?

   (c) Show that the previous conditions cannot be satisfied simultaneously.

6.2.2. Determine all extensions of the following default theories. (The number of extensions ranges from zero to two.) Use Theorem 6.1 to verify that each extension is indeed an extension. In each case prove that you have found all extensions.

(a) $\mathcal{W} = \{a, \neg a\}$, $\mathcal{D} = \left\{ \dfrac{:a}{b} \right\}$.

(b) $\mathcal{W} = \{a, b\}$, $\mathcal{D} = \left\{ \dfrac{a:b}{c} \right\}$.

(c) $\mathcal{W} = \{a, b, \neg c\}$, $\mathcal{D} = \left\{ \dfrac{a:b}{c} \right\}$.

(d) $\mathcal{W} = \{a\}$, $\mathcal{D} = \left\{ \dfrac{a:b}{b}, \dfrac{b:c}{c}, \dfrac{b:\neg c}{\neg c} \right\}$.

(e) $\mathcal{W} = \{b \to \neg(a \lor c)\}$, $\mathcal{D} = \left\{ \dfrac{:a}{a}, \dfrac{:b}{b}, \dfrac{:c}{c} \right\}$.

6.2.3. The purpose of this exercise is to prove that the minimality condition (c) in Definition 6.4 makes sense. Define $\Gamma^*(\mathcal{S})$ to be any set that satisfies (a) and (b) in the definition and define $\Gamma(\mathcal{S})$ to be the intersection of all $\Gamma^*(\mathcal{S})$'s. Prove that $\Gamma(\mathcal{S})$ satisfies (a), (b) and (c) in the definition.

6.2.4. The goal of this exercise is to prove (6.10). Suppose $\mathcal{E}$ and $\mathcal{E}'$ are both extensions of Δ and that $\mathcal{E} \subseteq \mathcal{E}'$.

(a) Prove by induction that when the construction in Theorem 6.1 is applied to $\mathcal{E}$ and $\mathcal{E}'$, we always have $\mathcal{E}_i \supseteq \mathcal{E}'_i$.

(b) Use the previous step and the theorem to conclude that $\mathcal{E} \supseteq \mathcal{E}'$. Then conclude that $\mathcal{E} = \mathcal{E}'$.

6.2.5. Prove (6.11). Recall that, if $\mathcal{T}$ is inconsistent, then $\mathcal{T} \models \alpha$ for all formulas α.
Hint. Use (6.10) for the inconsistency claim. Prove the consistency claim by contradiction: Assume $\mathcal{E}$ is inconsistent and prove that $\mathcal{E} = \mathcal{W}$.

*6.2.6. This exercise refers to the discussion in the Aside preceding the exercises. Recall that $\beta = \beta_1 \land \cdots \land \beta_m$. Let $\mathcal{E}$ be a set of FOL or propositional logic formulas with $\mathrm{Th}(\mathcal{E}) = \mathcal{E}$.

(a) Let $\mathcal{D}_1 = \{(a : b, b') \to c\}$, let $\mathcal{D}_2 = \{(a : b \land b') \to c\}$, and let $\mathcal{W} = \{ a, ((\neg b) \lor (\neg b')) \}$. Determine the extensions of $(\mathcal{D}_1, \mathcal{W})$ and $(\mathcal{D}_2, \mathcal{W})$. (You needn't prove your results.) Which extensions contain c?

(b) Suppose that the β_i are distinct propositional letters and that $m > 1$. Show that $\neg\beta \in \mathrm{Th}(\neg\beta)$ and that $\neg\beta_i \notin \mathrm{Th}(\neg\beta)$ for all i.

(c) Prove $\neg\beta \notin \mathcal{E}$ implies that $\neg\beta_i \notin \mathcal{E}$ for $1 \leq i \leq m$.
Hint. State and prove the contrapositive.

(d) Prove: If one may use $(\alpha : \beta) \to \gamma$ to include γ in an extension, then one may use $(\alpha : \beta_1, \ldots, \beta_m) \to \gamma$ to do so, but not necessarily conversely.

*6.2.7. Let $\mathcal{E}$ be a consistent extension of $\Delta = (\mathcal{D}, \mathcal{W})$, let $\mathcal{B} \subseteq \mathcal{E}$, and let
$\Delta' = (\mathcal{D}, \mathcal{W} \cup \mathcal{B})$. Prove that $\mathcal{E}$ is an extension of Δ'.
Hint. You can apply Definition 6.4. The hard part is the minimality condition.
You can also use Theorem 6.1, but it has a difficult part, too.

Normal Default Theories

To illustrate the nonmonotonicity of default logic, suppose that $\mathcal{W} = \{a, \neg b\}$
and $\mathcal{D} = \{(a : b) \longrightarrow b\}$. Using only a from $\mathcal{W}$ and the default in $\mathcal{D}$, we arrive
at the unique extension $\text{Th}(\{a, b\})$. Hence b is true. Using all of $\mathcal{W}$ and $\mathcal{D}$, we
arrive at the unique extension $\text{Th}(\{a, \neg b\})$. Hence it was incorrect to conclude
that b is true.

What can be done to avoid looking at all of Δ? The requirement that $\neg \beta \notin$
$\mathcal{E}$ in the definition of an extension means that we have to look at the entire
knowledge base $\mathcal{W}$ and whatever we have derived from defaults. Given the
definition, the need to use all of $\mathcal{W}$ is inevitable. People have created default
theories where not all of $\mathcal{D}$ need be used. Among these, Reiter's "normal"
default theories are the first and best known. For them, what you don't know
about $\mathcal{D}$ won't hurt you. In other words, an extension built using part of $\mathcal{D}$
can be augmented to obtain an extension that uses all of $\mathcal{D}$. Not only that,
every normal default theory has at least one extension.

Definition 6.5   Normal Defaults

A default of the form $(\alpha : \beta) \longrightarrow \beta$ is called a *normal default*. If all defaults
in $\mathcal{D}$ are normal, then $(\mathcal{D}, \mathcal{W})$ is called a *normal default theory*.

The following theorem contains the remarks in the opening paragraph and a
bit more.

Theorem 6.2   Properties of Normal Default Theories

Let $\mathcal{D}' \subseteq \mathcal{D}$ be sets of normal defaults, let $\Delta = (\mathcal{D}, \mathcal{W})$, and let
$\Delta' = (\mathcal{D}', \mathcal{W})$. The following are true.

(a) The theory Δ has at least one extension.

(b) For every extension $\mathcal{E}'$ of Δ', there is an extension $\mathcal{E} \supseteq \mathcal{E}'$ of Δ such
that

$$\text{GD}(\mathcal{E}', \Delta') \subseteq \text{GD}(\mathcal{E}, \Delta) \tag{6.12}$$

where

$$\text{GD}(\mathcal{E}, \Delta) = \left\{ \frac{\alpha : \beta}{\gamma} \in \mathcal{D} \;\middle|\; \alpha \in \mathcal{E} \text{ and } \neg \beta \notin \mathcal{E} \right\}.$$

(This is the general definition of GD( ). For normal theories, $\beta = \gamma$.)

We call the set GD the "generating defaults" set because it, together with $\mathcal{W}$, produces $\mathcal{E}$:

$$\mathcal{E} = \text{Th}\Big(\mathcal{W} \cup \big\{ \gamma \mid ((\alpha : \beta) \to \gamma) \in \text{GD}(\mathcal{E}, \Delta) \big\}\Big). \qquad (6.13)$$

What (6.12) tells us is that $\mathcal{E}$ is generated by a set of defaults that include those used to generate $\mathcal{E}'$. See Reiter's paper [20] for proofs of Theorem 6.2 and (6.13).

Theorem 6.2 allows us to construct a proof method for normal default theories similar to SLD-resolution—the basis of Prolog. Indeed, Prolog can be used to implement such a method.

How does this method work? For simplicity, let's examine a pseudo-Prolog example. Unlike ordinary Prolog, the pseudo-Prolog knowledge base will grow as the proof progresses (and shrink upon backtracking). Suppose we are currently trying to establish **pred(a,X)**. It may be possible to establish it by ordinary FOL from the knowledge base. If not, we look for a default rule $(\alpha : \beta) \to \beta$ where β can be unified with **pred(a,X)**. We must verify that $\neg\beta$ is not provable from the FOL part of the knowledge base and we must establish β. Once this is done, the fact **pred(a,X)**, possibly with X replaced by another term, is added to the FOL knowledge base. During backtracking, it must be removed.

To see why this works, let $\Delta = (\mathcal{D}, \mathcal{W})$ be the original normal default theory. Let D' consist of those defaults $(\alpha : \beta) \to \beta$ for which β was added to the knowledge base. The pseudo-Prolog reasoning produces an extension $\mathcal{E}'$ of $\Delta' = (\mathcal{D}', \mathcal{W})$ in which the original query is true. In fact, $\text{GD}(\mathcal{E}', \Delta') = \mathcal{D}'$—all the defaults in $\mathcal{D}'$ are generating defaults for $\mathcal{E}'$. Theorem 6.2 shows that $\mathcal{E}'$ is contained in some extension $\mathcal{E}$ of Δ.

Exercises

6.2.C. Define a normal default and a normal default theory.

6.2.D. What does it mean to say that what you don't know about $\mathcal{D}$ won't hurt you if $\mathcal{D}$ contains only normal defaults? Give an answer in words and then give an answer in terms of one extension's containing another.

6.2.8. Using the definition of $\Gamma(\mathcal{S})$, prove (6.13) when $\mathcal{E}$ is an extension of Δ. In other words, prove that if $\mathcal{E}$ is an extension as defined in Definition 6.4, then the right side of (6.13) is a formula for $\mathcal{E}$.

6.2.9. How does default logic deal with the four problems listed for nonmonotonic reasoning? (See p. 207.) What about normal default theories?

6.2.10. Sir Galahad, a knight of the Round Table, is facing a dragon which is guarding Lady Millicent, a damsel. The knights of the Round Table are acquainted with the following facts of medieval life:

A dragon will kill a knight if and only if the dragon is dangerous.

Dragons guarding damsels are always green.

Fire-breathing dragons are always dangerous.

Dragons guarding damsels normally breathe fire.

Green dragons normally do not breathe fire.

Green dragons are normally not dangerous.

(a) Write the rules and facts in the notation of a normal default theory.

(b) What are the possible extensions? Explain.

(c) What can you conclude about the dragon's color and ability to breathe fire in the various extensions? About Sir Galahad's getting killed? Why?

Prolog and Default Reasoning

Prolog treats negation as failure; that is, the formula $\neg p(\ldots)$ is considered to be true if the attempt to prove $p(\ldots)$ fails. (You can see the discussion on page 169 for further explanation; however, it's not needed here.) In this section, we'll briefly discuss this from two points of view. First, how Prolog's negation as failure can be viewed as default reasoning. Second, how Prolog's negation as failure might be used to implement default reasoning.

We claim that treating negation of a predicate as failure can be translated into a normal default theory. To do this, we add defaults of the form

$$\frac{:\ \neg P_n(X_1, \ldots, X_n)}{\neg P_n(X_1, \ldots, X_n)}, \tag{6.14}$$

for all n and all n argument predicates P_n. Why does this work? The explanation is closely related to the discussion at the end of the previous section on normal defaults. Try to explain why (6.14) works before reading further.

$*$ $*$ $*$ Stop and think about this! $*$ $*$ $*$

Suppose we want to establish $\neg P(a_1, \ldots, a_n)$. According to (6.14), we can assume it, provided it does not lead to an inconsistency. Recall that a consistent FOL knowledge base together with any formula α is inconsistent if and only if $\neg\alpha$ can be proved from the knowledge base. In our particular case, the fact that $\neg P(a_1, \ldots, a_n)$ is not inconsistent means that $\neg(\neg P(a_1, \ldots, a_n))$ cannot be deduced. Canceling the double negation and putting it all together, (6.14) tells us that we may assume $\neg P_n(a_1, \ldots, a_n)$, provided $P(a_1, \ldots, a_n)$ cannot be deduced from the knowledge base. This is precisely negation as failure.

In the previous section, we saw that the assumptions must be added to the knowledge base of the normal theory before continuing. Prolog doesn't do this—it doesn't add $\neg P_n(a_1, \ldots, a_n)$ to the knowledge base. In general, failing to add assumed formulas can lead to errors. Consider

$$\mathcal{W} = \{a \vee b\} \quad \text{and} \quad \mathcal{D} = \left\{ \frac{: \neg a}{\neg a}, \frac{: \neg b}{\neg b} \right\}.$$

We can assume either $\neg a$ or $\neg b$. Assuming both leads to an inconsistency because $\mathcal{W}$ contains $a \vee b$. Prolog's knowledge base does not allow general FOL statements. In particular, it doesn't allow $a \vee b$, so we can't conclude from this example that what Prolog is doing can produce inconsistencies. Clearly we need a solid grasp of the concepts and tools to decide if Prolog's treatment of negation is consistent with default logic. What Prolog does is okay, but let's not prove it here.

Let $\mathbb{T}(\alpha)$ be the translation of the FOL formula α to Prolog. If β is the negation of a predicate, define the translation of

$$\mathbb{T}\bigl((\alpha : \beta) \to \gamma\bigr) \quad \text{to be} \quad \mathbb{T}(\gamma) \; :- \; \textbf{not} \; \mathbb{T}(\neg\beta)) \, , \; \mathbb{T}(\alpha).$$

*6.3 Other Modifications of Logic

In this section, we'll briefly explore the two other approaches to adapting FOL mentioned earlier in the chapter: circumscription and modal logic. Although defeasible reasoning logically belongs in this section, we'll postpone it until Section 6.6, after we talk about semantic nets.

Circumscription

In default reasoning, the rule $(\alpha : \beta) \rightharpoonup \gamma$ means roughly that it's okay to assume β in proving γ as long as everything remains consistent. We're then told to assume as much as we can without causing inconsistency. This usually leads to more than one possible extension and so we're faced with the problem of what to believe.

John McCarthy's circumscription [14] attempts to deal with this problem by calling attention to the abnormal situation in which $\neg\beta$ is true rather than the normal situation in which β is true. It then seeks to minimize abnormality. If this were all, it would be no different from default reasoning. Circumscription makes two other changes. First, it eliminates default rules, using instead FOL statements like $(\alpha \wedge (\neg p)) \rightarrow \gamma$ where the predicate p indicates abnormality. Second, it insists that the circumscribed interpretation be such that p is not true unless it is true in all possible interpretations. *Interpretation* here refers to the semantics of the FOL language which was defined informally in Definition 3.6.

The approach to circumscription that we've just sketched requires working with the semantics of FOL. Since proof techniques for FOL are based on syntax, it's reasonable to ask if and how the concept of circumscription can be formulated purely syntactically. This can be done using *second-order predicate logic*: that is, FOL with the added feature that predicates can be quantified. Unfortunately, it is impossible to find complete proof methods for second-order predicate logic—they've been proved not to exist. Nevertheless, one could hope for something reasonable: Circumscription is only a special case of second-order logic, and besides, a proof method that sometimes failed might be acceptable in practice.

What is the syntactic form of circumscription? Let $p(X)$ be the predicate whose truth we want to limit and let $\mathcal{A}(p)$ be the conjunction of all the formulas in the knowledge base. (Hence, the knowledge base must be finite.) Add the following statement, where the variable ϕ stands for a predicate:

$$\forall\phi\Big(\big(\mathcal{A}(\phi) \wedge \forall X\big(\phi(X) \rightarrow p(X)\big)\big) \rightarrow \forall X\big(p(X) \rightarrow \phi(X)\big)\Big). \qquad (6.15)$$

This looks rather confusing and intimidating. What does it say? Since the formula inside the $\forall\phi$ is an implication, it only tells us something when the left side is true.

- $\mathcal{A}(\phi)$ tells us that ϕ is a predicate for which all formulas in the knowledge base are true.
- $\forall X\big(\phi(X) \rightarrow p(X)\big)$ tells us that $p(X)$ is true whenever $\phi(X)$ is true.

The conclusion can be rewritten in the equivalent form

$$\forall X \left((\neg\phi(X)) \to (\neg p(X)) \right).$$

In this form, it says that $p(X)$ is false whenever $\phi(X)$ is false.

Why does this ensure that p is not true except when necessary? The left side of (6.15) tells us that ϕ is any predicate that can be used to replace p in the knowledge base and that is true whenever FOL tells us that p is true. Imagine we have such a predicate and we use it in place of p because it works just as well. Might this new predicate do better than p; that is, might it be false sometimes when p is true? No, the right side of (6.15) says that this is impossible.

Thus, by describing properties that p must have when compared to all possible predicates, (6.15) creates limits on when p can be true.

Example 6.6   Applications of Circumscription

Suppose that our knowledge base contains the following FOL formulas:

$$\forall X \left((\text{bird}(X) \wedge \neg p(X)) \to \text{flies}(X) \right) \qquad\qquad 1$$

$$\forall X \left(\text{penguin}(X) \to p(X) \right) \qquad\qquad 2$$

$$\text{bird}(b) \qquad\qquad 3$$

$$\text{bird}(c) \qquad\qquad 4$$

$$\text{penguin}(c) \qquad\qquad 5$$

Using FOL, the only simple predicate conclusions we could reach about b and c are formulas 3–5 and, by using 2 and 5, $p(c)$.

Now suppose we circumscribe the predicate p using (6.15). Then $\mathcal{A}(p)$ is the conjunction of 1–5. By the previous paragraph, $\mathcal{A}(\phi) \models \phi(c)$. Because of this, we should try letting $\phi(X)$ be a predicate that is true when $X = c$ and false otherwise. For this ϕ, $\mathcal{A}(\phi)$ is true. Does $\phi(X)$ imply $p(X)$? Yes, as we now show. We need to check only the situation in which $\phi(X)$ is true. This only happens when $X = c$; but it was noted in the first paragraph that $\mathcal{A}(p) \models p(c)$, and so $p(c)$ is true.

Now for the right side of (6.15). By (6.15), $p(X)$ must be false whenever $\phi(X)$ is false. Since ϕ was chosen to be true if and only if $X = c$, it follows that $p(X)$ must be false except possibly when $X = c$. On the other hand, we know that $\mathcal{A}(p) \models p(c)$, and so the predicate p must be true for c and nowhere else.

As a result of circumscription, we can use formulas 1 and 3 and $\neg p(b)$ to conclude flies(b).

The previous example is a standard example in nonmonotonic reasoning. Here's another.

$$\text{quaker(nixon)} \hspace{12cm} 6$$

$$\text{republican(nixon)} \hspace{11cm} 7$$

$$\forall X \Big(\big(\text{quaker}(X) \wedge (\neg p_1(X))\big) \rightarrow (\text{dove}(X)) \Big) \hspace{5cm} 8$$

$$\forall X \Big(\big(\text{republican}(X) \wedge (\neg p_2(X))\big) \rightarrow \neg\text{dove}(X) \Big) \hspace{4cm} 9$$

You should show:

- By circumscribing p_1 we are able to conclude dove(nixon).
- By circumscribing p_2 we are able to conclude $\neg$dove(nixon).

Which conclusion is correct? It seems best to circumscribe both p_1 and p_2. This presents difficulties that are beyond the scope of the present discussion. ■

Modal and Autoepistemic Logics

Modal logics enrich the language of FOL (or propositional logic) by adding one or more *modal operators*. The operators are added to the language $\mathcal{L}$ and Definition 3.5 (p. 107) is expanded to allow for them. One possible addition to Definition 3.5 is:

(e) If $\heartsuit$ is a modal operator and α is a formula, then $(\heartsuit\alpha)$ is a formula.

If the operator $\heartsuit$ is intended to capture belief, you can think of $\heartsuit\alpha$ as saying "α is believed."

In FOL, manipulation and interpretation of formulas built using connectives are based on our knowledge of propositional logic. This in turn relies on the notions of T and F. Someone defining a modal logic is now faced with the problem of what to do with formulas containing modal operators.

One possible approach is to use the *axiomatic method* of proof (p. 140). In this method, we provide schemes that can be used to reason about formulas. A *scheme* looks like a formula, except that it contains arbitrary formulas. We've already used something like this in defining formulas; for example, if α is a formula then so is $(\neg\alpha)$. In the axiomatic approach to ordinary logic, a common axiom is

$$\big(\alpha \wedge (\alpha \rightarrow \beta)\big) \rightarrow \beta, \quad \text{called } \textit{modus ponens}.$$

In modal logic, the axiom system will depend on what meaning the designer intends to capture by the modal symbols. If we think of $\heartsuit\alpha$ as meaning that a person believes α and if we want to reason about beliefs in a rational manner, we might introduce the following axioms:

- $\big(\heartsuit(\alpha \to \beta)\big) \to \big((\heartsuit\alpha) \to (\heartsuit\beta)\big)$ If one believes that α implies β, then it is true that a belief in α implies a belief in β.

- $(\heartsuit\alpha) \to (\heartsuit\heartsuit\alpha)$ If α is believed, then one believes that α is believed.

- $(\neg\heartsuit\alpha) \to (\heartsuit\neg\heartsuit\alpha)$.   If α is not believed, then one believes that α is not believed.

In FOL, we state an axiom system and then attempt to prove that it is sound and complete. The notions of soundness and completeness require a notion of what formulas mean, that is, a semantics. Without any semantics, an axiom system is rather uncertain: How do we know that it is powerful enough to prove what we want but not so powerful that it allows us to reach incorrect conclusions? Without semantics, such "knowledge" is simply a belief. In some areas, belief suffices. In nonmonotonic reasoning, a paper proposing a reasoning method is usually followed by papers that point out unsuspected deficiencies in the method. Consequently, it's important to provide as much of a theoretical foundation as possible.

Modal logic was introduced by C. I. Lewis in the 1930s because of problems with implication. In FOL, the truth of $\alpha \to \beta$ depends only on the truth of α and β. The "normal" interpretation of implication says that $\alpha \to \beta$ is true only when β "can be deduced" from α. Consider the following two statements.

- If X is a unicorn, then X is a fish.
- If X is a unicorn, then X has hooves.

Since there are no unicorns, both statements are true in FOL. Only the latter statement is true with the normal interpretation of implication. The solution proposed for this problem was to introduce the modal operator $\Box$, which is read "it is necessary that." The statement $\Box\alpha$ means that in all possible worlds (which must be defined), α is true. Then

- $\Box$(if X is a unicorn, then X is a fish) is false because we can imagine worlds with unicorns; but

- $\Box$(if X is a unicorn, then X has hooves) is true because the concept of a unicorn, essentially a horse with a horn, includes the notion that unicorns have hooves.

This is called *strict implication*.

Note: Modal logics need not deal with belief; for example, *temporal logics* deal with situations where time is important.

6.4   Rule Systems

> *A rule base which contains say, 500 rules, might only contain some 10,000 words, perhaps less. Yet textbooks are seldom less than 100,000 words. What has happened to the 90,000 words which are omitted from expert systems? Were they superfluous to the textbook? Is expertise in an expert system more refined, more formalist even that that contained in a textbook? My suspicion is that all the faults which flow from the textbook tradition are simply magnified by the expert systems methodology.*
>
> —Philip Leith (1990)

We've already noted the distinction between $\alpha \rightarrow \beta$ and $(\alpha :) \rightarrow \beta$; namely, the former allows us to conclude $\neg\alpha$ from $\neg\beta$ while the latter does not. This is the essence of rule systems.

Basic Concepts of Monotonic Systems

A basic rule system consists of a language $\mathcal{L}$ consisting of atomic formulas $\mathcal{A}$, which were defined in Definition 3.5, and rules of the form

$$\forall X_1 \ldots \forall X_k \left((\alpha_1 \wedge \cdots \wedge \alpha_n) \rightarrow \beta \right) \tag{6.16}$$

where $n \geq 0$, $\alpha_1, \ldots, \alpha_n, \beta \in \mathcal{A}$, and the only variables in $\alpha_1, \ldots, \alpha_n, \beta$ are $X_1, \ldots, X_k$. Rule systems being used in research and commercial expert systems allow more general rules. Nevertheless, they share some features that distinguish them from FOL. In FOL, quantifiers and connectives provide the foundation for constructing and reasoning about complex formulas. In rule systems, both the construction and the reasoning are severely limited:

- The connectives $\rightarrow$ and $\wedge$ are always present, but their use is limited to the form given in (6.16). For example, we cannot say $p \rightarrow (q \rightarrow r)$. The connective $\neg$ is usually allowed, but its meaning differs from that in FOL.

- Usually the only quantifiers allowed are as specified in (6.16).

- Reasoning with the rule (6.16) is done as follows. If the α_i are known to be true, then β is also true. In particular, when $n = 0$ the rule has the form $\longrightarrow \beta$, and so β is known to be true *a priori*. We are not allowed to conclude $(\neg\beta) \longrightarrow (\neg\alpha)$ from $\alpha \longrightarrow \beta$ as we are in FOL.

One type of proof method in FOL, called the *axiomatic method*, is based on reasoning with rules that are often called axioms. Unlike rule systems, these rules are "metarules" about the language. An example of such a metarule is

For all formulas α and β, if α and $\alpha \rightarrow \beta$ are true, you may conclude that β is true. $\qquad(6.17)$

Note that this is a rule that holds for *all* formulas α and β. In contrast, in a rule such as (6.16), $\alpha_1, \ldots, \alpha_n, \beta$ are specific formulas. Thus an axiom like (6.17) represents an infinitude of rules in a rule system.

Given these limitations of rule systems, why are they used? Here are some arguments in favor of rule systems:

- By limiting the expressiveness of the language, it's possible to find more efficient algorithms than have been found for FOL.

- The rule system interpretation of if–then statements is closer to natural language than the FOL interpretation is. This closeness facilitates the design and maintenance of knowledge bases.

- Unlike FOL, rule systems isolate inconsistencies.

- Rule systems can incorporate procedural statements, which FOL cannot.

- Rule systems can incorporate quantitative features, which FOL cannot.

As a result, rule systems are a popular format for designing expert systems and can be found in nearly every expert system shell. These shells frequently include a numeric approach to uncertain reasoning, usually based on either certainty factors (Chapter 8) or fuzzy logic (Chapter 9).

Forward versus Backward Chaining

You may have noticed that rule systems look a lot like Prolog. In fact, Prolog is a rule system. Suppose that $\alpha_1, \ldots, \alpha_n$ and β are atomic formulas, that is, predicates with terms as arguments. The rule

$$(\alpha_1 \wedge \cdots \wedge \alpha_n) \rightarrow \beta \qquad\qquad (6.18)$$

is used just like a Prolog rule—we can conclude that β is true if the α_i's are true. When rule systems use Prolog's method of depth-first search for answering queries, it's referred to as *backward chaining*. The pros and cons of depth-first search were already discussed in connection with Prolog (p. 122).

Rule systems also use what is known as *forward chaining*. Instead of working backward from the desired conclusions, we work forward from the given facts. In other words, if $\alpha \rightarrow \beta$ is a rule and α is a fact or a conjunction of facts, then the system adds β to its collection of facts. This is called *firing* the rule $\alpha \rightarrow \beta$. Actually, things are not this simple. We should avoid firing rules that only give known facts. If rules were purely declarative, we could simply avoid firing rules twice. Unfortunately, most rule systems allow procedural code and so it makes sense to fire a rule more than once.

Negation

Rule systems often allow negation and deal with it the same way Prolog does—negation as failure. (We discussed this for Prolog on page 169 and in connection with default logic on page 223.)

If a rule system allows negation in the consequence of a rule (head of a clause), new problems arise. As long as the heads of clauses cannot contain negations, there is a consistent FOL interpretation of the knowledge base: Interpret every predicate as always true. This is consistent because $\alpha \rightarrow \beta$ is true in FOL whenever β is true. If negation is allowed in the heads of clauses, we can no longer use such a simple interpretation because any clause whose head has a negation will now have its head interpreted as false. Here's a simple set of three pseudo-Prolog rules that illustrate the problem:

$$p(X) \; :- \; q(X). \qquad \neg p(X) \; :- \; q(X). \qquad q(a).$$

This knowledge base is inconsistent and both p(a) and ¬ p(a) can be deduced.

Example 6.7   Implementing Negation in Prolog

It's actually a relatively simple matter to implement a form of predicate negation in Prolog. First, define a new operator (function), say **neg**, to indicate this negation. Next introduce rules for negation:

$$
\begin{aligned}
&\texttt{negate(neg Literal, Literal) :- !.} \\
&\texttt{negate(Literal, neg Literal).}
\end{aligned}
\tag{6.19}
$$

If **negate** is applied with its first argument instantiated to either an atomic formula or its negative, then **negate** will instantiate its second argument to the negative of the literal. To see this, note that the first clause strips off a **neg** if one is present and prevents the use of the second clause with a cut operator. The second clause simply adds a **neg** regardless of the nature of **Literal**. This is very far from being a full FOL implementation of negation. For example, suppose we are given

$$
\texttt{p(X) :- q(X).} \qquad\qquad \texttt{neg p(a).}
$$

In FOL, we can deduce **neg q(a)**, but the Prolog inference engine will not do so. When faced with the query "**?- neg q(a).**" it looks for clauses whose heads can be unified with **neg q(a)**. Since there are none, the query fails. Basically, all (6.19) does is allow Prolog to use negations in front of predicates and to realize that the negation of a negation is the original predicate. Nevertheless, this can be a useful alternative to negation as failure. ∎

Limiting the Effects of Contradictions

In the various forms of logic, consistency plays an important role; however, we cannot design an algorithm to check that a general FOL knowledge base is consistent. (This is the semidecidability of FOL mentioned earlier.) Unfortunately, if the knowledge base is inconsistent, then *everything* is a consequence of the knowledge base—from a false hypothesis you can deduce anything.

Aside. Why is this so? Recall that $S \models \alpha$ means that, for every interpretation in which S is true, α is also true. Equivalently, we can state this in contrapositive form: For every interpretation in which α is not true, S is also not true. The statement that S contains a contradiction means that there is no interpretation in which S is true. It follows that $S \models \alpha$ in this case, regardless of what the formula α is—it may even use only predicates and terms that don't appear in S!

In contrast, the limited reasoning powers of rule systems allow them to limit the effects of inconsistencies. As a result, errors in one part of the knowledge base will not affect unrelated parts of the knowledge base. This is important in practical applications because a large knowledge base may easily contain contradictions, especially if it is produced by more than one person.

If a reasoning system is to be practical for large applications, then it *must* localize the effect of knowledge-base contradictions.	(6.20)

What about Prolog? Since it's based on FOL, it apparently doesn't localize contradictions. Not so, because, as recently noted, an ordinary Prolog knowledge base is always consistent. Suppose Prolog is extended to allow a form of negation in the heads of clauses, as in (6.19). Inconsistencies are possible. Does Prolog localize them? Yes. Prolog with negation is a rule system rather than a full implementation of FOL negation.

Nonmonotonicity

So far, we've only described statements such as "If α then β." Rule systems might allow nonmonotonic statements such as

$$\text{"If } \alpha \text{ then } \beta \text{ unless } \gamma \text{."} \qquad (6.21)$$

Such a statement could be translated into the default logic rule

$$(\alpha : \neg\gamma) \to \beta.$$

Since negation is regarded as failure when using backward chaining, the rule will apply if α can be deduced and γ cannot be deduced.

When using forward chaining, the interpretation of (6.21) is different: The rule can fire if α is known to be true and γ is not *known* to be true. The condition on γ is much weaker than the requirement that γ cannot be deduced because it doesn't preclude the possibility that γ might be deducible. Why would someone want such a rule? The situation is quite common in procedural code because once something has been done, it's "known." A generic example of such a rule could be phrased "If α then do β unless γ has been done." An example of such a rule is "If going to bed then set the alarm clock unless the alarm clock has been set."

Exercises

6.4.A. What is a rule system? Why is Prolog a rule system?

6.4.B. What are some of the advantages of a (monotonic) rule system over FOL?

6.4.C. Why is a rule system able to isolate the effects of an inconsistency in its knowledge base? Why is this important?

6.4.D. What is backward chaining? forward chaining?

6.4.E. How are nonmonotonic rules interpreted differently in forward and backward chaining?

6.4.F. Why is a Prolog knowledge base always consistent?

6.5   Semantic Nets

> *Information is generally organized into hierarchies with a twofold goal: devise a conceptually clean model of the represented world, and give rise to a compact storage organization which may also enable easier navigation through and search of information.*
>
> —Maurizio Lenzerini, Daniele Nardi, and Maria Simi (1989)

> *Symbolic logic deals primarily with bricks and mortar, semantic nets more with principles of architecture.*
>
> —Fritz Lehmann

While rule systems depart from FOL by limiting the reasoning power to a "more natural interpretation" of if–then statements, semantic nets depart from FOL by focusing on relationships between concepts and representing them graphically. A semantic net is a directed graph (Definition 2.1 (p. 36)) in which each vertex contains information about a concept and each edge describes a relationship between two concepts.

Definition 6.6 Semantic Net

A *binary relation* R on a set V, often just called a *relation*, is simply a binary predicate. Binary relations are sometimes written in infix notation; for example, "$x \; R \; y$" means "$R(x,y)$ is true." A *semantic net* is a set V of concepts and one or more binary relations $R_1, R_2, \ldots$ on V.

The situation is represented graphically by letting V be a set of vertices and joining u and v with an edge labeled R_i if $u \; R_i \; v$. The edges are called *links*.

A common alternative representation replaces each labeled directed edge in the definition by two unlabeled directed edges connecting at a vertex whose label is that of the original edge:

Parse trees, which you have probably seen in connection with computer languages, are related to semantic nets. In fact, semantic nets were originated by researchers in natural language understanding. In this case, links indicate grammatical (i.e., syntactic) and semantic relations between parts of a sentence—agent, object, manner, et cetera. Some researchers still limit the use of the term to such situations while others have adopted the broader definition given above. A variety of semantic nets have been developed for a variety of tasks. We'll explore two types.

- **Frames**: Developed independently of the semantic net concept, frames can be viewed as general data structures for representing semantic nets. Frames use the same method for data representation as object-oriented programming: A frame has "slots" for data, may contain other frames, and may "inherit" properties from other frames. In fact, frames inspired the development of object-oriented languages even though frames were created to facilitate reasoning, not to design programs.

- **Inheritance Systems**: These are semantic nets in which attribute information tends to be "transitive" (defined below). For example, the information that a cat is a feline is a carnivore could be represented by the vertices "cat," "feline," and "carnivore" together with the edges cat $\rightarrow$ feline and feline $\rightarrow$ carnivore, with the edges labeled "is-a." Through inheritance, we can conclude that a cat is a carnivore.

We'll look at some default reasoning with a simple inheritance system. Here's a definition of an inheritance system.

Definition 6.7   Inheritance System

We say that a binary relation $x \prec y$ is a *partial order* if the following conditions hold:

- We never have $x \prec x$.
- Whenever $x \prec y$ and $y \prec z$, we have $x \prec z$.

A partial order is also called a *transitive order*. If the concepts V in a semantic net have a partial order, we say that the net is an *inheritance system* (with respect to the partial order).

In drawing the graph associated with a partial order, we frequently omit edges that can be deduced by transitivity. Thus, if (u, v) and (v, w) are edges, we may omit (u, w). The edge (u, w) is understood to be present by inheritance.

Example 6.8   Some Simple Inheritance Systems

The biological classification of organisms is a traditional example of an inheritance system. The concepts (vertices) are types of organisms and the ordering relation is "is a" as in "a lizard is a reptile" and "a reptile is a vertebrate." We'll denote the ordering by IS-A. Since IS-A is transitive, we have an inheritance system and we can draw conclusions such as "a lizard IS-A vertebrate."

In a tournament, the entities are the players and the ordering according to who wins each match. This ordering is not transitive. For example, it may happen that Alice beats Bob, Bob beats Cathy, and Cathy beats Alice. Transitivity would have required that Alice beat Cathy. Thus this is not an inheritance system.

A semantic net may have more than one relationship (edge). A common type of relationship indicates an attribute as in "a reptile has four limbs." This can be converted to IS-A: "A reptile IS-A four-limbed creature." Because IS-A is transitive, it follows that "a lizard has four limbs." We may still want to distinguish between the two types of edges, since one indicates a classification and the other indicates attributes.

Unfortunately, if we add "a snake IS-A reptile," we would be able to conclude "a snake has four limbs." One solution to this problem is to not declare that reptiles have four limbs, but rather declare it for each type of reptile that has four limbs. Since snakes are practically the only exception, it would be better if we could override the inheritance by declaring that a snake has no limbs. In IS-A terms, "not(a snake IS-A four limbed creature)." Since the ability to override default information is a key attribute of nonmonotonic reasoning, inheritance systems provide a method for implementing such reasoning. ■

Exercises

6.5.A. What is a binary relation?

6.5.B. What are semantic nets? inheritance systems?

6.5.C. Give an example of an inheritance system that is not found in the text.

Frames

A *frame* is something that contains *slots*. An attribute is associated with each slot. For example, the frame for a matrix might contain three slots whose attributes are:

(a) number of rows,

(b) number of columns, and

(c) entries in the matrix.

For any particular matrix, each slot (a) and (b) contains a positive integer and slot (c) contains a list of real numbers. Frames are related to one another in two ways:

- Containment: Usually a slot is simply filled with a value or values, but it may contain a frame; for example, the frame for **automobile** may contain the frame for **internal combustion engine**.

- Specialization: The frame for **computerized classroom** is a specialization of the frame for **classroom** which is a specialization of the frame for **room**. A specialization may inherit slots (**door**, **blackboard**), delete or override others (**desk**), and introduce new ones (**computer terminal**).

A frame is often represented pictorially as a tabular array, as shown in Figure 6.1.

We've seen that defaults play a major role in nonmonotonic reasoning. They first appeared when Minsky introduced frames in 1975. How do frame defaults work? The default value for the **fuel type** of an **automobile**'s **internal combustion engine** subframe may be **gasoline**. This leads to nonmonotonicity since we would normally conclude that the fuel is gasoline, but might override the default value with **diesel** for some engines.

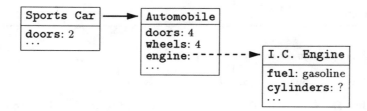

Figure 6.1   Part of an automobile frame system. **I.C.** stands for **Internal Combustion**. A dashed arrow points to a frame that is part of another frame and a solid arrow to a superframe of a frame. The values of attributes are inherited unless the frame contains a value that supersedes the inherited one. A question mark indicates that the value is not specified by the frame.

Example 6.9   Converting Frames to Rule Systems

The description of frames can be phrased in terms of rule systems or default logic. One possibility is to introduce predicates for the various frames and slots. This leads to a mess. The key idea for a cleaner translation is to introduce predicates that express the *relationships* of a frame system. Here's a possible (partial) set of predicates.

predicate	meaning
`frame(X)`	`X` is a frame.
IS-A`(X,Y)`	Frame `Y` is a special case of frame `X`.
`slot(X,S)`	`S` is a slot of `X`.
`default(X,S,V)`	`V` is the default value of slot `S` of `X`.

Some of the relationships in Figure 6.1 are described by the following Prolog facts

> frame(automobile).
>
> slot(automobile, doors).
>
> default(automobile, doors, 4).
>
> slot(automobile, engine).
>
> default(automobile, engine, ic_engine).
>
> IS-A(automobile, sports_car).
>
> default(sports_car, doors, 2).

In addition to the facts, the knowledge base must have some axioms or rules so that conclusions can be drawn. One FOL axiom is the transitivity of IS-A:

$$\forall X \, \forall Y \, \forall Z \Big(\big(\text{IS-A}(X,Y) \wedge \text{IS-A}(Y,Z) \big) \rightarrow \text{IS-A}(X,Z) \Big),$$

which can easily be rewritten as a rule for a rule-based system.

   This quick sketch omits quite a bit. For example, there must be axioms that allow defaults to be inherited and to be overridden. Also, a complete translation of a real frame system would use additional predicates. ■

Manipulating Simple Inheritance Systems

We can take two approaches to the manipulation of inheritance systems. First, we can interpret the system in terms of another reasoning tool, as was illustrated in Example 6.9. Second, we can develop methods directly without converting to another system. We'll explore this approach for semantic nets in which IS-A edges provide a partial order. (Remember that this may involve some rewriting, as in "a reptile IS-A four legged creature" rather than "a reptile has four legs.") The nets we'll consider will be directed graphs with two types of vertices and four types of edges.

- The edges are absolute ($\longrightarrow$) and default ($\cdots\!\!\blacktriangleright$), each of which may be negated. (We'll temporarily ignore default edges.)

- The two types of vertices are constants and predicates. No edge may point toward a constant and only absolute edges and their negations may point away from a constant.

- An edge $r \longrightarrow s$ means $s(r)$ if r is a constant and $\forall X\big(r(X) \rightarrow s(X)\big)$ if r and s are predicates.

- An edge $r \not\longrightarrow s$ means $\neg s(r)$ if r is a constant. If r is not a constant, it means $\forall X\big(r(X) \rightarrow \neg s(X)\big)$, which is equivalent to $\forall X\big(\neg r(X) \lor \neg s(X)\big)$. From this symmetric form, it follows that, when r and s are not constants,

$$s \not\longrightarrow r \text{ and } s \not\longrightarrow r \text{ are equivalent.} \tag{6.22}$$

Although we've translated $\longrightarrow$ and $\not\longrightarrow$ into FOL terms, full FOL reasoning with them is not allowed in inheritance theory. All that's allowed is (6.22) and transitivity based on $\longrightarrow$. This is like reasoning in a rule system. Here are the rules of inference:

(a) **Symmetry**: If $x \not\longrightarrow y$ and x is not a constant, we may add the edge $y \not\longrightarrow x$ to the digraph.

(b) **Positive chain**: If $x_i \longrightarrow x_{i+1}$ for $1 \leq i < k$ are edges in the digraph, we may add the edge $x_1 \longrightarrow x_k$ to the digraph.

(c) **Negative link:** If $x_i \longrightarrow x_{i+1}$ for $1 \leq i < k$, $y_j \longrightarrow y_{j+1}$ for $1 \leq j < m$, and $x_k \not\longrightarrow y_m$ are edges in the digraph, we may add the edge $x_1 \not\longrightarrow y_1$ to the digraph if y_1 is not a constant.

These inference rules can be justified by FOL interpretations. They can also be justified from the inheritance net viewpoint. In that case, we adopt them as axioms (or definitions) based on the intended meaning of the digraph. Which approach is better?

- In either case, we must introduce axioms or definitions. To convert to FOL, we must define the meaning of the digraph in FOL terms and then use the axioms and definitions of FOL.

- In converting to FOL, we make available the machinery of FOL, *but this is trap*. Notice that the allowed rules of inference are limited, so any FOL result must be checked to see that it uses only those rules. If we get sloppy on this, we'll make mistakes.

On balance, it seems preferable to justify (a)–(c) as axioms based on what we want the digraph to mean. Let's do that now.

Symmetry: An edge $x \not\longrightarrow y$ means that no x is a y. Suppose that $y \not\longrightarrow x$ is not true. Then there must be some u that is a y and also an x. This contradicts the meaning of $x \not\longrightarrow y$, and so we've justified (a). Note that this is *not a proof*, because we've shown that (a) should be assumed given our understanding of what $x \not\longrightarrow y$ should mean. This is comparable to the geometry assumption (axiom) that says, based on our understanding of what points and lines mean, precisely one straight line can be drawn through two distinct points.

Positive chain: Since $x \longrightarrow y \longrightarrow z$ means that every x is a y and every y is a z, every x is a z. This forms the basis of (b).

Negative link: We can justify (c) in one step, but it's probably easier to see if we do it in parts. First, $p \longrightarrow q \not\longrightarrow r$ says that every p is a q and no q is an r. It follows that we should conclude p is an r; that is, $p \not\longrightarrow r$. This argument can be expanded to justify (c). Do it!

$$*\quad * \quad * \qquad \text{Stop and think about this!} \qquad * \quad * \quad *$$

Alternatively, we can deduce (c) using the p-q-r result with (a) and (b): For simplicity, assume that neither x_1 nor y_1 is a constant. From (b), $x_1 \longrightarrow x_k$ and $y_1 \longrightarrow y_m$. It follows with $p = x_1$, $q = x_k$, and $r = y_m$ that $x_1 \not\longrightarrow y_m$. By (a), $y_m \not\longrightarrow x_1$. Since $y_1 \not\longrightarrow y_m$, we have $y_1 \not\longrightarrow x_1$. Apply (a).

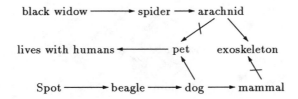

Figure 6.2 An inheritance network without default rules. All the vertices are predicates except for "Spot," which refers to some constant—a particular thing. See Example 6.10 for a discussion of possible inferences.

Example 6.10 Reasoning with $\longrightarrow$ and $\nrightarrow$

Figure 6.2 shows a small inheritance network. Here are some of the conclusions that (a)–(c) allow us to reach.

- Using (a), we can reverse the directions of the two negative edges if we wish.

- Using (b), we can reach conclusions such as "a black widow spider has an exoskeleton," "Spot is a dog," and "beagles live with humans."

- Using (c) with $m = 1$, we can conclude things like "spiders are not pets," "dogs do not have exoskeletons," and "pets are not arachnids." (Actually, the first is not correct in the real world because some people keep tarantulas as pets.)

- Using (c) and the paths from Spot to pet and black widow to pet, we can conclude that "Spot is not a black widow spider." This conclusion can also be reached by using the paths from Spot to mammal and black widow to mammal, where the latter ends with a negative edge from exoskeleton to mammal. ■

We've interpreted "Spot" and "beagle" in Figure 6.2 as a constant and a predicate, respectively; however, they could both be considered constants. Simply let "beagle" be the set of all X that are beagles. For consistency, let "Spot" be the singleton set {Spot}. Then $x \to y$ simply means that $x \subseteq y$. This process of converting a predicate (like beagle) into a constant is called *reification*.

What we've seen so far is just a subset of FOL. Now we come to an important distinction. Like rule systems, inheritance nets localize contradictions.

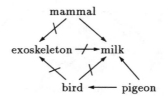

Figure 6.3   Additional edges and vertices for Figure 6.2. See Example 6.10 for a discussion.

Example 6.11   Limiting the Effects of Inconsistencies

Let's include information about milk giving and about birds in Figure 6.2 by adding the edges and vertices

$$\text{mammal} \longrightarrow \text{milk} \quad \text{and} \quad \text{bird} \not\longrightarrow \text{milk},$$

because we believe that mammals give milk and birds do not. Actually, pigeons secrete a cheeselike substance, called pigeon's milk, into their crops and regurgitate it for nestlings. Figure 6.3 shows the new parts that we've added to the digraph in Figure 6.2 plus two facts related to exoskeletons.

The figure contains a contradiction because we can conclude pigeon $\not\longrightarrow$ milk, contradicting pigeon $\longrightarrow$ milk. In FOL, the existence of a contradiction allows us to deduce that anything whatsoever is true. Because of the limited rules of inference in an inheritance net, this does not happen. Since the contradiction involves pigeons, any contradictory result must involve the mention of pigeons. (This is true in FOL as well as inheritance nets.) From the rules (a)–(c), we see that the derivation of a contradiction will involve a path through "pigeon." In other words, a contradiction will involve statements about pigeons or about things that are pigeons, such as Tweety in Tweety $\longrightarrow$ pigeon. Thus, the contradiction does not affect other parts of the digraph. ■

While localization of contradictions is important, we need more—a way to eliminate the contradiction and still preserve the structure of the net. That's the purpose of default edges. In particular, we should replace bird $\not\longrightarrow$ milk with bird $\cdot\!/\!\!\longrightarrow$ milk and introduce appropriate interpretations and rules for reasoning with such edges. The edge $x \cdots\!\!\longrightarrow y$ corresponds to the default rule $x \longrightarrow y$. The edge $x \cdot\!/\!\!\longrightarrow y$ corresponds to the statement "most x are not y" or to the statement "normally, x's are not y's."

What reasoning rules should be used when the nets contain default edges as well as absolute edges? Let's revisit Example 6.1 (p. 209) to get some guidance.

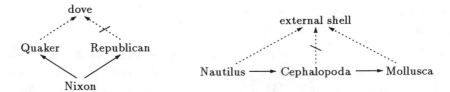

Figure 6.4 The information from Example 6.1 (p. 209) as an inheritance system. See Example 6.12 for a discussion.

Example 6.12   Nixon and Nautiloids Revisited Again

The information about Nixon and the Mollusca phylum that was given in Example 6.1 is represented pictorially in Figure 6.4.

In the Nixon diamond on the left in Figure 6.4, there are two directed paths from Nixon to dove. If we treat the default edges as solid, we could conclude that Nixon is a dove by using the left path and that he is not by using the right path. There are two options. Either accept one of the contradictory conclusions or accept neither.

In the Mollusca digraph, there are various paths that could allow us to conclude that nautiloids and/or cephalopods have or do not have external shells. Unlike the Nixon example, there are correct choices: Nautiloids usually have shells and cephalopods usually do not.   ■

How can the issues raised in Example 6.12 be dealt with in the context of inheritance systems? Let's sketch the basic ideas.

As a first step, pretend that all edges are absolute; that is, replace $\dashrightarrow$ with $\longrightarrow$ and $\cdot/\!\!\!\rightarrow$ with $\not\rightarrow$. Then use the rules of reasoning with $\longrightarrow$ and $\not\rightarrow$. If no contradiction is present, your conclusions should be accepted with two modifications. First, any chain of reasoning that used a default edge should state its conclusion with a default edge. Second $x \cdot/\!\!\!\rightarrow y$ does not imply $y \cdot/\!\!\!\rightarrow x$.

When contradictions arise, as in Figure 6.4, the reasoning process is more complicated. There are three basic principles:

1. If more than one path leads from A to B, the one with the more restrictive default rules is preferred. While this resolves the problem with the nautiloids, it needs to be stated more carefully and precisely to cover all cases.

2. If more than one path leads from A to B and if contradictory paths are not "preempted" by more restrictive ones by the previous principle, we are faced with an ambiguous situation. A conservative reasoner

would avoid drawing any conclusion. This arises in the Nixon example.

3. We should not add edges based on inferences that use default rules. This contrasts with what we said about reasoning with absolute rules (page 239). To see why this principle is needed, suppose we didn't know that nautiloids normally have shells. This means that the edge

$$\text{Nautilus} \dashrightarrow \text{external shell}$$

would be missing in Figure 6.4. Using the modified digraph, we would infer

$$\text{Nautilus} \not\dashrightarrow \text{external shell}. \qquad (6.23)$$

If someone later added

$$\text{Nautilus} \dashrightarrow \text{external shell} \qquad (6.24)$$

because she knew more than we, it would conflict with the inference (6.23). The rule of preferring the more specific default would lead a user to use (6.24) and so reach the correct conclusion. On the other hand, if the inference process had been allowed to add an edge for (6.23) before the fact (6.24) had been added, there would be no way to resolve the conflict.

Exercises

6.5.1. Show that, if edges are added using the positive chain rule (p. 239), then $\longrightarrow$ becomes a transitive relation.

6.5.2. For this exercise, use the FOL meaning of $\not\dashrightarrow$ given before (6.22). Suppose that $x \not\dashrightarrow y$ and $y \not\dashrightarrow z$. Should we be allowed to conclude either $x \longrightarrow z$ or $x \not\dashrightarrow z$? If the answer is "yes," prove it. If not, give a counterexample.

6.5.3. Select some topic and construct an inheritance network for it like those in Figures 6.3 and 6.4. Your network should have at least three of the four types of edges and in the neighborhood of ten vertices. Explain all edges that require specialized knowledge or are not self-evident. List at least two conclusions that can be drawn from your network.

6.6   Defeasible Reasoning

> *A simpler system is probably a better account of actual human*
> *reasoning, provided of course that it otherwise adequately reflects*
> *what we observe in ordinary practice.*
>
> —Donald Nute (1992)

The ideas and pitfalls encountered in previous sections can provide the basis for constructing still other methods for nonmonotonic reasoning. In this section, we'll briefly explore defeasible reasoning. Let's begin with some observations drawn from previous discussions.

- A solid algorithmic foundation is important. This lack was a problem for default logic. On the other hand, the complexity of nonmonotonic reasoning makes a solid theoretical foundation important, too.

- Negation plays an important role in nonmonotonic systems. Thus we must face what negation means—FOL negation, Prolog negation as failure, or something else as in Example 6.7.

- A reasoning system with a direction, like the edges in nets and the arrows in rules, helps localize the effects of contradiction.

- From our study of inheritance systems, it appears that generality is important—among competing default rules, those whose conditions are the most specific are the most believable. For example, a default rule about attributes of cephalopods is more believable than one about attributes of mollusks when we're studying a cephalopod.

Defeasible reasoning, or *defeasible logic*, is the name given to a collection of nonmonotonic reasoning methods that focus on algorithms with a solid theoretical foundation. These methods have a rule-system flavor. Defeasible reasoning methods use generality (and possibly other methods) to defeat chains of reasoning whose rules can have exceptions.

The Syntax of Defeasible Reasoning

A simple defeasible reasoning system is essentially a rule system. We'll define a system with three types of rules: absolute, defeasible, and defeating.

Definition 6.8   The Syntax of Defeasible Reasoning

As in FOL, we are given constants, variables, functions, and predicates. We use them to construct

(a) *terms*, which are the same as in FOL (Definition 3.4 (p. 106));

(b) *atomic formulas*, which are the same as in FOL—predicates whose arguments are terms (Definition 3.5(a));

(c) *literals*, which are atomic formulas and their negations;

(d) conjunctions of literals.

Instead of formulas, we have facts and rules:

(e) A *fact* is a literal that contains no variables.

(f) A *rule* is a string of the form $\forall X \ldots \forall Y (\alpha \Rightarrow l)$ where

- all variables have been universally quantified (so none are free);
- α is a conjunction of one or more literals;
- l is a literal and
- $\Rightarrow$ is one of the symbols $\longrightarrow$, $\cdots\!\!\rightarrow$, and $\cdot/\!\!\rightarrow$.

When we refer to a conjunction of literals, we allow for the possibility that there will be only one literal (and hence no $\wedge$ symbol).

Facts and rules using $\longrightarrow$ should be thought of as ordinary facts and rules; that is, they are always true. We'll call them "absolute." Rules using $\cdots\!\!\rightarrow$ are called "defeasible." They are presumed true unless there is reason to believe otherwise. Rules using $\cdot/\!\!\rightarrow$ are somewhat like the negative of those using $\cdots\!\!\rightarrow$ and are called "defeaters." Facts normally have no variables. The universal quantifiers are often omitted because they're understood to be present. Here's a simple example of each type of rule:

bird(tweety)	Tweety is a bird.	a fact
bird(X) $\longrightarrow$ wb(X)	Birds are warm-blooded.	rule 1
bird(X) $\cdots\!\!\rightarrow$ fly(X)	Birds can normally fly.	rule 2
bird(X) $\wedge$ sick(X) $\cdot/\!\!\rightarrow$ fly(X)	A sick bird may be unable to fly.	rule 3

A defeater $\alpha \cdot/\!\!\rightarrow \beta$ is like the defeasible rule $\alpha \cdots\!\!\rightarrow \neg\beta$. For example, we might say

bird(X) $\wedge$ sick(X) $\cdots\!\!\rightarrow \neg$fly(X)	A sick bird normally can't fly.	rule 4

There is a significant difference between rules 3 and 4. Both prevent us from concluding that a sick bird is likely to be able to fly, but rule 4 does more. It allows us to conclude that a sick bird probably can't fly. If we think that lots of sick birds can fly and lots can't, we'd put rule 3 in our knowledge base. If we think sick birds usually can't fly, we'd put rule 4 in our knowledge base instead.

The Laws of Defeasible Reasoning

What are the laws for manipulating the various rules of defeasible reasoning? Prolog can be used to manipulate rules and facts that contain →. In this manipulation, a negated term is treated like a term. For example, ¬bird(rover) can be thought of as `neg_bird(rover)`, where `neg_bird` is a new predicate. Defeasible rules are used in the same manner, unless there is a good reason for believing that the rule is not applicable. The notion of what constitutes a "good reason" forms the heart of defeasible reasoning and will take some time to explicate fully.

There are two reasons why we would not use a defeasible rule even though its body is satisfied. The first reason is the existence of a defeater. For example, if Tweety is sick, rule 2 and the fact `bird(tweety)` suggest that Tweety can fly, but rule 3 defeats this. The second reason is that we can also derive the negation of the rule's conclusion (head in Prolog). This brings us to the situation discussed earlier: Of two defeasible rules, the one with the more specific preconditions is favored.

How can the reasons in the preceding paragraphs be made explicit enough to be implemented in an algorithm? Four things require explication:

- how to negate a literal (copied from FOL),
- how to use facts and absolute rules (copied from monotonic rule systems),
- how to use defeasible rules and facts, and
- how to "defeat" a possible defeasible derivation.

Except for consistency and the notion of defeating, the rules for doing these things are fairly straightforward. To simplify the discussion, we'll assume consistency of the facts and absolute rules:

> **Consistency Assumption:** If the facts and absolute rules are used to form an FOL knowledge base, with ¬ interpreted as negation and → as implication, the FOL knowledge base is consistent. $\qquad$ (6.25)

Since descriptions are simpler without quantifiers, we'll assume that any rule in the knowledge base having universal quantifiers has been replaced by a (possibly infinite) set of rules in which all possible variable-free terms have been substituted for the variables.

Negation: If β is $p(\ldots)$ for some predicate p, then $\neg\beta$ is $\neg p(\ldots)$. If β is $\neg p(\ldots)$ for some predicate p, then $\neg\beta$ is $p(\ldots)$. The latter is simply the FOL fact that $\neg\neg\alpha$ is equivalent to α. (It's necessary to state this explicitly because FOL is not used as a basis for defeasible reasoning.)

Absolute Derivation: Call a literal β *absolutely derivable* if the fact β is in the knowledge base or if the rule $\alpha \longrightarrow \beta$ is in the knowledge base and α is a conjunction of absolutely derivable literals. This recursive definition is essentially a statement of the way monotonic rule systems operate.

Defeasible Derivation: This definition is more complicated than that of absolute derivation because it's possible to contradict a conclusion—a process called "defeating."

Call the literal β *defeasibly derivable* if one of the following is true:

- β is absolutely derivable;
- $\alpha \longrightarrow \beta$ is in the knowledge base, α is a conjunction of defeasibly derivable literals, and $\alpha \longrightarrow \beta$ is not defeated by method (a) or (b) below; or
- $\alpha \dashrightarrow \beta$ is in the knowledge base, $\alpha \dashrightarrow \beta$ is not defeated, and α is a conjunction of defeasibly derivable literals.

Defeating: Here are the ideas behind defeating. An absolute rule can be defeated by a fact or a contrary absolute rule. These also defeat a defeasible rule. In addition, a defeasible rule $\alpha \dashrightarrow \beta$ can be defeated by a defeasible rule $\gamma \dashrightarrow \neg\beta$ or a defeater $\gamma \nrightarrow \beta$ unless the hypothesis α is "more restrictive" than the hypothesis γ. Stating what "more restrictive" means is a bit tricky. It's contained in (ii) below.

Now for the explicit formal definition. Let α be a conjunction of the defeasibly derivable literals $\alpha_1, \ldots, \alpha_n$. The rule $\alpha \dashrightarrow \beta$ is *defeated* by all three of the following and the rule $\alpha \longrightarrow \beta$ is defeated by (a) and (b).

(a) $\neg\beta$ is fact.

(b) $\gamma \longrightarrow \neg\beta$ is in the knowledge base and γ is a conjunction of defeasibly derivable literals.

(c) γ is a conjunction of defeasibly derivable literals γ_j such that

   (i) either $\gamma \dashrightarrow \neg\beta$ or $\gamma \nrightarrow \beta$ is in the knowledge base and

   (ii) at least one γ_j is not absolutely derivable from the *rules* in the knowledge base and the additional facts $\alpha_1, \ldots, \alpha_n$.

Condition (ii) is somewhat confusing. Let's unwind it. Suppose that the hypothesis γ were less restrictive than the hypothesis α. Then we should be able to derive γ from α and our general knowledge. (General knowledge can be thought of as rules. Facts are normally specific knowledge.) Condition (ii) simply states that this doesn't happen, where "derive" is interpreted as "derive absolutely." Since this doesn't happen, the rule in (i) is at least as believable as $\alpha \dashrightarrow \beta$. The next example should help clarify the application of the definition and also help explain why it's appropriate.

Example 6.13   Nixon and the Mollusks One More Time

The rules for the Nixon diamond can be read from Figure 6.4 (p. 243). Perhaps the claim that most Republicans are not doves is too strong. Let's weaken it a bit. The resulting rules for our knowledge base are

quaker(X) ⋯▸ dove(X).	rule 1
republican(X) ·/▸ dove(X).	rule 2
quaker(nixon).	fact 1
republican(nixon).	fact 2

There's no rule that would let us conclude absolutely that Nixon is not a dove; however, rule 1 and fact 1 potentially let us conclude defeasibly that he is a dove. Is rule 1 defeated? The only rule that might defeat it is rule 2 since it satisfies (i) in the definition of defeating. It also satisfies (ii) because $\alpha = \alpha_1 = $ quaker(nixon), $\gamma = \gamma_1 = $ republican(nixon), and γ_1 is not absolutely derivable from α_1 and the rules in the knowledge base—we also need fact 2.

The mollusk rules from Figure 6.4 are

nautiloid(X) ⟶ cephalopod(X)	rule 1
cephalopod(X) ⟶ mollusk(X)	rule 2
nautiloid(X) ⋯▸ external-shell(X)	rule 3
cephalopod(X) ⋯▸ ¬external-shell(X)	rule 4
mollusk(X) ⋯▸ external-shell(X)	rule 5

Suppose we add the fact cephalopod(octopus). We can now conclude absolutely that an octopus is a mollusk. What can we say about its shell? The only rules that are useful in this connection are 4 and 5. Rule 5 does not defeat rule 4 because we can use rule 2 to derive the hypothesis mollusk(octopus) from the hypothesis cephalopod(octopus). Thus, we can use rule 4 to conclude defeasibly that an octopus does not have an external shell. On the other hand, you should be able to see that rule 4 defeats rule 5. Hence we cannot use rule 5. ∎

Concluding Remarks

We've seen why defeasible reasoning is not simply an extension of FOL. Is it a rule system? That depends on your definition. Some people think of rule systems as having relatively simple methods for manipulating the rules—something on the order of complexity of simple Prolog. Others allow more complex manipulations. Whether you call defeasible reasoning a rule system depends on your position on complexity.

How reasonable are the assumptions behind defeasible logic? Probably the most controversial issue is the way in which defeat is defined. In deciding whether a rule $\alpha \dashrightarrow \beta$ is defeated, we do not concern ourselves with how $\alpha = \alpha_1 \wedge \cdots \wedge \alpha_n$ was established. If we take into account the rules that were used to establish the α_i, we might be able to show that some rule $\gamma \dashrightarrow \beta$ defeats our rule. Essentially, this means that once we've decided to accept something, we don't question it every time we use it. Defenders argue that human reasoning seems to work this way and that it gives a computationally practical algorithm. Attackers argue that humans do eventually reexamine accepted "facts" and also point out that the goal is to develop a solid reasoning system, not to imitate a human being.

Aside. I've cheated a bit in this section. First, there's the consistency assumption (6.25). As I noted at the time, it's not essential. The ability of rule systems to isolate the effects of inconsistencies can be utilized to eliminate it. In doing this, you must be more careful with the definitions of derivability and defeat. Second, I chose the method of presentation I thought would best convey the motivation behind defeasible reasoning. Unfortunately, this is probably not the best approach for proving properties of defeasible reasoning. As a result, the treatment you'll find in the literature differs from the one given here.

Exercises

In the following exercises, *defeasible reasoning* always refers to the definition introduced in this section.

6.6.A. What is the syntax of defeasible reasoning?

6.6.B. When is something absolutely derivable?

6.6.C. Without explaining defeating, explain when something is defeasibly derivable.

6.6.D. What are the basic ideas behind defeating? How is this related to a major criticism of defeasible logic?

Each of the following exercises has four parts:

(a) Translate the prose into defeasible logic syntax.

(b) Use defeasible reasoning to answer the queries.

(c) Translate the prose into default logic syntax using normal defaults.

(d) Find all extensions.

6.6.1. (*Answer follows*) Here are the rules:

> Mammals usually don't fly.   Bats normally fly.
> Vampires are mammals.   Bats are mammals.

Given that Dracula is a vampire, what should we believe about Dracula's ability to fly? Given that Tea Tray is a bat, should we believe that Tea Tray flies?

6.6.2. Here are the rules:

1. College students are adults.
2. Adults normally work full time.
3. If someone works full time, he/she works.
4. College students often work.
5. College students normally do not work full time.
6. College students over 30 often work full time.

For each of the following should we believe that they work? work full time? don't work full time? Why? What happens if we remove rule 4?

> Mary, who is an adult.
> John, who is a college student.
> Bill, who is a 35-year-old college student.
> Karen, who is a 25-year-old college student.

6.6.3. These rules are taken from [15]:

1. Utah residents are usually Mormons.
2. Mormons usually do not drink beer.
3. BYU alumni living in Utah are usually football fans.
4. Football fans usually drink beer.

Should we believe that BYU alumni living in Utah drink beer? Do you agree with the conclusion? Why or why not?

6.6.4. The statements for this exercise are the same those as for Exercise 6.2.10 (p. 223).

(a) Write the statements in defeasible reasoning notation.

(b) What can you conclude about the dragon's color and ability to breathe fire? Why?

(c) What can you conclude about Sir Galahad's getting killed? Why?

Answers

6.6.1. Here are the translations of the rules in the same order given in the problem. Universal quantifiers have been omitted. The meanings of the predicates should be obvious.

$$m(X) \cdot/\!\!\!\!\rightarrow f(X) \qquad b(X) \cdots\!\!\rightarrow f(X)$$
$$v(X) \rightarrow m(X) \qquad b(X) \rightarrow m(X)$$

An alternative version of the first rule is $m(X) \cdots\!\!\rightarrow \neg f(X)$. Given $v(d)$, we can certainly conclude $m(X)$. With the rules as stated, that is all we can conclude. With the alternative first rule, we can conclude that Dracula presumably doesn't fly. Given $b(tt)$, we can certainly conclude $m(tt)$. From $b(tt)$, we can defeasibly derive $f(tt)$, unless the rule is defeated. Since only the first rule could defeat the conclusion, we need only look at case (c) in the definition of defeating. Since we are trying to defeat $b(X) \cdots\!\!\rightarrow f(X)$ with $m(X) \cdot/\!\!\!\!\rightarrow f(X)$, $\alpha = b(X)$ and $\gamma = m(X)$. Since γ is derivable from $b(X)$ and the rules, our attempt at defeat fails. In other words, we can conclude that Tea Tray can presumably fly.

For default reasoning, we can construct normal default rules for the two default situations:

$$\mathcal{D} = \left\{ \frac{m(X) : \neg f(X)}{\neg f(X)}, \ \frac{b(X) : f(X)}{f(X)} \right\}.$$

The remaining rules together with one of the facts constitute either $\mathcal{W}_d$ or $\mathcal{W}_t$, say. For $\mathcal{W}_d$, there is a unique extension obtained by using the first rule in $\mathcal{D}$ to conclude $\neg f(d)$ and then taking the closure using Th. This contrasts with the defeasible case where the two versions for the first rule gave two different results. For $\mathcal{W}_t$, we can use the second rule in $\mathcal{D}$ to add $f(t)$ and the take the closure using Th. Another extension can be obtained by first concluding $m(t)$ and then proceeding as we did for Dracula. In contrast to the defeasible case, we have two extensions. Whether Tea Tray can fly or not depends on the extension.

Notes

*Current logic will interact with AI and computer science,
it will be forced to evolve into its next evolutionary stage,
and the new logic will do wonders for AI and computing.*

—Dov M. Gabbay (1992)

Logicians and philosophers have studied *nonstandard logics* for some time.
(Any logic not contained in FOL is called nonstandard.) Much of their research has been oriented toward logics with different levels of truth (as
in modal logics, which we barely touched on). AI researchers are relative newcomers. But after McCarthy introduced his idea of circumscription,
they entered the field in force and began to dominate it. Their orientation was toward nonmonotonic logics, which appear to be different from
logics with multiple truth levels; however, the two can be reconciled. By
1980, many AI researchers accepted the idea that some form of nonmonotonic reasoning would be essential to achieve the goals of AI. The field is in
flux.

For a mathematical treatment of a wide variety of topics, see the handbook [6]. For various aspects of qualitative nonmonotonic reasoning, see the
articles collected and introduced by Ginsberg [8]. Brewka [2] and Ramsay [19]
have written monographs on nonmonotonic logic. Pollock [17] discusses aspects of nonmonotonic reasoning in connection with his research on self-aware
machines. For a more detailed discussion of Pollock's defeasible reasoning,
see [18]. The results on default logic in the text are taken from Reiter's
original paper [20] which is quite readable and contains many more results
and examples as well as a discussion of algorithms for normal default theories. Marek and Truszczyński [13] treat default and autoepistemic logics and
their relationships to each other. You may find Geffner's thesis [7] interesting.

Several AI texts discuss rule systems and some types of semantic nets,
for example, [9], [25], [1] and [21]. Brewka [2] relates rule systems to nonmonotonic logic. In the other direction, FOL can be done pictorially. See [22]
for an introduction. The broadest in-depth introduction to semantic nets is
provided by the collection [11] edited by Lehmann. A more limited introduction is given by Carpenter and Thomason [3]. Some aspects of inheritance systems and other semantic nets are discussed in [12] and [24], respectively.

My material on defeasible reasoning was based mainly on [4] and [15].
The former is an introductory text containing an implementation of defeasible
reasoning in Prolog. The version is somewhat different from that presented

here—the rule $\alpha \longrightarrow \beta$ is not defeated by (b). Nute's article [15] provides the theoretical background that I omitted. Since it contains logic techniques that I haven't discussed, it may be difficult. Pollock [18] presents a more expressive defeasible reasoning language than Nute does, but is then faced with potentially slower algorithms.

Most of the methods proposed for nonmonotonic reasoning face computational problems, at least as far as worst-case behavior is concerned. Some approaches are undecidable; others are NP-hard. Papadimitriou and Sideri [16] discuss the computational complexity of finding extensions in default logic. Selman and Levesque [23] discuss the computational complexity of inheritance systems. How important are these computational problems? If we're willing to accept the possibility of occasional failure, *worst-case* behavior may not be too important. Unfortunately, typical behavior is often much more difficult to determine. Regardless, the results on computational complexity indicate that nonmonotonic methods must be used with care.

Biographical Sketch

Raymond Reiter (1939–)
Born in Toronto, he received his B.A. and M.A. at the University of Toronto, where he now teaches. In 1967 he received his Ph.D. in computer science from the University of Michigan at Ann Arbor. His research areas are mechanical theorem proving, the theory of computation, and their connections with artificial intelligence. In 1978, Reiter stimulated research on the closed-world assumption by defining it for databases and establishing some of its properties. It seldom happens that an idea is born mature—later researchers usually improve and extend it. Reiter's important 1980 paper on default logic is an exception. In fact, the default logic section of this chapter is little more than an abridged version of Reiter's paper.

References

1. T. J. M. Bench-Capon, *Knowledge Representation. An Approach to Artificial Intelligence*, Academic Press, London (1990).

2. G. Brewka, *Nonmonotonic Reasoning: Logical Foundations of Commonsense*, Cambridge University Press, Cambridge, Great Britain (1991).

3. B. Carpenter and R. Thomason, Inheritance theory and path-based reasoning: An introduction. In [10] 309–343.

4. M. A. Covington, D. Nute, and A. Vellino, *Prolog Programming in Depth*, 2d ed., in preparation.

5. L. Fariñas del Cerro and M. Penttonen (eds.), *Intensional Logics for Programming*, Clarendon Press, Oxford (1992).

6. D. M. Gabbay, C. J. Hogger, and J. A. Robinson, *Handbook of Logic in Artificial Intelligence and Logic Programming*, Clarendon Press, Oxford. Six volumes are planned: Vol. 1 *Logical Foundations* (1993), Vol. 2 *Deduction Methodologies* (1994), Vol. 3 *Nonmonotonic Reasoning and Uncertain Reasoning* (1994), Vol. 4 *Epistemic and Temporal Reasoning*, Vol. 5 *Logic Programming I*, and Vol. 6 *Logic Programming II*.

7. H. Geffner, *Default Reasoning: Causal and Conditional Theories*, MIT Press, Cambridge, MA (1992).

8. M. L. Ginsberg, *Readings in Nonmonotonic Reasoning*, Morgan Kaufmann, San Mateo, CA (1987).

9. M. Ginsberg, *Essentials of Artificial Intelligence*, Morgan Kaufmann, San Mateo, CA (1993).

10. H. E. Kyburg, Jr., R. P. Loui, and G. N. Carlson (eds.), *Knowledge Representation and Defeasible Reasoning*, Kluwer Academic Publishers, Dordrecht (1990).

11. F. Lehmann (ed.), *Semantic Networks in Artificial Intelligence*. A special issue of the journal *Computers and Mathematics with Applications* **23** nos. 2–9 (1992).

12. M. Lenzerini, D. Nardi, and M. Simi (eds.), *Inheritance Hierarchies in Knowledge Representation and Programming Languages*, John Wiley and Sons, New York (1989). Beware of misprints.

13. V. W. Marek and M. Truszczyński, *Nonmontonic Logic. Context-Dependent Reasoning*, Springer-Verlag, Berlin (1993).

14. J. McCarthy, Circumscription—a form of non-monotonic reasoning, *Artificial Intelligence* **13** (1980) 27–39, 171–172. Reprinted in [8] 145–152.

15. D. Nute, Basic defeasible logic. In [5] 125–154.

16. C. H. Papadimitriou and M. Sideri, On finding extensions in default logic, *Proceedings of the 1992 Congress on Database Theory*, Springer-Verlag, Berlin (1992) 276–281.

17. J. L. Pollock, *How to Build a Person*, MIT Press, Cambridge, MA (1989).

18. J. L. Pollock, How to reason defeasibly, *Artificial Intelligence* **57** (1992) 1–42.

19. A. Ramsay, *Formal Methods in Artificial Intelligence*, Cambridge University Press, Cambridge, Great Britain (1988).

20. R. Reiter, A logic for default reasoning, *Artificial Intelligence* **13** (1980) 81–132. Reprinted in [8] 68–93.

21. E. Rich and K. Knight, *Artificial Intelligence*, 2d ed., McGraw-Hill, New York (1991).

22. D. D. Roberts, The existential graphs. In [11] 639–663.

23. B. Selman and H. J. Levesque, The tractability of path-based inheritance. In [12] 83–95.

24. J. F. Sowa (ed.), *Principles of Semantic Networks*, Morgan Kaufmann, San Mateo, CA (1991).

25. P. H. Winston, *Artificial Intelligence*, 3d ed., Addison-Wesley, Reading, MA (1992).

7

Probability Theory

*The theory of probabilities is at bottom nothing but common
sense reduced to calculus.*
—Pierre Simon de Laplace (ca 1820)

*Probabilities are numbers, and number crunching is exactly what
AI was supposed* not *to be.*
—Glenn Shafer and Judea Pearl (1990)

Introduction

In this chapter we'll look at some basic probability theory that's needed for
AI work. Formally, probability theory is a part of real analysis, whose study
presupposes a thorough grounding in calculus. Fortunately, we don't need all
this power; in fact, for this chapter we don't even need calculus.

The next section introduces the core concept—the finite probability
space—and some related ideas. Then we examine conditional probability and
independence—tools that are essential for incorporating new information in a
probability-based expert system. The next chapter applies these tools to the
study of Bayesian networks.

Prerequisites: No previous chapters are required.

Used in: Much of the following material relies on this chapter.

7.1   Finite Probability Spaces

This section introduces some basic concepts—finite probability spaces, compound events, and random variables—and associated notation. Be sure that you understand the concepts and are comfortable with the notation before studying later sections.

The Notion of Probability

> *I consider the word probability as meaning the state of mind with respect to an assertion, a coming event, or any other matter on which absolute knowledge does not exist.*
>
> —Augustus DeMorgan (1838)

> *Illustratative examples will be provided to explain the empirical background ..., but the theory itself will be of a mathematical character. We shall no more attempt to explain the "true meaning" of probability than the modern physicist dwells on the "real meaning" of mass and energy or the geometer discusses the nature of a point.*
>
> —William Feller (1957)

When applying mathematics to physical problems, we usually make assumptions that are untenable—matter is continuous, lines exist, et cetera. We've become so inured to these assumptions that we tend to ignore them. Probability theory is no exception. Like all mathematical theories, probability theory is not about the physical world. Rather, it is a theory about an abstract situation based on mathematical concepts. How it can and should be applied to the real world in any particular situation is open to discussion and even argument, which persists to this day. These arguments are not about the *mathematical* foundations of the probability *theory*. In the world of mathematics, which we are pursuing in this chapter, such issues can be put aside. This is a common situation: Einstein said, "As far as the laws of mathematics refer to reality, they are not certain; and as far as they are certain, they do not refer to reality."

Let's look at a concrete example of probability theory and the real world. What does a statement such as

$$\text{"The probability of heads when this coin is tossed is } \tfrac{1}{2}\text{"} \qquad (7.1)$$

mean? It many seem obvious on the face of it; however, this is not the case.

To begin with, there is no specification as to how the coin is to be tossed—what if it were tossed by a magician? On the other hand, specifying precisely how the coin is to be tossed may make tossing it correctly impossible and may eliminate the element of randomness in the tossing. Let's decide not to worry about this problem.

Having met defeat at the hands of randomness, let's attack another adversary. What does it mean to say that the probability is $\frac{1}{2}$? Presumably it means that if the coin is tossed many times, the ratio of heads to total tosses will approach $\frac{1}{2}$. This is rather vague; after all, we could be extremely unlucky and get no heads at all. There doesn't seem to be an operational definition of "the probability is $\frac{1}{2}$." To avoid a philosophical morass, let's assume that we understand what "the probability is $\frac{1}{2}$" means.

Let's try another problem: How can we possibly *know* that the probability is $\frac{1}{2}$? The answer is simple—we can't. What we can do is convince ourselves after some thought and experimentation that this is probably the case. (A careful person might say that it seems fairly certain that the probability is close to $\frac{1}{2}$.) In the end, our "probability" is simply a matter of *belief based on our (supposed) knowledge of the situation*. What we think is correct can easily be wrong. For example, if a coin is spun like a top on a smooth surface, what is the probability it will land with heads up? Not $\frac{1}{2}$! When a coin's two faces are unevenly balanced, the heavier face tends to fall downward—try it with a new penny. Even if we carry out some experiments, we may be wrong because when we assign a probability we are trying to *extrapolate* based on past experience and learning.

So what does the previous discussion tell us? Here's a summary:

> In applications, a probability reflects our (*informed?*) belief about what we *expect* to occur in situations that we *think* are similar to the one being considered. (This statement is intentionally vague and provides no information about how to estimate probabilities in the physical world.) (7.2)

Suppose we change the question about (7.1) slightly:

I've just tossed a coin, looked at the result, and hidden it from you. What is the probability that the coin I just tossed came up heads?"

Since you haven't seen the coin, you'll probably answer "one-half." On the other hand, I would answer "zero" because I've seen the coin and it came up tails. Since we're both talking about the same event, we can't both be right—can we?

Yes, we can. The probability is based on information and your information differs from mine: *I know how the coin landed*, but *you only know that the toss was fair*. Thus I'm looking at one particular toss, but you're looking at some sort of average behavior. In other words, our different answers reflect our state of knowledge about the situation. It's common to overlook this point and find it confusing that different individuals can obtain different probabilities when their knowledge of the situation differs.

The idea that probabilities change with changing knowledge is crucial for expert systems. An expert system can use probability to record its knowledge of a situation and to reflect changing expectations as the system's knowledge about a particular situation increases through interaction with a user. Therefore, *it is essential to understand how to calculate such changes.*

Probability Spaces

The mathematical concept of a finite probability space is quite simple. Nevertheless, it seems to capture the essence of what we mean when we speak of probability.

Definition 7.1   Finite Probability Space

Let $\mathcal{E}$ be a finite set and let Pr be a function from $\mathcal{E}$ to the nonnegative real numbers ($\mathbb{R}^+$) such that

$$\sum_{e \in \mathcal{E}} \Pr(e) = 1.$$

We refer to $\mathcal{E}$ as the *event set* and $\Pr(e)$ as the *probability that the event e occurs* or, more briefly, the *probability* of e. The pair $(\mathcal{E}, \Pr)$ is called a *(finite) probability space*. The elements of $\mathcal{E}$ are called *simple events* or *elementary events*. We call Pr the *probability distribution*.

Intuitively, $\mathcal{E}$ is the collection of things that can happen. In any particular situation, exactly one of these things will happen (or occur). The value $\Pr(e)$ is our estimate of the fraction of the time that e will occur in the type of situation we're studying.

Example 7.1   Heads and Tails: Some Simple Probability Spaces

Let $\mathcal{E}_1 = \{H, T\}$ and $\Pr_1(H) = \Pr_1(T) = \frac{1}{2}$. This can be interpreted as the probability space associated with a single toss of a fair coin. If we replace $\Pr_1$ with $\Pr_1'$ and set $\Pr_1'(H) = 0$ and $\Pr_1'(T) = 1$, we obtain a new probability space—same event set, new probability distribution. We can interpret the new space as the probability space of the toss of a coin where it has been observed that the coin came up tails.

Let $\mathcal{E}_k$ be all 2^k k-long strings of H's and T's and let $\Pr_k(e) = 2^{-k}$ for each $e \in \mathcal{E}_k$. We can interpret this as the outcome of a sequence of k tosses of a fair coin. Suppose we are only interested in the total number of heads in a sequence of k tosses. One approach is to define a new probability space with $\mathcal{E}_k' = \{0, 1, \ldots, k\}$ and $\Pr_k'(e')$ equal to 2^{-k} times the number of sequences in $e \in \mathcal{E}_k$ that contain exactly e' heads. A more flexible approach is to keep the same probability space $(\mathcal{E}_k, \mathrm{Pr}_k)$ and introduce a function on $\mathcal{E}_k$ whose value at $e \in \mathcal{E}_k$ is the number of heads in e. Such functions are, unfortunately, called random *variables*. (We'll discuss them soon.) Sets of elementary events, called compound events, can also be used to keep track of the number of heads. ■

Example 7.2   Selecting an Elementary Event at Random

In computer simulations of situations that are described by probability theory, it is often important to select elementary events at random. This is done by using so-called random-number generators.

What is a random-number generator? Suppose that the computer uses m bits to represent the digits of a number in floating point. (This doesn't include exponent and sign bits.) A good random-number generator will produce each of the 2^m numbers

$$\frac{0}{2^m}, \frac{1}{2^m}, \frac{2}{2^m}, \frac{3}{2^m}, \ldots, \frac{2^m - 2}{2^m}, \frac{2^m - 1}{2^m}$$

nearly equally often when it's run many times.

How do we use a random-number generator to choose elementary events from $(\mathcal{E}, \mathrm{Pr})$? Let the elements of $\mathcal{E}$ be $e_1, \ldots, e_n$. Define

$$s_i = \begin{cases} 0, & \text{if } i = 0, \\ s_{i-1} + \mathrm{Pr}(e_i), & \text{if } 1 \le i \le n. \end{cases}$$

Note that $s_n = 1$ since it is the sum of the probabilities of all the e_i. Let x be the number produced by the random-number generator. If $s_{i-1} \le x < s_i$, choose e_i. I'll leave it as an exercise for you to show that the frequency with which e_i is chosen is approximately $\mathrm{Pr}(e_i)$. ■

Definition 7.2   Compound Events

Let $(\mathcal{E}, \mathrm{Pr})$ be a probability space. A subset of the finite set $\mathcal{E}$ is called a *compound event*. For a compound event $A \subseteq \mathcal{E}$, define

$$\mathrm{Pr}(A) = \sum_{e \in A} \mathrm{Pr}(e).$$

(By the mathematical convention that the value of a summation with no terms is zero, $\mathrm{Pr}(\emptyset) = 0$, where $\emptyset$ is the empty set.) We identify e with the subset $\{e\}$. If e occurs and $e \in A$, we say that A occurs. The compound event consisting of that subset of elements of $\mathcal{E}$ not in A is called the *negation*, or *complement*, of A and is denoted by $\neg A$ or $\overline{A}$ or $\mathcal{E} - A$.

Intuitively, the elements of $\mathcal{E}$, the simple events, are such that exactly one of them must occur and $\Pr(A)$ is the probability that the simple event which does occur is an element of A. Unlike simple events, more than one compound event can occur. In particular, if A occurs and $A \subseteq B \subseteq \mathcal{E}$, then B also occurs.

Aside. Although we write $\Pr(e)$ and $\Pr(A)$ using the same function symbol, Pr, these are different functions. The domain of the former is $\mathcal{E}$ while the domain of the latter is all subsets of $\mathcal{E}$. Both functions are denoted by the symbol Pr to emphasize that they both arise from $(\mathcal{E}, \Pr)$. When you get to random variables, you'll see yet another usage for Pr. (In computer language terminology, another "overloading" of Pr.)

Example 7.3   Heads and Tails: Compound Events

As mentioned above, we can define various compound events in $\mathcal{E}_k$ for keeping track of the number of heads—simply let A_i be the set of $e \in \mathcal{E}_k$ that contain exactly i H's. For example, $A_0 = \{TT \cdots T\}$, $A_k = \{HH \cdots H\}$,

$$A_1 = \{HT \cdots T, THT \cdots T, \ldots, T \cdots TH\}$$

and $A_i = \emptyset$ whenever $i > k$. ∎

Definition 7.3   Terminology for "and" and "or"

It's common to think of $\Pr(\text{statement})$ as the probability that "statement" is true. Since we think of $\Pr(A)$ as the probability that A occurs, and since $\wedge$ stands for "and," we read $\Pr(A \wedge B)$ as the probability that both A and B occur. Similarly, $\Pr(A \vee B)$ denotes the probability that at least one of A and B occurs.

This notation is suggestive of propositional logic. The following example shows that the notation is more than just suggestive.

Example 7.4   Logical Connectives in Probability

If A and B were statements rather than sets, it would be natural to think of $A \wedge B$ in $\Pr(A \wedge B)$ as a logical "and" since the symbol is the same. Is this double meaning just an unfortunate overloading of the operator $\wedge$? No, there's more to it.

A compound event often has a simple verbal description such as

A: The first three tosses of the coin were heads.
B: The coin is fake—it has heads on both sides.
C: There will be rain next week.
D: There will be sunshine next week.

The operations $\vee$, $\wedge$, and $\neg$ then produce other compound events whose description corresponds to what we do with logical connectives. For example, the description of the compound event $(\neg A) \wedge B$ is "the first three tosses of the coin weren't all heads and the coin is two-headed." The event $C \vee D$ is described by "there will be rain or sun next week."

This doesn't explain how the probability enters. Imagine that each $e \in \mathcal{E}$ is a possible "world" and $\Pr(e)$ is the probability that it occurs. If we select a world at random according to Pr, then $\Pr(A)$ is the probability that we selected a world in which A is true. This is the starting point for probabilistic logic, which won't be discussed in this text. ■

Recall that $\cup$ and $\cap$ stand for set union and intersection, respectively; i.e.,

$$A \cup B = \{\, c \mid c \in A \text{ or } c \in B \text{ or both} \,\}$$
$$A \cap B = \{\, c \mid c \in A \text{ and } c \in B \,\}.$$

Thus, if $C = A \cap B$, then $\Pr(C) = \Pr(A \wedge B)$; while if $D = A \cup B$, then $\Pr(D) = \Pr(A \vee B)$. (If $A \cap B = \emptyset$, the empty set, then we say that A and B are *disjoint sets*.) Even given the previous example, you might wonder why we've used $\Pr(A \wedge B)$ instead of $\Pr(A \cap B)$. At this point, either notation would be acceptable. But for sets of random variables, the distinction becomes important.

The next theorem states some important basic properties of probability spaces. Its proof is left as an exercise.

Theorem 7.1   Properties of Probability

Let $(\mathcal{E}, \Pr)$ be a probability space.

(a) If $A \subseteq B \subseteq \mathcal{E}$, then $0 \le \Pr(A) \le \Pr(B) \le 1$.

(b) If $A, B \subseteq \mathcal{E}$, then

$$\Pr(A \wedge B) + \Pr(A \wedge \neg B) = \Pr(A). \tag{7.3}$$

(c) If $A, B \subseteq \mathcal{E}$, then

$$\Pr(A \vee B) = \Pr(A) + \Pr(B) - \Pr(A \wedge B). \tag{7.4}$$

(d) If $A_i \subseteq \mathcal{E}$ for $1 \le i \le n$ and $A_i \cap A_j = \emptyset$ whenever $i \ne j$, then

$$\Pr(A_1 \vee A_2 \vee \cdots \vee A_n) = \Pr(A_1) + \Pr(A_2) + \cdots + \Pr(A_n). \tag{7.5}$$

Equation (7.4) is essentially a special case of what we call the *Principle of Inclusion and Exclusion*. When we have $A_i \cap A_j = \emptyset$ as in (7.5), we say that A_i and A_j are *disjoint events*. Equation (7.3) is a special case of (7.5)—set $n = 2$, $A_1 = A \cap B$, and $A_2 = A \cap \neg B$, and use

$$A_1 \cup A_2 = (A \cap B) \cup (A \cap \neg B) = A \cap (B \cup \neg B) = A \cap \mathcal{E} = A.$$

Exercises

7.1.A. Define the terms simple event, probability space, and compound event. Give examples of these concepts.

7.1.B. What do $\Pr(\neg A)$, $\Pr(A \wedge B)$, and $\Pr(A \vee B)$ denote? Give some simple equations that relate their values and the values of $\Pr(A)$ and $\Pr(B)$.

7.1.1. (*Answer follows*) Express each of the following events in terms of the events A, B, and C and the operations $\neg$, $\wedge$, and $\vee$.

  (a) A and B occur, but C does not.

  (b) At least one of the events A, B, and C occurs.

  (c) At least two of the events A, B, and C occur.

  (d) At most one of the events A, B, and C occurs.

7.1.2. Let A and B be compound events in a probability space. Let $C = A \cup B$ and $D = A \cap B$. Prove that

$$\Pr(A \vee B) = \Pr(C) \quad \text{and} \quad \Pr(A \wedge B) = \Pr(D).$$

7.1.3. Using the notation of Example 7.1, find subsets $A_0, \ldots, A_k$ of $\mathcal{E}_k$ so that the event A_i corresponds in a natural way to the simple event $i \in \mathcal{E}'_k$.

7.1.4. Describe each of the parts of Theorem 7.1 in words. For example, the first part states that the probability of every event lies between zero and one, and if the occurrence of one event A implies the occurrence of another event B, then B is at least as probable as A.

7.1.5. Show that the case $n = 2$ of (7.5) follows from (7.4).

7.1.6. Prove Theorem 7.1.

7.1.7. In this exercise, you will prove the Principle of Inclusion and Exclusion.

  (a) Using (7.4) and $A \cap (B \cup C) = (A \cap B) \cup (A \cap C)$, prove that

$$\Pr(A \vee B \vee C) = \Pr(A) + \Pr(B) + \Pr(C)$$
$$- \Pr(A \wedge B) - \Pr(A \wedge C) - \Pr(B \wedge C) \qquad (7.6)$$
$$+ \Pr(A \wedge B \wedge C).$$

  *(b) State and prove a generalization of (7.4) and (7.6) to an arbitrary number of sets.

7.1.8. The set of die faces is given by

$$D = \{\; \boxed{\cdot}\;,\; \boxed{\cdot\,^{\cdot}}\;,\; \boxed{\cdot\,^{\cdot}\,_{\cdot}}\;,\; \boxed{::}\;,\; \boxed{::\cdot}\;,\; \boxed{:::}\; \}$$

The black dots on a face are called "pips." Let $\mathcal{E} = D \times D$, the set of all two long sequences of dice. An element of $\mathcal{E}$ is called a "roll." We'll assume that the dice are fair; that is, $\Pr(e) = \frac{1}{36}$ for all $e \in \mathcal{E}$.

(a) Prove that $(\mathcal{E}, \Pr)$ is a probability space.

(b) Let A_k be the set of rolls such that the sum of the pips on a roll is k. List the elements of A_7.

(c) Determine $\Pr(A_k)$ for $0 \le k \le 12$.

7.1.9. What is best can be tricky. For example, suppose we ask three people to list three outcomes in order of preference. We might obtain

	1st	2nd	3rd
Alice:	A	B	C
John:	B	C	A
Pat:	C	A	B

Now most people prefer A to B, most B to C, and most C to A!

Can we do something similar with probability? Imagine making three dice using only the standard spot patterns 1–6; however, you need not use each pattern exactly once on a die. Make three dice such that, when they are rolled, the first will probably be higher than the second, the second will probably be higher than the third, and the third will probably be higher than the first. "Probably" means more than half the time.

Answers

7.1.1. There's often more than one way to express these events. I've tried to give the ones you're mostly likely to come up with.

(a) $A \wedge B \wedge (\neg C)$, where the parentheses can be omitted.

(b) $A \vee B \vee C$.

(c) $(A \wedge B) \vee (A \wedge C) \vee (B \wedge C)$.

(d) You might observe that this is the same as saying "At least two of the events $\neg A$, $\neg B$, and $\neg C$ occur." Then you could use the previous answer and simply replace A, B, and C with $\neg A$, $\neg B$, and $\neg C$. Alternatively, you could start from scratch and come up with the same formula.

Random Variables

Functions on a probability space provide a useful way of defining compound events and manipulating information about them. Such functions are called random variables:

Definition 7.4   Random Variable

Let $(\mathcal{E}, \Pr)$ be a probability space and let S be a set. A function $X : \mathcal{E} \to S$ is called an *S-valued random variable* or, simply a *random variable*. Often the term random variable is reserved for the case in which $S = \mathbb{R}$; however, we won't require that $S = \mathbb{R}$. Random variables are typically indicated with uppercase letters near the end of the alphabet.

A random variable X can be used to define a set of compound events, namely those sets on which X is constant. In other words,

$$X^{-1}(s) = \left\{ \, e \in \mathcal{E} \mid X(e) = s \, \right\}$$

is one such compound event, namely the set on which X equals s. Of course, if we never have $X(e) = s$ for any $e \in \mathcal{E}$, then $X^{-1}(s) = \emptyset$. For a given X, these compound events "partition" $\mathcal{E}$; that is, each $e \in \mathcal{E}$ lies in precisely one of the compound events. Instead of the notation $X^{-1}(s)$, we usually use $X = s$ to indicate the compound event. This is suggestive since the compound event is just the set of simple events e for which $X(e) = s$.

Aside. You have probably seen inverse functions in calculus. The idea is that an inverse function undoes what the function does. For example, the function $x \to e^x$ is the inverse of the function $x \to \log_e(x)$. However, our "inverse function," X^{-1} is not a function because it has a whole set of values corresponding to s. Why is this? In general, there is an ambiguity concerning how to undo a function. For example, if $f : x \to x^2$, what should $f^{-1}(9)$ be? If f^{-1} is to be a function, it must be 3 or -3, *but not both*. Thus, we're faced with a decision: either (a) restrict the domain of f and the range of f^{-1} or (b) decide not to insist that f^{-1} be a function. The first choice is appropriate in calculus and the second choice is appropriate in the present context.

One type of probability space along with its associated random variables is particularly useful in expert systems. Recall that, if $A_1, \ldots, A_k$ are sets, then their *(Cartesian) product* is

$$A_1 \times \cdots \times A_k = \left\{ \, (a_1, \ldots, a_k) \mid a_1 \in A_1, \ldots, a_k \in A_k \, \right\}.$$

Suppose that $\mathcal{E} = \mathcal{E}_1 \times \cdots \times \mathcal{E}_k$. The random variable $X_i : \mathcal{E} \to \mathcal{E}_i$ defined by $X_i(e_1, \ldots, e_k) = e_i$ is a *projection* onto the ith component.

Example 7.5   Heads and Tails: Some Random Variables

Recall that in Example 7.1 we defined $\mathcal{E}_k$ to be all 2^k long strings of H's and T's. This can be thought of as the Cartesian product of k copies of $\{H,T\}$. The projection onto the ith component is simply the result of the ith toss. There are two compound events associated with it, namely $X_i = H$ (the set of all sequences that give a head on the ith toss) and $X_i = T$.

As mentioned in Example 7.1, we can define the random variable $X(e)$ to be the number of H's in e. The compound event $X = i$ is the same set as the compound event A_i defined in Example 7.3. ∎

Definition 7.5   Probability of Random Variables

Let X be a random variable on the probability space $(\mathcal{E}, \mathrm{Pr})$. As noted above, $X = s$ is a compound event, so $\mathrm{Pr}(X = s)$—read "the probability that $X = s$"—is the probability of that compound event. By $\mathrm{Pr}(X)$ we mean the function whose domain is the range of X and whose value at s is $\mathrm{Pr}(X = s)$.

From the definition, it follows that we combine compound events described by random variables as we do compound events; for example,

$$(X = a) \wedge (Y = b) \ \text{ is the same as } \ A \wedge B, \tag{7.7}$$

where $A = X^{-1}(a)$ and $B = Y^{-1}(b)$. The expression on the left of (7.7) is read "X equals a and Y equals b," which conveys just what we should expect the expression to mean.

Aside. Now our symbol Pr stands for a whole new class of functions whose domains depend on the domains of the random variables involved. This should not cause confusion because all the uses of Pr relate to the same idea—the probability that "statment" in Pr(statement) is true, given the probability of the elementary events. If the statement has (implied) variables as in $\mathrm{Pr}(X)$, then the result is a function instead of a single value. In fact, had I used standard functional notation, I'd've written something awkward like $\mathrm{Pr}(X^{-1}(\ ))$.

To get some idea how these concepts might relate to expert systems, let's consider two very simplified examples.

Example 7.6   Medical Testing

Suppose that a physician is attempting to diagnose a disease for which there are three possible tests. The first two tests can have two outcomes, positive or negative, which we denote by P and N, respectively. The third test produces a numeric value; however, it does not make sense to consider a distinction as fine as that implied by considering all real numbers. In fact, medical opinion says it is reasonable only to distinguish three levels, namely low, medium, and high. Finally, the patient either has or does not have the disease, denoted by Y and N, respectively. Define

$$\mathcal{E} = \{P, N\} \times \{P, N\} \times \{L, M, H\} \times \{Y, N\}.$$

Suppose that one person in 500 has the disease. The values of $\Pr(e)$ could have been estimated by medical researchers who have studied a large number of people, say 20,000. In fact, we could take a different point of view; we could say that the probability space consists of these 20,000 people and $\Pr(e) = 1/20,000$ for each person e. We could then imagine four random variables (X_1, X_2, X_3, X_4), the first three giving the results of the tests and the third reflecting the presence or absence of the disease.

In the previous paragraph, we took a large sample of people so that the subsample with the disease would be large enough to provide reasonable estimates for our probabilities. Researchers don't do this. Instead, they first estimate $\Pr(X_4 = Y)$ for the population. Then they study $\Pr_Y(e)$ (resp. $\Pr_N(e)$) for a sample of people having (resp. not having) the disease. This involves the notion of conditional probability, which we'll discuss in the next section. Anticipating the notation and concepts, $\Pr(A \mid B)$ is the probability of A when we know that B has occurred. In that notation,

$$\Pr_Y(e) = \Pr(e \mid X_4 = Y)$$
$$\Pr(e) = \Pr(e \mid X_4 = Y)\Pr(X_4 = Y) + \Pr(e \mid X_4 = N)\Pr(X_4 = N). \tag{7.8}$$

This approach allows researchers to obtain reliable estimates using smaller samples. ∎

*Example 7.7   Credit Risk Assessment

Suppose we're attempting to build an expert system to assess someone's creditworthiness. Imagine taking a collection of 1,000 people and classifying them according to six coordinates (possible *outcomes*—values—are indicated in parentheses):

1. marital status (M or S),

2. age in years (a positive integer)

3. annual income in thousands of dollars (a positive integer)

4. liquid assets in thousands of dollars (a positive integer),

5. nonliquid assets in thousands of dollars (a positive integer), and

6. creditworthiness as assessed by an expert (A to F).

Never mind how we gather these people or get our expert to assess them—this is only an example to illustrate some concepts. At present we cannot form a *finite* probability space because there is an infinite number of positive integers. Therefore, we introduce some bounds in 2–5. We can take $\mathcal{E}$ to be the Cartesian product of the six sets of possible outcomes. Using the results of the expert's assessment, we can assign probabilities to each $e \in \mathcal{E}$, namely the number of people that correspond to e divided by 1,000. Of course, many of the elementary events will be assigned the probability 0 and quite a few of the remaining elementary events will be assigned the probability 1/1,000.

Here is an extremely simple expert system that uses the data to assess the creditworthiness of a loan applicant. It begins by determining the age, marital status, annual income, liquid assets, and nonliquid assets of the applicant. Next our system looks at that subset S of the 1,000 people who have the same values of these five coordinates and defines a probability for creditworthiness as follows. If A_w denotes the subset of A with creditworthiness $w \in \{A, B, C, D, E, F\}$, then the probability that our applicant has creditworthiness w will be taken to be $|A_w|/|A|$. If we wish, this can be rewritten in terms of the original probabilities: Let a_i be the applicant's answer to the ith question and let X_i be the projection onto the ith coordinate. Then

$$\frac{|A_w|}{|A|} = \frac{1,000 \Pr\big((X_1 = a_1) \wedge \cdots \wedge (X_5 = a_5) \wedge (X_6 = w)\big)}{1,000 \Pr\big((X_1 = a_1) \wedge \cdots \wedge (X_5 = a_5)\big)}. \tag{7.9}$$

(This is actually an application of conditional probability.)

There are serious problems with this approach. If there are no people in our sample of 1,000 who match our applicant in the first five components, our system cannot assign any probabilities to her creditworthiness. Worse, suppose exactly one person matches our applicant. Then our expert system would happily assign the same creditworthiness that the matched person was given and tell us that it was certain of that assignment. Why is that bad? First, the expert's assessment is not perfectly accurate, and second the expert may have used additional criteria that we failed to take into account in constructing $\mathcal{E}$.

To some extent, this can be remedied by a less "fine-grained" approach; for example, $Y_2(e) \in \{2, 3, 4, 5, 6, O\}$ might depend on whether the age was 20–29, ... , 60–69 or older. We might define a random variable that reflects some combination of liquid and nonliquid assets. And so on. In assessing the credit risk of the applicant, we now look at persons who have the same values as the applicant on these random variables.

Of course, considerably more work needs to be done to build a useful expert system, including using more data. ∎

Exercises

7.1.C. Define the concept of a random variable X and define $\Pr(X = s)$.

7.1.D. Explain how Cartesian products and projections are often associated with random variables.

7.1.10. Let A_k be as in Exercise 7.1.8. Interpret a random variable for which $X^{-1}(k) = A_k$.

7.1.11. If we replace A_i in (7.5) by $X_i = a_i$, what does the condition $A_i \cap A_j = \emptyset$ become for random variables?

7.1.12. Let $(\mathcal{E}, \Pr)$ be a probability space. Let $\mathcal{P}$ be a *partition of the set* $\mathcal{E}$; that is, a collection of pairwise disjoint subsets of $\mathcal{E}$ whose union is $\mathcal{E}$. A *propositional variable* is a random variable mapping $\mathcal{E}$ to $\mathcal{P}$. Explain why a random variable which is a projection can be thought of as a propositional variable.

7.1.13. In Exercise 7.1.8, a probability space for tossing two identical "fair" dice, each with faces having 1–6 pips was defined. In Exercise 7.1.10, a random variable for this space was defined.

(a) It should be clear how to define a similar space and random variable when the dice are still fair but may differ in the number of pips on the faces; for example, it might happen that one die has 3 pips on each of its faces. Do it. Call the random variable for this space Y to distinguish it from the random variable of Exercise 7.1.10, which was called X.

(b) Show that $\Pr(X = k) = \Pr(Y = k)$ for all k if one die has face counts 1, 2, 2, 3, 3, 4 and the other has 1, 3, 4, 5, 6, 8.

7.2   Conditional Probability and Bayes' Theorem

> *But if probability is a measure of the importance of our state of ignorance, it must change its value whenever we add new knowledge. And so it does.*
>
> —Thornton C. Fry (1928)

Suppose that we're interested in some situation such as a sequence of ten tosses of a fair coin. We can construct a probability space for this situation, namely the space $(\mathcal{E}_{10}, \Pr_{10})$ of Example 7.1 (p. 260). So far, our only information has been that the coin is fair. Now suppose that someone gives us some additional information; for example, that there were at least three heads. How can we construct a new probability space that is consistent with the old space and the new information?

Constructing such new spaces in which Pr is altered due to new information is a central problem in the design of expert systems based on probability. This is because an expert system keeps acquiring new information by asking the user questions. Conditional probability is the concept that incorporates new information.

Conditional Probability

Let A and B be compound events in some probability space. Suppose some simple event e occurred. Then $\Pr(B)$ is the probability that $e \in B$, given our initial knowledge of the situation as reflected by Pr. Intuitively, the conditional probability $\Pr(B \mid A)$ is the probability that $e \in B$ when we are given the additional information that $e \in A$.

Let's look at what we should do from a frequency viewpoint. We can approximate Pr by using frequencies as follows. Suppose we make a large number N of observations and that the simple event e occurs N_e times. Then $\Pr(e) \approx N_e/N$. Suppose we are told that the only simple events that occur are in A, but that otherwise everything is as expected. We can still use our observations to estimate what the conditional probability ought to be—we simply ignore all observations that are not in A. Suppose that $e \in A$ and N_A of the observations are in A. By the intuitive notion of probability and also by the previous discussion, we should have $\Pr(e \mid A) \approx N_e/N_A$. But then

$$\Pr(e \mid A) \approx \frac{N_e}{N_A} = \frac{N_e/N}{N_A/N} \approx \frac{\Pr(e)}{\Pr(A)}.$$

Here's the formal definition.

Definition 7.6 Conditional Probability

Let $(\mathcal{E}, \Pr)$ be a probability space and let $A \subseteq \mathcal{E}$ be such that $\Pr(A) \neq 0$. Define a new probability space $(\mathcal{E}, f)$ by

$$f(e) = \begin{cases} \Pr(e)/\Pr(A), & \text{if } e \in A, \\ 0, & \text{if } e \notin A. \end{cases}$$

For any $B \subseteq \mathcal{E}$, $\Pr(B \mid A)$, the *conditional probability of B given A*, equals $f(B)$. We say that the probabilities have been *conditioned* on A. Note that, if $A = \mathcal{E}$, then $\Pr(e \mid \mathcal{E}) = \Pr(e)$. This is as it should be because knowing that $e \in A$ gives us no additional information in this case.

If $\Pr(A) = 0$, define $\Pr(e \mid A)$ in any manner that is consistent with its being a probability distribution on $\mathcal{E}$. (Its value will never actually be used.)

The notion of conditional probability applies to random variables: Given random variables X and Y, note that $\Pr(X = a \mid Y = b)$ merely involves the compound events $X^{-1}(a)$ and $Y^{-1}(b)$. In accordance with the convention for expressions like $\Pr(X)$, $\Pr(X \mid Y)$ is a function whose domain is range$(X) \times$ range(Y).

The definition claims that $(\mathcal{E}, f)$ is a probability space. This needs to be proved, but the proof is very simple: $f \geq 0$ and

$$\sum_{e \in \mathcal{E}} f(e) = \sum_{e \in A} f(e) = \sum_{e \in A} \Pr(e)/\Pr(A) = \Pr(A)/\Pr(A) = 1.$$

Aside. Note that Pr is used in $\Pr(B \mid A)$, not f. In effect, $\Pr(\mid A)$ is merely another notation for the function $f( )$; however, it has the advantages that it (a) explicitly conveys the fact that we have conditioned Pr on A's occurring and (b) explicitly mentions the function Pr from which it was derived instead of some new function f. Our mutable function Pr has acquired yet another meaning, which, as always, is based on the original function for the probability space $(\mathcal{E}, \Pr)$.

In some definitions of conditional probability, the new space is taken to be A. This does not matter except for some notational issues since $f(e) = 0$ when $e \notin A$. In Example 7.7 we constructed a conditional probability and we took A to be our new space. (See (7.9) and the preceding discussion.)

You should have no difficulty proving the following theorem. When $C = \mathcal{E}$, we recover Theorem 7.1 from parts (b)–(e).

Theorem 7.2   Properties of Conditional Probability

Let $(\mathcal{E}, \Pr)$ be a probability space. If $C \subseteq \mathcal{E}$ has $\Pr(C) \neq 0$, then

(a) $\Pr(A \mid C) = \dfrac{\Pr(A \wedge C)}{\Pr(C)}$;   (7.10)

(b) if $A \subseteq B \subseteq \mathcal{E}$, then $0 \leq \Pr(A \mid C) \leq \Pr(B \mid C) \leq 1$;

(c) if $A, B \subseteq \mathcal{E}$, then

$$\Pr(A \wedge B \mid C) + \Pr(A \wedge \neg B \mid C) = \Pr(A \mid C);   \quad (7.11)$$

(d) if $A, B \subseteq \mathcal{E}$, then

$$\Pr(A \vee B \mid C) = \Pr(A \mid C) + \Pr(B \mid C) - \Pr(A \wedge B \mid C);   \quad (7.12)$$

(e) if $A_i \subseteq \mathcal{E}$ for $1 \leq i \leq n$ and $A_i \cap A_j = \emptyset$ whenever $i \neq j$, then

$$\Pr(A_1 \vee A_2 \vee \cdots \vee A_n \mid C)$$
$$= \Pr(A_1 \mid C) + \Pr(A_2 \mid C) + \cdots + \Pr(A_n \mid C). \quad (7.13)$$

(f) If $A \subseteq \mathcal{E}$, $B_1 \cup \cdots \cup B_n = \mathcal{E}$ and $B_i \cap B_j = \emptyset$ whenever $i \neq j$, then

$$\Pr(A) = \Pr(A \mid B_1) \Pr(B_1) + \cdots + \Pr(A \mid B_n) \Pr(B_n). \quad (7.14)$$

Example 7.8   Conditional Probability versus Implication

In Example 7.4 (p. 262), we discussed the interpretation of compound events in terms of propositional logic. Since it includes $\neg$, $\vee$, and $\wedge$, there are enough connectives to build any proposition. Now we have a new connective in probability theory, namely $A|B$. It's tempting to rewrite verbal expressions as follows:

$\Pr(A \mid B)$ means   the probability of "A given B"
which means   the probability of "A if B"
which means   the probability of "$B \rightarrow A$"
which means   the probability of "$B' \vee A$"

We would then incorrectly conclude that $\Pr(A \mid B) = \Pr(B' \vee A)$. See Exercise 7.2.5 for the correct equation. The lesson here is

Don't confuse conditional probability with logical implication.

Since this approach is wrong, how should $A|B$ be interpreted in propositional logic? It can't be done. Why not?

$A|B$ is not a subset of the event space $\mathcal{E}$ whereas logical connectives applied to subsets of $\mathcal{E}$ produce subsets of $\mathcal{E}$. But how do we *know* that $A|B$ can't be thought of as a subset of $\mathcal{E}$? Not much can be done with logical connectives—a function is determined by its truth table and it can be shown that none of the $2^4 = 16$ functions of two variables has the right behavior for $A|B$.

It *is* possible to extend propositional logic to allow for conditionals. Here are two ways we might approach the problem.

- Recall the possible-worlds interpretation discussed in Example 7.4. We have a problem with $A|B$ when B is false in a possible world: There is no contribution to $\Pr(A \mid B)$ in this case and yet we must assign a truth value to $A|B$ if it is to be a proposition. This difficulty could be overcome with a multivalued logic by extending propositional logic to three truth values, say "?," "T," and "F." The truth value "?" could be assigned to $A|B$ when B is false.

- The preceding idea sidesteps the fact that $(A|B) \notin \mathcal{E}$. We could attack this problem directly and define rules for manipulating pairs $A|B$. Propositional logic would be embedded in the larger structure by the mapping $A \mapsto A|\mathcal{E}$. This is similar to the way rational numbers are defined by considering pairs of integers. This approach has been pursued some in the AI community. ∎

The next few examples illustrate reasoning with and about conditional probabilities.

Example 7.9   Two-Headed Coin

Care is needed in specifying a probability space. Consider the following problem:

> I have two coins in my pocket. One is normal and the other has two heads. I pull out one coin and see a head on the side I'm looking at. What is the probability that I took out the normal coin?

It's not uncommon to answer "50%." The reasoning for this answer goes something like this. Since the coin is chosen at random, half the time the coin will be normal. But this is wrong.

To solve this problem, we must construct a probability space that accurately reflects the random aspect of the coin removal and then condition on the fact that I saw a head. Here's how it's done.

Label the sides of the coin h and h' such that the side labeled h contains a head. We can define $\mathcal{E} = \{n, t\} \times \{h, h'\}$, where n and t correspond to selecting the normal and two-headed coin, respectively, and h and h' correspond to the side examined. Since everything is random, define $\Pr(e) = \frac{1}{4}$ for all $e \in \mathcal{E}$. Define the compound events

$$N = \big\{(n, h), (n, h')\big\}, \qquad \text{selecting the normal coin,}$$
$$H = \big\{(n, h), (t, h), (t, h')\big\}, \qquad \text{seeing a head.}$$

We want to compute $\Pr(N \mid H)$. Since $N \cap H = \{(n, h)\}$, it follows from (a) in the theorem that the probability is $\frac{1}{4} / \frac{3}{4} = \frac{1}{3}$. ∎

Example 7.10   It's a Girl!

Consider the following three problems.

1. My colleague Bob has two children. I don't know their sexes. One day in the supermarket I meet Bob with a child he identifies as his daughter. What is the probability that both of Bob's children are girls?

2. At a PTA meeting, the woman sitting next to me raises her hand when the speaker asks how many people have two (but not three) children in the school. She raises her hand again when the speaker asks how many have a daughter in the school. What is the probability that both her children in the school are girls?

3. My colleague Ken has two children, but I don't know their sexes or ages. My wife is friendly with Ken's wife, Barbie, and knows about their children. Today, my wife gave me some clothes our daughter had outgrown, saying that Barbie should be able to use them. What is the probability that both of Ken and Barbie's children are girls?

In answering these questions, let's assume that the sex of children is split exactly 50:50 and that there is no biological tendency for some people to have girls and others to have boys. These assumptions are not exactly biologically correct—but this isn't a biology text.

The possible children are

$$\mathcal{E} = \big\{(F,F), (F,M), (M,F), (M,M)\big\},$$

where the method for choosing which child to list first is not yet specified. With most methods of listing, the probability of each elementary event will be $\frac{1}{4}$. We're asked for the conditional probability of the elementary event (F,F). The question becomes, what should we condition on?

First Problem: Without any probability space construction at all, the answer to the first problem appears to be $\frac{1}{2}$ since, having met one child, I don't know any more than I did before about the sex of the other.

Using $\mathcal{E}$, with the first child being the one who went to the store with Bob, we should condition on $C = \{(F,F), (F,M)\}$. This also gives $\frac{1}{2}$.

Second Problem: We might think that the answer is the same as the previous one, but that isn't so. The information given is different. We can see the difference by determining what compound event to condition on.

Use $\mathcal{E}$ and order the children by age. Since the only information is that there is at least one girl, we condition on $C = \mathcal{E} - \{(M,M)\}$. The calculation is straightforward and leads to a probability of $\frac{1}{3}$ for both children's being girls.

What would happen in the first problem if the children were ordered by age? In this case, $\mathcal{E}$ would be inadequate. We'd have to use something like

$$\mathcal{E} = \big\{(F,F), (F,M), (M,F), (M,M)\big\} \times \big\{1, 2\big\}$$

instead, where 1 and 2 indicate whether meeting the first or second child in the supermarket. You should be able to work out that this still gives an answer of $\frac{1}{2}$ for the first problem.

Third Problem: This is trickier.

- It seems that the only information I have is that Ken and Barbie have at least one daughter. This is just the second problem and the answer is $\frac{1}{3}$.

- On the other hand, my wife has implicitly told me that they have a daughter who is slightly smaller than my daughter. Since this conveys no information about their other child, it's like the first problem. Thus the answer is $\frac{1}{2}$.

Since the two approaches give different answers, at most one is correct? Which, if either, is correct?

I can't tell from the information provided, since it really depends on what my wife knows about Barbie's children. Knowing my wife, I believe she wouldn't give me the clothes unless she knew that they were likely to fit. The previous paragraph indicates that the answer is then $\frac{1}{2}$. (It's possible to raise additional objections that cast doubt on this answer.) ∎

The second problem in the previous example shows that it can be tricky to apply probability theory even when a problem is unambiguous. The third problem shows that unsuspected ambiguities can make the application even more difficult. In most discussions, I'll either remove or ignore such ambiguities. (Like most texts, this one presents a "sanitized" version of the real world.)

Example 7.11   Medical Testing

Suppose that there is a disease that one person in 1,000 has. Suppose that there is a screening test for the disease, that the test is always positive if the person has the disease, and that it is also positive five percent of the time when the person does not have the disease. (These are not unreasonable figures.) If the test is adminstered to people at random, what is the probability that a person who tests positive has the disease?

Let $\mathcal{E}$ be the set of people who are tested and let $\Pr(e)$ be uniform. A *uniform distribution* is a probability function for which $\Pr(e) = 1/|\mathcal{E}|$ for all $e \in \mathcal{E}$. (It is also called a *flat distribution*.)

Define the following subsets of $\mathcal{E}$:

D the people who have the disease,
H the people who do not have the disease,
P the people who test positive for the disease,
N the people who test negative for the disease.

Clearly $D \cup H = P \cup N = \mathcal{E}$ and $D \cap H = P \cap N = \emptyset$. The information we are given can be stated as follows:

$$\Pr(D) = 0.001$$
$$\Pr(P \mid D) = 1$$
$$\Pr(P \mid H) = 0.05.$$

We are asked to compute $\Pr(D \mid P)$. By (7.10),

$$\Pr(D \mid P) = \Pr(D \wedge P)/\Pr(P).$$

We are not given either of the probabilities on the right-hand side, but we can compute them as follows. First, by (7.10),

$$\Pr(D \wedge P) = \Pr(P \mid D)\Pr(D) = 1 \times 0.001 = 0.001. \qquad (7.15)$$

Second, since $D \cup H = \mathcal{E}$ and $D \cap H = \emptyset$, it follows from (7.14) that

$$\Pr(P) = \Pr(P \mid D)\Pr(D) + \Pr(P \mid H)\Pr(H).$$

Finally, since $\Pr(P \mid H)\Pr(H) = 0.05 \times 0.999 = 0.04995$, we have

$$\Pr(P) = 0.001 + 0.04995 = 0.05005$$

and

$$\Pr(D \mid P) = \frac{\Pr(D \wedge P)}{\Pr(P)} = \frac{0.001}{0.05005} \approx 0.02.$$

In other words, a positive test means that the person has a 2% chance of having the disease. This sort of low figure is the reason doctors often say "I'd like to do some additional tests." ∎

*Example 7.12   The Car and the Goats

The following problem concerning a then popular TV game show received considerable publicity in 1991 thanks to the column of Marilyn vos Savant in the *Parade* magazine supplement to Sunday newspapers.

> A TV game show host shows you three doors. He tells you that one door hides a car and the other two doors hide goats, then asks you to choose a door. You will win whatever is behind the door you choose. You pick the first door. The host opens the third door, revealing a goat, and asks you if you'd like to switch to the second door. What should you do?

In order to analyze the situation, we will have to describe it in probability theory terms.

We'll let $\mathcal{E} = P \times H$, where $P = \{1, 2, 3\}$ indicates the door that hides the prize and $H = \{2, 3, N\}$ indicates the door the host opens, given that you have selected door 1. (N indicates that the host opens neither door.) Note that we have built into $\mathcal{E}$ the assumption that you pick the first door. Let X_P and X_H be the projections onto the two coordinates. It seems reasonable to assume that the car has been placed at random; that is, $\Pr(X_P = k) = \frac{1}{3}$ for $k = 1, 2, 3$.

Since the host has opened the third door, we know that $X_P \neq 3$, so we should condition on that compound event as well as $X_H = 3$. Thus, we want to determine

$$\frac{\Pr(X_P = 2 \mid X_H = 3 \wedge X_P \neq 3)}{\Pr(X_P = 1 \mid X_H = 3 \wedge X_P \neq 3)}.$$

If this ratio equals 1, the car is equally likely to be behind either door, so there is no reason to switch or not to switch doors. If it exceeds 1, the car is more likely to be behind the second door, so you should switch doors. You should be able to show that

$$\Pr(X_P = k \mid X_H = 3 \wedge X_P \neq 3) = \Pr(X_H = 3 \mid X_P = k \wedge X_P \neq 3)$$
$$\times \frac{\Pr(X_P = k \wedge X_P \neq 3)}{\Pr(X_H = 3 \wedge X_P \neq 3)}$$

by using (7.10). We have (you give the reasons!)

$$\Pr(X_H = 3 \mid X_P = k \wedge X_P \neq 3) = \Pr(X_H = 3 \mid X_P = k)$$
$$\Pr(X_P = k \wedge X_P \neq 3) = \Pr(X_P = k) = \tfrac{1}{3}$$
$$\Pr(X_H = 3 \wedge X_P \neq 3) = \Pr\big(X_H = 3 \wedge (X_P = 1 \vee X_P = 2)\big)$$
$$= \Pr(X_H = 3 \wedge X_P = 1)$$
$$+ \Pr(X_H = 3 \wedge X_P = 2)$$
$$= \tfrac{1}{3}\Big(\Pr(X_H = 3 \mid X_P = 1)$$
$$+ \Pr(X_H = 3 \mid X_P = 2)\Big).$$

Thus

$$\Pr(X_P = k \mid X_H = 3 \wedge X_P \neq 3)$$
$$= \frac{\Pr(X_H = 3 \mid X_P = k)}{\Pr(X_H = 3 \mid X_P = 1) + \Pr(X_H = 3 \mid X_P = 2)}$$

and so

$$\frac{\Pr(X_P = 2 \mid X_H = 3 \wedge X_P \neq 3)}{\Pr(X_P = 1 \mid X_H = 3 \wedge X_P \neq 3)} = \frac{\Pr(X_H = 3 \mid X_P = 2)}{\Pr(X_H = 3 \mid X_P = 1)}, \qquad (7.16)$$

where we must remember that we have built into $(\mathcal{E}, \Pr)$ the assumption that you pick the first door.

Without knowing more about how the host acts, we cannot determine the two probabilities that appear on the right side of (7.16). Some possibilities for the host are as follows:

(a) He does not know the location of the car. If he reveals a car, he tells the contestant that he/she has lost. If he reveals a goat, he offers the contestant a chance to switch. In this case,

$$\Pr(X_H = a \mid X_P = k) = \Pr(X_H = a)$$

since the host's choice is not influenced by the position of the car. Hence (7.16) equals 1 and it does not matter if you switch dooors.

(b) He knows the location of the car and always opens a door with a goat. In this case, the numerator of (7.16) is 1 and the denominator is at most 1. (Its exact value depends on how the host chooses when neither door 2 nor door 3 conceals a car.) Thus the ratio is at least 1 and you should switch doors.

(c) We can add to the previous situation by assuming that the host sometimes does not open a door and that the probability of not opening a door depends on the location of the car. In this case, the numerator of (7.16) may be less than 1 and we cannot say anything about the ratio without further information. ∎

Exercises

7.2.A. What does $\Pr(A \mid B)$ mean? How can it be computed from the probability of compound events?

7.2.B. Are $\Pr(A \mid B)$ and $\Pr(B \rightarrow A)$ the same?

7.2.1. Prove Theorem 7.2.

7.2.2. Assuming that none of the probabilities involved is zero, show that

$$\Pr(A \wedge B \mid C) = \Pr(A \mid C) \Pr(B \mid C)$$
$$\text{if and only if}$$
$$\Pr(A \mid B \wedge C) = \Pr(A \mid C).$$

What probabilities can be zero without losing this result?

7.2.3. Fill in the reasons for the various steps in Example 7.12.

7.2.4. Show that the ratio in (7.16) can be greater than, equal to, and less than 1 for the host behavior described in (c). Do this by giving three specific rules for the host's behavior that lead to ratios that are greater than, equal to, and less than 1, respectively.

7.2.5. Using rules of propositional logic, we have $\Pr(B \rightarrow A) = \Pr(A \vee (\neg B))$.

(a) Show that, when $\Pr(B) > 0$,

$$\Pr(B \rightarrow A) = \Pr(\neg B) + \Pr(A \wedge B) = \Pr(\neg B) + \Pr(B) \Pr(A \mid B)$$

and thus $1 - \Pr(B \rightarrow A) = \Pr(B)\big(1 - \Pr(A \mid B)\big)$.

(b) Conclude that $\Pr(A \mid B) \leq \Pr(B \rightarrow A)$, with equality if and only if $\Pr(B) = 1$ or $\Pr(A \mid B) = 1$.

7.2.6. *Simpson's Paradox* is the observation that it is possible for the three inequalities

$$\Pr(B \mid T) > \Pr(B \mid \neg T)$$
$$\Pr\big(B \mid T \wedge M\big) < \Pr\big(B \mid (\neg T) \wedge M\big)$$
$$\Pr\big(B \mid T \wedge (\neg M)\big) < \Pr\big(B \mid (\neg T) \wedge (\neg M)\big)$$

to hold simultaneously. Explain why Example 5.2 (p. 193) is a special case of Simpson's Paradox.

7.2.7. Adam, Bill, and Chuck, three convicts with similar records, apply for parole and the board decides to grant parole to two of them. Adam learns of this, but doesn't know which of them are being released. Adam reasons that his chance of release is 2/3. A friendly guard knows who is being paroled but is not allowed to tell Adam if he is one of the lucky ones. He offers to tell Adam the name of someone—Bill or Chuck—who is being paroled. Adam thinks briefly and begs the guard not to tell him. He reasons as follows. If the guard says "Bill," then Chuck and I each have a 50/50 chance of release. If the guard says "Chuck," then Bill and I each have a 50/50 chance of release. Thus, my chance of release will go down from 2/3 to 1/2!

(a) Construct a probability space like that in Example 7.12, with $\mathcal{E} = P \times H$ where $P = \{A, B, C\}$ indicates the one not being paroled and $H = \{B, C\}$ indicates the one the guard would name. Assuming that each prisoner is equally likely not to be paroled and that the guard chooses randomly if both Bill and Chuck are to be paroled, what is the probability of the elementary events?

(b) Define a random variable X whose value indicates who is *not* being paroled and another Y whose value indicates what the guard says. Compute the probability that $X = A$ when the guard says nothing, when the guard says "Bill," and when the guard says "Chuck." Express all your work in correct probability notation. Do the same for $X = B$ and $X = C$.

(c) Explain Adam's error.

Bayesian Reasoning

What we've done in Examples 7.11 and 7.12 is to apply Bayes' Theorem together with Theorem 7.2 and some simple calculations. Various slightly different (but equivalent) equations go by the name of Bayes' Theorem:

Theorem 7.3   Bayes' Theorem

Let $(\mathcal{E}, \mathrm{Pr})$ be a probability space and let $A, H_1, \ldots, H_k \subseteq \mathcal{E}$ be compound events, none of which has zero probability. Then

$$\mathrm{Pr}(H_i \mid A) = \frac{\mathrm{Pr}(A \mid H_i)\,\mathrm{Pr}(H_i)}{\mathrm{Pr}(A)}. \qquad (7.17)$$

If, in addition, $\mathrm{Pr}(H_i \wedge A) \neq 0$ for all i, then

$$\frac{\mathrm{Pr}(H_i \mid A)}{\mathrm{Pr}(H_j \mid A)} = \frac{\mathrm{Pr}(H_i)}{\mathrm{Pr}(H_j)} \times \frac{\mathrm{Pr}(A \mid H_i)}{\mathrm{Pr}(A \mid H_j)}. \qquad (7.18)$$

If, in addition $H_1 \cup \cdots \cup H_k = \mathcal{E}$ and $H_i \cap H_j = \emptyset$ whenever $i \neq j$, then

$$\Pr(H_i \mid A) = \frac{\Pr(H_i)\Pr(A \mid H_i)}{\Pr(H_1)\Pr(A \mid H_1) + \cdots + \Pr(H_k)\Pr(A \mid H_k)} \quad (7.19)$$

$$= \left(\sum_{j=1}^{k} \frac{\Pr(H_j)}{\Pr(H_i)} \times \frac{\Pr(A \mid H_j)}{\Pr(A \mid H_i)}\right)^{-1}. \quad (7.20)$$

The proof is simply a matter of manipulating the definitions with a bit of algebra.

In applications, the H_i's in Bayes' Theorem are often competing hypotheses which partition $\mathcal{E}$, as in (7.19). The event A can then be thought of as evidence, which causes us to change from our prior probabilities $\Pr(H_i)$ for the various hypotheses to the new probabilities $\Pr(H_i \mid A)$ which take into account the evidence A. The usefulness of Bayes' Theorem is based on the fact that we often want information about the probability of a hypothesis given some evidence, but the information available is usually the probability of the hypothesis given the evidence. A weakness in Bayes' Theorem is the fact that assessing prior odds can be difficult. Since Bayes' Theorem uses evidence to reach conclusions about hypotheses, it can be regarded as a form of statistical reasoning—a subject we'll discuss briefly in Chapter 12.

Example 7.13   Medical Testing Revisited

Let's redo Example 7.11 using Bayes' Theorem with $H_1 = D$, $H_2 = H$, and $A = P$. From (7.19),

$$\Pr(D \mid P) = \frac{\Pr(D)\Pr(P \mid D)}{\Pr(D)\Pr(P \mid D) + \Pr(H)\Pr(P \mid H)}$$

$$= \frac{0.001 \times 1}{0.001 \times 1 + 0.999 \times 0.05} \approx 0.02. \quad \blacksquare$$

Definition 7.7   Odds and Likelihood Ratios

Ratios such as

$$\frac{\Pr(H_i)}{\Pr(H_j)} \quad \text{and} \quad \frac{\Pr(H_i \mid A)}{\Pr(H_j \mid A)}$$

are called *odds*. The first ratio is called the *prior odds* and the second the *posterior odds*, or the *odds given the evidence A*.

Ratios such as $\Pr(A \mid H_i)/\Pr(A \mid H_j)$ are called *likelihood ratios*. This is because they are the ratios of the likelihoods (i.e., probabilities) of the evidence A given various hypotheses H_i and H_j.

These terms are sometimes restricted to the case in which $k = 2$ and $H_2 = \neg H_1$.

If we consider A as evidence that Bayes' theorem uses to assess the probability of hypotheses, we could then ask, "What would be the effect of further evidence B?" In that case we apply (7.18) twice. First, we condition everything on A and use (7.18) (with A in (7.18) replaced by B). Then we apply (7.18) to one of the resulting ratios. Thus

$$\frac{\Pr(H_i \mid B \wedge A)}{\Pr(H_j \mid B \wedge A)} = \frac{\Pr(H_i \mid A)}{\Pr(H_j \mid A)} \times \frac{\Pr(B \mid H_i \wedge A)}{\Pr(B \mid H_j \wedge A)}$$

$$= \frac{\Pr(H_i)}{\Pr(H_j)} \times \frac{\Pr(A \mid H_i)}{\Pr(A \mid H_j)} \times \frac{\Pr(B \mid H_i \wedge A)}{\Pr(B \mid H_j \wedge A)}. \quad (7.21)$$

The three factors in (7.21) have simple interpretations:

- The first factor is the prior odds. It reflects the lack of information.

- The second factor is a likelihood ratio. It reflects the information obtained from the fact that A occurred.

- The last factor is also a likelihood ratio. It reflects the information from the fact that B occurred, given that A also occurred.

Ideally, we'd like to have $\Pr(B \mid H_k \wedge A) = \Pr(B \mid H_k)$ so that we could eliminate A from the last factor in (7.21), thus uncoupling the effects of A and B. This leads us to the notion of independence, which is discussed in the next section.

Example 7.14   Soft Evidence

Sometimes evidence is not of the form "A has occurred." Instead, we may be 80% sure that A occurred and 20% sure that $\neg A$ occurred. This can be handled in a straightforward manner.

Call this unsure evidence α—a Greek letter to distinguish it from compound events. If A actually occurred, the correct probability of H would be $\Pr(H \mid A)$, which can be computed from Bayes' Theorem. Similarly, we have $\Pr(H \mid \neg A)$ if A did not occur. Looking at things from a frequency viewpoint, you should be able to see that it's reasonable to set

$$\Pr(H \mid \alpha) = \Pr(A \mid \alpha)\Pr(H \mid A) + \Pr(\neg A \mid \alpha)\Pr(H \mid \neg A), \quad (7.22)$$

where $\Pr(A \mid \alpha)$ is the probability we assign to A given the evidence α. (The value is 0.8 for the case given in the previous paragraph.)

Can we prove (7.22)? To do so, we would need to define $\Pr(B \mid \alpha)$. Looking back at the definition for conditional probability, we can see that this requires definitions for $\Pr(B \wedge \alpha)$ and $\Pr(\alpha)$, which are not compound events. There is a simpler approach. Let $\Pr_\alpha$ be the probability we assign to events after receiving the evidence α. Then

$$\Pr_\alpha(H) = \Pr_\alpha(A)\Pr_\alpha(H \mid A) + \Pr_\alpha(\neg A)\Pr_\alpha(H \mid \neg A) \quad (7.23)$$

follows from the fact that $\Pr_\alpha$ is a probability function. Now $\Pr_\alpha(B)$ is just what we said we wanted $\Pr(B \mid \alpha)$ to mean. Hence, (7.22) will follow from (7.23) if we can explain why we should have

$$\Pr_\alpha(H \mid A) = \Pr(H \mid A) \quad \text{and} \quad \Pr_\alpha(H \mid \neg A) = \Pr(H \mid \neg A).$$

The first of these equations states that the evidence α does not alter our assessment of the probability that H occurred given that A occurred. This is quite reasonable. For example, if I told you that I was 80% sure that the 8:10 bus already came, my information shouldn't alter your estimate of the conditional probability that you'll be late for class if you miss the 8:10 bus. Thus, we've shown that (7.22) is correct under rather broad conditions. ∎

Exercises

7.2.C. State one form of Bayes' theorem. Why is it useful?

7.2.D. What is a likelihood ratio?

7.2.8. Prove Bayes' Theorem (Theorem 7.3).

7.2.9. A bag contains k fair coins and one two-headed coin. Without looking, I select a coin at random and toss it n times. All n tosses are heads. What is the probability that the coin I selected was fair?

7.2.10. Testem University offers admissions based on scores on an exam given by National University Testing Service, which simply defines certain grades to be "pass" and others to be "fail." Of those applying to TU, 40% are capable of obtaining a degree there. Of those who are capable, 80% will pass the NUTS exam. Of those who are not capable, 30% will pass the NUTS exam. What percentage of those offered admission to TU are capable?

7.2.11. Jack tells his professor that he forgot to bring his project to hand in. From experience, the professor knows that students with finished projects forget and tell her so about once in 100 times. She also knows that about half of the students who haven't finished their projects will tell her they forgot. She thinks that about 90% of the students in this class completed their projects on time. What is the probability that Jack's excuse is valid?

7.3   Independence

The basic idea behind independence is that knowing certain information gives us no additional information about something else. For example, knowing the weather today gives you no information about how you will do on the quiz next week. Thus the random variable that gives your quiz score is independent of the random variable that gives today's weather. As a result, when you attempt to assess how well you will do on the quiz, you can ignore today's weather. People must often make assumptions of independence; otherwise, they would be unable to analyze a situation due to the lack of adequate data, the amount of computation required, or both. Of course, *assuming* independence doesn't make it so. This leads to the difficult problem: How much does assuming independence alter the results? There appear to be no useful results on this problem. Nevertheless, you may have to assume independence in some situations. In such cases, wave your arms and keep a good luck charm nearby.

Independence of Compound Events

Definition 7.8   Independence of Compound Events

Let $(\mathcal{E}, \mathrm{Pr})$ be a probability space and let $A_1, A_2 \subseteq \mathcal{E}$. If

$$\mathrm{Pr}(A_1 \wedge A_2) = \mathrm{Pr}(A_1)\,\mathrm{Pr}(A_2),$$

we say that A_1 and A_2 are *independent*. More generally, if, for every subset $\mathcal{A} = \{A_{i_1}, \ldots, A_{i_k}\}$ of $\{A_1, \ldots, A_n\}$, we have

$$\mathrm{Pr}(A_{i_1} \wedge \cdots \wedge A_{i_k} \mid B) = \mathrm{Pr}(A_{i_1} \mid B) \cdots \mathrm{Pr}(A_{i_k} \mid B), \qquad (7.24)$$

then we say that the A_i are *(mutually) independent events* given B.

If $\mathrm{Pr}(B) \neq 0$, an equivalent formulation of the statement that A and B are independent is $\mathrm{Pr}(A \mid B) = \mathrm{Pr}(A)$. On the other hand, if $\mathrm{Pr}(B) = 0$, then $\mathrm{Pr}(A \wedge B) = 0$. In other words,

A and B independent $\iff \mathrm{Pr}(A \wedge B) = \mathrm{Pr}(A)\,\mathrm{Pr}(B)$

$\iff$ either $\mathrm{Pr}(B) = 0$ or $\mathrm{Pr}(A \mid B) = \mathrm{Pr}(A)$.

Intuitively, $\mathrm{Pr}(A \mid B) = \mathrm{Pr}(A)$ says that knowing B occurred gives no information about the occurrence of A. Since $\mathrm{Pr}(A \wedge B) = \mathrm{Pr}(A)\,\mathrm{Pr}(B)$, is symmetric, it follows that $\mathrm{Pr}(B \mid A) = \mathrm{Pr}(B)$. This leads us to a simple fact:

> Knowing A occurred gives no information about the occurrence of B
> if and only if
> knowing B occurred gives no information about the occurrence of A.

It's easy to overlook this fact in intuitive probability. For example, many people might say that knowing Bill often wears blue jeans does not affect their estimate of the probability that he is a college student, but that knowing Bill is a college student does affect their estimate of the probability that he often wears blue jeans.

The next theorem clarifies the effect of independent evidence, which we alluded to after (7.21).

Theorem 7.4   Bayes Theorem with Independent Multiple Evidence

Suppose that $A_1, \ldots, A_n$ are independent given H_1 and are also independent given H_2, then

$$\frac{\Pr(H_1 \mid A_1 \wedge \cdots \wedge A_n)}{\Pr(H_2 \mid A_1 \wedge \cdots \wedge A_n)} = \frac{\Pr(H_1)}{\Pr(H_2)} \times \frac{\Pr(A_1 \mid H_1)}{\Pr(A_1 \mid H_2)} \times \cdots \times \frac{\Pr(A_n \mid H_1)}{\Pr(A_n \mid H_2)}. \quad (7.25)$$

Example 7.15   Further Medical Testing

Suppose we have the situation described in Example 7.11 (p. 276). The doctor decides to perform an additional test because the first test was positive. The second test gives a false positive only 0.5% of the time but gives a correct positive only 80% of the time. Since these two tests check for very different things, it's reasonable to assume that the tests are independent, given D and also given H.

We now want to know what the probability is that a person who tests positive on both tests has the disease. Let $H_1 = D$, $H_2 = H$, and let A_i be the event that the ith test was positive. By (7.25)

$$\frac{\Pr(D \mid A_1 \wedge A_2)}{\Pr(H \mid A_1 \wedge A_2)} = \frac{0.001}{0.999} \times \frac{1}{0.05} \times \frac{0.8}{0.005} = \frac{8 \times 10^{-4}}{2.4975 \times 10^{-4}}$$

and so, by (7.20) (p. 281),

$$\Pr(D \mid A_1 \wedge A_2) = \frac{1}{1 + 2.4975/8} \approx \frac{3}{4}.$$

Thus, about 3 out of 4 people who are positive for both tests have the disease.

What if the order of the tests had been reversed? The conclusions would have been the same, but the doctor would have missed about 20% of the people with the disease because a false negative would have led him to stop testing. Missing people with the disease is very bad. Thus, it is important that the number of false negatives on the first (screening) test be as low as possible, even if there are some false positives. What should the doctor do if the second test is negative? Perhaps he should order further testing, because 20% of the people with the disease will have a negative result on the second test. ■

Example 7.16   Independence and Cartesian Products

Independence arises in a natural manner when the probability space arises from products. Suppose that $(\mathrm{Pr}_0, \mathcal{E}_0)$, $(\mathrm{Pr}_1, \mathcal{E}_1)$, and $(\mathrm{Pr}_2, \mathcal{E}_2)$ are probability spaces. Let

$$\mathcal{E} = \mathcal{E}_0 \times \mathcal{E}_1 \times \mathcal{E}_2 \quad \text{and} \quad \mathrm{Pr}(e_0, e_1, e_2) = \mathrm{Pr}_0(e_0) \, \mathrm{Pr}_1(e_1) \, \mathrm{Pr}_2(e_2).$$

Let S_i be a subset of $\mathcal{E}_i$ and define

$$A_0 = S_0 \times \mathcal{E}_1 \times \mathcal{E}_2, \quad A_1 = \mathcal{E}_0 \times S_1 \times \mathcal{E}_2, \quad \text{and} \quad A_2 = \mathcal{E}_0 \times \mathcal{E}_1 \times S_2.$$

(One example of this is coin tossing where the subscript i refers to the ith toss.) It's left as an exercise for you to prove that A_1 and A_2 are independent given A_0 for this Cartesian product construction. The generalization to more than three A_i's should be fairly obvious. ∎

*Example 7.17   A Counterexample

Suppose that A_1 and A_2 are independent given B and are also independent given C. It is natural to ask if they are independent given $B \vee C$ and if they are independent given $B \wedge C$. To disprove such a statement, it suffices to construct a probability space where it is false. We will deal with the first case and leave the second as an exercise.

Here is an idea for constructing a counterexample given $B \vee C$. We could make $A_1 \cap B = \emptyset$ and $A_2 \cap C = \emptyset$. Then the given independences are trivial because the probabilities that must be equal are all zero. For example,

$$0 \leq \mathrm{Pr}(A_1 \wedge A_2 \mid B) \leq \mathrm{Pr}(A_1 \mid B) = 0 = \mathrm{Pr}(A_1 \mid B) \, \mathrm{Pr}(A_2 \mid B).$$

We also have $\mathrm{Pr}(A_1 \wedge A_2 \mid B \vee C) = 0$ because

$$\begin{aligned}
(A_1 \cap A_2) \cap (B \cup C) &= (A_1 \cap A_2 \cap B) \cup (A_1 \cap A_2 \cap C) \\
&\subseteq (A_1 \cap B) \cup (A_2 \cap C),
\end{aligned} \tag{7.26}$$

and the last expression is the union of two empty sets. On the other hand, $\mathrm{Pr}(A_1 \mid B \vee C)$ and $\mathrm{Pr}(A_2 \mid B \vee C)$ need not be zero. The left-hand side of Figure 7.1 should clarify this.

You might object that the counterexample is rather specialized since it requires that certain probabilities be zero. We can adjust the circles in Figure 7.1 to allow overlap, as shown on the right-hand side. Can this be made into a counterexample?

We need to have

$$\begin{aligned}
\mathrm{Pr}(A_1 \wedge A_2 \mid B) &= \mathrm{Pr}(A_1 \mid B) \, \mathrm{Pr}(A_2 \mid B), \\
\mathrm{Pr}(A_1 \wedge A_2 \mid C) &= \mathrm{Pr}(A_1 \mid C) \, \mathrm{Pr}(A_2 \mid C).
\end{aligned} \tag{7.27}$$

Suppose this is done so that $\mathrm{Pr}(A_1 \mid B)$ and $\mathrm{Pr}(A_2 \mid C)$ are small. The last expression in (7.26) will correspond to an event with small probability. All we need to do is keep $\mathrm{Pr}(A_1 \mid C)$ and $\mathrm{Pr}(A_2 \mid B)$ fairly large.

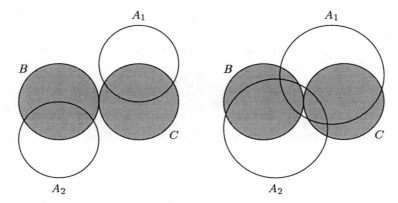

Figure 7.1 A counterexample for Example 7.17. The interior of a labeled circle corresponds to the elements of $\mathcal{E}$ that lie in the compound event associated with the label. The shaded area corresponds to $B \cup C$. On the left, $A_1 \cap B = A_2 \cap C = \emptyset$. The right-hand figure is the general case.

Let's try to assign some numbers so that what we described in the previous paragraph happens. To simplify calculations, we'll assume that $B \cap C = \emptyset$ and that the situation is symmetric as shown in Figure 7.1. We begin with variables instead of numbers, using δ and ϵ to suggest small numbers:

$$\Pr(B) = \Pr(C) = x,$$
$$\Pr(A_1 \wedge B) = \Pr(A_2 \wedge C) = \delta,$$
$$\Pr(A_1 \wedge C) = \Pr(A_2 \wedge B) = y,$$

and

$$\Pr(A_1 \wedge A_2 \wedge B) = \Pr(A_1 \wedge A_2 \wedge C) = \epsilon.$$

For these to make sense as probabilities, we need $\epsilon < \delta$ and x and y cannot be too big. By the definition of conditional probability and the definition of independence, the requirement that A_1 and A_2 be independent given B states that $(\epsilon/x) = (y/x)(\delta/x)$. Thus (7.27) is equivalent to

$$\epsilon = \delta \times (y/x). \qquad (7.28)$$

This gives us ϵ in terms of the other three parameters, x, y, and δ. Let D be any compound event. Since B and C have no events in common,

$$\Pr(D \wedge (B \vee C)) = \Pr\big((D \wedge B) \vee (D \wedge C)\big) = \Pr(D \wedge B) + \Pr(D \wedge C).$$

Setting $D = \mathcal{E}$, A_i, and $A_1 \wedge A_2$ in turn, we obtain

$$\Pr(B \vee C) = 2x,$$
$$\Pr(A_i \wedge (B \vee C)) = y + \delta,$$

and

$$\Pr((A_1 \wedge A_2) \wedge (B \vee C)) = 2\epsilon.$$

In order not to have independence given $A \vee B$, we need only have

$$\frac{2\epsilon}{2x} \neq \left(\frac{y+\delta}{2x}\right)^2.$$

Using (7.28) to eliminate ϵ, we see that we need

$$\frac{y\delta}{x^2} \neq \frac{(y+\delta)^2}{4x^2}.$$

Multiplying by $4x^2$ and subtracting $4y\delta$ results in $0 \neq (y-\delta)^2$. Thus we need only require that $y \neq \delta$.

Putting all this together, and remembering that our numbers must be probabilities, we see that we require

$$0 < y < x < 1, \quad y \neq \delta \quad \text{and} \quad \epsilon = \delta y / x.$$

Numeric examples are now easily constructed; for example, $x = \frac{1}{2}$, $y = \frac{1}{4}$, $\delta = \frac{1}{8}$, and $\epsilon = \frac{1}{16}$. You should be able to assign probabilities to the various shaded regions in the right-hand part of Figure 7.1. ■

Exercises

7.3.A. What does it mean to say that the compound events A and B are independent given C? Explain by an equation as well as intuitively.

7.3.B. How does independence simplify the use of Bayes' Theorem?

7.3.1. Assuming that $0 < \Pr(B) < 1$, show that A and B are independent events if and only if $\Pr(A \mid B) = \Pr(A \mid \neg B)$.

7.3.2. Suppose n people each toss a fair coin. What is the probability that all people get the same result? that all but one person get the same result?

7.3.3. Suppose that $A_1, A_2, \ldots, A_n$ are mutually independent given B.

 (a) Prove that any subset of the A_i's is mutually independent given B.

 (b) Suppose that, for each i, C_i equals A_i or $\neg A_i$. Prove that $C_1, C_2, \ldots, C_n$ are mutually independent given B.
    Hint. Induct on the number of complemented A_i's.

 (c) Prove that $\mathcal{A} = \{A_1 \cup A_2, A_3, \ldots, A_n\}$ are mutually independent given B.
    Hint. Use (7.12).

 (d) Let $T_1, T_2, \ldots, T_k$ be disjoint subsets of $\mathcal{A}$; that is, $T_i \cap T_j = \emptyset$ whenever $i \neq j$. Let T_i be the union of those A_j's that are in T_i. Prove that $T_1, \ldots, T_k$ are independent given B.

 (e) Let $1 \le t < n$ be an integer. Define

$$D = A_1 \cap \cdots \cap A_t \cap B.$$

   Prove that $A_{t+1}, \ldots, A_n$ are mutually independent given D.
   Hint. If it's hard, try doing the case $t = 1$ first.

7.3.4. Prove Theorem 7.4.

7.3.5. This exercise refers to Example 7.16.

 (a) Prove the claim that A_1 and A_2 are independent given A_0.

 (b) State and prove a claim where $(\mathrm{Pr}_0, \mathcal{E}_0)$ is missing and A_2 and A_3 are independent.

 (c) State an independence result for $\mathcal{E}_0 \times \cdots \times \mathcal{E}_n$ that says $A_1, \ldots, A_n$ are mutually independent given A_0.

 (d) Prove the independence result stated in (c). Conceptually, the proof is not hard, but all the sets and indices that are floating around may make it difficult to write down a proof. Try to write a careful proof by induction.

7.3.6. A petroleum geologist is exploring an area that has about one chance in a thousand of yielding oil. She finds two independent pieces of geological evidence favoring oil. Each piece has about a 20% chance of occurring if no oil is present and about a 95% chance of occurring if oil is present. Given this evidence, what is the probability that oil is present?

7.3.7. A doctor is checking a patient's symptoms. Of a list of four symptoms for the dreaded lurgy, the patient has three. Suppose that the symptoms are independent for people who have the dreaded lurgy and are also independent for people free of the disease. Suppose that a patient with dreaded lurgy has a 3/4 chance of exhibiting each of the symptoms and that a random person who does not have dreaded lurgy has 1 chance in 50 of exhibiting each of the symptoms. If 1 person in 1,000 has dreaded lurgy, what is the probability that the patient has dreaded lurgy?

7.3.8. The doctor in the previous exercise consults a specialist. She tells him that the figures he has for people who don't have dreaded lurgy are not quite right. There is a rare disease that only 1 person in 100,000 has. Patients with this meta-lurgy should also be excluded when figuring the 1 in 50 chances of exhibiting the symptoms. People with meta-lurgy always have the three symptoms the current patient has, but, like people without dreaded lurgy, have only 1 chance in 50 of having the symptom the patient does not have. What is the probability that the patient has dreaded lurgy? has the rare meta-lurgy? has neither?

7.3.9. Suppose you're using two independent sensors to locate an object. The object is in one of three regions, say R_1, R_2, and R_3. From experience (or otherwise), you have estimates of $p_{i,j,k}$, the probability that sensor i will report that the object is in region R_j when it is actually in region R_k. Based on the object's previous position, you believe the chance of its being in R_2 is about 60% and the chance of its being in each of the other regions is about 20%. Both sensors report that the object is in R_1. What are your new estimates for the probability of the object's being in each of the three regions?

7.3.10. Construct probability spaces to show that the following are FALSE in some situations:

(a) Suppose that $\Pr(B \wedge C) \neq 0$. If A_1 and A_2 are mutually independent given B and A_1 and A_2 are mutually independent given C, then A_1 and A_2 are mutually independent given $B \wedge C$.

(b) If A_1 and A_2 are mutually independent given B, A_1 and A_3 are mutually independent given B, and A_2 and A_3 are mutually independent given B, then A_1, A_2, and A_3 are mutually independent given B.

Independence of Random Variables

At the end of Definition 7.6, we saw that $\Pr(X \mid Y)$ is a function from $\text{range}(X) \times \text{range}(Y)$ to $\mathbb{R}$. This is a special case of a more general situation:

> Whenever an equation involves probability functions whose arguments contain random variables $Y_1, Y_2, \ldots$, the equation stands for the entire set of equations obtained by replacing each Y_i by $Y_i^{-1}(s_i)$ for all possible choices of s_i in the range of Y_i. (7.29)

For example, if X_i is either H or T, indicating the results of the ith toss of a coin, then

$$\Pr(X_1 \wedge X_2) = \Pr(X_1)\Pr(X_2) \qquad (7.30)$$

is actually a set of four equations corresponding to the four values HH, HT, TH, and TT that $X_1 X_2$ can assume. Equation (7.30) is read "the probability of X_1 and X_2 equals the probability of X_1 times the probability of X_2."

Definition 7.9   Notation

(a) To avoid having to write $\wedge$ so often, we shorten $\Pr(X_1 \wedge X_2 \wedge X_3)$ to $\Pr(\{X_1, X_2, X_3\})$. We refer to this as the *joint distribution* of X_1, X_2 and X_3.

(b) We will always use script letters such as $\mathcal{X}$ to denote sets of random variables.

(c) When we write a set of random variables as an index of summation, we mean that the sum is to extend over all values in the ranges of the random variables in the set.

There's an unfortunate notational consequence of (a):

$$\Pr(\{X\} \cup \{Y\}) = \Pr(\{X, Y\}) = \Pr(X \wedge Y)$$

It is *very important* to distinguish $\cup$ and $\cap$ from $\vee$ and $\wedge$ in probability statements involving sets of random variables. With a little care, you can keep this straight because

Inside $\Pr(\ )$ we use set union only for sets of random variables and we denote such sets by script letters such as $\mathcal{X}$ and $\mathcal{Y}$ or by the set notation $\{\ldots\}$.

Note that, in this notation, $\Pr(\mathcal{X} \mid \mathcal{Y}) = \Pr(\mathcal{X} \cup \mathcal{Y})/\Pr(\mathcal{Y})$.

Example 7.18   Manipulating Notation

Let $\mathcal{X}$ and $\mathcal{Y}$ be disjoint sets of random variables. We'll show that

$$\sum_{\mathcal{Y}} \Pr(\mathcal{X} \cup \mathcal{Y}) = \Pr(\mathcal{X}). \qquad (7.31)$$

We need to show that each simple event that contributes to the right side also contributes to the left side exactly once and vice versa. To see this, we switch from random variables to compound events. This is done by looking at the values that the random variables can take on. Let a_i (resp. b_i) be a value for the ith random variable in $\mathcal{X}$ (resp. $\mathcal{Y}$). Let A be the compound event in which $X_i = a_i$ for all $X_i \in \mathcal{X}$. Let $k = |\mathcal{Y}|$ and let $B(b_1, \ldots, b_k)$ be the compound event in which $Y_i = b_i$ for $1 \le i \le k$. We can now rewrite the claimed equality (7.31) as

$$\sum_{b_1, \ldots, b_k} \Pr\Big(A \wedge B(b_1, \ldots, b_k)\Big) = \Pr(A).$$

Each elementary event $e \in \mathcal{E}$ belongs to *exactly one* of the sets $B(b_1, \ldots, b_k)$, namely the one for which $b_i = Y_i(e)$ for all i. Equation (7.31) now follows from (d) of Theorem 7.1 (p. 263) with the A_i's equal to $A \cap B(b_1, \ldots, b_k)$ because

$$\bigcup_{b_1, \ldots, b_k} \Big(A \cap B(b_1, \ldots, b_k)\Big) = A \cap \Big(\bigcup_{b_1, \ldots, b_k} B(b_1, \ldots, b_k)\Big) = A \cap \mathcal{E} = A.$$

Looking over this proof, you may be wondering where I used the fact that $\mathcal{X}$ and $\mathcal{Y}$ are disjoint. You'll have to hunt a bit for that because I was careless about one point. Try to find it.

$*$ $\quad$ $*$ $\quad$ $*$ $\quad$ Stop and think about this! $\quad$ $*$ $\quad$ $*$ $\quad$ $*$

Here's the answer. I selected some values for the X_i's and fixed them at the a_i's. Then I allowed the Y_j's to vary over all possible values. I can't do this if one of the X_i's is one of the Y_j's.

Note that I said "one of the X_i's *is* one of the Y_j's", not "one of the X_i's *equals* one of the Y_j's." This is an important point. The convention (7.29) requires that each random variable be assigned a value, but it does not require two differently named random variables, say X and Y, to be assigned the same value even if $X(e) = Y(e)$ for *all* $e \in \mathcal{E}$. Of course, if we assign them different values, we're likely to end up with a compound event that is the empty set. For example, suppose that X and Y are random variables that take on the values 3 and 4 and that $X(e) = Y(e)$ for all $e \in \mathcal{E}$. Then $\Pr(X|X)$ is a function on $\{3,4\}$ that is identically equal to 1, while $\Pr(X|Y)$ is a function on $\{3,4\} \times \{3,4\}$ whose value at (a,b) equals 1 when $a = b$ and equals 0 otherwise. ∎

Now we'll define independence for random variables (actually, for sets of random variables). This definition is stronger than the definition of independence for compound events (Definition 7.8): Since it involves a function whose arguments take on all possible choices in the ranges of the random variables, one independence statement for random variables becomes several independence statements for compound events.

Definition 7.10   Independence of Sets of Random Variables

Let $(\mathcal{E}, \Pr)$ be a probability space and let $\mathcal{X}_1, \ldots, \mathcal{X}_n$ and $\mathcal{W}$ be pairwise disjoint sets of random variables on $\mathcal{E}$. We say that the $\mathcal{X}_i$'s are *(mutually) independent* given $\mathcal{W}$ if

$$\Pr(\mathcal{X}_1 \cup \cdots \cup \mathcal{X}_n \mid \mathcal{W}) = \Pr(\mathcal{X}_1 \mid \mathcal{W}) \cdots \Pr(\mathcal{X}_n \mid \mathcal{W}). \qquad (7.32)$$

Example 7.19   A Simple Independence Example

Let's return to our old friend, a sequence of n independent coin tosses with $\mathcal{E} = \{H, T\}^n$. Let $X_i(e_1, \ldots, e_n) = e_i$. If $\mathcal{Y}_1, \ldots, \mathcal{Y}_k$ and $\mathcal{W}$ are disjoint subsets of $\{X_1, \ldots, X_n\}$, then the $\mathcal{Y}_i$'s are independent given $\mathcal{W}$. This claim follows from Exercise 7.3.5 (p. 289). Decompose the coin-toss sequence space into pieces where the ith piece $\mathcal{E}_i$ contains those tosses that are referred to by $\mathcal{Y}_i$. The disjointness of the $\mathcal{Y}_i$'s ensures that the $\mathcal{E}_i$'s all refer to different positions. Let $\mathcal{E}_0$ be the remaining tosses. Then a particular assignment of values to $\mathcal{Y}_i$ will pick out values for the positions that $\mathcal{E}_i$ keeps track of. Since $\mathcal{W}$ is disjoint from the $\mathcal{Y}_i$'s, an assignment to $\mathcal{W}$ will pick out values for *some* of the positions that $\mathcal{E}_0$ keeps track of. The remaining details are left for you.

Let U_k be the number of heads on the first k tosses and let V_k be the number of heads on the last k tosses. If $k + j \leq n$, then U_k and V_j are independent. This follows from Exercise 7.3.5(b). Simply write $\mathcal{E}$ as a product of two spaces, one that contains the first k tosses and one that contains the remaining $n - k$ tosses (and hence the last j since $j \leq n - k$). The details are left for you. ∎

Independence was defined as one probability equalling a product of probabilities. It can also be defined as one conditional probability equalling another. For $n = 2$, this definition is $\mathcal{X}_1$ and $\mathcal{X}_2$ are independent given $\mathcal{W}$ if and only if conditioning $\mathcal{X}_1$ on $\mathcal{X}_2 \cup \mathcal{W}$ gives the same values as conditioning it on just $\mathcal{W}$. In terms of equations, this is

$$\Pr(\mathcal{X}_1 \mid \mathcal{X}_2 \cup \mathcal{W}) = \Pr(\mathcal{X}_1 \mid \mathcal{W}).$$

You should be able to prove this using Exercise 7.2.2 (p. 279).

The following notation and properties of independence are sometimes useful in studying Bayesian-based reasoning in expert systems.

Theorem 7.5   Independence Properties of Random Variables

Let $\mathcal{W}, \mathcal{X}, \mathcal{Y}, \mathcal{Z}$ be pairwise disjoint sets of random variables on a probability space. Let $I(\mathcal{X}, \mathcal{Y}, \mathcal{Z})$ indicate that $\mathcal{X}$ and $\mathcal{Z}$ are independent given $\mathcal{Y}$; that is, $\Pr(\mathcal{X} \cup \mathcal{Z} \mid \mathcal{Y}) = \Pr(\mathcal{X} \mid \mathcal{Y}) \Pr(\mathcal{Z} \mid \mathcal{Y})$ or, equivalently, $\Pr(\mathcal{X} \mid \mathcal{Y} \cup \mathcal{Z}) = \Pr(\mathcal{X} \mid \mathcal{Y})$. Then

(a) $I(\mathcal{X}, \mathcal{Z}, \mathcal{Y})$ implies $I(\mathcal{Y}, \mathcal{Z}, \mathcal{X})$;

(b) $I(\mathcal{X}, \mathcal{Z}, \mathcal{Y} \cup \mathcal{W})$ implies $I(\mathcal{X}, \mathcal{Z}, \mathcal{Y})$;

(c) $I(\mathcal{X}, \mathcal{Z}, \mathcal{Y} \cup \mathcal{W})$ implies $I(\mathcal{X}, \mathcal{Z} \cup \mathcal{W}, \mathcal{Y})$;

(d) $I(\mathcal{X}, \mathcal{Z}, \mathcal{Y})$ and $I(\mathcal{X}, \mathcal{Z} \cup \mathcal{Y}, \mathcal{W})$ together imply $I(\mathcal{X}, \mathcal{Z}, \mathcal{Y} \cup \mathcal{W})$.

Proof: We will prove (c). The other parts are left as an exercise. First, I'll translate I notation into Pr notation: In conditional probability notation, we must show that

$$\Pr(\mathcal{X} \mid (\mathcal{Y} \cup \mathcal{W}) \cup \mathcal{Z}) = \Pr(\mathcal{X} \mid \mathcal{Z}) \tag{7.33}$$

implies

$$\Pr(\mathcal{X} \mid \mathcal{Y} \cup (\mathcal{Z} \cup \mathcal{W})) = \Pr(\mathcal{X} \mid \mathcal{Z} \cup \mathcal{W}).$$

Since the left sides are equal, it suffices to show that (7.33) implies

$$\Pr(\mathcal{X} \mid \mathcal{Z}) = \Pr(\mathcal{X} \mid \mathcal{Z} \cup \mathcal{W});$$

which is equivalent to

$$\frac{\Pr(\mathcal{X} \cup \mathcal{Z})}{\Pr(\mathcal{Z})} = \frac{\Pr(\mathcal{X} \cup \mathcal{Z} \cup \mathcal{W})}{\Pr(\mathcal{Z} \cup \mathcal{W})}. \tag{7.34}$$

Similarly, (7.33) can be written as

$$\frac{\Pr(\mathcal{X} \cup \mathcal{Y} \cup \mathcal{W} \cup \mathcal{Z})}{\Pr(\mathcal{Y} \cup \mathcal{W} \cup \mathcal{Z})} = \frac{\Pr(\mathcal{X} \cup \mathcal{Z})}{\Pr(\mathcal{Z})}. \tag{7.35}$$

If we can eliminate $\mathcal{Y}$ from (7.35), the proof will be done. We can do this by first clearing fractions and then summing over all possible values for $\mathcal{Y}$. To see why this summation works and how it uses the fact that $\mathcal{Y}$ and $\mathcal{X} \cup \mathcal{Z} \cup \mathcal{W}$ are disjoint, see Example 7.18. ■

Exercises

7.3.C. Explain the notation $\Pr(\{X\} \cup \{Y\}) = \Pr(\{X, Y\}) = \Pr(X \wedge Y)$.

7.3.D. What does it mean for the random variables X and Y to be independent given Z?

7.3.E. What does $\Pr(\mathcal{X})$ mean when $\mathcal{X}$ is a set of random variables?

7.3.F. What does it mean for the sets $\mathcal{X}$ and $\mathcal{Y}$ of random variables to be independent given $\mathcal{Z}$?

7.3.G. Explain the notation $I(\mathcal{X}, \mathcal{Y}, \mathcal{Z})$.

7.3.11. Show that when $\mathcal{X} = \emptyset$ we should take $\Pr(\mathcal{X}) = 1$ to make equations work out correctly. Do it in at least the following two distinct ways:

    (a) Look at the corresponding compound event, using the fact that an intersection of no sets is taken to be the "universe."

    (b) Look at $\sum_{\mathcal{Y}} \Pr(\mathcal{X} \cup \mathcal{Y}) = \Pr(\mathcal{X})$ and evaluate the sum.

7.3.12. Let $\mathcal{Y}_i \subseteq \mathcal{X}_i$ for $1 \leq i \leq n$. Show that if $\mathcal{X}_1, \ldots, \mathcal{X}_n$ are independent given $\mathcal{W}$, then so are $\mathcal{Y}_1, \ldots, \mathcal{Y}_n$. Following the idea in the previous exercise, use $\Pr(\mathcal{X}_i \mid \mathcal{W}) = 1$ when $\mathcal{X}_i = \emptyset$.

7.3.13. Using independence terminology, express the four parts of Theorem 7.5 in words.

7.3.14. Using conditional probability terminology, express the four parts of Theorem 7.5 in words.

7.3.15. This exercise deals with breaking records as in sports and weather. Let $X_1, X_2, \ldots$ be a sequence of measurements (e.g., annual rainfall). Thus X_j is the measurement at time j. Although it is probably incorrect, assume that the X_j's are mutually independent and have the same probability distributions. For simplicity also assume that $\Pr(X_i = X_j) = 0$ whenever $i \neq j$. This can't be correct for finite probability spaces, but it can be close to correct.

(a) Show that the probability that X_j is the largest of $X_1, \ldots, X_t$ equals $1/t$.

(b) What is the probability that X_t is a new record?

(c) Show that the probability that X_{t+k} is a new record, given that X_t was the previous record, is $\dfrac{t}{(t+k)(t+k-1)}$.

7.3.16. Use $I(\mathcal{X}, \mathcal{Y}, \mathcal{Z})$ if and only if $I(\mathcal{Z}, \mathcal{Y}, \mathcal{X})$ to rearrange each of the I's in the remaining three parts of Theorem 7.5. Then, using conditional terminology, describe the new forms in words.

7.3.17. Complete the proof of Theorem 7.5.

Notes

There is what may seem a glaring omission in this chapter: Expectation and variance are not discussed. Since these concepts aren't really needed until later, I've postponed them until Section 12.1, which you can read now if you wish.

Neapolitan [12] discusses in some detail the controversy about relating probability to the real world. Another discussion is given by Shafer [14]. A more philosophical discussion is given by Cohen [1]. All give further references.

This chapter contains only the aspects of probability theory that are needed for expert systems. Some other aspects of probability theory will be discussed in a later chapter. In addition, there are many texts on the subject, ranging from useful elementary introductions such as that by Gnedenko and Khinchin [7] to highly mathematical texts. The classic mathematical introduction is the two-volume work by Feller [5].

Unfortunately, there is little introductory material devoted to those aspects of probability theory that are relevant to design and study of expert systems. The texts by Neapolitan [12], Pearl [13], and Kruse, Schwecke, and Heinsohn [10] are possible sources. In addition, some important papers in the area have been collected by Shafer and Pearl [15], who have written introductory material for the various parts of the collection.

Further discussion of the car and the goats example can be found in [6] and [11].

Various researchers have made proposals on how to incorporate $A|B$ into logic. The paper by Dubois and Prade [3] is a combination survey and research article. See also [8]. More recently, a special journal issue has appeared on conditional event algebras [2]. For further discussions of the relation between implication and conditional probability, see [4].

Bayes' Theorem plays an important role in some systems for uncertain reasoning. My approach to it has been traditional: I began with the necessary concepts and machinery from probability theory. It's possible to approach Bayes' Theorem from a viewpoint that is somewhat more congenial to expert systems: Start with a set of axioms that uncertain reasoning must obey and deduce Bayes' Theorem and probability theory as a consequence. Smith and Erickson [17] discuss Jaynes's way of doing this.

A point not emphasized in this chapter is the fact that Bayes' formula is not connected with causality. For example, it makes sense to talk about $\Pr(Y \mid X)$

- when X can cause Y,

- when Y can cause X, and

- when the relationship is more complicated as when X and Y have a common cause.

Jaynes [9] provides an interesting application of this lack of connection in the theory of diffusion.

Researchers have been working on various ideas for combining concepts from probability theory and logic. A couple of ideas related to conditional probability were discussed in Example 7.8 (p. 273). Another research topic is the possibility of approaching nonmonotonic logic through probability. We can cast ordinary logic in a probabilistic mode by using 0/1 valued probabilities. If we attempt to formulate a nonmonotonic logic by allowing arbitrary probabilities, many of the nice aspects of logic are lost. However, if we allow probabilities to differ from 0 and 1 by at most infinitesimals, we *can* develop an approach to nonmonotonic logic.

References

1. L. J. Cohen, *An Introduction to the Philosophy of Induction and Probability*, Clarendon Press, Oxford (1989).

2. D. Dubois, I. R. Goodman, and P. G. Calabrese (eds.), Special issue on conditional event algebra, *IEEE Transactions on Systems, Man and Cybernetics* **24** no. 12 (1994) 1665–1766.

3. D. Dubois and H. Prade, The logical view of conditioning and its application to possibility and evidence theories, *Intl. J. Approximate Reasoning* **4** (1990) 23–46.

4. E. Eells and B. Skyrms (eds.), *Probability and Conditionals: Belief Revision and Rational Decision*, Cambridge University Press, Cambridge, Great Britain (1994).

5. W. Feller, *An Introduction to Probability Theory and Its Applications*, Vol. 1 (3d ed.) and Vol. 2 (2d ed.), John Wiley and Sons, New York (1968, 1971).

6. L. Gillman, The car and the goats, *American Mathematical Monthly* **99** (1992) 3–7.

7. B. V. Gnedenko and A. Ya. Khinchin, *An Elementary Introduction to the Theory of Probability*, translated from the fifth Russian edition by L.F. Boron, Dover Publications, New York (1962).

8. I. R. Goodman, H. T. Nguyen, and E. A. Walker, *Conditional Inference and Logic for Intelligent Systems: A Theory of Measure-Free Conditioning*, North-Holland, Amsterdam (1991).

9. E. T. Jaynes, Clearing up mysteries—the original goal. In [16] 1–27.

10. R. Kruse, E. Schwecke, and J. Heinsohn, *Uncertainty and Vagueness in Knowledge Based Systems*, Springer-Verlag, Berlin (1991).

11. J. P. Morgan, N. R. Chaganty, R. C. Dahiya, and M. J. Doviak, Let's make a deal: The player's dilemma, *The American Statistician* **45** (1991) 284–287.

12. R. E. Neapolitan, *Probabilistic Reasoning in Expert Systems: Theory and Applications*, John Wiley and Sons, New York (1990).

13. J. Pearl, *Probabilisitic Reasoning in Intelligent Systems*, Morgan Kaufmann, San Mateo, CA (1988).

14. G. Shafer, What is probability? In D. C. Hoaglin and D. S. Moore (eds.), *Perspectives on Contemporary Statistics*, MAA Notes No. 21, Mathematical Association of America (1992).

15. G. Shafer and J. Pearl (eds.), *Readings in Uncertain Reasoning*, Morgan Kaufmann, San Mateo, CA (1990).

16. J. Skilling (ed.), *Maximum Entropy and Bayesian Methods*, Kluwer Academic Publishers, Dordrecht (1989).

17. C. R. Smith and G. Erickson, From rationality and consistency to Bayesian probability. In [16] 29–44.

8

Bayesian Networks

All science is concerned with the relationship of cause and effect.
—Laurence J. Peter and Raymond Hull (1969)

Introduction

Causality is the boon of our existence and the bane of our mathematics. We rely on causality in our interactions with reality, but we find it difficult to capture in our abstractions.

Example 3.5 (p. 97) discusses the impossibility of capturing causality in the context of standard logic. Although implication is the obvious choice for doing so, it can't be used. One problem is directionality:

(a) Since causes precede effects in time, it's essential to have a unidirectional process to model causality. In other words, if A is a cause of B, then B occurs later than A.

(b) In FOL, $\alpha \rightarrow \beta$ is equivalent to $(\neg\beta) \rightarrow (\neg\alpha)$, so there is no directionality to model causality.

This problem was overcome somewhat by rule systems for which the interpretation of "if–then" statements is unidirectional; that is, "if α then β" does not imply "if not β then not α." Semantic nets also incorporate directionality by using a *directed* graph. This suggests that some sort of directed graph structure might be useful for describing causality.

The concept of conditional probability is suggestive of causality, but as we saw in the previous chapter, conditional probability and Bayes' Theorem have no direct connection with causality (p. 296). Overcoming this by imposing a directed graph structure on the event space $\mathcal{E}$ is the idea behind Bayesian nets.

Bayesian Nets

A Bayesian (or causal) network has a structural part reflecting causal relationships and a probability part reflecting the strengths of the relationships. A user of a Bayesian network expert system specifies observations, each observation being information about the value of a random variable. The user may then ask for the probability of another random variable. For example, given certain geological observations ("instantiations" of random variables), what is the probability that drilling for oil will bring in a well? A user might ask other questions. For example, given some symptoms and medical test results, what are the most probable diagnoses and what are their probabilities? Like everything in AI, Bayesian networks have pros (+) and cons (−):

— In the general case, the only known algorithms require prohibitive amounts of data and calculation.

+ There is a solid theoretical base—probability theory and causality—rather than an uncertain heuristic one.

— Bayesian networks may be the wrong tool for the problem. (See the next chapter for some alternatives.)

+ Since results are given in terms of probabilities, they have a readily understood meaning.

— Numerical values give a false sense of accuracy since they are inevitably based on an inaccurate knowledge base.

The first objection is circumvented by making assumptions that limit the structure of the causal connections, the form of the probabilities, or both.

The last objection, regarding knowledge-base accuracy, would be weakened considerably if we had a practical method for doing a *sensitivity analysis* of a Bayesian network, that is, for estimating the effect changes in the values of data will have on predictions. Of course, this could be done by conducting a *huge* number of experiments with the network. That's seldom practical. Some empirical results indicate that predictions are often rather insensitive to numerical changes in the data. General theoretical results are needed.

In the next section, we get a first look at Bayesian networks. This introduction provides enough background for a more detailed perspective. The first application of Bayesian networks is to the "diagnosis problem" for some simple networks. Then, two sections are devoted to singly connected networks. The final application is a discussion of the limitations of certainty factors, using an axiomatic approach as we did for fair elections in the Arrow Impossibility Theorem (p. 194).

> Note: You can read the discussions of the three applications—diagnosis, singly connected networks and certainty factors—independently of one another.

Prerequisites: The concepts from probability theory introduced in Chapter 7 are used throughout. Basic graph theory definitions from Section 2.1 are also needed.

Used in: No chapters require this one; however, the introductory discussion in Section 9.1 refers to the basic concepts introduced in Section 8.1. Although it's not necessary, you may want to read Section 8.1 before Chapter 9.

Exercises

8.A. What are the two major components of Bayesian networks?

8.B. Give at least one positive and one negative aspect of Bayesian networks.

8.1 Bayesian Networks

> *Mathematicians are like Frenchmen; whenever you say something to them, they translate it into their own language, and at once it is something entirely different.*
> —Johann Wolfgang Goethe (1829)

Human reasoning about reality appears to be qualitative and heavily dependent on causality. AI reasoning methods can break down because they fail to come to grips with causality. Human reasoning methods can break down because they fail to come to grips with numerical aspects of uncertainty. A computer-based system incorporating both causality and numerical uncertainty might prove quite useful for dealing with situations where human reasoning has difficulty. Probability is the standard mathematical tool for dealing with uncertainty. A tool for representing a collection of causally related events is the directed acyclic graph—DAG. The combination of these concepts leads to a Bayesian network. In it, conditional probability and independence are linked with the notion of causality.

The goal of this section is to introduce the concept of a Bayesian network and, in Theorem 8.2, to establish its most important property.

Is all this concern with causality really necessary? The next example shows that there is a need for something like causality when we lack certainty.

Example 8.1   Causality Is Needed

Imagine two rules

"If B then A."   and   "If C then A."

which are taken to be true with some probability. If we learn that A is more likely to be true, then we might expect to increase the chances of both B and C. This is quite reasonable when A can cause B and C. It is also quite reasonable if B and C could each cause A.

Now suppose that we learn that B is likely to be true. This is evidence for A which is in turn evidence for C, so the probability of both A and C should increase. Again, this is reasonable if A can cause B and C. If B and C could each cause A, it is no longer reasonable to expect C's probability to increase, as the following illustration shows.

As I look out the window it appears that (A) "the sidewalk outside my front door is damp this morning," but I can't tell for sure because it is still fairly dark out. My opinion is evidence E_1 for A. This could have been caused by (B) "rain last night," (C) "the automatic lawn sprinkler system was on," or maybe something else. I might estimate $\Pr(A \mid E_1)$, $\Pr(B \mid E_1)$, and $\Pr(C \mid E_1)$. Now I get some evidence E_2 for B: The radio announcer says that we had rain showers last night. With this new evidence, $\Pr(B \mid E_1 \cup E_2)$ and $\Pr(A \mid E_1 \cup E_2)$ should be larger than the probabilities conditioned on just E_1, but $\Pr(C \mid E_1 \cup E_2)$ should be less! ■

We've just seen that the direction of causality in if–then rules may be important if we want to combine evidence from two or more rules. The easiest way to keep track of this in a rule-based setting is to require that causes always be on the same side of rules. Another approach is to describe causality with DAGs, which is the purpose of this section.

Directed Acyclic Graphs

At this time, you may wish to review the definition of directed graphs (digraphs) and cycles given in Section 2.1. Recall that:

Definition 8.1   Directed Graphs

Let V be a set. A *directed graph*, or *digraph*, is V together with a subset E of $V \times V$. We refer to V as the *vertices* of the digraph and to E as the *edges* of the digraph. We denote the digraph by (V, E).

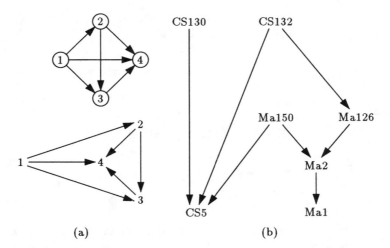

Figure 8.1 Two directed acyclic graphs (DAGs): (a) Two representations of a DAG which has $V = \{1, 2, 3, 4\}$ and has $(u, v) \in E$ whenever $u < v$; (b) some dependencies among imaginary mathematics and computer science courses, with an edge from B to A if A is a prerequisite for B.

A digraph is represented pictorially as follows. The vertices $v \in V$ are indicated by their names, possibly circled. An edge $(u, v) \in E$ is represented by a line or a curve connecting the representations of u and v, with an arrowhead indicating the direction from u to v. We usually abuse terminology and refer to such a representation as if it were the digraph. Since all that matters for a digraph is V and E, the shapes and positioning of the representations of the vertices as well as the shapes and crossings of the curves representing the edges are irrelevant. Figure 8.1 contains pictorial representations of two digraphs.

Definition 8.2 Acyclic Digraphs and Related Concepts

Recall from Section 2.1 that a cycle is a list of distinct vertices $v_1, \ldots, v_n$ such that (v_n, v_0) and all the (v_i, v_{i+1}) are edges. A digraph without directed cycles is called *acyclic* and is referred to as a *directed acyclic graph* or a *DAG*.

Let (V, E) be a DAG and let $v \in V$. We define

$$\mathcal{C}(v) = \big\{\, u \in V \mid (u, v) \in E \,\big\};$$
$$\mathcal{D}(v) = \big\{\, w \in V \mid \text{there is a path from } v \text{ to } w \,\big\};$$
$$\mathcal{A}(v) = \big\{\, x \in V \mid x \neq v \text{ and } x \notin \mathcal{C}(v) \cup \mathcal{D}(v) \,\big\}.$$

We refer to $\mathcal{C}(v)$ as the *parents* or *causes* of v and refer to $\mathcal{D}(v)$ as the *descendants* of v. For any subset W of V, define $\mathcal{C}(W)$ (resp. $\mathcal{D}(W)$) to be the union of $\mathcal{C}(w)$ (resp. $\mathcal{D}(w)$) over all $w \in W$.

Since a DAG has no cycles, $\mathcal{C}(v) \cap \mathcal{D}(v) = \emptyset$. Thus $\{v\}$, $\mathcal{A}(v)$, $\mathcal{C}(v)$, and $\mathcal{D}(v)$ are a partition of V—pairwise disjoint sets whose union is V.

Example 8.2   Notation

In Figure 8.1,

$$\mathcal{C}(\mathrm{Ma2}) = \{\,\mathrm{Ma126}, \mathrm{Ma150}\,\};$$
$$\mathcal{D}(\mathrm{Ma2}) = \{\,\mathrm{Ma1}\,\};$$
$$\mathcal{A}(\mathrm{Ma2}) = \{\,\mathrm{CS5}, \mathrm{CS130}, \mathrm{CS132}\,\};$$
$$\mathcal{D}(\mathrm{CS132}) = \{\,\mathrm{CS5}, \mathrm{Ma126}, \mathrm{Ma1}\,\}. \quad \blacksquare$$

With DAGs we have a potential tool for representing causality, which is an important aspect of everday reasoning. The idea is that (u, v) will be an edge if and only if u is an "immediate cause" of v. The digraph will be acyclic because the time of a cause is always earlier than the times of its effects. Unfortunately, the notion of an immediate cause is not immediately useful because

- It's not entirely clear what the concept means.

- Even if we knew what it meant, it's not clear what its mathematical implications are besides the DAG.

In other words, we have a somewhat vague, but apparently useful, concept whose translation into mathematics is unclear. This type of situation occurs often. How can it be handled? One approach begins by deciding what the implications of the concept—immediate cause—are in the mathematical framework we hope to use—probability theory. Next these mathematical consequences are taken as the basis of a definition of a mathematical concept. Finally, based on the original belief that the mathematical concept captures relevant properties of reality, we should be able to translate back and forth between reality and mathematics.

The idea in the previous paragraph has been the impetus for many areas of mathematics: measuring the earth inspired *geo*-metry, physical ideas inspired calculus, gambling inspired probability theory, and so forth.

Before formulating the mathematical consequences of immediate causes and connecting them with probability spaces, let's look at an example adapted from some of Judea Pearl's work.

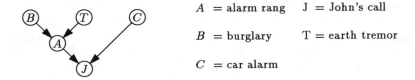

A = alarm rang   J = John's call

B = burglary   T = earth tremor

C = car alarm

Figure 8.2 The DAG for the burglar alarm example. Each edge connects an effect to one of its immediate causes, with the arrowhead indicating the effect.

Example 8.3   A Burglar Alarm: Introduction

Sally's home, which is south of San Francisco, is equipped with a burglar alarm. Upon returning to her office, Sally finds a message that her neighor John called five minutes ago to say that her burglar alarm was ringing. Unfortunately, John is somewhat unreliable—he might have heard the neighbor's car alarm. Sally knows that her alarm can be triggered by a minor earth tremor or a malfunction, as well as by burglars, but she thinks that tremors don't normally affect car alarms. Figure 8.2 shows the result of using a DAG to represent direct causes. (I omitted alarm malfunctions for reasons that will be discussed later.) Note that $\mathcal{C}(v)$ corresponds to the immediate causes of v and $\mathcal{D}(v)$ corresponds to the events on which v has a, possibly indirect, effect. ∎

The following simple theorem, which gives us the ability to order vertices in a convenient manner, will be useful in proving a key theorem about Bayesian networks.

Theorem 8.1

Let (V, E) be a DAG and let $u \in V$ be given. We may order the vertices, say as $v_1, v_2, \ldots, v_n$ so that

(a) for all $v \in V$, all vertices in $\mathcal{D}(v)$ follow v in the list and

(b) a vertex follows u in the list if and only if it lies in $\mathcal{D}(u)$.

Proof: Let's prove (a) by induction on the number of vertices. It is trivial when there is only one vertex. Since $|V| = n$ is finite and (V, E) contains no cycles, there must be some vertex $v_n \in V$ such that there is no edge of the form (v_n, v). Look at the DAG (V', E') where

$$V' = V - \{v_n\} \quad \text{and} \quad E' = E \cap (V' \times V'),$$

that is, all vertices except v_n and all edges in E that don't involve v_n. By the induction assumption, there is an ordering $v_1, \ldots, v_{n-1}$ of V' that satisfies (a). Since there is no edge of the form (v_n, v), the ordering $v_1, \ldots, v_n$ satisfies (a). This proves (a).

Now we can show that part (b) follows from (a). The set of vertices in $\mathcal{D}(u)$ together with the edges joining them form a DAG and so have an ordering $d_1, \ldots, d_k$ by (a). The set of vertices which are neither u nor in $\mathcal{D}(u)$ together with the edges joining them form a DAG and so have an ordering $v_1, \ldots, v_{n-k-1}$ by (a). It is easily verified that

$$v_1, \ldots, v_{n-k-1}, u, d_1, \ldots, d_k$$

is a ordering of the desired type. ∎

Bayesian Networks

At last, it's time to define a Bayesian network. The following definition is more restrictive than necessary, but is quite adequate for many purposes.

Definition 8.3   Bayesian Network

Let $(\mathcal{E}, \mathrm{Pr})$ be a probability space with $\mathcal{E} = \mathcal{E}_1 \times \cdots \times \mathcal{E}_k$. Let X_i be the projection onto $\mathcal{E}_i$. Let $(\mathcal{X}, E)$ be a DAG with $\mathcal{X} = \{X_1, \ldots, X_k\}$. We call $(\mathcal{X}, E, \mathrm{Pr})$ a *Bayesian network* (or *net*) if, for all $X_i \in V$ and all $\mathcal{W} \subseteq \mathcal{A}(X_i)$, X_i and $\mathcal{W}$ are independent given $\mathcal{C}(X_i)$. In the I notation of Theorem 7.5 (p. 293), $I(\{X_i\}, \mathcal{C}(X_i), \mathcal{W})$. In terms of Pr, if $\mathrm{Pr}\big(\mathcal{W} \cup \mathcal{C}(X_i)\big) \neq 0$, then

$$\mathrm{Pr}\big(X_i \mid \mathcal{W} \cup \mathcal{C}(X_i)\big) = \mathrm{Pr}\big(X_i \mid \mathcal{C}(X_i)\big). \tag{8.1}$$

Somewhat imprecisely in prose, if we know about the causes of X_i, nothing other than X_i itself or its descendants can give us any more information about X_i. A Bayesian network may also be referred to as a *causal network*, as a *belief network*, or as an *influence diagram*. I prefer "Bayesian" to "causal" because it alludes to the nature of the probabilistic reasoning associated with the network. "Belief" is unfortunate terminology because "belief functions" are a different, but related, topic.

There is no need to mention $\mathcal{E}$ in $(\mathcal{X}, E, \mathrm{Pr})$ because

$$\mathcal{E} = \mathrm{range}(X_1) \times \cdots \times \mathrm{range}(X_k). \tag{8.2}$$

Since an assignment of values to $X_1, \ldots, X_k$ gives a compound event that consists of just one elementary event, $(e_1, \ldots, e_k)$, we know Pr completely when we know $\mathrm{Pr}(\mathcal{X})$. Thus, the probability space in the definition is given completely by (8.2) and (8.1).

As discussed earlier, although the concept of causality is not defined mathematically, one of its important consequences is defined instead. This definition is equation (8.1). It expresses the idea that only immediate causes are relevant in determining the nature of an effect. Since immediate causation is represented by a DAG, (8.1) merges the qualitative representation (the

DAG) with the numerical consequences (independence). In doing this, we've dismissed causality from *mathematical* consideration.

Causality plays a role only by providing the means for translating a real-world situation into a Bayesian net. In fact, it's possible to create a Bayesian network whenever (8.2) holds—causality doesn't enter into it at all. (See Exercise 8.1.2.) Have we lost too much by reducing causality to (8.1)? Perhaps. Does this mathematical representation lead to useful tools? Perhaps. These sorts of issues were discussed in connection with FOL and Prolog, and we'll discuss them briefly for Bayesian nets later in this chapter.

Example 8.4   A Burglar Alarm: A Possible Bayesian Network

We can think of the burglar alarm situation introduced in Example 8.3 in terms of a Bayesian network by interpreting each of the vertices as a random variable. We can do this simply by using the corresponding letter to stand for a random variable that takes on the value y or n, for "yes" and "no." In these terms, we're interested in $\Pr(T = y \mid J = y)$.

At this time, look at each vertex in Figure 8.2, find the three sets $\mathcal{C}$, $\mathcal{D}$, and $\mathcal{A}$, and check for yourself that it is reasonable to believe that the figure would indeed represent a Bayesian network if accurate probabilities could be assigned to the various combinations of events; that is, verify that the conditions in Definition 8.3 seem likely to hold in the real world.

$$*\qquad*\qquad*\qquad\text{Stop and think about this!}\qquad*\qquad*\qquad*$$

We cannot expect to *prove* that the figure represents a Bayesian network because it is incomplete—there are no probabilities. Nevertheless, we can reasonably argue to ourselves that certain independencies would hold. For example, the chances that the car alarm went off are not affected by whether the burglar alarm went off or the house was burgled or an earth tremor occurred. (You might object to the last independency, but we're using Sally's perception and she thinks it doesn't matter.) We have just argued that

$$\Pr(C \mid \mathcal{W}) = \Pr(C) \text{ whenever } \mathcal{W} \subseteq \{A, B, T\}. \; \blacksquare$$

A direct verification that a probability space for the last example makes the DAG in Figure 8.2 into a Bayesian network could require a fair bit of hand calculation. We would have to be given the values for $\Pr(\{A, B, C, J, T\})$. Then we would have to check many independencies to verify the conditions in Definition 8.3. This can be avoided by the following procedure.

First argue that $(\mathcal{E}, E)$ should be the DAG of a Bayesian network by considering what causes what. Then estimate $\Pr$ by estimating $\Pr(X \mid \mathcal{C}(X))$. By Theorem 8.2, the only necessary constraints are that $0 \le \Pr(X \mid \mathcal{C}(X)) \le 1$ for all X and that $\sum_X \Pr(X \mid \mathcal{C}(X)) = 1$ for all X.

This procedure has three benefits. First, it reduces the amount of information that must be obtained and stored. Second, it greatly reduces the number of tests that must be performed on the information. Third, the numerical information is all local—conditional probabilities relating an effect to its immediate causes.

Theorem 8.2   Bayesian Network Probabilities

For the purposes of this theorem, call a function f a random variable function if

(i) its arguments are (sets of) random variables, but the domain is the values the random variables can assume and

(ii) its range is the nonnegative real numbers.

In other words, f behaves like Pr on its arguments.

Let $\mathcal{X} = \{X_1, \ldots, X_k\}$ be a set of functions, each of which has a finite range.

(a) If $(\mathcal{X}, E, \mathrm{Pr})$ is a Bayesian network, then

$$\mathrm{Pr}(\mathcal{X}) = \prod_{\substack{X \in \mathcal{X} \\ \mathrm{Pr}(\mathcal{C}(X)) \neq 0}} \mathrm{Pr}(X \mid \mathcal{C}(X)), \qquad (8.3)$$

where the notation in the product indicates that the product is taken over all $X \in \mathcal{X}$ for which $\mathrm{Pr}(\mathcal{C}(X)) \neq 0$.

(b) Conversely, if $(\mathcal{X}, E)$ is a DAG, and if $f(X \mid \mathcal{C}(X))$ is a random variable function such that $\sum_X f(X \mid \mathcal{C}(X)) = 1$, then

$$\mathcal{E} = \mathrm{range}(X_1) \times \cdots \times \mathrm{range}(X_k)$$

$$\mathrm{Pr}(\mathcal{X}) = \prod_{X \in \mathcal{X}} f(X \mid \mathcal{C}(X)) \qquad (8.4)$$

defines a probability space for which $(\mathcal{X}, E, \mathrm{Pr})$ is a Bayesian network. Furthermore, $\mathrm{Pr}(X \mid \mathcal{C}(X))$ is either 0 or $f(X \mid \mathcal{C}(X))$.

It's tempting to read through the theorem quickly and continue on. *Don't do that!* You should be able to say what a theorem means. Can you do it?

      *   *   *   Stop and think about this!   *   *   *

Since $\mathrm{Pr}(\mathcal{X})$ determines $(\mathcal{E}, \mathrm{Pr})$, the products in (a) and (b) completely determine $(\mathcal{E}, \mathrm{Pr})$. From (a), we see that probabilities in *any* Bayesian network can be obtained from a knowledge of the values of the conditional probabilities $\mathrm{Pr}(X_i \mid \mathcal{C}(X_i))$ simply by multiplying them together almost like independence. If we really had independence, we'd have $\mathrm{Pr}(\mathcal{X}) = \prod_{X \in \mathcal{X}} \mathrm{Pr}(X)$. Instead, we have to condition X on its immediate causes. From (b), we see that the *only* constraint that needs to be imposed on the conditional probabilities is that, for each i, the resulting numbers make sense as conditional probabilities.

From the theorem, it suffices to estimate the conditional probability of each X_i given its causes. Usually, this is more natural than estimating $\Pr(\mathcal{X})$ directly and involves the determination of fewer numbers. For example, the theorem tells us that Figure 8.2 requires the estimation of

$$\Pr(J \mid \{A,C\}), \ \Pr(A \mid \{B,T\}), \ \Pr(B), \ \Pr(C), \text{ and } \Pr(T).$$

When we take into account the fact that

$$\Pr(X = n \mid \mathcal{W}) = 1 - \Pr(X = y \mid \mathcal{W}),$$

this is a total of $4 + 4 + 1 + 1 + 1 = 11$ numbers. The direct approach requires $2^5 - 1 = 31$ numbers to estimate $\Pr(\mathcal{X})$ directly, followed by a verification that these estimates satisfy Definition 8.3. (Of course, they would only satisfy it approximately since they are estimates, so we would be faced with the question of how close is close enough.)

Example 8.5   A Burglar Alarm: A Bayesian Network

Let's finally make the burglar alarm example into a Bayesian network and, in the process, show that M can be eliminated. Sally estimated the following probabilities:

$$
\begin{array}{ll}
\Pr(A = y \mid B = y \wedge T = y) = 0.99, & \Pr(J = y \mid A = y \wedge C = y) = 0.6, \\
\Pr(A = y \mid B = y \wedge T = n) = 0.99, & \Pr(J = y \mid A = y \wedge C = n) = 0.6, \\
\Pr(A = y \mid B = n \wedge T = y) = 0.15, & \Pr(J = y \mid A = n \wedge C = y) = 0.6, \\
\Pr(A = y \mid B = n \wedge T = n) = 4\delta, & \Pr(J = y \mid A = n \wedge C = n) = 0, \\
\multicolumn{2}{c}{\Pr(B = y) = 0.1\delta, \ \Pr(C = y) = 12\delta, \ \Pr(T = y) = 3\delta,}
\end{array}
$$

where δ is some very small number. Sally doesn't know what value to assign to δ. How did Sally get these values and what is δ? For those probabilities without δ, she thought about past experience and her expectations. The others presented a bit of a problem. Sally thought that her burglar alarm went off for no apparent reason about 4 times a year, but she needed the probability that it would go off in some much shorter time period. How long a time period should she allow? 5 minutes? 30 minutes? She couldn't decide, but she reasoned it would be some small fraction δ of a year. In a similar fashion, she set $\Pr(B = y) = 0.1\delta$ because she thought that people were burgled about once every 10 years in her neighborhood.

After you think about the situation a bit, you should be able to convince yourself that, in the real world, the value of δ shouldn't matter. What happens in the Bayesian network model? When the calculations are carried out, which we won't do, it turns out that the value of δ is unimportant as long as it is quite small. ∎

We can imagine building a Bayesian network by consulting an expert. First we discover (or invent) the relevant random variables for the network and determine which are direct causes of which others. Having constructed the DAG, we elicit the conditional probabilities that Theorem 8.2 requires.

Exercises

8.1.A. What does DAG stand for? Define a DAG.

8.1.B. Define a Bayesian network. What do $\mathcal{A}(X)$, $\mathcal{C}(X)$, and $\mathcal{D}(X)$ denote?

8.1.C. How does a Bayesian network encompass the notion of causality? (The answer involves both the graphical structure and the probabilistic structure.)

8.1.D. What is the content and importance of Theorem 8.2?

8.1.1. For each vertex in Figure 8.2, find the three sets $\mathcal{C}$, $\mathcal{D}$, and $\mathcal{A}$. Interpreting the vertices in the figure as suggested in Example 8.5, give a commonsense argument that the requirements for a Bayesian network are satisfied in the case of T and every subset $\mathcal{W}$ of $\mathcal{A}(T)$.

8.1.2. Let $\mathcal{X} = \{X_1, \ldots, X_k\}$ be random variables on a probability space $(\mathcal{E}, \Pr)$ for which
$$\mathcal{E} = \operatorname{range}(X_1) \times \cdots \times \operatorname{range}(X_k).$$
Let $E = \big\{ (X_i, X_j) \,\big|\, 1 \le i < j \le k \big\}$. Prove that $(\mathcal{X}, E, \Pr)$ is a Bayesian network.

8.1.3. In Example 8.4, we argued for the condition in Definition 8.3 when $X_i = C$. Do this for the four other random variables: A, B, J, and T.

8.1.4. Two types of causes are necessary causes and sufficient causes. Let $\mathcal{Z} \subseteq \mathcal{C}(X)$. We say

(i) $\mathcal{Z}$ is a *necessary cause* of X if, whenever X occurs, we know that $\mathcal{Z}$ *must* have occurred to (help) produce X and

(ii) $\mathcal{Z}$ is a *sufficient cause* of X if, whenever $\mathcal{Z}$ occurs it *must* produce X.

(a) Show that (i) is equivalent to
$$\Big(\Pr(X = \mathrm{T} \mid \mathcal{C}(X)) > 0\Big) \implies \Big(Z = \mathrm{T} \text{ for all } Z \in \mathcal{Z}\Big).$$

(b) State and prove a similar result for (ii).

8.1.5. We want to represent the following situation with a Bayesian network. If I have enough money, I will probably buy a car or a boat, but not both. Here is a proposed DAG. There is a random variable M that indicates whether I have enough money. It is connected by an edge to B and by an edge to C, which are random variables indicating whether or not I buy a boat or a car.

(a) Explain why this Bayesian network is incorrect.

(b) Construct a correct network.

Proof of Theorem 8.2

To prove the theorem we must complete the following four steps:

1. Verify that the definition of a Bayesian network implies (8.3).
2. Verify that the function Pr defined in (8.4) does in fact give a probability space.
3. Verify the equality of the conditional probabilities and f.
4. Verify that $(\mathcal{X}, E, \mathrm{Pr})$ defined in (b) is a Bayesian network.

From now on, we'll assume that $X_1, \ldots, X_k$ have been ordered according to Theorem 8.1 where X_t plays the role of u and the subscripts on the X_i's reflect this ordering. Let $\mathcal{X}_0 = \emptyset$ and $\mathcal{X}_i = \{ X_j \mid j \leq i \}$ for $i > 0$. Since $\mathrm{Pr}(\mathcal{X}_0)$ is the probability of the compound event in which no constraints are imposed on the random variables, we have $\mathrm{Pr}(\mathcal{X}_0) = 1$.

Step 1: Suppose that $\mathrm{Pr}(\mathcal{X}_i) \neq 0$. Using the definition of conditional probability, it is easily shown that

$$\mathrm{Pr}(\mathcal{X}_i) = \prod_{j \leq i} \mathrm{Pr}(X_j \mid \mathcal{X}_{j-1}). \tag{8.5}$$

By the ordering of the vertices, $\mathcal{D}(X_j) \cap \mathcal{X}_j = \emptyset$; and so, from Definition 8.3 (p. 306), $\mathrm{Pr}(X_j \mid \mathcal{X}_{j-1}) = \mathrm{Pr}(X_j \mid \mathcal{C}(X_j))$. Using this in (8.5), we obtain

$$\mathrm{Pr}(\mathcal{X}_i) = \prod_{j \leq i} \mathrm{Pr}(X_j \mid \mathcal{C}(X_j)).$$

If $\mathrm{Pr}(\mathcal{X}) \neq 0$, we set $i = k$ to prove (8.3). Otherwise, there is an i such that $\mathrm{Pr}(\mathcal{X}_i) = 0$ and $\mathrm{Pr}(\mathcal{X}_{i-1}) \neq 0$. This implies that $\mathrm{Pr}(X_i \mid \mathcal{C}(X_i)) = 0$, from which it follows that both sides of (8.3) are zero. This completes Step 1.

Step 2: Since the function defined in (8.4) takes on only nonnegative real values, all that is needed for Step 2 is to show that summing the function over $\mathcal{E}$ gives 1. Define

$$g(\mathcal{X}_i) = \prod_{j \leq i} f(X_j \mid \mathcal{C}(X_j)). \tag{8.6}$$

Note that $\mathcal{X}_0 = \emptyset$ and so $g(\mathcal{X}_0)$ is the empty product, which equals 1 by standard mathematical convention. Since X_i appears only in the $j = i$ term in the product in (8.6),

$$\sum_{X_i} g(\mathcal{X}_i) = \left(\prod_{j<i} f(X_j \mid \mathcal{C}(X_j)) \right) \sum_{X_i} f(X_i \mid \mathcal{C}(X_i)).$$

Since the last summation is 1 by assumption, we've shown that

$$g(\mathcal{X}_0) = 1 \quad \text{and} \quad \sum_{X_i} g(\mathcal{X}_{i+1}) = g(\mathcal{X}_i) \quad \text{if } i \geq 0. \tag{8.7}$$

It follows by induction on i that $\sum_{X_1,\ldots,X_i} g(\mathcal{X}_{i+1}) = 1$ for all i. Setting $i = k$ completes Step 2; that is, we have a probability space.

A Summation: We now study a summation that arises in Steps 3 and 4. From the previous paragraph, $g(\mathcal{X}_{k+1}) = \Pr(\mathcal{X})$. Summing over $X_{i+1}, \ldots, X_k$ gives

$$g(\mathcal{X}_i) = \Pr(\mathcal{X}_i). \tag{8.8}$$

Suppose that $\mathcal{Y} \subseteq \mathcal{X}_i$ and that $\mathcal{Y} \cap \mathcal{C}(X_i) = \emptyset$. Using (8.8) and some definitions, we have

$$\sum_{\mathcal{Y}} \Pr(\mathcal{X}_i) = \sum_{\mathcal{Y}} g(\mathcal{X}_i) = \sum_{\mathcal{Y}} \prod_{j \leq i} f(X_j \mid \mathcal{X}_j)$$

$$= f(X_i \mid \mathcal{C}(X_i)) \sum_{\mathcal{Y}} \prod_{j<i} f(X_j \mid \mathcal{C}(X_j))$$

$$= f(X_i \mid \mathcal{C}(X_i)) \sum_{\mathcal{Y}} g(\mathcal{X}_{i-1}).$$

Thus

$$\sum_{\mathcal{Y}} \Pr(\mathcal{X}_i) = f(X_i \mid \mathcal{C}(X_i)) \sum_{\mathcal{Y}} \Pr(\mathcal{X}_{i-1}). \tag{8.9}$$

Step 3: Let $\mathcal{Y} = \mathcal{X}_{i-1} - \mathcal{C}(X_i)$, that is, those X_j for which $j < i$ and $X_j \notin \mathcal{C}(X_i)$. If $\Pr(\mathcal{C}(X_i)) \neq 0$, then

$$\Pr(X_i \mid \mathcal{C}(X_i)) = \frac{\sum_{\mathcal{Y}} \Pr(\mathcal{X}_i)}{\sum_{X_i} \sum_{\mathcal{Y}} \Pr(\mathcal{X}_i)}.$$

First, interchange the sums in the denominator and evaluate the sum over X_i, then apply (8.9) to the numerator to obtain

$$\Pr(X_i \mid \mathcal{C}(X_i)) = \frac{f(X_j \mid \mathcal{C}(X_j)) \sum_{\mathcal{Y}} \Pr(\mathcal{X}_{i-1})}{\sum_{\mathcal{Y}} \Pr(\mathcal{X}_{i-1})} = f(X_j \mid \mathcal{C}(X_j)).$$

This completes Step 3.

Step 4: To verify that we have a Bayesian network, we'll take $i = t$. (Recall that X_t equals the u of Theorem 8.1.) We must show that if $\mathcal{W} \subseteq \mathcal{A}(X_t)$, then

$$\Pr(X_t \mid \mathcal{C}(X_t) \cup \mathcal{W}) = \Pr(X_t \mid \mathcal{C}(X_t)).$$

We can easily see that $\mathcal{A}(X_t) \subset \mathcal{X}_{t-1}$ and so $\mathcal{W} \subset \mathcal{X}_t$. Let $\mathcal{Y}$ consist of all X_j with $j < i$ that are not in $\mathcal{C}(X_t) \cup \mathcal{W}$. Then

$$\Pr(\{X_t\} \cup \mathcal{C}(X_t) \cup \mathcal{W}) = \sum_{\mathcal{Y}} \Pr(\mathcal{X}_t).$$

Since $\mathcal{Y} \subseteq \mathcal{X}_{t-1}$ and $X_t \notin \mathcal{X}_{t-1}$, it follows that

$$\sum_{\mathcal{Y}} \Pr(\mathcal{X}_t) = f(X_t \mid \mathcal{C}(X_t)) \sum_{\mathcal{Y}} \Pr(\mathcal{X}_{t-1})$$

by (8.9), and so

$$\sum_{X_t} \sum_{\mathcal{Y}} \Pr(\mathcal{X}_t) = \left(\sum_{X_t} f(X_t \mid \mathcal{C}(X_t)) \right) \left(\sum_{\mathcal{Y}} \Pr(\mathcal{X}_{t-1}) \right).$$

Combining these with the definition of conditional probability, we have

$$\begin{aligned}
\Pr(X_t \mid \mathcal{C}(X_t) \cup \mathcal{W}) &= \frac{\Pr(\{X_t\} \cup \mathcal{C}(X_t) \cup \mathcal{W})}{\sum_{X_t} \Pr(\{X_t\} \cup \mathcal{C}(X_t) \cup \mathcal{W})} \\
&= \frac{f(X_t \mid \mathcal{C}(X_t)) \sum_{\mathcal{Y}} \Pr(\mathcal{X}_{t-1})}{\left(\sum_{X_t} f(X_t \mid \mathcal{C}(X_t)) \right) \left(\sum_{\mathcal{Y}} \Pr(\mathcal{X}_{t-1}) \right)} \\
&= \frac{f(X_t \mid \mathcal{C}(X_t))}{\sum_{X_t} f(X_t \mid \mathcal{C}(X_t))},
\end{aligned}$$

which equals $\Pr(X_t \mid \mathcal{C}(X_t))$ since $f(X_t \mid \mathcal{C}(X_t)) = \Pr(X_t \mid \mathcal{C}(X_t))$. ∎

Exercises

8.1.6. Prove (8.5). Remember to treat the case when probabilities are zero.

8.1.7. Fill in the steps leading to (8.9) more carefully.

8.1.8. The purpose of this exercise is to show that if the vertices of a Bayesian network can be divided into two sets that have no edges connecting them, we can regard it as two separate causal networks.
Let $(\mathcal{X}, E)$ and $(\mathcal{Y}, F)$ be DAGs.

(a) Prove that $(\mathcal{X} \cup \mathcal{Y}, E \cup F)$ is a DAG and show how to construct a picture of it from the pictures of the DAGs $(\mathcal{X}, E)$ and $(\mathcal{Y}, F)$.

(b) Let $(\mathcal{X}, E, \Pr_1)$ and $(\mathcal{Y}, F, \Pr_2)$ be Bayesian networks with $\mathcal{X} \cap \mathcal{Y} = \emptyset$. Define $\Pr(\mathcal{X} \cup \mathcal{Y}) = \Pr_1(\mathcal{X}) \Pr_2(\mathcal{Y})$. Prove that $(\mathcal{X} \cup \mathcal{Y}, E \cup F, \Pr)$ is a Bayesian network.
Hint. The most involved part is verifying (8.1). It can be done with summation arguments or by clever use of Theorem 7.5 starting with $I(\mathcal{Y}, \emptyset \mathcal{X})$ and the separate Bayesian network (8.1).

(c) Let $(\mathcal{X} \cup \mathcal{Y}, E \cup F, \Pr)$ be a Bayesian network. Prove that

$$\Pr(\mathcal{X} \cup \mathcal{Y}) = \Pr(\mathcal{X}) \Pr(\mathcal{Y})$$

and that $(\mathcal{X}, E, \Pr)$ and $(\mathcal{Y}, F, \Pr)$ are Bayesian networks.

8.2   Some Types of Bayesian Networks

> *The chief drawback with using probability theory is the complexity that sometimes results, and the need to assess an often surprisingly large number of conditional probabilities.*
>
> —Stephen R. Watson

In this section we'll take a first look at the three types of Bayesian networks that are explored in this chapter and indicate the nature of the principal results. First, a bit of terminology.

Definition 8.4   Notation

- Similar to $\mathcal{C}(X)$, the immediate causes of X, define the immediate results of X by

$$\mathcal{R}(X) = \big\{ \, Y \mid (X, Y) \text{ is an edge} \, \big\}.$$

- A random variable is said to be *instantiated* if it has been assigned a value.

- A set of random variables with the superscript asterisk (or star), as in $\mathcal{X}^*$, will indicate that each of the random variables is to be instantiated. If $U \in \mathcal{X}^*$, we'll usually denote the value by u^*; that is, we'll be setting $U = u^*$ which is the compound event $U^{-1}(u^*)$.

- A set of random variables with the superscript plus, as in $\mathcal{X}^+$, will indicate that each of the random variables in the set is to be instantiated to T (true).

- A set of random variables with the superscript minus, as in $\mathcal{X}^-$, will indicate that each of the random variables in the set is to be instantiated to F (false).

For example, if $\mathcal{X}^+ = \{X_2\}$, $\mathcal{X}^- = \{X_1, X_4\}$, and $\mathcal{R}^+ = \{R_3, R_5\}$, then the notation $\Pr(\mathcal{R}^+ \mid \mathcal{X}^+ \wedge \mathcal{X}^-)$ means

$$\Pr\!\Big((R_3 = \text{T}) \wedge (R_5 = \text{T}) \,\Big|\, (X_2 = \text{T}) \wedge (X_1 = \text{F}) \wedge (X_4 = \text{F}) \Big).$$

We'll usually make the following compromise It's not always necessary, but it simplifies the text:

> **Compromise:** Assume that the only information provided by the user is the instantiation of certain random variables.

Recall that two sets are called *disjoint* if their intersection is empty. The problem for the expert system can now be phrased somewhat vaguely as follows:

> **Problem:** Given a Bayesian network and disjoint sets $\mathcal{Y}^*$ and $\mathcal{Z}$ of random variables, "study" $\Pr(\mathcal{Z} \mid \mathcal{Y}^*)$.   (8.10)

The situation is still too complicated. No reasonable algorithms are available in the general case. Nor is any likely to be found because the general problem is NP-hard. (See [4].) As a result,

> **Compromise:** To obtain a reasonable worst-time algorithm, we restrict the structure of the DAG in some manner.

Bipartite Multiple-Diagnosis Problems

In some situations there may be many causes that could produce any given result; for example, in medical diagnosis many diseases could cause a particular symptom. Furthermore, there may be more than one underlying cause that is producing the observed results. Here's a definition of the problem.

Definition 8.5   Diagnosis Problem and Abductive Inference

Given a Bayesian network $(\mathcal{X}, E, \Pr)$ in which the range of each variable is $\{\text{T}, \text{F}\}$, let

$$\mathcal{C} = \{\, X \in \mathcal{X} \mid \mathcal{C}(X) = \emptyset \,\}$$
$$\mathcal{R} = \{\, X \in \mathcal{X} \mid \mathcal{D}(X) = \emptyset \,\}.$$

Let $\mathcal{R}^+$ and $\mathcal{R}^-$ be disjoint subsets of $\mathcal{R}$. The *diagnosis problem* is to partition $\mathcal{C}$ into $\mathcal{C}^+$ and $\mathcal{C}^-$ so that $\Pr(\mathcal{C}^+ \wedge \mathcal{C}^- \mid \mathcal{R}^+ \wedge \mathcal{R}^-)$ is as large as possible. The process of looking for probable causes based on their effects is referred to as *abductive inference*.

In the medical arena, we may think of $\mathcal{R}$ as the set of possible symptoms. By observation and tests, it has been determined that that patient has those symptoms in $\mathcal{R}^+$ and does not have those in $\mathcal{R}^-$. Symptoms in $\mathcal{R}$ that are in neither $\mathcal{R}^+$ nor $\mathcal{R}^-$ have not been examined for this patient. The set $\mathcal{C}$ consists of the causes (diseases) that could produce the symptoms. The solution to the diagnosis problem is the most probable explanation for the symptoms. Obviously, we may also wish to find a second best explanation, third best, and so on.

There are three major difficulities with general diagnosis:

- No reasonable algorithms are available for Bayesian networks with arbitrary DAGs. This problem will be overcome by restricting attention to "bipartite" DAGs.

- If $|\mathcal{C}(X)| = n$, the values of $\Pr(X \mid \mathcal{C}(X))$ are 2^n parameters that must be estimated. The available data often makes such estimates impossible even when n is only 4 or 5. This problem will be overcome by using "noisy-OR" functions, thereby reducing the number of parameters from 2^n to n.

- Instead of a specific probability, we want to find those sets of diseases that are most likely to produce the observed symptoms. *A priori*, this means looking at all 2^d subsets of the set of d diseases that are being built into the expert system—too large a number even for fairly small d. The previously mentioned restrictions plus a minor compromise will make this problem more tractable.

What's a noisy-OR? Let $\mathcal{X} = \{X_1, \ldots, X_k\}$ be a set of independent random variables. If Y is a random variable that equals T when any $X_i = $ T and that equals F otherwise, then Y is the OR of the X_i's. You can think of $\mathcal{X}$ as the possible causes for Y, any of which can cause Y. In this case, the probability that Y occurs (i.e., $Y = $ T), given that X_i occurs, is 1. If this probability can be less than 1 and if the X_i's act independently as causes of Y, the result is a noisy-OR.

Definition 8.6   Noisy-OR

Let $\mathcal{X}$ be a set of random variables with range$(X) = \{$T, F$\}$ for each $X \in \mathcal{X}$. The random variable Y is called the *noisy-OR* of $\mathcal{X}$ if range$(Y) = \{$T, F$\}$ and

$$1 - \Pr(Y = \text{T} \mid \mathcal{X}^*) = \prod_{\substack{X \in \mathcal{X}^* \\ X = \text{T}}} (1 - A_X), \qquad (8.11)$$

where $0 < A_X \leq 1$.

The definition may seem a bit mysterious at first. But the idea is fairly simple when looked at correctly. Think of A_X as the probability that $Y = \text{T}$, given that $X = \text{T}$ and all other variables in $\mathcal{X}$ equal F. If the random variables independently try to make $Y = \text{T}$, what is the probability that $Y = \text{F}$? Since Y is supposed to mimic an OR, $\Pr(Y = \text{F})$ is just the product of the probabilities that each of the X's independently fails to make $Y = \text{T}$. When $X = \text{F}$, it can never cause $Y = \text{T}$, so the probability is 1 in this case and hence can be ignored in the product. When $X = \text{F}$, the probability that it forces $Y = \text{T}$ is just A_X, and so the probability that it fails is $1 - A_X$. You should be able to combine these observations to produce (8.11).

Note that the number of factors in the product (8.11) depends on the instantiation of $\mathcal{X}$—one factor for each instantiation to T. When all instantiations are F, the product is empty and so equals 1 by the usual mathematical convention.

Definition 8.7   Bipartite Diagnosis Problem

Consider a Bayesian network $(\mathcal{X} \cup \mathcal{Y}, E, \Pr)$ such that

(a) The range of each $Z \in \mathcal{X} \cup \mathcal{Y}$ is $\{\text{T}, \text{F}\}$.

(b) For each $X \in \mathcal{X}$, $\Pr(X = \text{T})$ is much less than $\frac{1}{2}$.

(c) $E \subseteq \mathcal{X} \times \mathcal{Y}$.

(d) Each $Y \in \mathcal{Y}$ is the noisy-OR of $\mathcal{C}(Y)$; that is, for each $(X, Y) \in E$ there exists $\Pr(X \rightarrow Y) > 0$, known as the probability that X causes Y, such that

$$1 - \Pr(Y = \text{T} \mid \mathcal{C}(Y)) = \prod_{\substack{X \in \mathcal{C}(Y) \\ X = \text{T}}} \left(1 - \Pr(X \rightarrow Y)\right). \qquad (8.12)$$

Call the diagnosis problem for this type of Bayesian network the *bipartite diagnosis problem*; that is,

> Given disjoint subsets $\mathcal{R}^+$ and $\mathcal{R}^-$ of $\mathcal{Y}$, find disjoint sets $\mathcal{X}^+$ and $\mathcal{X}^-$ such that $\mathcal{X}^+ \cup \mathcal{X}^- = \mathcal{X}$ and $\Pr(\mathcal{X}^+ \wedge \mathcal{X}^- \mid \mathcal{R}^+ \wedge \mathcal{R}^-)$ is as large as possible. $\qquad (8.13)$

We can now state the assumptions we'll make so that the multiple-diagnosis problem becomes computationally feasible.

Compromises: The DAG is bipartite, X is the noisy-OR of $\mathcal{C}(X)$, and the prior probability of each cause is not large.

Singly Connected DAGs

There is another approach to limiting the structure of the DAG. Instead of insisting that there be no intermediate causes (the bipartite case), we could assume that the interconnections are rather sparse:

> **Definition 8.8**   (Singly) Connected DAGs
>
> A DAG (V, E) is *connected* if, between every two vertices, there is *at least* one undirected path. It is *singly connected* if, between every two vertices there is *exactly one* undirected path. A singly connected DAG is also called a *polytree*.

We'll make the following assumptions so that we can obtain a reasonable algorithm without assuming bipartiteness and noisy-ORs.

Compromises: The DAG is singly connected and $\mathcal{Z}$ in (8.10) contains only a single variable.   (8.14)

There are methods for converting a general DAG into a singly connected one. Unfortunately, those methods tend to be computationally prohibitive. Nothing is lost in assuming that the network is connected because Exercise 8.1.8 (p. 313) tells us we may as well assume that it is. This is what you should expect intuitively, for if there is no path between certain sets of vertices, then they cannot exert any influence over one another.

MYCIN and Certainty Factors

Certainty factors were introduced as an ad hoc device by Shortliffe and Buchanan [16] in their medical diagnosis program MYCIN. Since that time, certainty factors have been misunderstood, misused, redefined, and employed in some form in a variety of expert system shells. The ad hoc nature of certainty factors and the misuse of expert shells based on them may have resulted in poor expert systems.

Certainty factors have also been investigated from an axiomatic viewpoint. As a result of this mathematical investigation, they have been put on a clear probabilistic foundation and their meaning and limitations have been elucidated. This investigation is a "textbook example" of the importance of taking a rigorous mathematical approach whenever feasible.

Deduction, Abduction, and Induction

In FOL, deduction in its simplest form is the process of starting with $\forall X\big(p(X) \to q(X)\big)$ and $p(a)$ and concluding $q(a)$. If the assumptions are true, so is the conclusion. In contrast, abduction starts with $\forall X\big(p(X) \to q(X)\big)$ and $q(a)$ and concludes $p(a)$—a conclusion that need not be true even if the assumptions are true. Some defeasible rules $\forall X\big(q(X) \dashrightarrow p(X)\big)$ arise in this manner. (Note the reversal of p and q.) Since an abductive conclusion need not be valid and since there are competing abductive conclusions, we need some method for choosing amongst them. In the context of Bayesian nets, Definition 8.5 (p. 315) provides such a criterion.

Inductive reasoning is a process in which we extrapolate from a variety of specific examples $p(a_i) \to q(a_i)$ either to the FOL generalization $\forall X\big(p(X) \to q(X)\big)$ or to the defeasible generalization $\forall X\big(p(X) \dashrightarrow q(X)\big)$. *This is not the same as a mathematical proof by induction.* In a mathematical proof by induction, the conclusion is valid, whereas in inductive reasoning, the generalization need not be valid. We can think of inductive reasoning as a form of learning: The inductively derived generalization is a hypothesis put forward by the learner and will be tested against the knowledge base or future experience or both. Inductive reasoning is also called *inductive inference*.

Another form of learning is analogical reasoning. If there is some transformation that maps p to p' and q to q' and if we believe $p \to q$, then we expect $p' \to q'$ to be true as well.

Exercises

8.2.A. What does it mean for a random variable to be instantiated? Explain the notations $\mathcal{X}^-$, $\mathcal{X}^+$, and $\mathcal{X}^*$.

8.2.B. Define and explain a noisy-OR.

8.2.C. What is the (multiple) diagnosis problem? What is abductive inference? What is the bipartite diagnosis problem?

8.2.1. Suppose that Y is a noisy-OR of U and V. Write out $\Pr(Y = \text{F} \mid \{U, V\})$ in tabular form. (There should be four rows, one for each possible instantiation of $\{U, V\}$, and they will involve A_U and A_V.)

8.2.2. Define a noisy-AND that is like the noisy-OR

8.2.3. Let (V, E) be a DAG. We want to determine how many parameters must be specified to make the DAG into a Bayesian network in which the range of every random variable is $\{\text{T}, \text{F}\}$.

    (a) Using Theorem 8.2 and $\Pr(X = \text{T} \mid A) + \Pr(X = \text{F} \mid A) = 1$, show that we must specify

$$\sum_{X \in V} 2^{|\mathcal{C}(X)|}$$

    probabilities. Recall that $|A|$ is the number of elements in the set A.

    (b) Suppose that each $X \in V$ for which $\mathcal{C}(X) \neq \emptyset$ is a noisy-OR of $\mathcal{C}(X)$. Show that fewer than $|V| + |E|$ parameters are required and determine the exact number.

    (c) Suppose that V is a union of two disjoint sets $\mathcal{X}$ and $\mathcal{Y}$, each of which has n elements, that $E \subseteq \mathcal{X} \times \mathcal{Y}$, and that $|\mathcal{C}(Y)| = k$ for all $Y \in \mathcal{Y}$. Show that the ratio of the result in (a) to that in (b) is $(2^k + 1)/(k + 1)$ and tabulate it for $2 \le k \le 7$. What does this show?

8.3   Multiple Diagnosis in Bipartite Networks

> *Although we know that doctors do so, we do not understand just how they weigh the evidence that favors and that opposes various hypotheses or courses of action; this is an important unsolved problem for both AI and cognitive psychology.*
>
> —Peter Szolovits and Stephen C. Pauker (1978)

> *Make three correct guesses consecutively and you will establish a reputation as an expert*
>
> —Laurence J. Peter (1969)

Some aspects of our formulation the bipartite diagnosis problem in Definition 8.7 (p. 317) deserve a bit more consideration.

- The limitation of the ranges to $\{\text{T}, \text{F}\}$ is quite reasonable since many problems fall into this category, with T indicating that something is present and F indicating that it is absent.

- The assumption that the prior probability of X is small is often reasonable. It merely states that the prior probability of X is small. In the medical diagnosis case, this means that a person visiting a doctor is not, *a priori*, likely to have any specific disease.

- The assumption that $E \subseteq \mathcal{X} \times \mathcal{Y}$ means that we can think of $\mathcal{X}$ as the possible causes and $\mathcal{Y}$ as the possible effects (or results). It rules out the possibility of intermediate-level causes. The assumption can be eliminated at the cost of more computation.

- The "real-world" content of (8.12) is that each cause $X \in \mathcal{X}$ that is present (i.e., $X = \text{T}$) acts independently of other causes to produce an effect. In other words, there are no synergistic or inhibiting interactions. This independence assumption is a reasonable approximation in many cases, but is sometimes unreasonable.

- Asking for the set of causes that give the largest probability may not be what we want to do. Some causes are much more dangerous than others and should not be ruled out unless they give an extremely low probability.

- In terms of conditional probability, $\Pr(X \rightarrow Y)$ equals

$$\Pr\left(Y = \text{T} \;\middle|\; X = \text{T} \wedge \left(X' = \text{F for all other } X' \in \mathcal{C}(Y)\right)\right). \qquad (8.15)$$

Some researchers have suggested that an attempt to elicit the conditional probability $\Pr(Y = \text{T} \mid X = \text{T})$ from an expert is likely to produce $\Pr(X \rightarrow Y)$ instead. Thus $\Pr(X \rightarrow Y)$ would be a more natural parameter than $\Pr(Y = \text{T} \mid X = \text{T})$.

Example 8.6

Imagine a medical diagnosis situation. We let $\mathcal{Y}$ be a list of possible symptoms and let $\mathcal{X}$ be a list of possible causes. We observe the presence or absence of various symptoms in $\mathcal{Y}$ and want to select the probable cause or causes from among the diseases in $\mathcal{X}$. When the diseases act independently of one another to produce the symptoms, we have (8.12). In some medical situations, there can be several diseases acting to produce the observed symptoms; however, a doctor will not normally propose more causes than are needed to produce the observed symptoms. "Irredundant covers" will provide the tool for finding such sets of causes.

In a typical diagnosis situation, the presence or absence of all possible symptoms is not known at the beginning of the diagnosis. Rather, such information is elicited stepwise through questioning, observation, and medical tests. Thus, it is important to have an approach that can incorporate further information about the symptoms $\mathcal{Y}$ without starting the calculations over. The algorithm for finding irredundant covers is just such an incremental process.

This formulation of the diagnosis problem does not allow for the possibility that a disease X may suppress a symptom Y. An ad hoc fix can be made within the system: Introduce a new symptom $\overline{Y}$, say, and a new probability $\Pr(X \rightarrow \overline{Y})$. This solution is not entirely satisfactory because we now have some diseases acting to produce Y and others acting independently to produce $\overline{Y}$. ∎

Now let's see what can be said about the bipartite diagnosis problem. A simple way to visualize the problem is to imagine we've observed that a patient has certain symptoms and does not have certain others (the information about $\mathcal{R}$). We want to know what combination of diseases provides the most probable explanation for the symptoms (which X's should equal T). The following theorem, whose proof is left as an exercise, provides the connection between (8.13) and the graph theory concept of "covers."

Theorem 8.3

Let $\mathcal{C}^+(Y) = \mathcal{C}(Y) \cap \mathcal{X}^+$ and

$$\mathcal{X}_1 = \left\{\, X \in \mathcal{X} \mid \Pr(X \rightarrow R^-) = 1 \text{ for some } R^- \in \mathcal{R}^- \,\right\}.$$

In words, $\mathcal{X}_1$ are those causes X which, if present, *must* produce some result in $\mathcal{R}^-$; that is, a result we know has *not* occurred. Every solution to (8.13) satisfies

$$\mathcal{X}_1 \subseteq \mathcal{X}^-, \quad \mathcal{X}^+ \subseteq \mathcal{C}(\mathcal{R}^+), \tag{8.16}$$

and

$$\text{for each } Y \in \mathcal{R}^+, \mathcal{C}^+(Y) \neq \emptyset. \tag{8.17}$$

$(\mathcal{C}(\mathcal{W}) = \cup_{W \in \mathcal{W}} \mathcal{C}(W)$ was defined in Definition 8.2 (p. 303).)

We'll conclude the probabilistic studies of the simple diagnosis model by developing a useful formula for computing $\Pr(\mathcal{X}^+ \wedge \mathcal{X}^- \mid \mathcal{R}^+ \wedge \mathcal{R}^-)$. Using Bayes' Theorem and the properties of Bayesian networks, we have

$$\Pr(\mathcal{X}^+ \wedge \mathcal{X}^- \mid \mathcal{R}^+ \wedge \mathcal{R}^-) = \frac{\Pr(\mathcal{X}^+ \wedge \mathcal{X}^-)}{\Pr(\mathcal{R}^+ \wedge \mathcal{R}^-)} \Pr(\mathcal{R}^+ \wedge \mathcal{R}^- \mid \mathcal{X}^+ \wedge \mathcal{X}^-)$$

$$= \frac{1}{\Pr(\mathcal{R}^+ \wedge \mathcal{R}^-)} \prod_{X \in \mathcal{X}^+} \Pr(X = \text{T}) \prod_{X \in \mathcal{X}^-} \Pr(X = \text{F})$$

$$\times \prod_{Y \in \mathcal{R}^+} \Pr(Y = \text{T} \mid \mathcal{R}^+ \wedge \mathcal{R}^-) \prod_{Y \in \mathcal{R}^-} \Pr(Y = \text{F} \mid \mathcal{R}^+ \wedge \mathcal{R}^-)$$

Rearranging a bit and using (8.12) to expand the last two products, we have

$$\Pr(\mathcal{X}^+ \wedge \mathcal{X}^- \mid \mathcal{R}^+ \wedge \mathcal{R}^-) \left(\Pr(\mathcal{R}^+ \wedge \mathcal{R}^-) \prod_{X \in \mathcal{X}} \frac{1}{\Pr(X = \text{F})} \right)$$

$$= \prod_{X \in \mathcal{X}^+} \frac{\Pr(X = \text{T})}{\Pr(X = \text{F})}$$

$$\times \prod_{Y \in \mathcal{R}^+} \left(1 - \prod_{X \in \mathcal{C}^+(Y)} \left(1 - \Pr(X \rightarrow Y) \right) \right)$$

$$\times \prod_{Y \in \mathcal{R}^-} \left(\prod_{X \in \mathcal{C}^+(Y)} \left(1 - \Pr(X \rightarrow Y) \right) \right). \tag{8.18}$$

The second factor on the left side of (8.18) depends only on $\mathcal{X}$, $\mathcal{R}^+$, and $\mathcal{R}^-$, which are constant in the search for $\mathcal{X}^+$ and $\mathcal{X}^-$. Thus, we can locate the maximum of $\Pr(\mathcal{X}^+ \wedge \mathcal{X}^- \mid \mathcal{R}^+ \wedge \mathcal{R}^-)$ by locating the maximum of (8.18). The right side of (8.18) has the nice feature that it involves only those X which are in $\mathcal{X}^+$.

Irredundant Covers

A naive approach to applying Theorem 8.3 requires a considerable amount of computation: If $|\mathcal{C}(\mathcal{R}^+)| = m$, we'd have to check out all $2^m - 1$ nonempty subsets of $\mathcal{C}(\mathcal{R}^+)$. This may be a large number in problems of practical interest. We'll avoid this problem as follows:

- First, we look at the related problem of finding "irredundant covers."

- Then we *assume* that the solution to our bipartite diagnosis problem is among the solutions of the irredundant covering problem.

- Finally, we evaluate (8.18) for the various irredundant covers.

Definition 8.9   Covers

Let P and Q be disjoint sets, $V = P \cup Q$, and $E \subseteq P \times Q$. A *cover* of the DAG (V, E) is a subset C of P such that $\mathcal{D}(C) = Q$. In other words, for every $q \in Q$, there is a $p \in C$ such that $(p, q) \in E$. If no proper subset of C is a cover, we call C an *irredundant cover*. It is also called a *minimal cover*.

To use covers, we form a sub-DAG of the one for the one in the bipartite diagnosis problem (Definition 8.7):

Theorem 8.4

Let $(\mathcal{X} \cup \mathcal{Y}, E, \Pr)$ be a bipartite network. Define a DAG (V', E') by

$$V' = \mathcal{X}' \cup \mathcal{R}^+ \quad \text{and} \quad E' = (\mathcal{X}' \times \mathcal{R}^+) \cap E,$$

where $\mathcal{X}'$ is set of all elements of $\mathcal{C}(\mathcal{R}^+)$ that are not in $\mathcal{X}_1$. It follows that any set $\mathcal{X}^+$ satisfying (8.13) lies in $\mathcal{X}'$ and is a cover of the DAG (V', E').

The proof is left as an exercise.

Limit your attention to irredundant covers. While there's no *guarantee* that the solution to (8.13) is an irredundant cover, the solutions to many real-world problems are irredundant covers. Why is this?

We have a possible "explanation" for the given sets $\mathcal{R}^+$ and $\mathcal{R}^-$ if and only if we have a cover $\mathcal{X}^+$. Saying that $\mathcal{X}^+$ is an *irredundant* cover is the same as saying that $\mathcal{X}^+$ and $\mathcal{X}^-$ provide an "explanation" and no smaller choice for $\mathcal{X}^+$ does. In many cases, it's reasonable to look for solutions among the irredundant covers. That's *Ockham's Razor* (also spelled Occam's Razor):

> *Do not multiply entities beyond what is strictly necessary to explain phenomena.*   —William of Ockham

For example, in medical diagnosis, physicians don't normally propose that the patient has more conditions ($\mathcal{X}^+$) than are needed to explain the symptoms ($\mathcal{R}^+$, $\mathcal{R}^-$). Of course, if the model is accurate and irredundant covers are the right choice, then the highest scoring explanation must be an irredundant cover. This should follow from (8.18). See Exercise 8.3.7 for more discussion.

Let's recapitulate. We've seen that it's often reasonable to look for solutions to (8.13) among the irredundant covers of (V', E') and we have a formula (8.18) for computing the probability of the observations given a particular cover. Since the use of (8.18) is straightforward and not computationally expensive, all that's missing is a reasonable algorithm for finding irredundant covers. We now develop such an algorithm.

Let (V, E) be the DAG for which we want irredundant covers, where

$$V = P \cup Q, \quad E \subseteq P \times Q, \quad \text{and} \quad Q = \{\, q_1, \ldots, q_n \,\}.$$

Define $\mathcal{D}_k = (V_k, E_k)$ by

$$V_k = P \cup Q_k \quad \text{and} \quad E_k = E \cap (P \times Q_k) \quad \text{where} \quad Q_k = \{\, q_1, \ldots, q_k \,\}.$$

The idea of the algorithm is to produce irredundant covers for $\mathcal{D}_k$ from those for $\mathcal{D}_{k-1}$. Since $\mathcal{D}_n = (V, E)$, we are done when $k = n$. The key is the following.

Lemma

Suppose that S is an irredundant cover for $\mathcal{D}_k$. Then either

- S is an irredundant cover for $\mathcal{D}_{k-1}$ or
- S contains exactly one element in $\mathcal{C}(q_k)$ and removal of this element gives an irredundant cover for $\mathcal{D}_{k-1}$.

Proof: Suppose that S is not an irredundant cover for $\mathcal{D}_{k-1}$. Since it is a cover for $\mathcal{D}_{k-1}$, it follows that we must be able to remove at least one element of S and still have a cover. Let $x \in S$ be such that $S - \{x\}$ is a cover for $\mathcal{D}_{k-1}$. Since S is an irredundant cover for $\mathcal{D}_k$, $S - \{x\}$ cannot be a cover for $\mathcal{D}_k$. It follows that no element of $S - \{x\}$ "causes" q_k; that is, $(S - \{x\}) \cap \mathcal{C}(q_k) = \emptyset$. Hence we must have $x \in \mathcal{C}(q_k)$ and we can't have $y \in \mathcal{C}(q_k)$ for any $y \in (S - \{x\})$. We've just shown that

- any element of S whose removal leaves a cover for $\mathcal{D}_{k-1}$ must be in $\mathcal{C}(q_k)$ and

- there is no other element of S that is in $\mathcal{C}(q_k)$.

It follows that there is just one element that can be removed from S and still leave a cover for $\mathcal{D}_{k-1}$. This says that no other elements can be removed from $S - \{x\}$, which means it must be an irredundant cover for $\mathcal{D}_{k-1}$. ∎

The importance of this lemma is that we can turn it around: Every irredundant cover for $\mathcal{D}_k$ is either an irredundant cover for $\mathcal{D}_{k-1}$ or is obtained from an irredundant cover for $\mathcal{D}_{k-1}$ by adding an element of $\mathcal{C}(q_k)$. This form of the lemma is the key idea behind the following incremental algorithm given by Peng and Reggia [14]. [Bracketed comments help in proving correctness.]

Algorithm 8.1   Irredundant Covers

Given a digraph $(P \cup Q, E)$ where $E \subseteq P \times Q$ and $Q = \{q_1, \ldots, q_n\}$, the following algorithm produces the irredundant covers of $\{q_1, \ldots, q_k\}$ in Σ_k.

1. **Initialize:** Set $\Sigma_0 = \{\emptyset\}$ and $k = 1$.

2. **Split Σ:** Set Λ_k to $\{\, S \in \Sigma_{k-1} \mid S \cap \mathcal{C}(q_k) = \emptyset \,\}$. Set $\Sigma'_k = \Sigma_{k-1} - \Lambda_k$. [$\Sigma'_k$ consists of all irredundant covers in Σ_{k-1} that are irredundant covers for $\mathcal{D}_k$ and Λ_k consists of those that are not.]

3. **Adjoin $\mathcal{C}(q_k)$:** Set Λ'_k to be the set of all sets of the form $S \cup \{p\}$ where $S \in \Lambda_k$ and $p \in \mathcal{C}(q_k)$. [We add a unique element of $\mathcal{C}(q_k)$.]

4. **Create new Σ:** Set Σ_k to Σ'_k together with all sets in Λ'_k that do not contain any set of Σ'_k. [The lemma is the key to the fact that this gives us the irredundant covers.]

5. **Test Loop:** If $k = n$, stop; otherwise, increment k and go to Step 2.

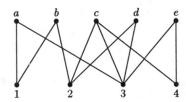

Figure 8.3 The digraph for Example 8.7. To avoid congestion in the figure, arrowheads have been omitted. They all point downward from $\{a, b, c, d, e\}$ to $\{1, 2, 3, 4\}$.

The incremental nature of Algorithm 8.1 is quite important for applications. Frequently, we gradually increment $\mathcal{R}^+$ (and $\mathcal{R}^-$) by acquiring additional information. Each time we add an element to $\mathcal{R}^+$, we are essentially just going from $\mathcal{D}_n$ to $\mathcal{D}_{n+1}$. Thus the algorithm allows us to produce irredundant covers without the need to start over. Another aspect of the algorithm that can be useful is the fact that an irredundant cover at one step never gives a smaller cover at a later step. As a result, we can design the algorithm to produce only those irredundant covers that do not exceed some predetermined cardinality.

How long does the algorithm take? If the number of diseases that can produce each symptom is not too large and if we also limit our search to irredundant covers of small cardinality, the algorithm can be quite efficient. On the other hand, with no restrictions the worst-case behavior can be prohibitive.

Example 8.7   Using the Algorithm

Let's apply the algorithm to find the irredundant covers in the digraph given by $P = \{a, b, c, d, e, f\}$, $Q = \{1, 2, 3, 4\}$, and the "causes"

$$\mathcal{C}(1) = \{a, b\}, \quad \mathcal{C}(2) = \{b, c, d\}, \quad \mathcal{C}(3) = \{a, c, d, e\}, \quad \mathcal{C}(4) = \{c, e\}.$$

The DAG is shown in Figure 8.3. Here are the steps in the algorithm.

Step 1. $\Sigma_0 = \{\emptyset\}$ and $k = 1$.

Step 2. $\Lambda_1 = \{\emptyset\}$ and $\Sigma_1' = \emptyset$.

Step 3. $\Lambda_1' = \{\{a\}, \{b\}\}$.

Step 4. $\Sigma_1 = \{\{a\}, \{b\}\}$.

Step 5. $k = 2$.

Step 2. $\Lambda_2 = \{\{a\}\}$ and $\Sigma_2' = \{\{b\}\}$.

Step 3. $\Lambda_2' = \{\{a, b\}, \{a, c\}, \{a, d\}\}$.

Step 4. $\Sigma_2 = \{\{b\}, \{a, c\}, \{a, d\}\}$.

Step 5. $k = 3$.

Step 2. $\Lambda_3 = \{\{b\}\}$ and $\Sigma_3' = \{\{a, c\} \ \{a, d\}\}$.

Step 3. $\Lambda_3' = \{\{b,a\}, \{b,c\}, \{b,d\}, \{b,e\}\}$.

Step 4. $\Sigma_3 = \{\{a,b\}, \{a,c\}, \{a,d\}, \{b,c\}, \{b,d\}, \{b,e\}\}$.

Step 5. $k = 4$.

Step 2. $\Lambda_4 = \{\{a,b\}, \{a,d\}, \{b,d\}\}$ and $\Sigma_4' = \{\{a,c\}, \{b,c\}, \{b,e\}\}$.

Step 3. $\Lambda_4' = \{\{a,b,c\}, \{a,b,e\}, \{a,d,c\}, \{a,d,e\}, \{b,d,c\}, \{b,d,e\}\}$.

Step 4. $\Sigma_4 = \{\{a,c\}, \{b,c\}, \{b,e\}, \{a,d,e\}\}$.

Step 5. Stop.

We now have the irredundant covers in Σ_4. If the probabilistic part of the network were given, we could use (8.18) (p. 322) to determine which irredundant cover was the most probable explanation for the observed effects, which were so imaginatively called 1, 2, 3, and 4. ■

Further Remarks

Splitting the bipartite diagnosis problem into a structural part (irredundant cover algorithm) and a numerical part (probability calculations using (8.18)) has important repercussions:

+ We might replace (8.12) because it doesn't reflect reality; for example, we might allow one disease to interfere with another in producing a symptom. This means that (8.18) would no longer be valid; however, the irredundant cover algorithm would be unchanged. As long as the replacement for (8.18) was computationally practical, we could proceed to evaluate the irredundant covers numerically.

+ We might use the irredundant covers as the basis of a qualitative expert system by replacing $\Pr(X \to Y)$ with statements such as "always" and "low likelihood"—a much more feasible goal than obtaining accurate estimates. (KMS.HT, an expert system shell implementing this idea, is discussed briefly in [14].)

− A negative consequence of this separation is avoidable work, especially when $\mathcal{R}^+$ and $\mathcal{R}^-$ are built up by instantiating elements of $\mathcal{R}$ one by one. The algorithm for finding irredundant covers has no mechanism for rejecting those covers in $\mathcal{D}_k$ that will lead only to covers in $\mathcal{D}_n$ with low probabilities. (An approach to this problem is discussed in [14].)

How are the values of $\Pr(X \to Y)$ to be determined? One possibility is to consult experts—but their values would probably be quite crude. Another possiblity is to examine data. Equation (8.15) suggests that we must look for situations in which only one disease is present, but this is not the case. Taking logarithms in (8.12), we have

$$\log\Big(\mathrm{Pr}(Y = \text{\sc f} \mid \mathcal{C}(Y))\Big) = \sum_{\substack{X \in \mathcal{C}(Y) \\ X = \text{\sc t}}} \log\Big(1 - \mathrm{Pr}(X \rightarrow Y)\Big).$$

Estimates of the left-hand side can be obtained by examining known situations. Statistical methods can then be used to estimate the logarithms on the right-hand side.

Exercises

8.3.A. What is a cover? an irredundant cover?

8.3.B. Explain how irredundant covers help in the bipartite diagnosis problem.

8.3.C. What is the key idea behind the algorithm for finding irredundant covers?

8.3.1. The last paragraph of Example 8.6 suggests a method for incorporating suppression of symptoms.

    (a) Describe a plausible situation in which this method could lead to erroneous results.

    (b) Suggest general guidelines for deciding when the method proposed in the example is likely to be acceptable.

8.3.2. Prove Theorem 8.3 and provide a verbal interpretation of (8.16) and (8.17). (A start on interpreting (8.16) was made in the text by explaining $\mathcal{X}_1$.)

8.3.3. Prove Theorem 8.4.

8.3.4. Prove that the algorithm for producing all irredundant covers is correct.

8.3.5. Reverse the order of 1–4 in Figure 8.3. Use the algorithm to compute the irredundant covers with the reversed ordering. (Of course, you should get the same answer as that arrived at in the text, but the intermediate steps will differ.)

8.3.6. Draw a picture of the Bayesian network whose nonempty $\mathcal{C}$'s are given below. Apply the algorithm to find all of its irredundant covers.

$$\begin{aligned}
\mathcal{C}(1) &= \{\, a, c, d, i \,\} & \mathcal{C}(2) &= \{\, a, d, e \,\} & \mathcal{C}(3) &= \{\, d, f, h, i \,\} \\
\mathcal{C}(4) &= \{\, b, e, g \,\} & \mathcal{C}(5) &= \{\, b, f, g, h \,\} & \mathcal{C}(6) &= \{\, c, d, e \,\}.
\end{aligned}$$

8.3.7. Let $\mathcal{X}_+ \cup \mathcal{X}_- \cup \{X_0\}$ be a partition of $\mathcal{X}$ such that $\mathcal{X}^+ = \mathcal{X}_+$ is a possible explanation for the observations $\mathcal{R}^+$ and $\mathcal{R}^-$.

(a) Compute (8.18) with $\mathcal{X}^+ = \mathcal{X}_+ \cup \{X_0\}$ and divide it by the same product with $\mathcal{X}^+ = \mathcal{X}_+$. Explain why the highest probability covers will be irredundant if this ratio is always less than 1. Also, show that the ratio equals

$$\frac{\Pr(X_0 = \mathrm{T})}{\Pr(X_0 = \mathrm{F})} \prod_{Y \in \mathcal{R}_- \cap \mathcal{R}(X_0)} \Big(1 - \Pr(X \to Y)\Big)$$

$$\prod_{Y \in \mathcal{R}_+ \cap \mathcal{R}(X_0)} \frac{1 - \big(1 - \Pr(X_0 \to Y)\big) \prod_{X \in \mathcal{C}^+(Y)}\big(1 - \Pr(X \to Y)\big)}{1 - \prod_{X \in \mathcal{C}^+(Y)}\big(1 - \Pr(X \to Y)\big)},$$

where $\mathcal{C}^+(Y)$ is calculated using $\mathcal{X}^+ = \mathcal{X}_+$.

(b) This is a rather open-ended question: What sorts of conditions are likely to make the ratio less than 1? You might consider things such as the size of $\Pr(X_0 = \mathrm{T})$, the bounds on $\Pr(X \to Y)$, and $\big|\mathcal{R}_\pm \cap \mathcal{R}(X_0)\big|$.

8.4   An Algorithm for Singly Connected Networks

> *I presume that to the uninitiated the formulae will appear cold and cheerless.*
> —Benjamin Peirce (1882)

Unlike the preceding section, this section includes some rather complicated formulas involving probabilities. The fact is, you may have some difficulty understanding this section if you concentrate on the formulas. To simplify matters, I've postponed all proofs until the next section. In addition, I suggest that on your first reading you ignore the expressions that are given for the various λ's and π's and focus on understanding the general structure of the algorithm and the way in which information is propagating through the network. To help you ignore the expressions, I'll not give the probabilistic meanings of the λ's and π's until page 340.

Let $(\mathcal{X}, E, \Pr)$ be a *singly connected* Bayesian network. At each vertex and at each edge we have two real-valued functions, a π and a λ:

- at the vertex $X \in \mathcal{X}$, the functions λ_X and π_X, both with domain range(X);
- at the edge (X, W), the functions $\lambda_{X,W}$ and $\pi_{X,W}$, both with domain range(X).

Of course, such a function can be thought of as a vector of real numbers where the length of the vector is the size of the function's domain, namely |range(X)| for λ_X, π_X, $\lambda_{X,W}$, and $\pi_{X,W}$.

This presentation of Kim and Pearl's [8] algorithm for singly connected Bayesian networks has two parts. The first part consists of the algorithm and the second part consists of a subalgorithm for propagating "activation." A vertex will be referred to as being "active" or "inactive." When all vertices are inactive, we'll say that the network is inactive.

In the following, remember that when a random variable appears as an index of summation, it means that we have to sum over all allowed instantiations of the variable. Also, remember the notation

$$\mathcal{R}(X) = \Big\{ Y \in \mathcal{X} \,\Big|\, (X, Y) \in E \Big\}.$$

[Bracketed comments will later help in proving correctness.]

Algorithm 8.2   Singly Connected Bayesian Networks

These are the steps that must be taken to trigger the propagation rules so that the network will calculate the desired probabilities. They begin by getting the network to compute probabilities with no instantiations $(\mathcal{W}^* = \emptyset)$; then, in the penultimate step, instantiations are introduced.

1. **Initialize λ:** Set all π's to "unknown" and all λ's to 1. [Since nothing is instantiated, the values of λ's are all correct.]

2. **Initialize π:** For each X with $\mathcal{C}(X) = \emptyset$, activate X, telling it that it has received a new π. When the network is inactive, go to the next step. [Since $\mathcal{C}(X) = \emptyset$, π_X will be computed. The remaining values of the π's will be properly computed throughout the network since they propagate downward. No λ messages will be sent since the λ's are constant.]

3. **Instantiate $\mathcal{W}^*$:** For each $X \in \mathcal{W}^*$, do the following:

   - Let x^* be the value to which it is to be instantiated and set

   $$\lambda_X(b) = \begin{cases} 1 & \text{if } b = x^*, \\ 0 & \text{otherwise.} \end{cases}$$

   - Activate X, telling it that it has received a new π and wait until the network is inactive

   When done with $\mathcal{W}^*$, proceed to the next step.

4. **Compute Answers:** For whatever $X \notin \mathcal{W}^*$ you wish, calculate $\Pr(X \mid \mathcal{W}^*)$ using

   $$\Pr(X \mid \mathcal{W}^*) = \frac{\lambda_X \pi_X}{\sum_X \lambda_X \pi_X}.$$

After Step 4, the expert system may decide to request additional information to distinguish between competing alternatives. The new information provides newly instantiated variables. We need not start at the beginning but can simply add these new instantiations by going back to Step 3 as if we hadn't finished with $\mathcal{W}^*$.

Now come the messy formulas—the propagation rules. In using the following formulas, remember that the value of a product over an empty set is 1.

Algorithm 8.3   Propagation Rules

A vertex becomes active either through outside intervention or automatically by receiving a "message" from another vertex. An active vertex performs some computations, sends messages, then becomes inactive. The following describes the action vertex X takes when active. The appropriate action is assumed to be automatically carried out when activation occurs.

1. If some π_Y for $Y \in \mathcal{C}(X)$ is unknown, become inactive and stop calculation. Otherwise do the following.

2. Compute $\pi_X = \sum_{\mathcal{C}(X)} \Pr(X \mid \mathcal{C}(X)) \prod_{U \in \mathcal{C}(X)} \pi_{U,X}$.

3. If X is uninstantiated, compute λ_X using $\lambda_X = \prod_{Y \in \mathcal{R}(X)} \lambda_{X,Y}$.

4. Send, to each $Y \in \mathcal{R}(X)$ that did not activate X, a message as follows. [If Y activated X, then the value of $\pi_{X,Y}$ does not change. The proof is left as an exercise.]

    - If X is uninstantiated, send

    $$\pi_{X,Y} = \pi_X \prod_{\substack{Z \in \mathcal{R}(X) \\ Z \neq Y}} \lambda_{X,Z} \left/ \sum_X \left(\pi_X \prod_{\substack{Z \in \mathcal{R}(X) \\ Z \neq Y}} \lambda_{X,Z} \right) \right.$$

    - If X is instantiated, send

    $$\pi_{X,Y}(b) = \begin{cases} 1 & \text{if } b = x^*, \text{ the instantiated value of } X, \\ 0 & \text{otherwise.} \end{cases}$$

5. If λ_X was the constant function both before and after its recalculation in Step 3, become inactive and stop calculation. Otherwise, send, to each $U \in \mathcal{C}(X)$ such that U did not cause the activation of X, the message

    $$\lambda_{U,X} = \sum_X \left(\lambda_X \sum_{\mathcal{C}(X)-\{U\}} \Pr(X \mid \mathcal{C}(X)) \prod_{\substack{Z \in \mathcal{C}(X) \\ Z \neq U}} \pi_Z \right)$$

    to U. [If U caused the activation of X, then $\lambda_{U,X}$ is unchanged. If λ_X is constant, then $\lambda_{U,X} = \lambda_X$. The proofs are left as an exercise.]

6. Become inactive and stop calculation.

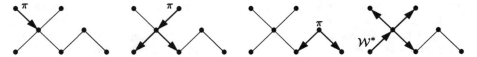

Figure 8.4 All DAG arrows are presumed to be directed downward along the edges. The arrows in the DAGs show the propagation of messages through the Bayesian network as a result of initializing π or instantiating $\mathcal{W}^*$ at the marked vertex. Downward messages are π's and upward messages are λ's. The first three DAGs show the results of propagations caused by initializing π in Step 2 and the last shows the results of a single instantiation in Step 3.

Now for a brief pause, brought to you by a subtle notational point about sums indexed by random variables. Recall that when random variables appear as an index of summation, it means we should sum over all possible instantiations of the random variables. What, then, is the meaning of an expression like $\sum_{\mathcal{Z}} f(\mathcal{X})$ when $\mathcal{Z} = \emptyset$? An empty sum is interpreted to be 0; however,

$$\text{if } \mathcal{Z} = \emptyset, \text{ then } \sum_{\mathcal{Z}} f(\mathcal{X}) = f(\mathcal{X}). \tag{8.19}$$

Why is this? A sum over $\mathcal{Z} = \{Z_1, \ldots, Z_k\}$ is really a shorthand notation for k nested sums, the ith sum ranging over all possible values for Z_i. In other words, $|\mathcal{Z}|$ determines the *number* of summations and the ranges of the Z_i's determine the *ranges* on the sums. For example,

$$\sum_{\{Z_1, Z_2\}} f(\mathcal{X}) = \sum_{z_1 \in \text{range}(Z_1)} \left(\sum_{z_2 \in \text{range}(Z_2)} f(\mathcal{X}) \right).$$

If $\mathcal{Z} = \emptyset$, then there are no summations present in $\sum_{\mathcal{Z}} f(\mathcal{X})$—it's just as if we'd written $f(\mathcal{X})$. This is relevant to Steps 2 and 5 in the previous algorithm.

Example 8.8   Propagation in the Algorithm

How do the algorithms cause calculations to be propagated through a network? Figure 8.4 shows a Bayesian network. All edges in the DAG are directed downward, but I've omitted them because I want to use arrows to show how computations propagate. This is shown by heavy edges with arrows. The vertex to which a rule is applied is labeled with the rule. ∎

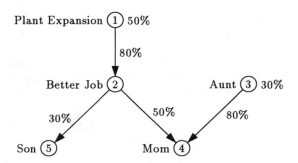

Figure 8.5 The nodes in this Bayesian network correspond to (1) plant expanding, (2) uncle's getting a better job, (3) aunt's getting better, (4) my mom's getting news and (5) my cousin's getting news. See the text for further details.

Example 8.9   Calculations with the Algorithm

My aunt needs an operation and my uncle is trying to find a better job.

The surgeon says that there is about a 30% chance that the operation will lead to improvement this week. If it's good news, my aunt will probably call my mom. If it's bad, she won't want to call anyone. In the same week the biggest employer in town is making a major decision on plant expansion. My guess is there's a 50% chance the plant'll expand. Mom said that my uncle told her he thought there was an 80% chance of a better job if the expansion occurs and that he'd know right away. If he has good news he'll probably call my mom and maybe even call his son, my cousin—the one he's fighting with. My estimates of the three phone calls' probabilities are shown on the edges of the DAG in Figure 8.5.

With vertices 1–5, we'll associate random variables V_1–V_5 but we'll often simply refer to them by number. Each random variable will have range $\{\text{T}, \text{F}\}$, where true means good news. We'll frequently be writing vectors of numbers. The first component will correspond to T and the second to F. For example, $\Pr(5 \mid 2 = \text{T}) = (0.3, 0.7)$ tells us that $\Pr(V_5 = \text{T} \mid V_2 = \text{T}) = 0.3$.

We need to compute $\Pr(X \mid \mathcal{C}(X))$. The two job possibilities were treated like a noisy-OR, as were the two sources of good news for Mom. Here are the results:

$$\Pr(1) = (0.2, 0.8),$$
$$\Pr(3) = (0.3, 0.7),$$
$$\Pr(2 \mid 1 = \text{T}) = (0.8, 0.2), \qquad \Pr(2 \mid 1 = \text{F}) = (0, 1),$$
$$\Pr(5 \mid 2 = \text{T}) = (0.3, 0.7), \qquad \Pr(5 \mid 2 = \text{F}) = (0, 1),$$
$$\Pr(4 \mid 2 = \text{T} \wedge 3 = \text{T}) = (1 - \text{next component}, 0.5 \times 0.2) = (0.9, 0.1),$$
$$\Pr(4 \mid 2 = \text{F} \wedge 3 = \text{T}) = (0.8, 0.2),$$

$$\Pr(4 \mid 2 = \text{T} \wedge 3 = \text{F}) = (0.5, 0.5),$$
$$\Pr(4 \mid 2 = \text{F} \wedge 3 = \text{F}) = (0, 1).$$

Now let's apply the two algorithms to the networks. A number like SC1 refers to the first step in the singly connected network algorithm, while PR3 refers to the third step in the propagation rules. We begin at SC2 with $X = V_1$:

PR1–3. We have $X = V_1$, so $\pi_1 = \Pr(V_1) = (0.5, 0.5)$ since the empty product equals 1. Lambda does not change.

PR4. Activate 2 by setting $\pi_{1,2} = \pi_1 = (0.5, 0.5)$.

PR1–3. We have $X = V_2$, so

$$\pi_2 = \sum_{1=\text{T,F}} \Pr(2 \mid 1)\pi_{1,2} = (0.8, 0.2) \times 0.5 + (0, 1) \times 0.5 = (0.4, 0.6).$$

Lambda does not change.

PR4. Activate 4 and 5 by setting $\pi_{2,4} = \pi_2$ and $\pi_{2,5} = \pi_2$.

PR1. We have $X = V_4$, so we stop because $V_3 \in \mathcal{C}(X)$.

PR1–3. We have $X = V_5$, so

$$\pi_5 = \sum_{2=\text{T,F}} \Pr(5 \mid 2)\pi_{2,5} = (0.3, 0.7) \times 0.4 + (0, 1) \times 0.6 = (0.12, 0.88).$$

We now return to SC2 and set $X = V_3$:

PR1–4. Since $X = V_3$, it follows that $\pi_3 = \Pr(V_3) = (0.3, 0.7)$, λ_3 does not change, and $\pi_{3,4} = \pi_3$.

PR1–3. Since $X = V_4$,

$$\pi_4 = \sum_{2,3} \Pr(4 \mid 2 \wedge 3)\pi_{2,4}\pi_{3,4}$$
$$= \Big((0.9, 0.1) \times 0.3 + (0.5, 0.5) \times 0.7\Big) \times 0.4$$
$$+ \Big((0.8, 0.2) \times 0.3 + (0.0, 1.0) \times 0.7\Big) \times 0.6$$
$$= (0.392, 0.608).$$

This completes the initialization. If there are no instantiations, we can proceed directly to SC4 with $\mathcal{W}^* = \emptyset$. Since all lambdas are 1, it turns out that $\Pr(V_i) = \pi_i$ for all i. For example, the probability that Mom received good news is 0.392. Of course, with the rough guesses that were made on the numbers in the network, we'd better not claim anything more accurate than about a 40% chance of receiving good news.

Now let's see what happens when we make some instantiations. Here's a typical example: After spending the weekend with my cousin, who received no news, there is a call on my answering machine, "This is Mom on Saturday morning. I just got some good news." What are the new probabilities? To find out we'd have to instantiate $V_5 = \text{F}$ and $V_4 = \text{T}$ and do a good deal more calculation. Let's not do this beyond indicating how lambda calculations work when $V_5 = \text{F}$ is instantiated by doing SC3 with $X = 5$. Then $\lambda_5 = (0, 1)$ and propagation gives

PR1–4. We have $X = V_5$. Nothing changes and $\mathcal{R}(5) = \emptyset$.

PR5. Since $\lambda_5 = (0, 1)$, it is not the constant function. We have

$$\lambda_{2,5} = \sum_{5 = \text{T,F}} \lambda_5 \Pr(5 \mid 2) = (0.7, 1.0).$$

PR1–3. We have $X = V_2$. π_2 is unchanged but $\lambda_2 = (0.7, 1.0)$.

PR4. We activate 4 and set

$$\pi_{2,4} = \pi_2 \lambda_{2,5}/C = (0.4 \times 0.7, 0.6 \times 1.0)/C = (0.28, 0.6)/C,$$

where $C = 0.28 + 0.6 = 0.88$. Thus $\pi_{2,4} = (0.318, 0.682)$ to the nearest 0.001. The remaining calculations are similar to what we've already done. ∎

Exercises

There are some exercises concerned with modifications of the algorithm. Since these exercises require material in the next section, they have been placed there, starting with Exercise 8.5.8 (p. 344).

8.4.1. Which rules in the algorithm have summations that might involve summing over an empty set of random variables? (There are two.) Write out those equations in that case.

8.4.2. Prove the bracketed claims in Steps 4 and 5 of the Propagation Rules. (You may find it easier to do this exercise after the experience of studying the next section.)
   Hint. For the case of λ_X constant, interchange the order of summation.

8.4.3. In this exercise, consider the amount of work in propagating the information throughout the Bayesian network during initialization. You may wish to look at Exercise 8.2.3 (p. 320).

   (a) Roughly how does the total amount of work change if the network doubles in size?

   (b) Roughly how does the amount of work change on a π (resp. λ) propagation through X if the network is changed so that $C(X)$ is doubled?

   (c) Roughly how does the amount of work change on a π (resp. λ) propagation through X if the ranges of X and all vertices adjacent to it are doubled?

*8.4.4. Suppose that

- a singly connected network has $ncr + 1$ vertices,
- $|\text{range}(X)| = t$ for all X,
- $|\mathcal{C}(X)| = c$ for nr vertices and $|\mathcal{C}(X)| = 0$ for the rest and
- $|\mathcal{R}(X)| = r$ for nc vertices and $|\mathcal{R}(X)| = 0$ for the rest.

(a) Obtain the best upper and lower bounds you can for the number of times π information is transmitted in Step 4 of Algorithm 8.3 when the network is initialized. Also obtain bounds on the amount of work involved, that is the number of additions and multiplications that are needed. You may wish to look at Exercise 8.2.3 (p. 320).

(b) Obtain bounds for λ information similar to those obtained in (a) for π information.

*8.5   Some Theorems and Proofs

> *Victory at all costs, victory in spite of all terror, victory however long and hard the road may be; for without victory there is no survival.*
>
> —Winston Churchill (1940)

We'll begin with two theorems for Bayesian networks. The first is for general Bayesian networks and the second for singly connected ones. Next, the theorems will be used to derive the formulas in the previous section and to prove the correctness of the algorithm. The first theorem is also needed briefly in the discussion of certainty factors.

Two Theorems

Theorem 8.5   Independence in Bayesian Networks

Let $(\mathcal{X}, E, \text{Pr})$ be a Bayesian network and let $\mathcal{S}$, $\mathcal{T}$, and $\mathcal{V}$ be three disjoint sets whose union is $\mathcal{X}$. Assume that

$$E \subseteq \Big((\mathcal{S} \cup \mathcal{V}) \times (\mathcal{S} \cup \mathcal{V})\Big) \cup \Big(\mathcal{V} \times \mathcal{T}\Big) \cup \Big(\mathcal{T} \times \mathcal{T}\Big),$$

that $\mathcal{S}' \subseteq \mathcal{S} \cup \mathcal{V}$, and that $\mathcal{T}' \subseteq \mathcal{T} \cup \mathcal{V}$. Then

$$\text{Pr}(\mathcal{S}' \cup \mathcal{T}' \mid \mathcal{V}) = \text{Pr}(\mathcal{S}' \mid \mathcal{V}) \, \text{Pr}(\mathcal{T}' \mid \mathcal{V}). \qquad (8.20)$$

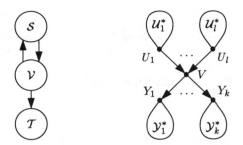

Figure 8.6 The left-hand picture illustrates the possible types of edges *between* the three sets $\mathcal{S}$, $\mathcal{T}$, and $\mathcal{V}$ in Theorem 8.5. (Any edges are allowed within the sets, with the restriction that the result must be a DAG.) The right-hand picture illustrates the locations of the various vertices in Theorem 8.6.

Figure 8.6 illustrates the assumptions made about edges in this theorem and in the next one.

Proof: It suffices to limit $\mathcal{S}'$ and $\mathcal{T}'$ to subsets of $\mathcal{S}$ and $\mathcal{T}$, respectively. It also suffices to prove (8.20) for $\mathcal{S}' = \mathcal{S}$ and $\mathcal{T}' = \mathcal{T}$ because we can then apply Theorem 7.5 (p. 293) to obtain the general case. The proof of this special case of (8.20) is simply a matter of summing

$$\Pr(\mathcal{X}) = \prod_{X \in \mathcal{X}} \Pr(X \mid \mathcal{C}(X)) = \prod_{X \in \mathcal{T}} \Pr(X \mid \mathcal{C}(X)) \prod_{X \notin \mathcal{T}} \Pr(X \mid \mathcal{C}(X))$$

judiciously. In doing so, we use

$$X \notin \mathcal{T} \quad \text{implies} \quad \mathcal{C}(X) \cap \mathcal{T} = \emptyset.$$

After some steps which are left as an exercise,

$$\Pr(\mathcal{T} \mid \mathcal{V}) = \frac{\left(\sum_{\mathcal{S}} \prod_{X \notin \mathcal{T}} \Pr(X \mid \mathcal{C}(X)) \right) \prod_{X \in \mathcal{T}} \Pr(X \mid \mathcal{C}(X))}{\left(\sum_{\mathcal{S}} \prod_{X \notin \mathcal{T}} \Pr(X \mid \mathcal{C}(X)) \right) \left(\sum_{\mathcal{T}} \prod_{X \in \mathcal{T}} \Pr(X \mid \mathcal{C}(X)) \right)}$$

$$= \frac{\prod_{X \in \mathcal{T}} \Pr(X \mid \mathcal{C}(X))}{\sum_{\mathcal{T}} \prod_{X \in \mathcal{T}} \Pr(X \mid \mathcal{C}(X))}.$$

The same final expression is also obtained for $\Pr(\mathcal{T} \mid \mathcal{S} \cup \mathcal{V})$. This proves (8.20) for $\mathcal{S}' = \mathcal{S}$ and $\mathcal{T}' = \mathcal{T}$, subject to your filling in the steps I left for you. ∎

The key to dealing with singly connected Bayesian networks is provided by the following theorem. It will be convenient to have a notation for the "results" of X:

$$\mathcal{R}(X) = \left\{ Y \in \mathcal{X} \mid (X, Y) \in E \right\}.$$

Theorem 8.6

Let $(\mathcal{X}, E, \Pr)$ be a singly connected Bayesian network, let $\mathcal{W}^* \subset \mathcal{X}$, let $\mathcal{C}(V) = \{U_1, \ldots, U_l\}$, and let $\mathcal{R}(V) = \{Y_1, \ldots, Y_k\}$. Define

$$\mathcal{Y}_i^* = \left\{ W \in \mathcal{W}^* \;\middle|\; \text{the undirected path from } V \text{ to } W \text{ contains } Y_i \right\},$$

$$\mathcal{U}_i^* = \left\{ W \in \mathcal{W}^* \;\middle|\; \text{the undirected path from } V \text{ to } W \text{ contains } U_i \right\},$$

$\mathcal{U}^* = \mathcal{U}_1^* \cup \cdots \cup \mathcal{U}_l^*$, and $\mathcal{Y}^* = \mathcal{W}^* - \mathcal{U}^*$. Then

$$\Pr(\mathcal{W}^* \mid V) = \Pr(\mathcal{U}^* \mid V)\Pr(\mathcal{Y}^* \mid V) \tag{8.21}$$

$$\Pr(\mathcal{Y}^* \mid V) = \begin{cases} \displaystyle\prod_{i=1}^{k} \Pr(\mathcal{Y}_i^* \mid V), & \text{if } V \notin \mathcal{W}^*, \\[2mm] \displaystyle\Pr(V^* \mid V)\prod_{i=1}^{k} \Pr(\mathcal{Y}_i^* \mid V), & \text{if } V \in \mathcal{W}^*, \end{cases} \tag{8.22}$$

and

$$\Pr(\mathcal{C}(V) \mid \mathcal{U}^*) = \prod_{i=1}^{l} \Pr(U_i \mid \mathcal{U}_i^*). \tag{8.23}$$

In the above equations we make an exception to the usual rule that all occurrences of a variable must be set to the same value: A variable need not be set to the value it is assigned in a starred set. In particular, $\Pr(V^* \mid V)$ means $\Pr(V = v^* \mid V = v)$ which is 1 when $v = v^*$ and 0 otherwise.

Proof: Consider (8.21) and (8.22) when $V \in \mathcal{W}^*$. If V is not equal to its instantiated value, then both sides of these equations are zero and so equality holds. On the other hand, when V equals its instantiated value, $\Pr(V^* \mid V) = 1$ and V^* can be removed from $\mathcal{W}^*$ and $\mathcal{Y}^*$ without changing the probabilities containing them. In other words, when $V = V^*$, we can reduce it to the uninstantiated case. We now prove (8.21) and (8.22) when $V \notin \mathcal{W}^*$.

Let $\mathcal{V} = \{V\}$, let $\mathcal{T}$ be all $X \in \mathcal{X}$ such that a path from X to V passes through some Y_i, and let $\mathcal{S}$ be the rest of $\mathcal{X}$. Then $\mathcal{U}^* \subseteq \mathcal{S}$ and $\mathcal{Y}^* \subseteq \mathcal{T}$, and so (8.21) follows from (8.20).

To prove (8.22), remove one factor $\Pr(\mathcal{Y}_i^* \mid V)$ at a time. Let $\mathcal{V} = \{V\}$, let $\mathcal{T}$ by all $X \in \mathcal{X}$ such that a path from X to V passes through Y_i and let $\mathcal{S}$ be the rest of $\mathcal{X}$. Then $\mathcal{Y}_{i+1}^* \cup \cdots \cup \mathcal{Y}_k^* \subseteq \mathcal{S}$ and $\mathcal{Y}_i^* \subseteq \mathcal{T}$. Applying (8.20) with $i = 1, \ldots, k - 1$ successively, we have

$$\Pr(\mathcal{Y}_1^* \cup \cdots \cup \mathcal{Y}_k^* \mid V) = \Pr(\mathcal{Y}_1^* \mid V)\Pr(\mathcal{Y}_2^* \cup \cdots \cup \mathcal{Y}_k^* \mid V)$$

$$= \cdots = \Pr(\mathcal{Y}_1^* \mid V) \cdots \Pr(\mathcal{Y}_k^* \mid V).$$

Suppose that $U_i \notin \mathcal{U}_i$ for all i. A similar argument with $\mathcal{V} = \mathcal{U}^*$, $\mathcal{S}' = \{U_{i+1}, \ldots, U_l\}$, and $\mathcal{T}' = \{U_i\}$ yields

$$\Pr(\mathcal{C}(V) \mid \mathcal{U}^*) = \prod_{i=1}^{l} \Pr(U_i \mid \mathcal{U}^*).$$

(You should provide $\mathcal{S}$ and $\mathcal{T}$ for this.) Finally, with $\mathcal{Z}$ equal to the part of $\mathcal{C}(U_i)$ that is not in $\mathcal{U}^*$,

$$\Pr(U_i \mid \mathcal{U}^*) = \sum_{\mathcal{Z}} \Pr(U_i \mid \mathcal{U}^* \cup \mathcal{Z}) \Pr(\mathcal{Z})$$

$$= \sum_{\mathcal{Z}} \Pr(U_i \mid \mathcal{C}(U_i)) \Pr(\mathcal{Z}), \qquad \text{by (8.1) (p. 306)},$$

$$= \Pr(U_i \mid \mathcal{U}_i^*).$$

This proves (8.23) when no U_i is being conditioned. Suppose that some U_i's are in $\mathcal{U}^*$. If they are not equal to their instantiated values, both sides of (8.23) equal zero. On the other hand, if they equal their instantiated values, then the left side is unchanged if they are removed from $\mathcal{C}(V)$ and the right side is unchanged if those $\Pr(U_i \mid \mathcal{U}_i^*)$ are removed. Then the previous argument for the uninstantiated case applies. ∎

Exercises

8.5.1. Express the theorems verbally using words like "cause," "effect," and "independent." Then argue that the conclusions of the lemma and the theorem follow from our "commonsense" understanding of your statements. (Of course, this does not obviate the need for a mathematical proof; however, it makes it reasonable to expect one to be forthcoming if the mathematical definitions have captured the relevant aspects of cause and independence.)

8.5.2. Fill in the details in the proof of (8.20).

8.5.3. Fill in the details of the proof of (8.23).

The Algorithm for Singly Connected Networks

Let α_V and $\alpha_{V,W}$ be some unspecified nonzero constants that depend on the state of the network. The subscripts are used merely to distinguish between constants associated with different vertices of the DAG. In the notation of Theorem 8.6, we make the following definitions. (See the right-hand side of Figure 8.6 (p. 337).)

- $\pi_V = \Pr(V \mid \mathcal{U}^*)$, that is, the probability of V given $\mathcal{U}^*$—those random variables in $\mathcal{W}^*$ that are reached from V by an undirected path through $\mathcal{C}(V)$.

- $\pi_{U_i,V} = \Pr(U_i \mid \mathcal{U}_i^*)$, that is, the probability of U_i given $\mathcal{U}_i^*$—those random variables in $\mathcal{W}^*$ that are reached from V by an undirected path through U_i.

- $\lambda_V = \alpha_V \Pr(\mathcal{Y}^* \mid V)$, that is, some unspecified nonzero constant times the probability, given V, of $\mathcal{Y}^*$—those random variables in $\mathcal{W}^*$ that are not reached from V by an undirected path through $\mathcal{C}(V)$. (The λ notation is used because of the connection with "likelihood ratios.")

- $\lambda_{V,Y_i} = \alpha_{V,Y_i} \Pr(\mathcal{Y}_i^* \mid V)$, that is, some unspecified nonzero constant times the probability, given V, of $\mathcal{Y}_i^*$—those random variables in $\mathcal{W}^*$ that are reached from V by an undirected path beginning with the edge (V, Y_i).

In fact, these π's and λ's are just precisely the π's and λ's that appeared in the algorithm. Now that we have their definitions and have just proved the necessary tools, we'll be able to derive the formulas involving λ's and π's that appeared in the algorithms.

Suppose that V is not instantiated; that is, $V \notin \mathcal{W}^*$. According to (8.22), we may take

$$\lambda_V = \prod_{i=1}^{k} \lambda_{V,Y_i}, \qquad (8.24)$$

and then $\alpha_V = \prod \alpha_{V,Y_i}$. Applying Bayes' Theorem to (8.21), we have

$$\Pr(V \mid \mathcal{W}^*) = \left(\frac{\Pr(\mathcal{U}^*)}{\Pr(\mathcal{W}^*)\alpha_V} \right) \lambda_V \pi_V.$$

Since $\sum_V \Pr(V \mid \mathcal{W}^*) = 1$, we can determine the parenthesized factor and eliminate it:

$$\Pr(V \mid \mathcal{W}^*) = \frac{\lambda_V \pi_V}{\sum_V \lambda_V \pi_V} \quad \text{when } V \notin \mathcal{W}^*. \qquad (8.25)$$

It follows from (8.24) and (8.25) that, once we determine how to compute λ_{V,Y_i} and π_V, we'll have the algorithm for noninstantiated variables. It will turn out that π_V will be computed from π values at $\mathcal{C}(V)$ and that λ_{V,Y_i}

will be computed at Y_i. These values will be "passed" to V from the various vertices that generate them.

Formulas for π and λ

Let's begin with π_V. By summing over disjoint events and then using Bayes' Theorem, we have

$$
\begin{aligned}
\pi_V = \Pr(V \mid \mathcal{U}^*) &= \sum_{\mathcal{C}(V)} \Pr\Big(V \cup \mathcal{C}(V) \,\Big|\, \mathcal{U}^*\Big) \\
&= \sum_{\mathcal{C}(V)} \Pr(V \mid \mathcal{C}(V) \cup \mathcal{U}^*)\,\Pr(\mathcal{C}(V) \mid \mathcal{U}^*) \\
&= \sum_{\mathcal{C}(V)} \Pr(V \mid \mathcal{C}(V))\,\Pr(\mathcal{C}(V) \mid \mathcal{U}^*), \qquad \text{by (8.1) (p.\,306),} \\
&= \sum_{\mathcal{C}(V)} \Pr(V \mid \mathcal{C}(V)) \prod_{i=1}^{l} \Pr(U_i \mid \mathcal{U}_i^*), \qquad \text{by (8.23),} \\
&= \sum_{\mathcal{C}(V)} \Pr(V \mid \mathcal{C}(V)) \prod_{X \in \mathcal{C}(V)} \pi_{X,V}, \qquad \text{by definition.}
\end{aligned}
$$

Thus, if each vertex S sends to each vertex T for which $(S,T) \in E$ the value $\pi_{S,T}$, we can compute the π's using

$$
\pi_V = \sum_{\mathcal{C}(V)} \Pr(V \mid \mathcal{C}(V)) \prod_{X \in \mathcal{C}(V)} \pi_{X,V} \tag{8.26}
$$

As a refresher on notation, let's see what this is computing. The result is a function whose domain is range(V). The summation extends over all values in the ranges of $\mathcal{C}(V)$. In other words, if $\mathcal{C}(V) = \{U_1, \ldots, U_l\}$, then the summation is over all l-tuples

$$
\text{range}(U_1) \times \cdots \times \text{range}(U_l).
$$

For each particular l-tuple $(u_1, \ldots, u_l)$, $\Pr(V \mid \mathcal{C}(V))$ is a function whose domain is range(V) because $\mathcal{C}(V)$ has been instantiated to $(u_1, \ldots, u_l)$ and V has not been instantiated. The product on the right side of (8.26) is a number; in fact, it equals $\prod_{i=1}^{l} \pi_{U_i,X}(u_i)$.

The formula for π_{V,Y_i} is given by

$$
\pi_{V,Y_i} = \pi_V \prod_{j \neq i} \lambda_{V,Y_j} \,\Big/\, \sum_{V} \Big(\pi_V \prod_{j \neq i} \lambda_{V,Y_j} \Big), \tag{8.27}
$$

the proof of which is left as an exercise.

Now let's consider the λ value that V will pass to U_i, that is, $\lambda_{U_i,V}$. Let $\overline{\mathcal{U}}_i^*$ be the union of all the $\mathcal{U}_j^*$'s except for $\mathcal{U}_i^*$. We're going to have a long series of equations to compute $\Pr(\mathcal{Y}^* \cup \overline{\mathcal{U}}_i^* \mid U_i)$, hence the λ's. The plan is to separate $\mathcal{Y}^*$ and $\overline{\mathcal{U}}_i^*$ using (8.21) and then separate the various $\mathcal{U}_j$'s in $\overline{\mathcal{U}}_i^*$ using (8.23). In the process, we'll be using Bayes' Theorem in two ways:

$$\Pr(A \mid B)\Pr(B) = \Pr(B \mid A)\Pr(A)$$

and

$$\Pr(A \wedge B \mid C) = \Pr(A \mid B \wedge C)\Pr(B \mid C).$$

By Bayes' Theorem and summing over disjoint events, we have

$$\Pr(\mathcal{Y}^* \cup \overline{\mathcal{U}}_i^* \mid U_i)\Pr(U_i)$$
$$= \Pr(\mathcal{Y}^* \cup \overline{\mathcal{U}}_i^* \cup \{U_i\}) = \sum_V \sum_{\mathcal{C}(V)-\{U_i\}} \Pr(\mathcal{Y}^* \cup \overline{\mathcal{U}}_i^* \cup \mathcal{C}(V) \cup \{V\})$$
$$= \sum_V \sum_{\mathcal{C}(V)-\{U_i\}} \Pr(\mathcal{Y}^* \cup \overline{\mathcal{U}}_i^* \cup \mathcal{C}(V) \mid V)\Pr(V)$$
$$= \sum_V \sum_{\mathcal{C}(V)-\{U_i\}} \Pr(\mathcal{Y}^* \mid V)\,\Pr(\overline{\mathcal{U}}_i^* \cup \mathcal{C}(V) \mid V)\Pr(V),$$

by (8.21) with $\mathcal{U}^*$ taken to be the present $\overline{\mathcal{U}}_i^* \cup \mathcal{C}(V)$,

$$= \Pr(\overline{\mathcal{U}}_i^*)\sum_V \left(\Pr(\mathcal{Y}^* \mid V)\sum_{\mathcal{C}(V)-\{U_i\}} \Pr(V \mid \overline{\mathcal{U}}_i^* \cup \mathcal{C}(V))\Pr(\mathcal{C}(V) \mid \overline{\mathcal{U}}_i^*) \right),$$

by Bayes' Theorem,

$$= \frac{\Pr(\overline{\mathcal{U}}_i^*)}{\alpha_V}\sum_V \left(\lambda_V \sum_{\mathcal{C}(V)-\{U_i\}} \Pr(V \mid \mathcal{C}(V))\Pr(\mathcal{C}(V) \mid \overline{\mathcal{U}}_i^*) \right),$$

by the definition (8.1) of a Bayesian network (p. 306),

$$= \frac{\Pr(\overline{\mathcal{U}}_i^*)}{\alpha_V}\sum_V \left(\lambda_V \sum_{\mathcal{C}(V)-\{U_i\}} \Pr(V \mid \mathcal{C}(V))\Pr(U_i)\prod_{j \neq i} \pi_{U_j} \right),$$

by (8.23) with $\mathcal{U}^*$ replaced with $\overline{\mathcal{U}}_i^*$. Putting the start and finish together, you should see that we can define $\alpha_{U_i,V} = \alpha_V / \Pr(\overline{\mathcal{U}}_i^*)\Pr(U_i)$ and

$$\lambda_{U_i,V} = \sum_V \left(\lambda_V \sum_{\mathcal{C}(V)-\{U_i\}} \Pr(V \mid \mathcal{C}(V))\prod_{j \neq i} \pi_{U_j} \right). \qquad (8.28)$$

Now that we've discussed the necessary calculations for an uninstantiated vertex, how do we handle instantiation? The answer is fairly simple. Suppose that $X \in \mathcal{W}^*$ is to be instantiated to x^*. Only a few of the earlier calculations are incorrect.

- Equation (8.25) may be invalid because its derivation involves factors of $\Pr(V)$, which could be zero. On the other hand, there is no reason to use (8.25) if we have assigned a value to V by instantiation.

- The derivations of $\pi_{X,Y}(b)$ and $\lambda_X(b)$ are incorrect, but they are easily derived directly from their definitions: They will be one or zero depending on whether $b = x^*$ or not.

We can now prove that the algorithm is correct. There are two keys to the proof:

- The λ's are initialized correctly for the case of no instantiation.
- Information need not be sent backward: If X activates Y, the values Y would send back to X are simply the values of $\pi_{X,Y}$ and $\lambda_{X,Y}$ that X already possesses.

These observations were made as bracketed notes in the algorithms.

As a result of the λ's being correctly initialized, the π's will be correct after Step 2 in the main algorithm. Thus we need only analyze the effect of instantiation. Since information never needs to be sent backward, as the activation due to an instantiation moves outward from its start, it leaves behind correct values of the $\pi_{X,Y}$'s and the $\lambda_{X,Y}$'s. Thus, after each instantiation's calculations are complete, all values are correct.

Exercises

8.5.4. Write down (8.26) when $\mathcal{C}(V) = \emptyset$.

8.5.5. Write down (8.27) when $\mathcal{R}(V) = \{Y_1\}$.

8.5.6. Write down (8.28) when $\mathcal{C}(V) = \{U_1\}$.

8.5.7. Prove (8.27).

Further Remarks

The independence result (8.20) is not the most general. Pearl's approach uses "d-separation." Had I introduced it, the derivation of (8.28) would have been slightly shorter, but I think that (8.20) is more easily understood than d-separation.

It would have involved less calculation to restrict the ranges of $X \in \mathcal{W}^*$ to $\{x^*\}$ on page 330 and leave the rest of the Bayesian network unchanged. Why was this not done? It gives an incorrect model! The assertion that X is instantiated to x^* leads to conditioning on $X = x^*$, which can affect the probabilities of the various instantiations of $\mathcal{C}(X)$. On the other hand, the suggestion made at the beginning of the paragraph leads to a situation in which the probabilities for $\mathcal{C}(X)$ are unaffected since $\Pr(X = x^* \mid \mathcal{C}(X)) = 1$, regardless of the instantiation of $\mathcal{C}(X)$ in that probability function.

Exercises

8.5.8. The purpose of this exercise is to allow more parallelism in the algorithm.

    (a) How can the algorithm be modified to allow a vertex to receive more than one message at the same time?

    (b) Illustrate the modification on the DAG in Figure 8.4 when all the π's are started simultaneously and all the λ's are started simultaneously after the π's have finished. Make the assumption that, at a single time step, each active vertex sends out all the messages it can and becomes inactive. To make it clear what is going on, you should draw a separate picture for each time step, in contrast to the pictures in Figure 8.4 where all the time steps for a single Outside Action have been shown on one DAG.

    (c) Illustrate the modification on the DAG in Figure 8.4 when all the π's and all the λ's are started at the same time. Make the assumption that, at a single time step, each active vertex sends out all the messages it can and becomes inactive. To make it clear what is going on, you should draw a separate picture for each time step.

8.5.9. Suppose that we've applied the algorithm to a network. Now we want to change the function $\Pr(X)$ in the light of new evidence.

    (a) Suppose that $\mathcal{C}(X) = \emptyset$. Describe a method for doing this without restarting the algorithm from the beginning.

    *(b) Prove your result.

    *(c) Can you find an algorithm when $\mathcal{C}(X) \neq \emptyset$?

8.6 Certainty Factors

> *There is only one thing certain and that is that nothing is certain.*
>
> —G. K. Chesterton (1874–1936)

In this section we partially develop D. Heckerman's approach [6] to certainty factors: First we'll list *desiderata* (properties I—and, I hope, you—want them to have). Next we'll deduce information about certainty factors from the desiderata. In more mathematical terms, the desiderata would be called axioms.

What Are Certainty Factors?

Certainty factors are used in a diagnosis situation. Thus we (a) have a Bayesian network, (b) are given information about certain vertices (random variables), and (c) want to "push" the information back up the network. In other words, information about Y is used in obtaining information about $C(Y)$ but not vice versa. Because the direction information is being moved is always toward causes, people working with certainty factors usually *reverse the direction of the edges* in the DAG from the normal Bayesian network usage. (In fairness, I should note that certainty factors entered AI *before* Bayesian networks.)

Since a certainty factor concerning some hypothesis (vertex of a DAG) should provide some measure of the certainty of our belief in the hypothesis, we have the first desideratum:

> Desideratum 1:   A *certainty factor* is a function whose range is the interval $[-1, +1]$. The certainty factor $\mathrm{CF}(H, E)$ is intended to measure the *change* in belief in H given the evidence E with 0 indicating no change, $+1$ indicating that H is certain, and -1 indicating that $\neg H$ is certain. As $\mathrm{CF}(H, E)$ increases, so does the change in belief in H given E. The value of $\mathrm{CF}(H, E)$ does not depend on the prior belief in H.

From our viewpoint, H is a random variable and E is a set of random variables all of which are being instantiated. It's known that, if certainty factors conform to the usual rules of probability, then $|\mathrm{range}(H)| = 2$. We won't prove this; instead, we'll assume that

> All random variables discussed here have the same range, namely $\{\mathrm{T}, \mathrm{F}\}$. (See [6] for a proof.)

We'll temporarily make another assumption: All evidence consists of instantiation of random variables. In keeping with previous notation, we'll use the notation $\mathrm{CF}(X^+, \mathcal{Y}^*)$ and $\mathrm{CF}(X^-, \mathcal{Y}^*)$ for certainty factors.

The idea that evidence for $X = \mathrm{T}$ is the same as evidence against $X = \mathrm{F}$ leads to our second desideratum:

> Desideratum 2: $\mathrm{CF}(X^+, \mathcal{Y}^*) = -\mathrm{CF}(X^-, \mathcal{Y}^*)$.

An Interpretation of Certainty Factors

Since certainty factors measure changes in belief, we can think of them in terms of conditional probability and so have the key desideratum:

> **Desideratum 3**: It should be possible to somehow compute $\Pr(X \mid \mathcal{Y}^*)$ from $\Pr(X)$ and $\mathrm{CF}(X^+, \mathcal{Y}^*)$

How can this be translated into something usable? First, we should determine what's needed to compute the conditional probability. Then we must insist that the certainty factor and $\Pr(X)$ provide that information. Finally, we can examine the impact of the two desiderata on this. The results are summarized in the following theorem.

> **Theorem 8.7**   Nature of Certainty Factors
>
> Let $\lambda(X^+, \mathcal{Y}^*)$ be the *likelihood ratio* $\Pr(\mathcal{Y}^* \mid X^+) \, / \, \Pr(\mathcal{Y}^* \mid X^-)$. If CF is a certainty factor, then
>
> $$\mathrm{CF}(X^+, \mathcal{Y}^*) = F\big(\lambda(X^+, \mathcal{Y}^*)\big), \tag{8.29}$$
>
> where F in an increasing function from $[0, \infty]$ to $[-1, 1]$ such that $F(0) = -1$ and $F(1/x) = -F(x)$. It follows that $F(\infty) = 1$ and $F(1) = 0$.

Proof: By the usual manipulation of conditional probabilities, we have

$$\frac{\Pr(X^+ \mid \mathcal{Y}^*)}{\Pr(X^- \mid \mathcal{Y}^*)} = \frac{\Pr(X^+)}{\Pr(X^-)} \frac{\Pr(\mathcal{Y}^* \mid X^+)}{\Pr(\mathcal{Y}^* \mid X^-)} = \frac{\Pr(X^+)}{\Pr(X^-)} \lambda(X^+, \mathcal{Y}^*). \tag{8.30}$$

The ratio on the left-hand side is a probabilistic measure of belief given $\mathcal{Y}^*$, and the ratio on the right-hand side is a probabilistic measure of prior belief—the *prior odds*. The last sentence in Desideratum 1 asserts that since CF measures *change* in belief, it does depend on prior information. In other words, the contribution of $\mathrm{CF}(X^+, \mathcal{Y}^*)$ to the right side of (8.30) must not change as $\Pr(X^+)/\Pr(X^-)$ varies. Since the likelihood ratio does not change as we vary the prior odds, the likelihood ratio is a function of the certainty factor but not of the prior odds. Thus, $\lambda = f(\mathrm{CF})$ for some function f. Applying f^{-1} to both sides and calling it F, we obtain (8.29). Since the range of λ is $[0, \infty]$ and the range of CF is $[-1, +1]$, the domain and range of F must be as stated in the theorem. The penultimate sentence of Desideratum 1 tells us that F is an increasing function. Since $\Pr(X^+ \mid \mathcal{Y}^*) = 0$ requires that $\lambda = 0$, Desideratum 1 gives us $F(0) = -1$. Since interchanging the roles of X^+ and X^- changes λ to its reciprocal, Desideratum 2 tells us that $F(1/x) = -F(x)$. The last two results now follow from these results: $F(1) = F(1/1) = -F(1)$ and so $F(1) = 0$. $F(\infty) = F(1/0) = -F(0) = 1$. They could also have been deduced from Desideratum 1. ∎

Are there restrictions on F besides those in the theorem? No. If we describe how to combine certainty factors, and if that method of combining is consistent with the rules for combining λ, then F would be determined. Alternatively, we could start with a definition of certainty factors and derive the rules for combining them.

The details of this paragraph are left as an exercise. To illustrate the previous ideas, let's define

$$\mathrm{CF}(X^*, \mathcal{Y}^*) = \begin{cases} \dfrac{\Pr(X^* \mid \mathcal{Y}^*) - \Pr(X^*)}{\Pr(X^* \mid \mathcal{Y}^*)\big(1 - \Pr(X^*)\big)}, & \text{if } \Pr(X^* \mid \mathcal{Y}^*) \geq \Pr(X^*), \\[2mm] \dfrac{\Pr(X^* \mid \mathcal{Y}^*) - \Pr(X^*)}{\Pr(X^*)\big(1 - \Pr(X^* \mid \mathcal{Y}^*)\big)}, & \text{if } \Pr(X^* \mid \mathcal{Y}^*) \leq \Pr(X^*). \end{cases} \tag{8.31}$$

Then it can be shown that

$$F(x) = \begin{cases} \dfrac{x-1}{x}, & \text{if } x \geq 1, \\[3mm] x - 1, & \text{if } x \leq 1. \end{cases} \tag{8.32}$$

Suppose that $\mathcal{Y}^*$ and $\mathcal{Z}^*$ are independent pieces of evidence that bear on X. In this case, it can be shown that

$$\lambda(X^+, \mathcal{Y}^* \cup \mathcal{Z}^*) = \lambda(X^+, \mathcal{Y}^*)\lambda(X^+, \mathcal{Z}^*). \tag{8.33}$$

(Remembering that evidence means effects, you should be able to derive (8.33) from (8.20) (p. 336).) This leads to a rule, depending on F, for combining certainty factors. For the case of (8.32), with $y = \mathrm{CF}(X^+, \mathcal{Y}^*)$ and $z = \mathrm{CF}(X^+, \mathcal{Z}^*)$, we have

$$\mathrm{CF}(X^+, \mathcal{Y}^* \cup \mathcal{Z}^*) = \begin{cases} y + z - yz, & \text{if } y \geq 0 \text{ and } z \geq 0, \\[2mm] \dfrac{y + z}{1 - \min\big(|y|, |z|\big)}, & \text{if } yz \leq 0, \\[2mm] y + z + yz & \text{if } y \leq 0 \text{ and } z \leq 0. \end{cases} \tag{8.34}$$

This is just the combining function given in MYCIN; however, (8.31) differs slightly from that in MYCIN because MYCIN's definition of CF is not consistent with (8.34).

Limits on Certainty Factor Validity

We noted earlier that Heckerman proved certainty factors cannot be defined in a consistent fashion if the random variables have ranges that contain more than two elements. Now let's look for other limitations on using certainty factors.

Let $(\mathcal{X}, E, \Pr)$ be the Bayesian network associated with the problem. Suppose that we have the situation shown on the left-hand side of Figure 8.7. The formula for parallel combination (8.34) provides a method for combining the evidence $\mathcal{Y}_1^*$ and $\mathcal{Y}_2^*$ and applying it to U_1. Unfortunately, we have no mechanism for applying $\mathcal{Y}_1^*$ to U_2. This is needed since, if $\mathcal{Y}_1^*$ makes U_1 more probable, then the competing explanation U_2 for X_2 will be less probable. We cannot simply take the evidence $\mathcal{Y}_1^*$ and apply it to U_2 because we need to know how it interacts with U_1 and how this interaction affects X_2. This was done in the algorithm on page 330 for singly connected Bayesian networks, but it cannot be done with certainty factors because they only move evidence upward. (Actually we do not even need X_1 since the evidence could be $\{U_1^*\}$ instead of $\mathcal{Y}_1^*$.) We've shown that each $X \in \mathcal{X}$ can have at most one cause if certainty factors are to work correctly.

From the result in the previous paragraph, it can be shown that a connected Bayesian network in which certainty factors can be used is singly connected and has precisely one X with $\mathcal{C}(X) = \emptyset$. (Details are left as an exercise.)

Even in networks that are this limited, we haven't dealt fully with the propagation of certainty factors. You can think of (8.34) as a method for combining certainty factors when parallel, independent evidence is available for X. (See the center diagram in Figure 8.7.) What happens if we have evidence $\mathcal{Y}^*$ for X and if X, in turn, is evidence for U as shown in the right-hand side of Figure 8.7? First, how do we express the fact that X is evidence for U numerically, and second, how do we move from this and $\mathrm{CF}(X^+, \mathcal{Y}^*)$ to $\mathrm{CF}(U^+, \mathcal{Y}^*)$? The answer to the first question is "Via likelihood ratios" since that's what certainty factors are related to. Using a Bayesian approach as in (8.30), we have

$$
\lambda(U^+, \mathcal{Y}^*) = \frac{\Pr(\mathcal{Y}^* \mid U^+)}{\Pr(\mathcal{Y}^* \mid U^-)} = \frac{\Pr(U^-)}{\Pr(U^+)} \frac{\Pr(\{U^+\} \cup \mathcal{Y}^*)}{\Pr(\{U^-\} \cup \mathcal{Y}^*)}
$$

$$
= \frac{\Pr(U^-)}{\Pr(U^+)} \frac{\sum_X \Pr(\{U^+, X\} \cup \mathcal{Y}^*)}{\sum_X \Pr(\{U^-, X\} \cup \mathcal{Y}^*)}
$$

$$
= \frac{\Pr(U^-)}{\Pr(U^+)} \frac{\sum_X \Pr(U^+) \Pr(X \mid U^+) \Pr(\mathcal{Y}^* \mid \{U^+, X\})}{\sum_X \Pr(U^-) \Pr(X \mid U^-) \Pr(\mathcal{Y}^* \mid \{U^-, X\})}
$$

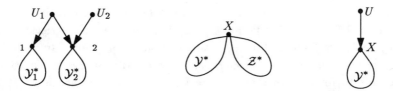

Figure 8.7   Pieces of various Bayesian networks.

$$= \frac{\sum_X \Pr(X \mid U^+) \Pr(\mathcal{Y}^* \mid X)}{\sum_X \Pr(X \mid U^-) \Pr(\mathcal{Y}^* \mid X)},$$

where the last equality follows from (8.20) (p. 336) with $\mathcal{T} = \mathcal{Y}^*$, $\mathcal{V} = \{X\}$, and $\mathcal{S} = \{U\}$ because we have a Bayesian network. The details are left as an exercise. You should then be able to show that

$$\lambda(U^+, \mathcal{Y}^*) = \frac{\Pr(X^+ \mid U^+)\lambda(X^+, \mathcal{Y}^*) + 1 - \Pr(X^+ \mid U^+)}{\Pr(X^+ \mid U^-)\lambda(X^+, \mathcal{Y}^*) + 1 - \Pr(X^+ \mid U^-)}. \tag{8.35}$$

Certainty factors are directly related to likelihood ratios, not probabilities. Therefore, in order to convert to certainty factors, we need to somehow replace the probabilities in (8.35) with likelihood ratios. This can be done using the two equations

$$\lambda^+ := \lambda(U^+, X^+) = \frac{\Pr(X^- \mid U^+)}{\Pr(X^- \mid U^-)}$$

$$\lambda^- := \lambda(U^+, X^-) = \frac{1 - \Pr(X^+ \mid U^+)}{1 - \Pr(X^+ \mid U^-)} \tag{8.36}$$

as follows. Clear denominators to obtain two linear equations in the Pr's, solve for the Pr's in terms of the λ's, and substitute into (8.35) to obtain

$$\lambda(U^+, \mathcal{Y}^*) = \frac{\lambda^+(1 - \lambda^-)\lambda(X^+, \mathcal{Y}^*) - \lambda^-(1 - \lambda^+)}{(1 - \lambda^-)\lambda(X^+, \mathcal{Y}^*) - (1 - \lambda^+)}. \tag{8.37}$$

To summarize, certainty factors can be used in a causal network only if the following hold:

- Each random variable has $\{\text{T}, \text{F}\}$ as range.
- The network is singly connected with $|\mathcal{C}(X)| \leq 1$ for all X. (Such a network is called a "rooted tree" and the root U has $\mathcal{C}(U) = \emptyset$.)
- The goal is to determine $\Pr(X)$ for some X given information about some random variables in $\mathcal{D}(X)$—the vertices in the network "below" X. (Note that the only information discussed here was instantiation. We can allow probabilistic information, too.)

As part of the network information, we must be given the certainty factor information defined in (8.36). Heckerman [6] derives these restrictions on the use of certainty factors starting with different assumptions. In fact, he derives the Bayesian network properties, which I assumed. His assumptions are preferable to mine since they assume less, but my approach is shorter and still illustrates the power of reasoning from the properties of a function to restrictions on the nature and applicability of the function.

Exercises

8.6.A. Describe the idea of a certainty factor.

8.6.B. Explain: "Under reasonable assumptions, certainty factors are essentially likelihood ratios." (This is the content of Theorem 8.7.)

8.6.C. For what sorts of Bayesian nets do certainty factors provide a valid means of Bayesian inference?

8.6.1. After the definition (8.31) and up through (8.34), a considerable amount of detail has been omitted. Fill it in.

8.6.2. Derive results like those from (8.31) to (8.34) for the simpler function

$$F(x) = \frac{x-1}{x+1}. \tag{8.38}$$

8.6.3. Show that a connected Bayesian network in which certainty factors can be used is singly connected and has precisely one X with $C(X) = \emptyset$.

8.6.4. Fill in the details of the derivation of (8.37) and use it to derive certainty function equations for the certainty factor defined in the text and for the certainty factor defined by (8.38).

Notes

I've tried to adhere reasonably closely to the notation that appears in the literature, but have made some changes in the interests of clarity and notational consistency in the text.

A broader definition of Bayesian networks can be found in Neapolitan's text [10, p. 158]. He uses the notion of "propositional variables," which are a type of random variables that are more general than the projections I used. The texts by Neapolitan [10] and Pearl [13] discuss many topics more thoroughly than I did. They also discuss the diagnosis problem from somewhat different viewpoints. Kruse et al. [9] provide another point of view. There has been quite a bit of literature on certainty factors. A partial bibliography can be found in [6]. Shafer and Pearl [15] reprint a variety of articles related to this chapter. The journal *Statistical Science* is a forum for discussion of issues related to statistics. This is done by having one or more papers presenting viewpoints on or surveys of a subject. Other researchers then comment on the papers and the authors reply. The subject of graphical models of dependency is treated in the August 1993 issue (**8** 204–283).

My discussion of the bipartite diagnosis problem is based largely on the book by Peng and Reggia [14], where the algorithm for irredundant covers appears. This book and Neapolitan's [10] discuss abduction (the diagnosis problem) for more general DAGs. Hobbs et al. [7] discuss another method for carrying out abductive inference. It lacks the firm foundation that probability theory provides for Bayesian nets. On the other hand, it can be used in reasoning situations where the data and computational demands of Bayesian nets preclude their use.

Charniak [3] gives a nontechnical introduction to Bayesian nets. For further discussion of Bayesian nets, see the texts [10] and [13]. They both discuss techniques for extending the algorithm to cases where the DAG is multiply connected. One method eliminates edges by considering all instantiations of carefully selected variables. Another eliminates edges by merging vertices that share causes and/or effects. A different approach, which leads to an undirected graph called the "moral graph," has been developed by Spiegelhalter and Lauritzen. See [10, Ch. 7] for information. Kim and Pearl [8] have implemented some singly connected Bayesian networks in CONVINCE. Less restrictive Bayesian networks are used in the expert system shell HUGIN by Andersen et al. [1].

Further research needs to be done on Bayesian networks. One problem is to find more efficient algorithms for handling complicated networks. Approximate methods are also being studied. Bonissone [2] lists methods for dealing with Bayesian networks and provides some references. In a sense, this research is doomed to failure because it is NP-hard to find even crude approximations [5]. In another sense, success is possible because researchers may find algorithms that make many commonly encountered situations computationally tractable. This is no different from any other area of AI: All the big problems from nonmonotonic reasoning (Chapter 6) to neural nets (Chapter 11) are NP-hard, but researchers still look for better and/or more general algorithms.

In addition to approximate methods for solving networks, there's the difficulty of obtaining accurate estimates for $\Pr(X \mid \mathcal{C}(X))$, which I mentioned earlier. Available estimates for these probabilities are likely to be crude. When approximations are involved, it is useful to have empirical or, if possible, theoretical results on the extent to which the approximations cause the answers to deviate from their correct values—this is called sensitivity analysis. *Sensitivity analysis* provides estimates on how much conclusions change when the problem is changed slightly. It's a standard aspect of numerical analysis. Sensitive problems—those in which slight changes in data produce large changes in answers—are called *ill-conditioned*. Practically nothing is known about the sensitivity of AI techniques.

Fault-tree analysis, which is usually considered to be outside AI, is related to Bayesian network theory. In this case, there is just one vertex, "system failure," with $\mathcal{R}(X) = \emptyset$ and there are many causes that are combined with noisy-ORs and noisy-ANDs. The Bayesian network is, in fact, usually not singly connected (and hence not a tree). For an introduction, see Page and Perry [11, 12].

I defined abductive inference in a probabilistic setting in Definition 8.5 (p. 315). The nonprobabilistic form is the process of concluding "A" from "B" and "If A then B." Of course, this in not a valid deduction in the propositional calculus. If the abduction is usually true, it could be stated as a defeasible rule in a nonmontonic reasoning system.

Biographical Sketch

Judea Pearl (1936–)

Born in Tel Aviv, he received a bachelor's degree in electrical engineering from Technion, a master's degree in physics from Rutgers, and a doctorate in electrical engineering from the Polytechnic Institute of Brooklyn. Pearl is a professor of computer science at UCLA and Director of the Cognitive Systems Laboratory there.

His focus has been the use of probabilistic methods in reasoning. This includes work in heuristic search, Bayesian nets, and approaches to nonstandard logics using infinitesimal probabilities. Thanks to his work on algorithms for Bayesian nets, probabilistic reasoning in networks is on a secure theoretical foundation, something that the certainty factor approach lacked. As a result, researchers have become more interested in exploring such methods.

References

1. S. K. Andersen, K. G. Olesen, F. V. Jensen, and F. Jensen, HUGIN—A shell for building Bayesian belief universes for expert systems. In *Proceedings of the Eleventh International Joint Conference on Artificial Intelligence* (IJCAI-89), Morgan Kaufmann, San Mateo, CA (1989) 1080–1085. Reprinted in [15], 332–337.

2. P. P. Bonissone, Approximate reasoning systems: A personal perspective. In *AAAI-91*, Vol. 2, 923–929, MIT Press, Cambridge, MA (1991). (AAAI is also known as the *National Conference on Artificial Intelligence*.)

3. E. Charniak, Bayesian networks without tears, *AI Magazine* **12** (1991) 50–63.

4. G. F. Cooper, The computational complexity of probabilistic inference using Bayesian belief networks, *Artificial Intelligence* **42** (1990) 393–405.

5. P. Dagum and M. Luby, Approximating probabilistic inference in Bayesian belief networks is NP-hard, *Artificial Intelligence* **60** (1993) 141–153.

6. D. Heckerman, Probabilistic interpretations for MYCIN's certainty factors. In L. N. Kanal and J. F. Lemmer (eds.), *Uncertainty in Artificial Intelligence*, North-Holland, Amsterdam (1986) 167–196. Reprinted in [15], 298–312.

7. J. R. Hobbs, M. E. Stickel, D. E. Appelt, and P. Martin, Interpretation as abduction, *Artificial Intelligence* **63** (1993) 69–142.

8. J. H. Kim and J. Pearl, CONVINCE: A conversational inference consolidation engine, *IEEE Trans. on Systems, Man and Cybernetics* **17** (1987) 120–132.

9. R. Kruse, E. Schwecke, and J. Heinsohn, *Uncertainty and Vagueness in Knowledge Based Systems*, Springer-Verlag, Berlin (1991).

10. R. E. Neapolitan, *Probabilistic Reasoning in Expert Systems: Theory and Algorithms*, John Wiley and Sons, New York (1990).

11. L. B. Page and J. E. Perry, Reliability, recursion and risk, *Americam Mathematical Monthly* **98** (1991) 937–946.

12. L. B. Page and J. E. Perry, Direct-evaluation algorithms for fault-tree probabilities, *Computers Chem. Engng.* **15** (1991) 157–169.

13. J. Pearl, *Probabilisitic Reasoning in Intelligent Systems*, Morgan Kaufmann, San Mateo, CA (1988).

14. Y. Peng and J. A. Reggia, *Abductive Inference Models for Diagnostic Problem-Solving*, Springer-Verlag, Berlin (1990).

15. G. Shafer and J. Pearl (eds.), *Readings in Uncertain Reasoning*, Morgan Kaufmann, San Mateo, CA (1990).

16. E. H. Shortliffe and B. G. Buchanan, A model of inexact reasoning in medicine, *Math. Biosciences* **23** (1975) 351–379. A shortened version is reprinted in [15], 259–273.

9

Fuzziness and Belief Theory

Introduction

Randomness and uncertain judgment were once thought to require different mathematical formalisms. Since inanimate objects do not exercise judgment, the emergence of physical science fostered the development of a probability theory based on randomness. Applications of probability theory achieved some spectacular successes, such as the statistical theory of gases. Interest in theories based on uncertain judgment waned. When the sciences of human behavior—economics, sociology, and psychology—sought to incorporate uncertainty, they looked to probability theory since it was the only available mathematical tool. (AI is another science related to human behavior.) Some researchers have questioned the appropriateness of probability theory and proposed other alternatives.

In this chapter, we'll examine some objections to probability theory and introduce two alternative choices—fuzzy reasoning and belief theory (also called evidential reasoning) . Fuzzy methods, belief theory, Bayesian nets, and certainty factors are the four main methods currently used for quantitative uncertain reasoning. With this chapter, we come to the end of our exploration of uncertain reasoning, which we began in Chapter 5; thus we conclude the

chapter with a brief assessment of the strengths and weaknesses of some of
the qualitative and quantitative methods we've explored. In the next chapter, we begin a discussion of expert systems that are not based on reasoning
methods.

Since the sections of this chapter are practically independent of one another, you can pick and choose among them.

Prerequisites: Section 7.1 is needed. The concepts of conditional probability, Bayes' Theorem, and independence, which are found elsewhere in
Chapter 7, are needed to follow some of the discussions; however, those
concepts are not needed to understand or apply belief and fuzzy methods. The notation and terminology used for rule systems are also used in
fuzzy reasoning, but reading about rule systems in Section 6.4 isn't essential.

Used in: No chapters require this material.

9.1   Is Probability the Right Choice?

*It seems unfair to crow about the successes of a theory and to
sweep all its failures under the rug.*

—Richard Brauer (1963)

Probability theory was formulated to deal with uncertainty of occurrence. Is
probability theory also appropriate for situations involving vagueness, ambiguity, and so forth? This is the same issue that was raised for first-order logic
(FOL) beginning on page 175. The ideas discussed there relate to (a) ability to represent information, (b) ease of use, and (c) computational adequacy.
Probability applied to expert system design leads to Bayesian nets and approximations such as certainty factors. In the previous chapter, Bayesian nets
received mixed reviews for ease of use and computational adequacy. Now let's
explore probability theory's representational ability; that is, is probability
theory adequate for representing the situation we want to study?

Some researchers claim that probability is inappropriate for representing important forms of quantitative uncertainty. They champion alternative
methods as more appropriate. Other researchers attack these alternatives and
defend probability theory. To avoid choking on the dust clouds thus raised,
we need a clear view of what probability theory does and doesn't do. Then
we can look at the world and see how well probability fits.

How can we obtain an understanding of probability theory's capabilities
and limitations? Careful, deep study is one approach. Debate is another. The

former is time-consuming and the latter often fails because unexpressed pre-conceptions obscure the situation. The axiomatic method provides a third approach. You've seen it in action before:

- In Chapter 5, the axiomatic approach was used to show that fair elections are impossible. Axioms were given for fair elections and we showed that they were inconsistent.

- In Chapter 8, the axiomatic approach was used to explore the nature and limitations of certainty factors. Desired properties—essentially axioms—were given and their consequences were explored.

We can approach probability theory in a similar manner—list a set of axioms and prove they imply we *must* use probability theory. What does this gain us? The idea is to choose reasonable axioms for measuring, as we did for fair elections. Of course, these axioms relate to a measure of "uncertainty" or "plausibility"—not fair elections. Next we prove that the only measure satisfying the axioms obeys the laws of probability. It follows that we must either accept probability theory as the appropriate tool or reject at least one of the axioms. Thus, the axioms provide a clear focus for debate and/or study. We'll state some axioms that imply probability theory, but will omit the proof.

We'll work with propositions as in propositional logic and assign a measure of "plausibility" to propositions, conditioned on the truth (occurrence) of other propositions. The restriction to propositional logic is merely for simplicity—we could use predicate logic. Here are the assumptions.

Axiom 1: Our universe (or domain) $\mathcal{U}$ consists of a set of basic propositions and formulas that can be built from them using the connectives $\neg$, $\vee$, and $\wedge$ of propositional logic.

Axiom 2: There is a function $f : \mathcal{U} \times \mathcal{U} \to \mathbb{R}$, written $f(\alpha \mid \beta)$. We should think of this as the plausibility of α when we know that β is true.

Axiom 3: If α and α' are logically equivalent and likewise β and β', then $f(\alpha \mid \beta) = f(\alpha' \mid \beta')$.

Axiom 4: If $f(\alpha \mid \gamma) < f(\beta \mid \gamma)$, then $f(\neg\alpha \mid \gamma) > f(\neg\beta \mid \gamma)$.

Axiom 5: There is some function g such that
$$f((\alpha \wedge \beta) \mid \gamma) = g\Big(f(\alpha \mid \gamma), f(\beta \mid (\alpha \wedge \gamma))\Big).$$

These assumptions imply that f is "essentially" a conditional probability, that is, there is an increasing function ϕ such that $\phi(f(\alpha \mid \beta))$ is a conditional probability.

It follows that we must either accept the use of probability or reject one or more of the axioms. Which axiom, if any, is most open to intelligent attack? Since the later axioms are less controversial, let's begin with Axiom 5 and work backward.

Axiom 5:

This axiom arises from thinking about α and β sequentially—first α and then β. (This does not mean that there has to be some sense of time—it's our thoughts that are sequential, not the events.) Suppose that γ is true. The axiom says that if you know how plausible α is and how plausible β is given α, you should be able to compute the plausibility of $\alpha \wedge \beta$—a reasonable assumption.

Axiom 4:

This axiom is based on the idea that if β is more plausible than α when γ is true, then not having β should be less plausible than not having α—another reasonable assumption.

Axiom 3:

This axiom claims that equivalent descriptions (e.g., α and α') can be substituted for one another. "Equivalent" here refers to propositional logic, not the real world. For example, "Monday" and "the day after Sunday" are not equivalent in propositional logic, but are in the real world. We might object to the computational problem of determining equivalence, but that is not an objection to the idea of Axiom 3.

Axiom 2:

This axiom contains overt and hidden assumptions.

First, the overt assumption. Since the values of f must be real, Axiom 2 claims that the plausibility of a statement can be measured by a single real number. Is this reasonable? We might say that plausibility is a measure— and aren't measures, after all, real numbers? No. The use of only one real number implies a simple linear ordering and some things aren't that simple. For example, position is measured by three real numbers. An example closer to AI is "intelligence." We could circularly and uselessly define intelligence to be that which is measured by IQ. A more fruitful approach is to begin by admitting that there are too many aspects to intelligence for one number such as IQ to provide a meaningful overall measure. Perhaps the same is true of our concept of plausibility. Although this is a cogent objection, we won't pursue it.

The hidden assumption in Axiom 2 is the requirement that we can assign a plausibility to every statement. Consider the following scenario:

> I haven't seen Jim for a while because he's been busy in the lab. On Friday the 9th there's a message from him on my answering machine: "I should be free of the lab soon. How about getting together next Sunday for a game of squash?"   (9.1)

Knowing Jim, I may be willing to assign a probability of $\frac{3}{4}$ that he'll be free as predicted. Unfortunately "next" Sunday might be the 11th or the 18th. Thus, I may be willing to assign a plausibility of $\frac{3}{4}$ to the statement "Jim will be able to get together with me on either the 11th or the 18th." In other words,

$f(J_{11} \lor J_{18}) = \frac{3}{4}$, where the notation should be obvious. However, I may be unwilling to assign any plausibility to either J_{11} or J_{18}. A. P. Dempster and Glenn Shafer developed belief theory in response to such problems.

Axiom 1:

First, a spurious objection: Since Chapter 6 was devoted to a discussion of logics that overcome some limitations of FOL, isn't it a major step backward to limit attention to propositional logic? No. Most of the ideas in Chapter 6 built on FOL, adding ways of assigning truth values other than "certainly true" and "certainly false." Here we're concerned with assigning plausibility to FOL formulas, which is another way of going beyond certainly true and certainly false.

Actually, this idea is closely related to a serious problem. Inherent in the combination of Axioms 1 and 3 is a hidden assumption about the nature of the world. For example, suppose you have never met Marcel and wonder if he is bright. Let M be the statement that Marcel is bright. A mutual friend says that Marcel is reasonably bright, but he also points out that, in some ways, Marcel isn't particularly bright. It appears that both M and $\neg M$ are plausible and so $M \land (\neg M)$ has some degree of plausibility. This vagueness wreaks havoc with ordinary true/false logic where $\alpha \land (\neg \alpha)$ is always false.

The major objections we've raised to probability can be characterized by looking at the types of ambiguities that can occur. Using this approach, we can distinguish among the three main quantitative tools for dealing with ambiguity:

- **Probability:** Probability theory deals with clearly defined events (i.e., true/false events) whose plausibility it makes sense to discuss. The source of ambiguity is occurrence: we're uncertain what event will occur (or has occurred but isn't yet known).

- **Belief Theory:** We may object to the idea of assigning plausibility to some events, thereby introducing a further level ambiguity. Belief theory is designed to deal with evidence that points only to sets of events and does not allow us to directly assess the likelihood of individual events within a set. Instead of treating events directly, it focuses on evidence.

- **Fuzzy Methods:** The ambiguity may lie in the nature of the event rather than its occurrence: We may object that a concept is ambiguous and so cannot be described and manipulated using FOL. Many everyday concepts are like this; e.g., intelligent, ripe, fast. Fuzzy methods are designed for situations in which the events, rather than their occurrence, are the source of ambiguity.

In this chapter, we'll briefly explore the two alternatives to probability theory. How are they defined and used? What are their foundations? Are these newcomers viable alternatives when probability theory seems inapplicable?

Exercises

9.1.A. What is the objection to the axioms for probability that leads to belief theory?

9.1.B. What is the objection to the axioms for probability that leads to fuzzy theories?

9.1.C. What is the difference between the ambiguities treated by probability, belief theory, and fuzzy methods?

9.1.D. It has been said that probability describes ambiguity about what event will occur and fuzziness describes ambiguity about what the event is. Explain why this is or is not a reasonably accurate characterization.

9.2   Fuzziness

> *Because of its unorthodoxy, it has been ... controversial ....*
> *Eventually, though, the theory of fuzzy sets is likely to be*
> *recognized as a natural development in the evolution of scientific*
> *thinking, and the skepticism about its usefulness will be viewed,*
> *in retrospect, as a manifestation of the human attachment to*
> *tradition and resistance to innovation.*
> —Abraham Kandel and Mordechay Schneider (1989)

> *Great is a word of great vagueness.*
> —Martin Gardner (1983)

FOL and set theory are closely related—we can identify a predicate with the set of arguments for which it is true. For example, the predicate "less than" can be identified with $\{ (x, y) \mid (x, y \in \mathbb{R}) \wedge (x < y) \}$. You can see this idea in action in Chapter 7, where we often treated a compound event A as a predicate that was true for those elementary events $e \in A$. Here's another approach for propositional logic. Let $\mathcal{X}$ be the set of all possible interpretations of the propositional letters; that is, $\mathcal{X}$ corresponds to all possible worlds. The subset of $\mathcal{X}$ corresponding to a formula contains those interpretations for which the formula is true. As before, $\mathcal{X}$ corresponds to a tautology.

Since logic and set theory are closely related, developing the notion of fuzzy sets can help pave the way for a fuzzy approach to logic. Fuzzy logic has been most useful when restricted to reasoning analogous to rule systems. In this section we'll briefly explore the notion of fuzzy sets and fuzzy rule systems.

The Fuzzy Set Concept

A central concept in fuzzy reasoning is the notion of a "fuzzy set." Let $\mathcal{X}$ be the set of all objects under consideration. There's a natural correspondence between ordinary sets $A \subset \mathcal{X}$ and functions $\chi_A : \mathcal{X} \to \{0,1\}$; namely, $\chi_A(x) = 1$ whenever $x \in A$ and $\chi_A(x) = 0$ whenever $x \notin A$. (The function χ_A is called the *characteristic function* of the set A.) Membership in ordinary sets is "crisp"—a yes or no affair. Membership in fuzzy sets is vague. That leads to the following definition.

Definition 9.1 Fuzzy Sets are Membership Functions

Let $\mathcal{X}$ be the set of all objects under consideration. A *fuzzy set* A is a function $\mu_A : \mathcal{X} \to [0,1]$. We say that two fuzzy sets A and B are equal if and only if $\mu_A(x) = \mu_B(x)$ for all $x \in \mathcal{X}$. We call μ_A the *membership function* of A. The fuzzy set whose membership function is identically zero is called the empty set.

An ordinary set A is defined as a collection of objects and the characteristic function χ_A is then defined in terms of A. In contrast, the fuzzy set A has no such independent definition: μ_A *is* its definition. Nevertheless, we still use notation like $A \cap B$. This is simply the name of a fuzzy set given by the as yet undefined function $\mu_{A \cap B}$.

If $\mu_A(\mathcal{X}) = \{0,1\}$, then A can be thought of as an ordinary set. If there is some $x \in \mathcal{X}$ with $0 < \mu_A(x) < 1$, then A cannot be thought of as an ordinary set because x has a degree of membership in A that is neither 0% nor 100%.

In Chapter 7, we came to some understanding of what probability meant before launching into the subject. We need to do the same for fuzzy sets.

To reach such an understanding, let's take an often cited example. The universe $\mathcal{X}$ consists of people and the fuzzy set A consists of tall people. Young children are not in A and so $\mu_A(x) = 0$ when x is a young child. Many basketball stars are certainly in A and so have $\mu_A(x) = 1$. We might have $\mu_A(p) = 0.7$ for a person p who is somewhat tall. What does the value 0.7 mean? How is it determined or estimated?

It's tempting to take a frequentist approach, as was done for probability theory. In that case we'd say that $\mu_A(p) = 0.7$ because 70% of the people (whoever they are) would agree that p is tall and 30% would disagree. This tempting approach must be resisted. Here are some reasons.

(a) If we accept this approach, the study of fuzzy sets is reduced to probability theory. In that case, there would be nothing more to say except "Go back to Chapters 7 and 8."

(b) There is something inherently vague in the notion of tall. Thus, it's unnatural to force a person to either completely agree or completely disagree with the statement that p is tall. This forcing is required for the 70% versus 30% survey.

(c) The probability approach misses the point—it's the *concept* that's vague or uncertain, not people's responses. In other words, even if these people whose opinions I sought did not exist, *I personally* would consider the concept of tallness to be vague.

(d) The probability approach distorts the situation—I'm concerned with *my* reaction to a particular person's tallness, not some frequency information.

The first objection merely states that the approach is pointless. The second, that it is unnatural. However, (c) and (d) contain the germ of an alternative approach.

Since I perceive the vagueness of a concept, I should be the one to determine degree of membership; that is, I determine $\mu_A(p)$. Howls of protest can be heard from those who advocate an objective approach. They righteously proclaim "Probability theory, which is based on a solid objective foundation, is vastly superior to a method based on subjective opinion." Apparently they didn't read (or didn't accept) the first part of Section 7.1, especially (7.2) (p. 259). In many situations, empirical determination of a probability is impossible and so a subjective estimate is used. Thus, a subjective approach to fuzzy sets shouldn't be rejected out of hand. Instead, I can take the same approach I did with probability theory: Assuming an understanding of the basic notion, I'll determine the properties it should have. The problem of estimating membership functions can be postponed until I face the question of designing an actual fuzzy rule system. At that time, you'll see that the conceptual aspects are largely ignored in favor of an empirical approach.

Example 9.1   Temperature, Youth, and Age

Figure 9.1 shows some commonly used examples of fuzzy sets.

The graph on the left is my membership function for room temperature (degrees Fahrenheit) in the set "warm." It might be obtained by my reaction when the temperature of a room is at various values. When the temperature is too high or too low, I would not characterize it as warm. What about intermediate temperatures? Sometimes I might say something like

It's warm. Actually, it's somewhat more than warm, so I couldn't say that it's simply 100% warm. Since you insist on a number, I estimate 80% warm.

As already noted, I won't ask how the 80% was arrived at.

There's an important discrepancy between my description and my use of the set "warm." Can you spot the problem?

*   *   *   Stop and think about this!   *   *   *

The fuzzy set I claimed to be talking about contains numbers that measure temperature. My discussion involved deciding about particular situations, where the answer could be influenced by relative humidity, time of year, and many other factors.

This distinction may be more evident in the graphs on the right side of Figure 9.1. There, an age—not a person—is being described as young, mature, or old. In fact, I might consider one twenty-year old to be very mature

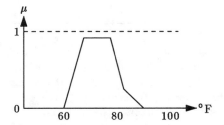 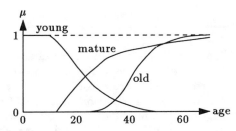

Figure 9.1 Some examples of fuzzy sets. The vertical axes measure μ and the horizontal axes list the elements of the set, which are real numbers here. The left-hand graph pictures a warm temperature and the right-hand graph pictures various age categories. See Example 9.1 for a discussion.

and another to quite immature. Thus, maturity depends on more than age. "Young" and "old" may also depend on more than age. In that case, what does it mean to say a twenty-year old has about 0.6 membership in the set "young"? I'll continue this discussion in the next example. ∎

Exercises

9.2.A. What real-world aspect of membership is not captured by ordinary sets but is captured by fuzzy sets?

9.2.B. Explain the statement "a fuzzy set is a membership function" and define a membership function.

9.2.C. Why does a probabilistic approach to fuzzy sets miss the point? In particular, what is wrong with saying $\mu_{young}(\text{Jane}) = 0.3$ because 30% of the people consider Jane to be young?

9.2.D. How do the uncertainties represented by fuzzy logic and probability theory differ?

9.2.1. Which of the following set descriptions require the notion of a fuzzy set? Justify your positive and negative answers.

    (a) daytime   (b) blue-colored   (c) sunny day
    (d) beautiful   (e) healthy   (f) deceased
    (g) A-student   (h) old person   (i) teenager

This is probably best done by a quick class discussion.

Properties of Fuzzy Sets

What are the basic properties of fuzzy sets? When do they agree with the corresponding concepts for ordinary sets? When do they differ? Basic notions for ordinary sets are subset, union, intersection, and complement. If we can describe these basic notions in terms of characteristic functions, we have a description that might be applied to fuzzy sets by replacing the characteristic function χ with the membership function μ. As a start, you should be able to easily prove the following theorem about ordinary sets.

Theorem 9.1   Characteristic Functions of Sets

Let A and B be sets with characteristic functions χ_A and χ_B.

(a) $A \subseteq B$ if and only if $\chi_A(x) \leq \chi_B(x)$ for all x.

(b) The characteristic function of the complement of A is
$\chi_{A'}(x) = 1 - \chi_A(x)$.

(c) The characteristic functions of the intersection and union of A and B are

- $\chi_{A \cap B}(x) = \min(\chi_A(x), \chi_B(x))$ and
- $\chi_{A \cup B}(x) = \max(\chi_A(x), \chi_B(x))$.

(d) The characteristic functions of the intersection and union of A and B are also computable by

- $\chi_{A \cap B}(x) = \chi_A(x)\chi_B(x)$ and
- $\chi_{A \cup B}(x) = \chi_A(x) + \chi_A(b) - \chi_A(x)\chi_B(x)$.

The two formulas for union and intersection are paired in a natural fashion. Recall that one of de Morgan's laws states that

$$A \cup B = (A' \cap B')',$$

which you should be able to prove easily. Using this law, (b), and the first formula in (c), we have

$$\chi_{A \cup B} = \chi_{(A' \cap B')'} = 1 - \chi_{A' \cap B'} = 1 - \min(\chi_{A'}, \chi_{B'})$$
$$= 1 - \min(1 - \chi_A, 1 - \chi_B) = 1 - 1 + \max(\chi_A, \chi_B)$$
$$= \max(\chi_A, \chi_B),$$

the second formula in (c). If min is replaced by product in this derivation, the second formula in (d) is obtained from the first formula in (d).

Should (c) or (d) be used as the basis of a definition for fuzzy sets? There are pros and cons to either choice. (Exercise 9.2.6 presents one argument.) The min and max of (c) are usually used. We'll use them, too.

Definition 9.2   Operations on Fuzzy Sets

Let μ_A and μ_B be the membership functions for two fuzzy sets.

(a) We say A is a subset of B whenever $\mu_A(x) \le \mu_B(x)$ for all x.

(b) The fuzzy complement of A is defined by $\mu_{A'}(x) = 1 - \mu_A(x)$.

(c) Fuzzy intersection is defined by $\mu_{A \cap B}(x) = \min(\mu_A(x), \mu_B(x))$.

(d) Fuzzy union is defined by $\mu_{A \cup B}(x) = \max(\mu_A(x), \mu_B(x))$.

Is this definition reasonable? This means, does it fit with our everyday interpretation of the concepts?

(a) If the notion of subset is understood in the sense of "less than," the definition is quite reasonable: It says that every x has at least as much membership in B as it has in A.

(b) We claim that the definition of complement is reasonable: It certainly agrees when $\mu_A(x)$ is 0 or 1. For intermediate values, it reflects the idea that the more something has the property described by A, the less it has the property described by A'. What about the particular choice of $1 - \mu_A(x)$ rather than, say $1 - \mu_A(x)^2$? It has the nice property that $\mu_{A''} = \mu_A$, just as $A'' = A$ for ordinary sets. We can describe these properties in terms of a decreasing function g on $[0,1]$ with $g(0) = 1$ and $g(g(t)) = t$: For any fuzzy set A, $\mu_{A'}(x) = g(\mu_A(x))$. It can be shown that $g(t) = 1 - t$ is the only g with these properties.

(c) Think of A and B as describing two properties and $\mu_A(x)$ and $\mu_B(x)$ as measuring the extent to which x possesses each of the properties. To what extent does x possess both properties? If $\mu_A(x) = \mu_B(x)$, then it seems reasonable that this number measures the extent to which x possesses both properties. What if $\mu_A(x) \ne \mu_B(x)$? In this case, it seems reasonable that the smaller number should measure the extent to which x possesses both properties.

(d) A similar argument can be used to defend the definition of $A \cup B$.

Another way of assessing the definition is to ask how many of the usual properties of complementation, intersection, and union it preserves. The following theorem partially answers this question.

Theorem 9.2   Properties of Fuzzy Set Operations

For all fuzzy sets A, B, and C, we have the following properties, just as for ordinary sets.

(a) $A'' = A$;

(b) $(A \cup B)' = (A') \cap (B')$ and $(A \cap B)' = (A') \cup (B')$;

(c) $A \cap B \subseteq A$ with equality if and only if $A \subseteq B$;

(d) $A \cup B \supseteq A$ with equality if and only if $A \supseteq B$;

(e) if $B \subseteq C$, then $B' \supseteq C'$, $A \cup B \subseteq A \cup C$, and $A \cap B \subseteq A \cap C$;

(f) $A \cup (B \cap C) = (A \cap B) \cup (A \cap C)$ and $A \cap (B \cup C) = (A \cup B) \cap (A \cup C)$.

Two important formulas that *do not carry over from ordinary sets* are $A \cap (A') = \emptyset$ (the empty set) and $A \cup (A') = \mathcal{X}$ (the set of everything).

The proof is left as an exercise. The formula $A \cap A' = \emptyset$ corresponds to the law of the excluded middle in propositional logic—a statement cannot be both true and false. In fuzzy reasoning, statements are often both somewhat true and somewhat false.

Example 9.2   Youth and Age (continued)

In Example 9.1, we concluded by asking what it means to say a twenty-year old has about 0.6 membership in the set "young."

One possible interpretation is that what is represented in Figure 9.1 (p. 363) is precisely what was intended: "Young" is a concept that depends only on age and not on aspects of an individual's personality.

What if we reject this and say that "young" depends on more than age? Then the membership function $\mu_{\text{young}}(20)$ must represent an average over twenty-year-old individuals. It seems reasonable to agree that the fuzzy sets represented in Figure 9.1 are obtained from other fuzzy sets this way. We'll soon see that this leads to problems—operations on the resulting sets don't quite fit Definition 9.2. Is this important? Yes and no.

- If we want a solid mathematical foundation whose concepts mirror the interpretations we have in mind, then we must be very careful in calling something a fuzzy set.

- If we're interested in rule systems for automatic control—the major application of fuzziness—then pragmatic considerations are more important than theoretical ones. In that case, we may be perfectly willing to ignore inconsistencies in our usage of fuzzy concepts and tools as long as the final result works.

We'll conclude this example with a demonstration that averaging usually destroys fuzziness.

We have one universe whose set elements are individual people. Call it $\mathcal{E}$. We have another universe $\mathcal{X}$ whose set elements are ages in years. Finally, we have an obvious function "age" from $\mathcal{E}$ to $\mathcal{X}$. Let $A(x) \subseteq \mathcal{E}$ be those people whose age is x. If B is a fuzzy set in the $\mathcal{E}$ universe, define B^* to be the fuzzy set in the $\mathcal{X}$ universe by letting $\mu_{B^*}(x)$ be the average of $\mu_B(e)$ over all $e \in B^*$. For example $\mu_{\text{young}^*}(20)$ is the average of $\mu_{\text{young}}(x)$ over all $x \in A(20)$. We'll denote this averaging by the notation

$$\mu_{B^*}(x) = \mathrm{E}\big(\mu_B(e) \mid \text{age}(e) = x\big). \tag{9.2}$$

(This notation is used in probability theory to denote an average that is called "expected value" and is studied in Chapter 12.)

Let B, C, and $D = B \cap C$ be three fuzzy sets based on $\mathcal{E}$. It's natural and desirable to identify the fuzzy set $B^* \cap C^*$ with the fuzzy set D^*. Unfortunately, they're usually different sets and all that can be said is that $(B \cap C)^* \subseteq (B^*) \cap (C^*)$. Let's prove this. Fix some $x \in \mathcal{X}$. Without loss of generality, we may suppose that $\mu_{B^*}(x) \leq \mu_{C^*}(x)$. We have

$$
\begin{aligned}
\mu_{B^* \cap C^*}(x) &= \mu_{B^*}(x) && \text{by Definition 9.2} \\
&= \frac{1}{|A(x)|} \sum_{e \in A(x)} \mu_B(x) \\
&\geq \frac{1}{|A(x)|} \sum_{e \in A(x)} \min\big(\mu_B(e), \mu_C(e)\big) \\
&= \frac{1}{|A(x)|} \sum_{e \in A(x)} \mu_{B \cap C}(e) && \text{by Definition 9.2} \\
&= \mu_{(B \cap C)^*}(x).
\end{aligned}
$$

This proves that $(B^*) \cap (C^*) \supseteq (B \cap C)^*$. Furthermore, we have equality throughout the previous chain of equalities and inequalities if and only if $\mu_B(e) \leq \mu_C(e)$ for all $e \in A(x)$. In other words, $(B^*) \cap (C^*) = (B \cap C)^*$ if and only if whenever $\mu_B(e)$ is less (resp. greater) than $\mu_C(e)$, the same is true for all e' that have the same age as e. This is unlikely to hold. For example, some twenty-year olds may be more young than they are mature and others may be more mature than they are young. In other words, we have

$$\mu_{\text{young}}(e) > \mu_{\text{mature}}(e) \quad \text{and} \quad \mu_{\text{young}}(e') < \mu_{\text{mature}}(e')$$

for some e and e' in $A(20)$ and so $\mu_{\text{young}^* \cap \text{mature}^*}(20) > \mu_{(\text{young} \cap \text{mature})^*}(20)$. ∎

In probability theory, our probability spaces were often Cartesian products with

$$\Pr((e_1, \ldots, e_n)) = \Pr_1(e_1) \cdots \Pr_n(e_n),$$

a probability function that made the various components behave independently of one another. The following example looks at a similar situation for fuzzy sets.

Example 9.3   Cartesian Products and Projections

The concepts of Cartesian product and projection are closely related. Recall that the Cartesian product of k sets is defined by

$$A_1 \times \cdots \times A_k = \{ (a_1, \ldots, a_k) \mid a_1 \in A_1, \ldots, a_k \in A_k \}.$$

If $S \subseteq A_1 \times \cdots \times A_k$, its projection onto the ith coordinate is given by

$$\text{proj}_i(S) = \{ p \mid p = x_i \text{ for some } (x_1, \ldots, x_k) \in S \}. \qquad (9.3)$$

How can these ideas be extended to fuzzy sets?

Let's discuss one approach. Suppose that $A_1 \subseteq \mathcal{X}_1$ and $A_2 \subseteq \mathcal{X}_2$. Notice that

$$A_1 \times A_2 = (A_1 \times \mathcal{X}_2) \cap (\mathcal{X}_1 \times A_2). \qquad (9.4)$$

This generalizes to the Cartesian product of k sets. Since we know how to form a fuzzy intersection, all we need do is decide what the fuzzy Cartesian product $A_1 \times \mathcal{X}_2$ means. Since $\mathcal{X}_i$ contains everything that can be in the ith component of a Cartesian product, it's reasonable to have the degree of membership of (x_1, x_2) in $A_1 \times \mathcal{X}_2$ depend only on the membership of x_1 in A_1; that is,

$$\mu_{A_1 \times \mathcal{X}_2}((x_1, x_2)) = \mu_{A_1}(x_1) \text{ and similarly } \mu_{\mathcal{X}_1 \times A_2}((x_1, x_2)) = \mu_{A_2}(x_2).$$

Using (9.4) for fuzzy sets, we now have

$$\mu_{A_1 \times A_2}((x_1, x_2)) = \min\Big(\mu_{A_1 \times \mathcal{X}_2}((x_1, x_2)),\ \mu_{\mathcal{X}_1 \times A_2}((x_1, x_2))\Big)$$
$$= \min\big(\mu_{A_1}(x_1),\ \mu_{A_2}(x_2)\big).$$

The extension to an arbitrary length product is obvious: The *fuzzy Cartesian product* of $A_1, \ldots, A_k$ is defined by

$$\mu_{A_1 \times \cdots \times A_k}((x_1, \ldots, x_k)) = \min\big(\mu_{A_1}(x_1), \ldots, \mu_{A_k}(x_k)\big). \qquad (9.5)$$

The fuzzy projection is given by

$$\mu_{\text{proj}_i(S)}(p) = \max\{ \mu_S((x_1, \ldots, x_k)) \mid x_i = p \}, \qquad (9.6)$$

that is, the maximum of the membership function of all $\vec{x}$'s whose ith component is p. (Actually, since the set of $\vec{x}$'s can be infinite, the maximum should really be a supremum—a technicality that doesn't matter here.) You should prove that (9.5) and (9.6) agree with the definitions for ordinary sets when μ is replaced by $\mathcal{X}$. We may write projection differently; for example, if $S \subseteq \mathcal{X} \times \mathcal{Y}$, we may write $\text{proj}_\mathcal{X}$ instead of proj_1. For so called "normal" fuzzy sets, the projection of $A_1 \times \cdots \times A_k$ onto the ith component is A_i. See Exercise 9.2.8 for details. ∎

Fuzzy Predicates

We're now ready to look at fuzzy predicates. As indicated at the start of the section, this will be done through the use of fuzzy sets. In fact, you can think of the fuzzy sets in previous examples as if they were predicates. For example, "young" is a predicate that takes a single argument, either an age or a person, depending on which version of "young" we're thinking of. Let's be a little more precise, beginning with correspondence between ordinary predicates and ordinary sets.

Let $\mathcal{C}_i$ be set of possible values for the ith argument of a predicate p that takes k arguments. We can associate a subset P of $\mathcal{X} = \mathcal{C}_1 \times \cdots \times \mathcal{C}_k$ with p in a simple way, namely

$$P = \left\{ \vec{c} \in \mathcal{X} \mid p(\vec{c}) \text{ is true} \right\}.$$

Conversely, any subset $Q \subseteq \mathcal{X}$ determines a predicate q by the rule that $q(\vec{c})$ is true if and only if $\vec{c} \in Q$. Suppose we have two predicates and we want to combine them; for example, $p(\vec{x}, \vec{y}) \wedge q(\vec{y}, \vec{z})$. Note that the predicates share some arguments but not others ($\vec{x}$ and $\vec{z}$). Thus p is associated with $P \subseteq \mathcal{X} \times \mathcal{Y}$ and q is associated with $Q \subseteq \mathcal{Y} \times \mathcal{Z}$. We associate $p(\vec{x}, \vec{y}) \wedge q(\vec{y}, \vec{z})$ with the set $(P \times \mathcal{Z}) \cap (\mathcal{X} \times Q)$. One way to think of this is by extending p and q to predicates whose arguments lie in $\mathcal{X} \times \mathcal{Y} \times \mathcal{Z}$. The truth of p is independent of the components in $\mathcal{Z}$ and the truth of q is independent of the components in $\mathcal{X}$. The sets associated with p and q are then $P \times \mathcal{Z}$ and $\mathcal{X} \times Q$, respectively. Here's a restatement of this with formulas for two other connectives:

$$\begin{aligned}
p(\vec{x}, \vec{y}) \wedge q(\vec{y}, \vec{z}) &\text{ is associated with } (P \times \mathcal{Z}) \cap (\mathcal{X} \times Q), \\
p(\vec{x}, \vec{y}) \vee q(\vec{y}, \vec{z}) &\text{ is associated with } (P \times \mathcal{Z}) \cup (\mathcal{X} \times Q), \qquad (9.7) \\
\neg p(\vec{x}, \vec{y}) &\text{ is associated with } (\mathcal{X} \times \mathcal{Y}) - P.
\end{aligned}$$

Because of the close relationship between sets and predicates, *we'll frequently denote a predicate and the associated set with the same symbol.*

The set operations can be extended to the fuzzy case by using fuzzy sets. This gives us a fuzzy logic. However, the logic is not uniquely determined by this procedure. The fuzzy interpretation of $p \rightarrow q$ is open to debate. Researchers have advocated a variety of fuzzy versions of $p \rightarrow q$. Why not simply rewrite it using the connectives $\neg$, $\wedge$, and $\vee$ and use Definition 9.2 (p. 365)? One way to rewrite it is $(\neg p) \vee q$. Another way of rewriting it is $(q \wedge p) \vee (\neg p)$. While these formulas are equivalent in propositional logic, they don't necessarily describe the same fuzzy sets. Thus we're faced with the problem of which rewrite to choose. Another approach is to develop the idea that $p \rightarrow q$ is a tautology if and only if $P \subseteq Q$. As you'll see later, there's a fairly natural interpretation of "if p then q" for fuzzy controllers.

The problem with $p \rightarrow q$ is just one of the difficulties that arise in developing a fuzzy logic. The words "not," "or," and "and" bear a heavy burden of

meaning based on everyday usage. FOL was formulated to provide a precise
mathematical formulation of that meaning (the semantics of FOL) so that
we could manipulate some aspects of language using the syntax. It's natural to assume that the fuzzy meanings will carry forward this program. *This
mistake has led to considerable misunderstanding.* A fuzzy logic must have its
own semantics that differs from that for FOL. To begin with, the notion of
truth must be changed since set membership takes on values other than 0 and
1 (corresponding to false and true in FOL). Thus a fuzzy logic is some sort
of "multiple-valued" logic. As in the case of nonmonotonic logics, there is no
consensus on what *the* multiple-valued logic or *the* fuzzy logic should be.

We must also be careful when manipulating fuzzy logic formulas because
not all the rules of ordinary set theory hold for fuzzy sets—see the last part
of Theorem 9.2 (p. 366) and Exercise 9.2.5. As the following example shows,
there are also problems interpreting the results of fuzzy manipulations.

Example 9.4   Interpreting Combined Predicates

Given that Mary is young (denoted by y below) and that John is much older
than Mary (denoted by mo), what can we say about John's age (denoted
by Ja)? It seems reasonable to simply project the conjunction onto the coordinate associated with John's age. Using x to indicate John's age and t to
indicate Mary's and using formulas (9.7) and (9.3), we have

$$\mu_{\mathrm{Ja}}(x) = \max_t \Big(\min \big(\mu_{\mathrm{y}}(t),\, \mu_{\mathrm{mo}}(x,t) \big) \Big). \tag{9.8}$$

Although this is a simple application of preceding formulas, the result is
counterintuitive. Since μ_{y} is a decreasing function of its argument and μ_{mo} is
a decreasing function of its second argument, the maximum will be achieved
when Mary is as young as possible. Presumably this is 0 (newborn), which is
certainly a young age ($\mu_{\mathrm{y}}(0) = 1$). Thus $\mu_{\mathrm{Ja}}(x) = \mu_{\mathrm{mo}}(x, 0)$. We claim that
even a toddler is certainly much older than a newborn so that $\mu_{\mathrm{mo}}(x, 0) = 1$
even for small values of x. Hence the membership value for John's being a
toddler would be 1, which is the same as that for an octogenarian!

Although we've shown that (9.8) contains practically no information on
John's age, the example's opening sentence gives us the impression that John
is more likely to be older than younger. What happened?

This impression may come from thinking in somewhat probabilistic
terms—Mary could be any age, not just newborn. What do we have instead of
(9.8) if we're dealing with probability? There are several changes. We would
probably obtain something like

$$\Pr(\mathrm{J} = x) = \sum_t \Pr(\mathrm{M} = t) \Pr\Big(\mathrm{J} = x \,\Big|\, (\mathrm{M} = t) \wedge (\mathrm{M}\ \text{is much older than J})\Big),$$

where J stands for John and $\mathrm{J} = x$ means his age is x. Some thought is
required for estimating the conditional probability because of its "much older"
condition.

Fuzzy memberships are not probabilities, so we must avoid thinking in probabilistic terms. Since not thinking is also a poor strategy, how might we think about this example? The projection (maximum) causes us to look for the choice of t that makes things as large as possible. In other words, "How possible is it that John's age is x?"—not "How probable is it?" ∎

Exercises

9.2.E. How are fuzzy union, intersection, and complement computed?

9.2.F. Assuming that Figure 9.1 represents fuzzy sets, indicate the intersection of young and mature in the figure. What age maximizes "young and mature"? Indicate the union of young and mature. What age minimizes "young or mature"?

9.2.G. How can a predicate be interpreted as a fuzzy set? Give an example.

9.2.2. Prove Theorem 9.1.

9.2.3. Defend $\mu_{A \cup B}(x) = \max(\mu_A(x), \mu_B(x))$. This is probably best done in discussion.

9.2.4. Prove Theorem 9.2, including the fact that two important formulas do not carry over to fuzzy sets.

9.2.5. Prove that all the following are valid for ordinary sets. Which are valid for fuzzy sets? Give a proof or counterexample in each case.

(a) $P' \cup Q = (Q \cap P) \cup (P')$

(b) $(A \cup B) \cap (A' \cup B) \subseteq B$

(c) $(A \cup B) \cap (A' \cup B) \supseteq B$

9.2.6. This is one way of justifying the definition of fuzzy intersection. When you work this problem, *do not assume* Definition 9.2(c); however, you may assume (a).

(a) Explain why each of the following is a reasonable requirement to place on fuzzy intersection.

(i) $B \cap A = A \cap B$.

(ii) $A \cap A = A$.

(iii) Whenever $B \supseteq C$, we have $A \cap B \supseteq A \cap C$.

(b) Define a fuzzy set D by $\mu_D(x) = \min(\mu_A(x), \mu_B(x))$. Using (i)–(ii), show that

$$\mu_D(x) \geq \mu_{A \cap B}(x) \geq \mu_{A \cap D}(x) \geq \mu_{D \cap D}(x) = \mu_D(x)$$

and then conclude that $\mu_{A \cap B}(x) = \min(\mu_A(x), \mu_B(x))$.
Hint. With $\mu_B(x) = 1$ for all x, use (iii) to prove $A \supseteq A \cap C$.

9.2.7. For the fuzzy set $S \subseteq \mathcal{X} \times \mathcal{Y}$ and $x \in \mathcal{X}$, define the set $S(x) \subseteq \mathcal{Y}$ by $\mu_{S(x)}(y) = \mu_S((x,y))$. Prove that $\text{proj}_2(S) = \bigcup_{x \in \mathcal{X}} S(x)$.

9.2.8. A fuzzy set S is called *normal* if there is an $x \in S$ with $\mu_S(x) = 1$.

   (a) Suppose that $A_1, \ldots, A_k$ are normal fuzzy sets. Prove that
       $A_i = \text{proj}_i(A_1 \times \cdots \times A_k)$.

   (b) Show by means of a counterexample that the previous result requires
       the normality assumption. (Do not give the trivial example in which one
       of the $A_i = \emptyset$.)

9.2.9. Suppose that $f : \mathcal{X} \to \mathcal{Y}$. Let $A \subseteq \mathcal{X}$ be a set. The set $f(A)$ is defined by

$$f(A) = \left\{\, y \in \mathcal{Y} \mid f(x) = y \text{ for some } x \in A \,\right\}.$$

The corresponding definition for fuzzy sets is called the *extension principle*.
The purpose of this exercise is to formulate it.

   (a) Let A be an ordinary set and let $A^* = \left\{\, (x, f(x)) \mid x \in A \,\right\}$. Prove that
       $f(A) = \text{proj}_2(A^*)$.

   (b) Let A be a fuzzy set and define A^* by

$$\mu_{A^*}((x,y)) = \begin{cases} \mu_A(x), & \text{if } y = f(x), \\ 0, & \text{if } y \neq f(x). \end{cases}$$

   Prove that this agrees with the previous definition when A is an ordinary
   set; that is, μ_A only takes on the values 0 and 1.

   (c) If A is a fuzzy set, define $f(A) = \text{proj}_2(A^*)$. Prove that $\mu_{f(A)}(y)$ equals
       the maximum of $\mu_A(x)$ over all x for which $f(x) = y$ and equals 0 if
       there are no such x.

9.2.10. You may have noticed that I didn't offer a fuzzy analog for the FOL formula
        $\alpha \to \beta$. In fact, several analogs have been proposed—another reason to
        beware of using a fuzzy analog of FOL. You'll look at two in this exercise.

   (a) Show that the FOL formulas $\alpha \to \beta$ and $(\beta \vee \neg\alpha)$ are equivalent; i.e.,
       one is true if and only if the other is. What fuzzy definition does this
       suggest for $\alpha \to \beta$; this is, what is the value of $\mu_{A \to B}(x)$?

   (b) If A (resp. B) is the set of X for which $\alpha(X)$ (resp. $\beta(X)$) is TRUE, show
       that $\forall X(\alpha(X) \to \beta(X))$ is equivalent to $A \subseteq B$. Use this to propose a
       definition for $\mu_{A \to B}(x)$ which is 0/1-valued.

*9.2.11. Adverbs like "very" and "slightly" provide the opportunity for creating new and arbitrary functions. Suppose some predicate p (such as "tall" or "mature") describes a fuzzy set A. The adverbs "very" and "slightly" can be used to modify p and produce new fuzzy sets $V(A)$ and $S(A)$ (such as "very tall" or "slightly mature"). People have proposed various functions v and s such that $V(A)$ and $S(A)$ could be automatically created from A by

$$\mu_{V(A)}(x) = v(\mu_A(x)) \quad \text{and} \quad \mu_{S(A)}(x) = s(\mu_A(x)).$$

One possibility is $v(t) = t^2$. Note that v and s are completely specified by defining them on $[0,1]$.

(a) What are some properties you think v and s should have. Why?

(b) Suggest and defend one or more choices of functions for v and for s.

Suppose our universe consists of integers $\mathbb{Z}$ and we want to add two of them. Unfortunately, we have only a vague notion of their values. The next two exercises explore probabilistic and fuzzy approaches.

9.2.12. Our elementary events will be $\mathcal{E} = \mathbb{Z} \times \mathbb{Z}$, a slight extension of the concept in Chapter 7, where $\mathcal{E}$ had to be finite. The probability function is $\Pr(x, y)$. Two particular numbers x and y correspond to the elementary event (x, y).

(a) What is the compound event corresponding to x and y having a particular sum and what is its probability?

(b) We'll say that the values of x and y are independent if the compound events $\{x\} \times \mathbb{Z}$ and $\mathbb{Z} \times \{y\}$ are independent for all x and y. Let

$$\Pr_1(x) = \Pr(\{x\} \times \mathbb{Z}) \quad \text{and} \quad \Pr_2(y) = \Pr(\mathbb{Z} \times \{y\}).$$

Prove that the probability that the sum is c is the sum over all x and y that sum to c of the product $\Pr_1(x) \Pr_2(y)$—in mathematical notation,

$$\Pr(\text{sum is } c) = \sum_{\substack{x,y: \\ x+y=c}} \Pr_1(x) \Pr_2(y).$$

9.2.13. Suppose that X and Y are sets of integers.

(a) Let Z be the set of sums $x + y$ where $x \in X$ and $y \in Y$. For example, if $X = Y = \{1, 2\}$, then $Z = \{2, 3, 4\}$. Defend the corresponding fuzzy definition

$$\mu_Z(c) = \max_{\substack{x,y: \\ x+y=c}} \Big(\min\big(\mu_X(x), \mu_Y(y)\big) \Big).$$

Hint. Why is set intersection like "and" and set union like "or"?

(b) Extend the previous ideas from addition to an arbitrary binary operation $g(x, y) \in \mathbb{Z}$. This is the basis of *fuzzy arithmetic*.

Fuzzy Rule Systems

> *Fuzzy Wuzzy was a bear.*
> *Fuzzy Wuzzy had no hair.*
> *Fuzzy Wuzzy wasn't very fuzzy,*
> *Was 'e?*
> —Anonymous

It should be clear by now that using and interpreting fuzzy sets can be a tricky business. This makes it particularly important to have a clear method for moving between the real world and the mathematics of fuzzy reasoning. Of course, this correspondence must be such that the operations (union, intersection, etc.) on fuzzy sets mirror what happens in the real world. It seems to me no such method currently exists and its lack will cause severe problems if we attempt to use fuzzy methods for complex reasoning.

For some types of reasoning, the difficulties can be overcome or ignored. These are the "shallow rule systems." In fact, they've been so commercially successful, that "fuzzy logic" has come to be synonymous with such systems despite the their rather limited logical reasoning capabilities.

First, a one-paragraph review of rule systems for those of you who skipped or forgot Section 6.4 (p. 229). Rule systems are often used to control various devices. A thermostat uses a very simple rule system for controlling a furnace. If T is the present temperature, the rule system is

> `If` $T < T_{\text{low}}$`, then furnace should be ON.`
> `If` $T > T_{\text{hi}}$`, then furnace should be OFF.`   (9.9)
> `If` $T_{\text{low}} \leq T \leq T_{\text{hi}}$`, then furnace may be in any state.`

The use of such a system is straightforward: Whenever the *antecedent* ("if" part) of a rule is true, its *consequent* ("then" part) is also taken to be true—or made true if it describes an action. The consequents of one rule can become the antecedents of another, a process known as *forward chaining*. The *depth* of a rule system is the length such a chain of reasoning can reach. The example (9.9) has depth 1 since the rules cannot be chained together. A rule system whose depth is small is called *shallow*. The general form of a rule is

$$\forall X_1 \ldots \forall X_k \Big((\alpha_1 \wedge \cdots \wedge \alpha_n) \rightarrow \beta \Big) \qquad (9.10)$$

where $\alpha_1, \ldots, \alpha_n, \beta$ are predicates whose only variables are $X_1, \ldots, X_k$ and rules are used when their hypotheses are true. In other words, whenever $\alpha_1, \ldots, \alpha_n$ are true for some values of $X_1, \ldots, X_k$, then β is also taken to be true—or is made true if it implies an action as in (9.9). When rules are stated, the universal quantifiers are usually omitted.

A more complicated thermostat might control a furnace whose rate of heat production can be set to levels other than just off and on. For example, the first rule in (9.9) might be replaced by

> If $T_{\text{low}} - T$ is large and positive,
>   then furnace should be ON high.
> If $T_{\text{low}} - T$ is medium and positive,
>   then furnace should be ON medium.
> If $T_{\text{low}} - T$ is small and positive,
>   then furnace should be ON low.

(9.11)

Terms like large, medium, small, high, and low should suggest fuzzy sets to you. We'll discuss the foundation of such controllers in this section.

How do fuzzy controllers overcome the difficulties raised by fuzziness? Here's a brief answer.

- **Restricted logic:** Rule systems allow only two logical connectives, $\wedge$ and $\rightarrow$. They can be interpreted in a way that fits fairly well with "commonsense" notions of what the rule systems should mean.

- **Shallowness:** Chains of reasoning can magnify discrepencies between how we imagine fuzziness should work and how it actually does work. Since controllers use very shallow rule systems—often the depth is only 1—they sidestep this problem.

- **Predicate visibility:** In a rule system of depth 1, the effects of all predicates are visible as either input information (temperature) or output action (furnace control). This makes it easier to adjust them, an important ability since we lack a solid foundation for constructing fuzzy sets. (Remember, predicates can be thought of as sets and vice versa. See p. 369.)

Fuzzy Controllers without Chaining

For the time being, let's assume that no chaining takes place; that is, the antecedents α_i of a rule refer to observations, and the consequent β refers to an action to be taken. Furthermore, we'll assume that the consequents all contain one and the same variable, say Z. This is the situation in the rule systems (9.9) and (9.11).

To interpret such a system, we must understand the meaning of the corresponding ordinary rule system better. Suppose it's time to take action. Then we must apply one of the rules to the present situation. This means $\alpha_1, \ldots, \alpha_n$ would be true in some rule (9.10) and we would act so as to make β true. In other words, we act so that at least one rule has both its antecedents and consequent true. If the jth rule is $(\alpha_{1,j} \wedge \alpha_{2,j} \wedge \cdots) \rightarrow \beta_j$ and there are m rules, we choose an action that makes the formula

$$\Big((\alpha_{1,1} \wedge \alpha_{2,1} \wedge \cdots \wedge \beta_1) \vee \cdots \vee (\alpha_{1,m} \wedge \alpha_{2,m} \wedge \cdots \wedge \beta_m)\Big)$$

(9.12)

true. To illustrate, add a third rule to (9.9) to cover intermediate temperatures. Then rewrite the system in the form (9.12) to obtain

$$
\Big(\big((T < T_{\text{low}}) \wedge \text{Furnace}(\texttt{ON}) \big) \vee \big((T > T_{\text{hi}}) \wedge \text{Furnace}(\texttt{OFF}) \big)
$$
$$
\vee \big((T_{\text{low}} \le T \le T_{\text{hi}}) \wedge \text{Furnace}(Z) \big) \Big),
\tag{9.13}
$$

where Z is arbitrary and "Furnace" and the temperature inequalities are predicates. You should convince yourself that the control system (9.9) is choosing the state of the furnace so that this formula is true.

We can think of (9.12) and (9.13) in a slightly different manner: Given the values of variables in the antecedents, these formulas determine a set of allowed values for the consequent variable Z. This can be done simply by substituting in the values of the antecedent variables and interpreting predicates as sets—something the next example will clarify.

We now have something that fits easily into a fuzzy context. Formula (9.12) with its connectives $\wedge$ and $\vee$ makes sense from a fuzzy viewpoint.

Example 9.5   Implementing These Ideas: Balancing a Stick

A classic example of a fairly simple fuzzy controller is the problem of balancing a stick in a vertical position by supporting it on a moving platform. For simplicity, let's do the problem in two dimensions rather than three—the stick can fall only to the left or right, not forward or backward. We'll have five fuzzy predicates to describe the stick's tilt:

$$
\begin{aligned}
T_L &= \text{moderate tilt to left}, \\
T_l &= \text{slight tilt to left}, \\
T_0 &= \text{nearly vertical}, \\
T_r &= \text{slight tilt to right}, \\
T_R &= \text{moderate tilt to right};
\end{aligned}
$$

three predicates to describe the stick's velocity:

$$
V_f = \text{falling slowly}, \qquad V_0 = \text{essentially still}, \qquad V_r = \text{rising slowly};
$$

and five predicates to describe the desired motion:

$$
\begin{aligned}
M_L &= \text{move support quickly to the left}, \\
M_l &= \text{move support moderately to the left}, \\
M_0 &= \text{do not move support}, \\
M_r &= \text{move support moderately to the right}, \\
M_R &= \text{move support quickly to the right}.
\end{aligned}
$$

There is no specification in the movement rules about how far to move. In all cases, carry out the movement for a fixed time interval.

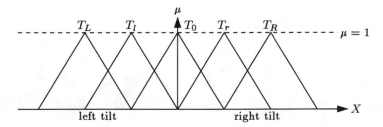

Figure 9.2   Fuzzy sets for stick tilt. The graph of each membership function has a triangular bulge above the $\mu = 0$ line. See Example 9.5 for a discussion.

Possible choices for the fuzzy sets associated with the tilt predicates are shown in Figure 9.2. The sets for the other predicates have a similar triangular shape. Why triangular? It's an easy shape to work with and captures the fact that the predicate fits less well as we move away from some central value. Fuzzy controllers are frequently such that the exact graphs of the membership functions are not important—only a crude estimate is needed. Of course, some additional thought or experimental work will be needed to establish a scale on the horizontal axes.

All rules have the form $\big(T_?(X) \wedge V_?(Y)\big) \rightarrow M_?(Z)$, where the variables X, Y, and Z indicate tilt, velocity, and movement, respectively. Since the subscript on M is determined by those on S and V, there are 5×3 possible rules. Depending on the implementation, some rules may be omitted because the situation is deemed hopeless (moderate tilt and falling) or ambiguous (vertical and falling doesn't indicate direction of fall). The model should be changed to avoid the ambiguity, but we won't do it. Here's a possible set of rules for leftward-tilted sticks:

$$T_L(X) \wedge V_0(Y) \rightarrow M_L(Z) \qquad\qquad \text{rule 1L}$$
$$T_L(X) \wedge V_r(Y) \rightarrow M_l(Z) \qquad\qquad \text{rule 2L}$$
$$T_l(X) \wedge V_0(Y) \rightarrow M_l(Z) \qquad\qquad \text{rule 3L}$$
$$T_l(X) \wedge V_f(Y) \rightarrow M_L(Z) \qquad\qquad \text{rule 4L}$$
$$T_l(X) \wedge V_r(Y) \rightarrow M_0(Z) \qquad\qquad \text{rule 5L}$$

There are corresponding rules for rightward-tilted sticks. Vertical sticks have only one rule because the information on falling does not allow us to decide which direction it's falling.

$$T_0(X) \wedge V_0(Y) \rightarrow M_0(Z). \qquad\qquad \text{rule V}$$

Suppose the stick is tilting very slightly leftward and is falling very slowly so that

$$\mu_{T_l}(X) = 0.4, \quad \mu_{T_0}(X) = 0.6, \quad \mu_{V_f}(Y) = 0.7, \quad \mu_{V_0}(Y) = 0.2$$

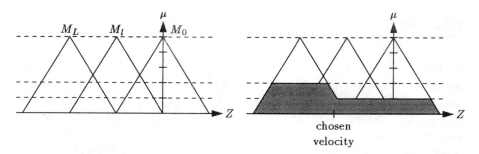

Figure 9.3   The first graph shows fuzzy sets for leftward movement. The shaded area in the second shows (9.14). Horizontal dashed lines indicate $\mu = 0.2$, 0.4, and 1. The chosen velocity on the second graph is the center of gravity.

and all other membership functions for X and Y are 0. Rules 3L, 4L, and V are the only ones whose antecedents have nonzero membership functions. They give the following fuzzy contributions to (9.12) (p. 375):

$$\text{rule 3L:}\quad \min\big(0.4,\ 0.2,\ \mu_{M_l}(Z)\big) = \min\big(0.2,\ \mu_{M_l}(Z)\big),$$
$$\text{rule 4L:}\quad \min\big(0.4,\ 0.7,\ \mu_{M_L}(Z)\big) = \min\big(0.4,\ \mu_{M_L}(Z)\big),$$
$$\text{rule V:}\quad \min\big(0.6,\ 0.2,\ \mu_{M_l}(Z)\big) = \min\big(0.2,\ \mu_{M_l}(Z)\big).$$

Combining these three, we obtain (9.12) for this situation:

$$\max\Big(\min\big(0.2,\ \mu_{M_l}(Z)\big),\ \min\big(0.4,\ \mu_{M_L}(Z)\big),\ \min\big(0.2,\ \mu_{M_0}(Z)\big)\Big). \quad (9.14)$$

Figure 9.3 shows the membership functions for the movement sets and the resulting set (9.14). I have yet to explain how a velocity is finally chosen. That's next. ∎

As you can see from Example 9.5, we're not done yet because our control system must choose a value for the output variable Z. Given a fuzzy set like that on the right side of Figure 9.3, how should we choose Z?   Making this choice is called *defuzzifying*.

In the ordinary set situation, μ_R only takes on the values 0 and 1 and we choose a value z for Z so that $\mu_R(z) = 1$. We may hope that there is only one such z so that our action is determined unambiguously. Perhaps the same idea can be used in a fuzzy system:

(a) Choose the element z for which $\mu_R(z)$ is largest.

Often, the value of z is not unique, so how do we choose? Furthermore, since $\mu_R(z)$ is usually nonzero for many values of z, why should we choose one of the largest? Here's two other possibilities.

(b) If R consists of real numbers, choose an average value; for example,

$$z = \frac{\int t\mu_R(t)\,dt}{\int \mu_R(t)\,dt}, \quad \text{the center of gravity.}$$

(c) Use μ_R to construct a probability space in which $\Pr(z)$ is proportional to $\mu_R(z)$. Then choose a random z using this distribution.

When we are dealing with an ordinary rule system that specifies a unique action, all three methods give that action.

How can we choose among the possibilities for fuzzy systems? I'm not aware of a good theoretical defense for any of these three options. Indeed, the entire concept of defuzzifying is foreign to the basic concept of a fuzzy set—how do you choose a representative "tall" individual? Nevertheless, if fuzzy methods are to be used in control systems, we *must* defuzzify in order to act. What method is likely to work well in practice?

We attacked option (a) almost as soon as we proposed it. Since the action of a control system usually involves setting a numerical variable in some device, option (b) could be used. In fact, it's the common choice. Using it in the stick-balancing example gives the velocity in Figure 9.3.

Option (c) seems a bad idea since we've been emphasizing that fuzziness is not probability. On the other hand, defuzzification is, by definition, nonfuzzy. The randomness of a probabilistic approach may increase the stability of a controller. The use of randomness to improve stability is a common technique that arises again in Chapter 11. Although it's probably been tried, I'm not aware of any attempts to use it for designing fuzzy controllers.

Note that both the input and output of a fuzzy controller are "crisp"; that is, they are actual numbers rather than a distribution. The input causes fuzziness through its interaction with the fuzzy rules and the fuzzy result is defuzzified to produce a crisp output.

More General Rule Systems

Thanks to the previous exploration, we now have some information about applying rules. In the previous study we

(a) started with some values of the variables in rule antecedents,

(b) combined rules to produce a fuzzy set for the variable in the consequents, and

(c) extracted a member of this set for use in setting a control.

In order to extend this procedure to chains of reasoning we eliminate (c) for intermediate consequents and extend (a) and (b) to allow fuzzy information about variables.

Let's do it. Suppose we have

$$\text{the fact that } X \in F \text{ and the rule } (\alpha(X) \rightarrow \beta(Z)). \qquad (9.15)$$

Based on earlier discussion, we replace the rule with $\alpha \wedge \beta$. In other words,

$$\mu_{\alpha \wedge \beta}(x, z) = \min\big(\mu_\alpha(x), \mu_\beta(z)\big).$$

How can we combine the fact $X \in F$ with this rule? Since we must use the fact AND the rule, we obtain

$$\mu_{F \wedge \alpha \wedge \beta}(x, z) = \min\big(\mu_F(x), \mu_\alpha(x), \mu_\beta(z)\big). \qquad (9.16)$$

Note that x can have any value whatsoever; that is, $x = x_0$ OR ... OR $x = x_n$. Since OR corresponds to set union, we take a maximum over all x:

$$\mu(z) = \max_x \Big(\min\big(\mu_F(x), \mu_\alpha(x), \mu_\beta(z)\big) \Big). \qquad (9.17)$$

Alternatively, we can view (9.17) as projecting (9.16) onto the z coordinate using (9.6) (p. 368). In the earlier situation, the value of x was given explicitly—it was input.

The previous paragraph holds the key for chaining in a fuzzy rule system: We take fuzzy information of the form (9.15) for all rules that talk about z, apply (9.17) to obtain a $\mu(z)$ for each of these rules, and compute the fuzzy union of these $\mu(z)$ results to obtain fuzzy information about z. Suppose we're given several rules $\alpha_i(\vec{X}) \rightarrow \beta_i(Z)$ and the fact $\vec{X} \in F$. Then (9.17) becomes

$$\begin{aligned}
\mu(z) &= \text{proj}_Z \Big((F \wedge \alpha_1 \wedge \beta_1) \vee \cdots \vee (F \wedge \alpha_m \wedge \beta_m) \Big) \\
&= \max_{\vec{x}} \Big(F \wedge \big((\alpha_1 \wedge \beta_1) \vee \cdots \vee (\alpha_m \wedge \beta_m) \big) \Big).
\end{aligned} \qquad (9.18)$$

This is actually a generalization of (9.12) (p. 375) with $\alpha_i = \alpha_{1,i} \wedge \alpha_{2,i} \wedge \cdots$, $\mu_F(\vec{x}) = 1$ when $\vec{x}$ equals the value of the input variables and $\mu_F(\vec{x}) = 0$ otherwise. To see this, you simply need to notice

- $\mu_{F \wedge (\ldots)}(\vec{x}) = 0$ when $\vec{x}$ differs from the input because $\mu_F(\vec{x}) = 0$ and

- $\mu_{F \wedge (\ldots)}(\vec{x}) = \mu_{(\ldots)}(\vec{x})$ when $\vec{x}$ equals the input values because $\mu_F(\vec{x}) = 1$.

We're now able to chain rules. I won't pursue the subject any further; however, I should point out that I've quietly slid some issues by you. One of these is discussed in Exercise 9.2.19.

Constructing Fuzzy Sets and Rules

So far, we've avoided two central problems any creator of a fuzzy rule system must face:

- How do we create rules?
- How do we create fuzzy set membership functions?

Creating rules: An obvious method is to understand the situation well enough to be able to construct the rules. What if understanding is in short supply? Researchers are working on methods for developing rules from observational data. For example, how would you back up a tractor trailer to park it at a loading dock without jackknifing? Ford researchers constructed a "neural net" to do this. By studying the neural net, they developed a system using just three fuzzy rules. Here's a rough description of it in ordinary English.

> If near jackknifing, then reduce steering angle.
>
> If far from dock, then back up toward dock.
>
> If near dock, then back up directly toward dock.

Neural nets have also been used to find clustering in data from which rules are extracted. (For information on neural nets, see the next couple of chapters.)

Creating membership functions: We faced a similar problem with probabilistic expert systems in the previous chapter. Theoretically, we could estimate the probabilities by measuring frequencies of events. Practically, this is often infeasible. As a result, the numbers in probabilistic expert systems are often crude estimates or educated guesses. Extensive testing of the particular expert system is done to adjust the numbers and to decide how much the inaccuracies of the numbers affect the performance of the system.

The situation with fuzzy systems is worse. Since there's no clear way to estimate a fuzzy set like "slight tilt to the left," designers guess. Usually they choose a membership function to have one of the following three simple shapes:

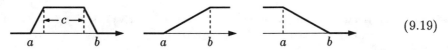

$$(9.19)$$

where the top of each graph is at $\mu = 1$. The leftmost graph is symmetric—an isosceles trapezoid. If $c = 0$, it becomes an isosceles triangle. Sometimes the graphs (9.19) are replaced by others without corners. Although shape is relatively unimportant, *overlap of fuzzy sets is very important*—it allows the results of several rules to be combined. If only one rule contributes to the output, the system behaves like an ordinary rule system. Look back at the stick-balancing controller in Example 9.5. The overlap of the tilt sets (Figure 9.2) and the overlap of the stick motion sets resulted in three rules' being applicable in the example given there.

How are the parameters a, b, and c chosen? We start out with guesses and then adjust the parameters to improve expert system performance—a "learning" process called "tuning." Fortunately, this can often be automated. How practical is tuning? For general rule systems, it's at least NP-hard; that is, prohibitively expensive—a problem we'll encounter again for neural networks. In practice, tuning has proved practical for *very* shallow rule systems of the sort used in controllers.

Exercises

9.2.H.   Suppose we have rules for a controller in which no chaining is required. Explain how the rules are used to choose an action given the input data.

9.2.I.   Describe how forward chaining can be done in fuzzy rule systems by using projections.

9.2.J.   What is meant by a shallow rule system? Why is it important that rule systems for simple controllers be shallow?

9.2.14.   This exercise refers to Example 9.5 and Figure 9.2.

   (a) Describe the position of the stick when $\mu_{T_r}(X) = 0.2$ and $\mu_{T_R}(X) = 0.8$

   (b) Describe the position of the stick when $\mu_{T_r}(X) = 0.2$ and $\mu_{T_R}(X) = 0.2$

   (c) What rules apply when $\mu_{T_r}(X) = 0.2$, $\mu_{T_R}(X) = 0.8$, $\mu_{V_r}(Y) = 0.6$, and $\mu_{V_0}(Y) = 0.3$? Construct a sketch like Figure 9.3.

9.2.15.   Modify the balancing-stick example to allow the stick to be caught if it falls beyond the point at which it can be balanced.
   Hint. You may want to introduce additional fuzzy sets and they may not have triangular graphs.

9.2.16. The purpose of this exercise is to construct a fuzzy traffic light controller for a freeway entrance ramp. At the entrance ramp there is a signal and a sign $\boxed{\substack{\text{TWO CARS} \\ \text{PER GREEN}}}$. The controller determines the length of time between greens. A sensor on the freeway provides information on the rate at which traffic is moving. The controller is used to reduce flow onto the freeway when additional cars would cause problems on the freeway.

(a) Using short, medium, and long to describe the duration of the red signal and jammed, crowded, and moving well to describe the traffic flow, write down fuzzy rules for the controller.

(b) Draw graphs like Figure 9.2 for the six fuzzy sets introduced in (a). (A triangular shape may be inappropriate for some graphs.)

9.2.17. Design a fuzzy controller for steering a car. (Speed is not an issue.) You have sensors that measure (a) the amount the car's front wheels are turned, (b) the deviation of the car's position from the center of the lane, and (c) the angular difference between the car's heading and the direction of the road. The following information may or may not be relevant. Road direction is measured a slight distance ahead of the car since that is more important than the road direction at the car's present position. This measurement is subject to a fair amount of inaccuracy.

9.2.18. Design a fuzzy controller for a car's speed. The following information is available:

- the current speed limit;
- the current speed of the car;
- the current acceleration of the car;
- the distance to the vehicle in front of the car, or a very large number if no vehicle is within a reasonable distance;
- if a vehicle is in front of the car, its speed relative to the car's.

9.2.19. All sets in this exercise are fuzzy. In Exercise 9.2.8 (p. 372) we defined a normal fuzzy set to be one in which $\mu_S(x) = 1$ for at least one x.

(a) Suppose $\emptyset \neq F_1 \subseteq \mathcal{X}_1$, $F_2 \subseteq \mathcal{X}_2$, and $C \subseteq \mathcal{X}_2 \times \mathcal{Z}$. Prove that

$$\text{proj}_{\mathcal{Z}}\big((F_1 \times F_2 \times \mathcal{Z}) \wedge (\mathcal{X}_1 \times C)\big) = \text{proj}_{\mathcal{Z}}\big((F_2 \times \mathcal{Z}) \wedge C\big)$$

when F_1 is normal but that it is sometimes false when $F_1 \neq \emptyset$ is not normal.

(b) Suppose we have two rules, $\alpha_1(X_1) \rightharpoonup \beta_1(Z)$ and $\alpha_2(X_2) \rightharpoonup \beta_2(Z)$, and two facts, $X_1 \in F_1$ and $X_2 \in F_2$. Using (a), discuss whether

$$\text{proj}_{\mathcal{Z}}\Big((F_1 \wedge \alpha_1 \wedge \beta_1) \vee (F_2 \wedge \alpha_2 \wedge \beta_2)\Big)$$

$$= \text{proj}_{\mathcal{Z}}\Big((F_1 \times F_2) \wedge \big((\alpha_1 \wedge \beta_1) \vee (\alpha_2 \wedge \beta_2)\big)\Big).$$

(c) The two sides of the previous equation suggest different ways of incorporating facts F_i about rule antecedents into the calculations. State in words what the two ideas are.

Hint. One way of doing this incorporates with each rule just those facts that are relevant to the rule.

*(d) The previous results in this exercise suggest three ways of handling fuzzy facts about rule antecedents. You should have found two in the previous part. The third method involves "normalizing" the facts. This means, given a fact F, replace it with the normal set F^* where $\mu_{F^*}(x) = \mu_F(x)/(\max_t \mu_F(t))$. Normalization is the method usually used, but arguments can be given for the other methods. Come to class prepared to defend normalization. Try to bring a specific example to illustrate your argument.

Hint. What happens if you combine two rules, one whose antecedent involves input data and another whose antecedent involves the consequent of another rule?

Further Remarks

The first industrial use of a fuzzy controller was by a Danish company in 1980. It used about fifty rules to regulate the controls of a cement kiln. Fuzzy reasoning has also been applied to other areas, including character recognition, speech recognition, medical diagnosis, financial planning, and information retrieval.

In spite of these successes, fuzzy logic hasn't really been tried! All the systems mentioned in the previous paragraph need is a way of combining competing rules based on the degrees to which the preconditions of the rules hold. Other methods having adjustable parameters could work as well as fuzzy logic. But isn't fuzzy logic the "correct" or "natural" approach? Doesn't it reflect reality? If this were so, then there should be little or no need to "tune" parameters. In practice, fuzzy systems usually undergo extensive tuning.

Given the ad hoc nature of fuzzy logic, does it have anything to offer "true" AI? That depends on what you think true AI is. Fuzzy methods will probably play vital roles in motor control and low-level processing of sensory input for robots. On the other hand, they probably won't play a significant role in complex reasoning: Even if the rules are adequate, tuning a complex system probably isn't feasible. See Chapter 11 for discussion of this problem in connection with neural nets.

I've emphasized fuzzy methods for controller design. How do these methods differ from the traditional approach?

- Traditional methods start with a mathematical model of the system—a technical and often difficult step that usually requires considerable understanding of the system. Since fuzzy methods rely on descriptions in ordinary language, much less technical skill and much less analysis is needed to design a fuzzy controller.

- Traditional analysis produces a nonfuzzy rule system, which often contains many more rules than a comparable fuzzy controller. Traditional methods need intermediate rules for intermediate situations. Fuzzy methods, in effect, construct intermediate rules by combining several rules as illustrated in Figure 9.3 (p. 378).

- Standard methods of control have a better theoretical foundation. As a result, it may be possible to say something about system stability but often only near equilibrium. No such theory is available for fuzzy systems.

- The net result of all of this is that the fuzzy controller is often available sooner than the traditional one and frequently functions better.

9.3   Dempster-Shafer Belief Theory

> *It is a capital mistake to theorize before you have all the evidence. It biases the judgement.*
> —Arthur Conan Doyle [Holmes] (1888)

> *Ignorance gives one a large range of probabilities.*
> —George Eliot (1876)

Dempster-Shafer belief theory is also referred to as the Dempster-Shafer theory of evidence, which may be preferable since "belief" has other meanings in AI. Our discussion here is restricted to conceptual issues.

Example 9.6   Sherlock Holmes's New Method

The following is taken from a manuscript discovered in October of 1993 and destroyed shortly thereafter in a major Los Angeles fire.

> His near failure in the "Affair of the Tickled Trout" forced Holmes to admit that his deductions could contain uncertainties. Acting on the suggestion of his brother Mycroft, Holmes studied and refined a method proposed by the Reverend Bayes in the *Philosophical Proceedings* of 1763. Since I do not understand the rarified mathematics he employed, I have omitted mention of the method in my accounts of Holmes's cases. Holmes plans to publish a manuscript on his Probabilistic Method of Evidence so as to set the record straight.

No such manuscript has been found. What might Holmes's method have been?

Suppose that Holmes has reduced the suspects in a burglary to B_1, B_2, B_3, B_4, and B_5. Then he discovers some badly decomposed cigar ash at the scene and is certain that it was dropped by the burglar. Since the ash's state makes identification uncertain, he can only say that, with probability $\frac{1}{3}$ it implicates one of B_1 and B_2, and with probability $\frac{2}{3}$ it implicates one of B_2, B_4, and B_5. (B_2 smokes two types of cigars.) As a result, he can rule out B_3. What other use can he make of this evidence?

Let the random variable E denote the evidence, Let $E = e_1$ denote the case that implicates B_1 and B_2 and let $E = e_2$ denote the other case. We don't actually know the value of E; we know only that the evidence exists. Let $\mathcal{E}$ denote the existence of the evidence and let B_i denote the event that B_i was the thief. We want $\Pr(B_i \mid \mathcal{E})$. Applying basic probability and then Bayes' Theorem (Theorem 7.3 (p. 280)), we have

$$\Pr(B_i \mid \mathcal{E}) = \sum_E \Pr(B_i \wedge E \mid \mathcal{E}) = \sum_E \Pr(B_i \mid E \wedge \mathcal{E}) \Pr(E \mid \mathcal{E})$$

$$= \sum_E \Pr(B_i \mid E) \Pr(E \mid \mathcal{E}) \tag{9.20}$$

$$= \sum_E \frac{\Pr(E \mid B_i) \Pr(B_i)}{\Pr(E)} \times \Pr(E \mid \mathcal{E}),$$

where $\Pr(B_i \mid E \wedge \mathcal{E}) = \Pr(B_i \mid E)$ since the value E conveys more information about B_i than the mere existence $\mathcal{E}$ does. Holmes gave the values of $\frac{1}{3}$ and $\frac{2}{3}$ for $\Pr(E = e_j \mid \mathcal{E})$. We have

$$\Pr(E) = \sum_{i=1}^{5} \Pr(E \mid B_i) \Pr(B_i).$$

The values of $\Pr(E = e_j \mid B_i)$ are 0 when B_i could not have led to E_j; that is, $\Pr(E = e_1 \mid B_i) = 0$ for $i = 3, 4, 5$ and $\Pr(E = e_2 \mid B_i) = 0$ for $i = 1, 3$. At the other extreme, $\Pr(E = e_j \mid B_i) = 1$ when (i, j) is $(1, 1)$, $(2, 4)$, or $(2, 5)$. Holmes may be unable to put values on $\Pr(E = e_i \mid B_2)$.

The values of $\Pr(B_i)$ are more troublesome. Bayes' method requires Holmes to put a value on $\Pr(B_i)$, but he may have believed that this violated his precept of working solely with the evidence. Instead, Holmes might have collected further evidence and combined all the information to see what emerged. How?

Suppose some independent evidence $\mathcal{F}$—a footprint partially destroyed by rain—is found. Associate the random variable F with it. The formal adjustment in (9.20) is simple: Allow the sum to run over values of F as well as E, replace E with $E \wedge F$, and replace $\mathcal{E}$ with $\mathcal{E} \wedge \mathcal{F}$. Since Holmes believed the cigar and footprint evidences are independent, we expect to have

$$\Pr(E \wedge F \mid \mathcal{E} \wedge \mathcal{F}) = \Pr(E \mid \mathcal{E}) \Pr(F \mid \mathcal{F}). \tag{9.21}$$

Suppose that $F = f_1$ implicates one of B_3 and B_5. Then $E = e_1$ and $F = f_1$ taken together are impossible since they eliminate all suspects. Hence, (9.21) is incorrect in this case since the left side is zero and the right side is not. What, then does it mean to say that the evidence is independent? We cannot know how Holmes dealt with such situations; however, it's possible he discovered the Dempster-Shafer approach. ∎

Before dealing with the problem of combining "independent" evidence, we need to define some basic concepts.

Definition 9.3   Belief and Plausibility

Let X be a finite set and let m be a function from the subsets of X to the nonnegative reals such that

$$m(\emptyset) = 0 \ \text{ and } \sum_{T\,:\,T\subseteq X} m(T) = 1,$$

where the notation $T : T \subseteq X$ means that the sum is over all T such that $T \subseteq X$. We call m a *basic probability assignment*. The *belief* and *plausibility* of S are defined by

$$\mathrm{Bel}(S) = \sum_{T\,:\,T\subseteq S} m(T) \ \text{ and } \ \mathrm{Pl(S)} = \sum_{\substack{T:\\ T\cap S\neq\emptyset}} \mathrm{m(T)} = 1 - \mathrm{Bel}(S'),$$

where the last equality is left as an exercise. The set X is called the *frame of discernment*.

Belief is also called *certainty*. The function m is sometimes written as a set of ordered pairs (p, S) where $S \subseteq X$ is such that $m(S) = p \neq 0$. Capital Greek letters such as Θ are usually used instead of X to denote frames of discernment.

Let's look at some extreme cases to gain a bit of insight. Suppose we are looking for one $x \in X$—the answer—and m reflects our knowledge about X.

- If $\mathrm{Bel}(S) = 1$, then we are certain that the answer is in S.

- If $\mathrm{Pl}(S) = 0$, then we are certain that the answer is not in S because $1 = \mathrm{Pl}(S) = \mathrm{Bel}(S')$ tells us that we're certain the answer is in S'.

- Finally, if $\mathrm{Bel}(S) = 0$ and $\mathrm{Pl}(S) = 1$, we're totally ignorant about whether the answer is in S: $\mathrm{Bel}(S) = 0$ says there is no evidence pointing to S and $\mathrm{Pl}(S) = 1$ says there is no evidence pointing to S'.

Example 9.7   Belief in Cigar Ashes

Here's how to think about the definition in terms of the previous example. Let $X = \{B_1, B_2, B_3, B_4, B_5\}$ and

$$m(S) = \begin{cases} \frac{1}{3}, & \text{if } S = \{B_1, B_2\}, \\ \frac{2}{3}, & \text{if } S = \{B_2, B_4, B_5\}, \\ 0, & \text{otherwise.} \end{cases}$$

Thus m is constructed directly from the "testimony" of evidence: There are two distinct cases suggested by the evidence, namely (i) that it comes from a cigar of the type smoked by B_1 and B_2 and (ii) that it comes from a cigar of the type smoked by B_2, B_4, and B_5. How certain can we be that the burglar is in some particular set S? If we have evidence that the thief belongs to a subset of S, that contributes to our certainty. Bel(S) is simply the sum of that evidence. On the other hand, plausibility is a measure of whether something could be true. The burglar could belong to S if there is evidence that he belongs to a set that shares a burglar with S.

Let's do a couple of calculations. In the first place, there's no evidence that points definitely to any of the burglars and so Bel($\{B_i\}$) = 0 for all i. More generally, Bel(S) = 0 unless S contains B_2 and also contains either B_1 or both B_4 and B_5. (Why is this so?) Plausibility is another matter. We have

$$\text{Pl}(B_1) = m(\{B_1, B_2\}) = \tfrac{1}{3},$$
$$\text{Pl}(B_2) = m(\{B_1, B_2\}) + m(\{B_2, B_4, B_5\}) = 1,$$
$$\text{Pl}(B_3) = 0,$$
$$\text{Pl}(B_4) = \text{Pl}(B_5) = m(\{B_2, B_4, B_5\}) = \tfrac{2}{3}.$$

You might practice computing Pl and Bel for other subsets of X. ∎

Example 9.8   Other Simple Examples of Evidence

Let's phrase (9.1) (p. 358) in belief theory terms. We can let the frame of discernment be $X = \{f_{11}, f_{18}, b\}$, where f_d indicates that Jim is free on day d and b indicates that Jim will be busy on both days. In this case $m(\{f_{11}, f_{18}\}) = \frac{3}{4}$, $m(\{b\}) = \frac{1}{4}$, and $m(S) = 0$ otherwise. In the alternative notation, the evidence consists of the two pairs

$$\left(\tfrac{3}{4}, \{f_{11}, f_{18}\}\right) \quad \left(\tfrac{1}{4}, \{b\}\right).$$

You should show that

$$\text{Bel}(S) = \begin{cases} 1, & \text{if } S = X = \{f_{11}, f_{18}, b\}, \\ \frac{3}{4}, & \text{if } S = \{f_{11}, f_{18}\}, \\ \frac{1}{4}, & \text{if } S = \{b\}, \\ 0, & \text{otherwise,} \end{cases}$$

and

$$\text{Pl}(S) = \begin{cases} 0, & \text{if } S = \emptyset, \\ \frac{1}{4}, & \text{if } S = \{b\}, \\ \frac{3}{4}, & \text{if } S = \{f_{11}, f_{18}\}, \\ 1, & \text{otherwise.} \end{cases}$$

A traditional example of evidence is the unreliable witness. Let X be the possible suspects in a crime. If the witness is correct, the criminal is in the set A. If p is the probability that the witness is correct, then the evidence is (p, A) and $(1 - p, X)$. In this case, you should be able to show that

$$\text{Bel}(B) = \begin{cases} 1, & \text{if } B = X, \\ p, & \text{if } B \supseteq A \text{ and } B \neq X, \\ 0, & \text{otherwise.} \end{cases}$$

What is the value of $\text{Pl}(B)$? ∎

Combining Independent Evidence

Suppose we have two sets of evidence that we regard as independent—like Holmes's cigar ash and footprint. How should we combine them? In the new terminology, we have two basic probability assignments, m_1 and m_2 on X, and we want to combine them to produce a new basic probability assignment, m_{12} on X. Here it is.

Definition 9.4   Dempster's Rule of Combination

Let m_1 and m_2 be basic probability assignments for two bodies of evidence E_1 and E_2, both having X as the frame of discernment. The basic probability assignment $m_{12} = m_1 \oplus m_2$ is defined by $m_{12}(\emptyset) = 0$ and

$$m_{12}(S) = \frac{1}{1 - K} \sum_{\substack{T, U: \\ T \cap U = S}} m_1(T) m_2(U), \quad \text{if } S \neq \emptyset \text{ and } K < 1, \quad (9.22)$$

where

$$K = \sum_{\substack{T, U: \\ T \cap U = \emptyset}} m_1(T) m_2(U).$$

If $K = 1$, we say that m_1 and m_2 are incompatible and $m_1 \oplus m_2$ is undefined. If m_{12} is the basic probability assignment for the combination of E_1 and E_2, we say that E_1 and E_2 are *independent* (bodies of) evidence.

K is chosen so that $\sum_S m_{12}(S) = 1$. We can think of K as the strength of the contradiction between the evidence provided by m_1 and m_2. Since m_{12} should reflect a consistent situation, K is distributed to the noncontradictory evidence in a proportional manner: "When you have eliminated the impossible, whatever remains, *however improbable*, must be the truth" (Holmes, *The Sign of Four*).

What does independence of evidence *mean*? This is like asking what any mathematical concept means—it's the wrong question. We should ask "In what real-life situations is it reasonable to assume independence of evidence?"

For independence of random variables, causality and Bayesian nets provide an answer: When a random variable X is conditioned on its immediate causes $\mathcal{C}(X)$, it is independent of many other random variables. See Definition 8.3 (p. 306) for a formal statement and Theorem 8.5 (p. 336) for a more general result. Can we find something analogous for independence of evidence?

Imagine a diagram indicating *causal* dependence of evidence. Trace the dependency paths of the evidence E_1 and E_2 backward. If they meet before reaching elements of the frame of discernment X, they are dependent. If they do not meet before X, they are independent. To illustrate, suppose I ask Chris and Liz if it's still raining. Each of them may actually know the answer or may make a guess. If I have prior estimates of their reliabilities in such a situation and I question them separately, then the evidence about the rain is independent. On the other hand, if Liz listens when I ask Chris, and this affects her answer, then the evidence is no longer independent.

If you find my description of independence of evidence a bit vague, that's fine. Researchers are not in agreement about when Dempster's Rule of Combination can and cannot be used. More detailed discussions can be found in the literature.

Example 9.9   A Warning

If we have no prior reason to prefer one alternative to another, it's natural to divide up the basic probability assignment among the alternatives. In other words, if $m(\{d, e, f\}) = p$, it's natural to decide that this contributes $p/3$ each to the posterior probabilities of d, e, and f. *This is wrong.*

To see why, suppose that

$$m_1(\{a\}) = m_1(\{b, c\}) = \tfrac{1}{2} \quad \text{and} \quad m_2(\{a, b\}) = m_2(\{c\}) = \tfrac{1}{2}.$$

Let $m_{12} = m_1 \oplus m_2$. You should be able to show that

$$m_{12}(\{a\}) = m_{12}(\{b\}) = m_{12}(\{c\}) = \tfrac{1}{3}. \tag{9.23}$$

If we divide up m_i as suggested to get a probability p_i, we obtain the following table

$x =$	a	b	c
$p_1(x)$	$\tfrac{1}{2}$	$\tfrac{1}{4}$	$\tfrac{1}{4}$
$p_2(x)$	$\tfrac{1}{4}$	$\tfrac{1}{4}$	$\tfrac{1}{2}$

(9.24)

Any reasonable combination of these probabilities would ascribe equal probability to each of a and c and a lesser probability to b. This disagrees with (9.23).

What happened? A belief theorist might react as follows. The probability is assigned to different possibilities for evidence; not to the consequences of the evidence. What do I mean? To answer this, let's first look at Sherlock's cigar ash problem. After examining the crime scene, he decided the ash must have been dropped by the burglar. Examining the ash, he decided that it was from one of two varieties of cigars and twice as likely to be from the second type. You're free to bring to bear whatever probability interpretations you want on all of this. For example, the 1:2 for types of cigars might even be given a frequentist interpretation. It's only *after* obtaining the 1:2 that we look at what the evidence implies about the suspects. In more general terms:

> The probabilities in belief theory apply directly to the evidence (subsets of the frame of discernment X), not to the elements of X as in traditional probability theory.

The table in (9.24) was derived by (improperly) assuming that $m_i(S)$ was obtained by summing the probability of x over all $x \in S$. In essence, we assumed that the prior probabilities of the elements of X were all equal and then conditioned on the evidence. As already noted, independent evidence is not the same as independent probabilities, so we should not expect that we could then use the rows of (9.24) as if they were independent. ■

Why redistribute K proportionally as done in (9.22)? There are various justifications. Here are some.

- Think about what combining evidence should mean. In particular, the result should be the same regardless of the order in which $m_1, m_2, \ldots$ are combined. This is equivalent to saying that $\oplus$ commutative and associative, a result stated in the next theorem.

- The notion of evidence can be interpreted using a combination of propositional logic and Bayes' Theorem. The next example shows that, if we do this, Dempster's Rule emerges as the way to combine independent evidence.

- In a some sense, this method of redistribution involves less of an assumption than other methods.

Theorem 9.3   Combining Evidence Is Commutative and Associative

Let m_1, m_2, and m_3 be basic probability assignments for the frame of discernment X. Dempster's Rule of combination is

(a) commutative; that is, $m_1 \oplus m_2 = m_2 \oplus m_1$, and

(b) associative; that is, $(m_1 \oplus m_2) \oplus m_3 = m_1 \oplus (m_2 \oplus m_3)$.

The proof is left as an exercise.

*Example 9.10   A Possible-Worlds Viewpoint

Let $\mathcal{L}$ be a propositional logic language. Choose a finite set $\mathcal{E}$ of formulas in $\mathcal{L}$. Let $\Pr$ make $\mathcal{E}$ into a probability space; that is, $\Pr \geq 0$ and $\sum_{\alpha \in \mathcal{E}} \Pr(\alpha) = 1$. For any formula $\beta \in \mathcal{L}$, define the random variables

$$B_\beta(\alpha) = \begin{cases} 1, & \text{if } \alpha \models \beta, \\ 0, & \text{otherwise;} \end{cases} \qquad P_\beta(\alpha) = 1 - B_{\neg\beta}(\alpha) = \begin{cases} 0, & \text{if } \alpha \models (\neg\beta), \\ 1, & \text{otherwise.} \end{cases}$$

There are various ways to think about $\alpha \models \beta$:

- It is the same as $\models (\alpha \rightarrow \beta)$; that is, $\alpha \rightarrow \beta$ is valid (a tautology).

- In any possible world where α is true, β is also true.

- If we believe that α is true and we are logical, we must believe that β is true, too.

Let $\text{Bel}(\beta) = \Pr(B_\beta = 1)$ and $\text{Pl}(\beta) = \Pr(P_\beta = 1)$. Thus, $\text{Bel}(\beta)$ is the probability that we *must* logically believe β, given that the formula for $\alpha \in \mathcal{E}$ which we choose to believe is selected according to our probability distribution. Saying that α does not entail $(\neg\beta)$ is the same as saying that α and β can both be true simultaneously. Thus $\text{Pl}(\beta)$ is the probability that β could be true.

Suppose that someone else chooses a set of formulas $\mathcal{E}'$ and a probability distribution $\Pr'$. Now we can choose a formula α and she can independently choose a formula α'. If they are inconsistent, it would be illogical to believe both of them If they are consistent, we can ask whether they tell us that we must believe β; that is, we can ask if $(\alpha \wedge \alpha') \models \beta$.

What we did in the last paragraph was look at the probability distribution on $\mathcal{E} \times \mathcal{E}'$ given by $\Pr^*((\alpha, \alpha')) = \Pr(\alpha)\Pr'(\alpha')$. Insisting on consistency of α and α' means conditioning on $\alpha \wedge \alpha'$ being satisfiable. Since only the truth values are relevant, we can often replace $\mathcal{E} \times \mathcal{E}'$ by a smaller set $\mathcal{E}''$ of formulas logically equivalent to the consistent $\alpha \wedge \alpha'$. In other words, there is a function f from consistent $\alpha \wedge \alpha'$ to logically equivalent formulas in $\mathcal{E}''$. The probability conditioned on consistency is given by

$$\Pr''(\alpha'') = \frac{1}{1-K} \sum \Pr(\alpha)\Pr'(\alpha'),$$

where the sum ranges over all (α, α') such that $f(\alpha \wedge \alpha') = \alpha''$ and K is the sum of $\Pr(\alpha)\Pr'(\alpha')$ over all inconsistent $\alpha \wedge \alpha'$. We'll soon see that this is a generalization of Dempster's Rule of Combination.

Why go through all this formalism? It gives another view of the operation $\oplus$: Two formulas are chosen independently at random and we condition on their being consistent. We can then ask if this seems to be a reasonable way to model a particular situation.

To recover standard belief theory as a special case, let $\mathcal{E}$ consist of all formulas of the form $\gamma(S) = \delta(S) \wedge (\neg\delta(S'))$ where

- S is a nonempty subset of X, the set of propositional letters,
- S' is $X - S$, the complement of S,
- $\delta(T)$ is the disjunction of the propositional letters in T.

We claim that assigning a value to $\Pr(\gamma(S))$ is equivalent to assigning that value to $m(S)$ in standard belief theory and that $\mathrm{Bel}(\delta(S))$ here corresponds to $\mathrm{Bel}(S)$ there. Here's a sketch of how to prove the claim. You should be able to show that

$$\gamma(S) \models \delta(T) \text{ if and only if } S \subseteq T \text{ and so } \mathrm{Bel}(\gamma(T)) = \sum_{S \subseteq T} \Pr(\delta(S)).$$

To verify Dempster's Rule, you need to observe that, if $S \cap T = \emptyset$, then $\gamma(S)$ and $\gamma(T)$ are inconsistent, and, otherwise, $\gamma(S) \wedge \gamma(T) = \gamma(S \cap T)$. The details are left for you to fill in. ∎

Further Remarks

There are two major problems with evidence theory.

Dempster's Rule of Combination is an essential feature of the theory since it's our prescription for accumulating evidence. Unfortunately, it's difficult to decide when it can be used. Researchers have proposed various viewpoints, some of which I've mentioned.

Since μ is defined on the nonempty subsets of X, its domain is extremely large even for relatively small X. This can cause problems in both data collection and data manipulation. Data collection is often facilitated by the *a priori* information that $\mu(S) = 0$ for many $S \subseteq X$. Researchers have looked at approximate methods of computation.

Exercises

9.3.A. How does belief theory sidestep the problem of assigning prior probabilities?

9.3.B. Define frame of discernment and basic probability assignment.

9.3.C. Define Bel, Pl, and $m_1 \oplus m_2$.

9.3.D. What does it mean mathematically to say that two sets of evidence are independent?

9.3.E. Prove that $\mathrm{Bel}(\emptyset) = \mathrm{Pl}(\emptyset) = 0$ and $\mathrm{Bel}(X) = \mathrm{Pl}(X) = 1$, where X is the frame of discernment.

9.3.1. Prove that $\displaystyle\sum_{\substack{T: \\ T \cap A \neq \emptyset}} m(T) = 1 - \mathrm{Bel}(A')$.

9.3.2. Prove Theorem 9.3.

9.3.3. Let m_1, m_2, m_{12} and K be as in Definition 9.4 and suppose that $K = 0$. Let Bel_x (resp. Pl_x) be the belief (resp. plausibility) function associated with m_x.

    (a) Prove that $\mathrm{Pl}_{12}(S) \leq \mathrm{Pl}_1(S)\mathrm{Pl}_2(S) \leq \min\big(\mathrm{Pl}_1(S), \mathrm{Pl}_2(S)\big)$.

    (b) Prove that $\mathrm{Bel}_{12}(S) \geq \max\big(\mathrm{Bel}_1(S), \mathrm{Bel}_2(S)\big)$.

    (c) Do you think these results are plausible given the intended interpretation of Bel and Pl that I discussed after Definition 9.3? Why?

*9.3.4. Fill in the details in the last paragraph of Example 9.10.

9.3.5. It is interesting to compare results obtainable from Bayes' Theorem with the Dempster-Shafer theory; however, it's not clear what conclusions to draw from this. The following is adapted from an example discussed by Shafer and others. Suppose it's a cold, wet day and I'm wondering if the sidewalks are icy. My friend George stops by and I ask him if the sidewalks are icy. He says they are. A few minutes later, Kramer comes in and, when asked, tells me the sidewalks aren't icy. Unfortunately either friend might answer a question without thinking carefully; however, if they do think carefully, their answers are accurate. The probability that George is careful is $P_G = 0.8$ while for Kramer the value is only $P_C = 0.3$. Let R_G and R_C be the probability that each is accurate when not being careful. Let S and N denote the events "slippery" and "not slippery." We begin with probability theory.

    (a) Assuming independence as needed, show that

$$\frac{\Pr(S \mid \text{evidence})}{\Pr(N \mid \text{evidence})} = \frac{\Pr(S)}{\Pr(N)} \times \frac{0.8 + 0.2R_G}{0.2(1 - R_G)} \times \frac{0.7(1 - R_C)}{0.3 + 0.7R_C}.$$

    (b) Assume that $\Pr(S) = \Pr(N) = \frac{1}{2}$. Why might it be reasonable to assume that $R_G = R_C = \frac{1}{2}$? In that case, what is $\Pr(S)$?

    (c) Assume that $\Pr(S) = \Pr(N) = \frac{1}{2}$. Why might it not be reasonable to assume that $R_G = R_C$? Show that the product of the last two factors in (a) can take on any value from 0 to ∞ depending on the values of the probabilities R_C and R_G. What does this say about $\Pr(S \mid \text{evidence})$?

    (d) Assume that $\Pr(S) = \Pr(N) = \frac{1}{2}$. Suppose that $R_G = R_C$. What are the possible values for $\Pr(S|\text{evidence})$?

We now switch to belief theory.

    (e) Explain why the following translation is correct and compute m_{GC}.

$$m_G(\{S\}) = 0.8 \qquad\qquad m_G(\{S, N\}) = 0.2$$
$$m_C(\{N\}) = 0.3 \qquad\qquad m_C(\{S, N\}) = 0.7$$

(f) Compute $\text{Bel}_{GC}(\{S\})$ and $\text{Pl}_{GC}(\{S\})$.

*(g) How might the previous result be interpreted as saying something about the range of probability for the sidewalks being slippery? What happened to the prior probabilities that we needed in (a)?

9.4 Looking Backward

> *When we enter a new field, very often new concepts are needed, and these new concepts usually come up in a rather unclear and undeveloped form. Later they are modified, sometimes they are almost completely abandoned and are replaced by better concepts which then, finally, are clear and well-defined.*

—Werner Heisenberg (1972)

Sophisticated reasoning systems are a perpetual source of visions and nightmares for AI researchers. Hopes of designing such systems have continually foundered on the rocks of reality only to be refloated by someone with another symbolic reasoning approach. A growing number of researchers have decided to look elsewhere, seeking solutions in "subsymbolic" methods (such as neural nets) or in ad hoc ("scruffy") methods (such as scripts). The following argument is often seen: Symbolic reasoning requires a huge number of calculations which our brains aren't powerful enough to carry out in a reasonable time. Furthermore, computers will also never be powerful enough to do so. In other words, symbolic reasoning is too hard, it's never been done and it never will be. Before turning to subsymbolic methods, let's look back over where we've been.

Long Chains of Deductions

Developers of first-order predicate logic (FOL) were partially motivated by a desire to formalize the reasoning used in mathematics. Thus it's no surprise that FOL handles long chains of reasoning flawlessly. (This is called "deep" reasoning as opposed to "shallow," which uses only short chains of implications.) In fact FOL is the only system we've studied that can safely do this. Nonmonotonic methods can sometimes give unexpected (and unwanted) results when defeasible assumptions accumulate in long chains of reasoning. Probabilistic methods are likely to require staggering amounts of data and computation for long chains of reasoning unless we make crude approximations of independence as is done with certainty factors. Since we can't predict what the cumulative effect of such approximations will be, we can't be sure how valid the results are. The more complex the chain of reasoning, the more questionable the conclusions. Fuzzy methods are even more inept in their handling of deep reasoning and belief theory is based on the assumption of independence.

Things are not as bleak as they seem. Most human reasoning is shallow; i.e., based on short chains. This can be seen by introspection, by reading "deductive" detective fiction, or by realizing that our brains simply don't have the processing speed needed to perform lengthy serial calculations on a regular basis. We substitute knowledge for reasoning. For example, you compute $\frac{d}{dx}\big(\sin(x^2)\big)$ using knowledge of the chain rule and derivatives of the sine and of powers—you don't start with the definition of the derivative. When driving a car, you make extensive use of unconscious knowledge that was gained through experience—you don't repeat the steps you went through in learning. In other words, human expert behavior depends heavily on knowing and seldom on deep reasoning.

Learning plays a fundamental role because it provides the knowledge on which rapid, shallow reasoning is based. AI researchers are beginning to make progress in the practical aspects of learning; however, there is little in the way of theoretical foundations. There are theories of learning in AI, but they are primarily negative ("This is too hard to learn") rather than positive ("Here's a method for learning"). I expect that the practical aspects of learning will soon reach the point where significant learning is routinely incorporated in commercial expert systems.

Robustness

Large bodies of information are almost certain to contain errors. Thus, it's important to know how sensitive a reasoning method is to errors in the knowledge base. There are various types of errors.

- Contradictions: These are the most obvious. As we've seen, a single contradiction in the knowledge base of an FOL reasoning system destroys it. This arises because $\mathcal{K} \models \alpha$ means that α is true for every interpretation which makes all formulas in the knowledge base $\mathcal{K}$ true. When a contradiction is present, there are no such interpretations of $\mathcal{K}$ and so the set of "every interpretation" is empty. Clearly α is true for every interpretation in this empty set! Since default logic includes FOL, it suffers from the same problem. This fragility ensures that no effective reasoning method will include the unbridled power of FOL. Most other qualitative methods we've discussed, including Prolog and semantic nets, are essentially rule-based. Such systems only apply rules whose antecedents or consequents contain information relevant to what the system is reasoning about. This limits the effects of contradictions. For example, knowing that all penguins fly and that no penguins fly will create problems only when we're reasoning about penguins.

- **Numerical errors**: It's usually impossible to obtain accurate estimates for quantitative systems. In fact, numerical values are frequently "educated" guesses and such guesses may contain relatively large errors. It's generally believed that systems are relatively insensitive to all but the grossest errors. Since this is based on empirical observation, it may or may not apply to the particular expert system you happen to be interested in. Although theoretical results are urgently needed, it's unlikely that any useful general results will be obtained in the foreseeable future. This prediction is based on the difficulties numerical analysts and control theorists have encountered in problems that seem more tractable.

- **Structural sensitivity**: Here are two examples of structural sensitivity:

  (a) Bayesian nets can misbehave if intermediate causes are inadvertently omitted.

  (b) Some qualitative methods may give different results when the knowledge base is modified so that reasoning chains lengthen.

Although these are both forms of structural sensitivity, they're really quite different. Omitting intermediate causes as in (a) is a genuine error in the knowledge base, something that would cause problems for any reasoning method. The situation in (b) is insidious: We make a change that seemingly should not alter the conclusions. Nevertheless it does owing to counterintuitive quirks in the reasoning method. The discovery of such quirks followed by their correction and/or defense is one of the facets of AI research. (Analogous research is found in many areas of science.) Another form of structural sensitivity arises when we make a somewhat arbitrary choice in the design of a reasoning method. For example, what would be the effect on fuzzy controllers if defuzzification were done probabilistically instead of by using the center of gravity?

Computational Feasibility

A tension exists between the pull to design general reasoning systems and the pull to design computationally feasible systems. You've encountered various compromises—Prolog is limited to a subset of FOL, certainty factors ignore the dependency problem in Bayesian nets, the link between the real world and the theoretical one is tenuous for the methods in this chapter. Methods that work well for small knowledge bases may bog down when applied to larger problems. Advances in computers have made it possible to move technology from supercomputers to desktops. However, I don't believe that future breakthroughs lie in that direction. I think they'll come from massive parallelism. Achieving this requires knowledge-intensive, modular, shallow reasoning methods that can carried out on very simple processors. Connectionists are exploring such ideas on the "subsymbolic" level and are also exploring the integration of symbolic with subsymbolic methods. I believe future researchers will also develop powerful parallel purely symbolic methods.

Psychological Validity

The previous hundred pages have dealt with methods for representing and manipulating numerical uncertainty. Some methods, such as Bayesian networks are "normative"; that is, if we accept some basic axioms concerning measures of uncertainty, then Bayesian nets are the one and only correct method of manipulation. This approach was discussed in Section 9.1 where objections to the axioms led to the alternative approaches of this chapter. Although these methods are often intended for use in expert systems, little attention has been paid to what experts do. Since we often obtain rules from an expert who interprets and manipulates them in a certain way, it behooves us to design methods that reflect the approaches of experts. If we fail to do so, our systems may produce numerical predictions that disagree with expert opinion. Little research has been done in this area. See [23].

Similar issues can be raised regarding the qualitative methods of representing and manipulating uncertainty that were discussed in Chapter 6.

Choosing a System

Numerical methods provide an entire range from certainly true to certainly false, in contrast to qualitative methods which abandon the middle ground, ranging only from certainly true (or false) to probably true (or false). This ability comes with a high price tag: considerable data collection and extensive computation are required. To reduce the costs, people invent data and rely on approximate methods such as certainty factors. Fuzzy methods seem to overcome such problems by declaring data invention to be a virtue and using simple numerical manipulations. This works well in very shallow systems, but you should probably not use it otherwise.

Qualitative methods eliminate the need for expensive numerical data and offer more sophisticated reasoning strategies. Although default and multivalued logics provide some ability to discriminate between levels of belief, they can't provide the fine discrimination of Bayesian nets and other numerical methods. We might hope that fuzzy logics would combine sophisticated reasoning with fine discrimination. But, as discussed on page 384, this is unlikely to happen.

What method of reasoning should you use?

Obviously, that depends on the knowledge you have and what you want to do with it. In other words, choose a method whose knowledge representation fits the task.

If they are adequate, qualitative methods are probably preferable because they require less data. Unless a higher level system is going to decide between alternatives, default logic cannot be used. You might use a method based on defeasible reasoning or semantic nets if you want to be presented with an answer rather than a choice.

What about quantitative methods? Fuzzy methods are suspect for all but the shallowest reasoning. Exact probabilistic methods are often impractical because of extensive dependencies and so approximate methods such as certainty factors are needed. The approximations of independence assumptions and crude data estimates may lead to a breakdown when the reasoning becomes complicated. People are poor quantitative reasoners and usually break down much sooner than these methods.

The previous discussion is based on the assumption that you have a single problem on which you must use a single system. It may be possible to break the problem down into stages and use different knowledge representations and reasoning methods for different stages—a hybrid system. Some expert systems have been constructed this way; for example, CHATKB (p. 24). I believe this sort of modular construction with different methods being used in different parts is the only way AI will construct large reliable systems and that it will soon be a standard feature of large reasoning systems.

Notes

The axiomatic approach to conditional probability was developed by R. T. Cox, G. Pólya, and E. T. Jaynes. An exposition is given by Smith and Erickson [21]. The axioms, in various forms, have been discussed by a variety of people. For example, see [8, p. 241] and [3, p. 112]. In a 1988 workshop [3], researchers discussed the pros and cons of various approaches to quantitative reasoning. Klir [10] discusses various forms of uncertainty.

I dismissed the knotty problem of the semantic content of the fuzzy membership function rather quickly. Various authors have attacked the problem, but no consensus has been reached. If you'd like further discussion, see [7], a special issue with papers devoted to the problem. For a discussion of the sociological and technological as well as theoretical aspects of fuzzy logic, see McNeill and Freiberger's book [16]. Some of my discussion on the foundations and potential of fuzzy logic was stimulated by Elkan's paper and the reactions to it [19]. If you want to see what AI researchers think and feel about fuzzy logic, read that symposium. The only work on stability that I'm aware of is [12]. Books on fuzzy methods are appearing at a rapid rate. There are [5], [9], [11], [22], [25], and [28], to mention a few. In addition, the text by Dougherty and Giardina [6] has several sections on fuzzy sets and logic.

There are far fewer books devoted to belief theory. Some of the important papers on the subject have been reprinted and annotated in [18, Ch. 7]. See also the papers in [24].

Paris [17] discusses the mathematical foundations of quantitative uncertainty more deeply than I've done in this chapter. His text is at a slightly higher level than mine. In his monograph for graduate students, López de Mántaras [15] discusses Bayesian nets, belief theory, and fuzzy methods. In particular, he provides more information about inference in fuzzy systems and discusses "possibility theory." Dubois and Prade [4] also discuss possibility theory and its connections with fuzzy reasoning.

Levesque [13] discusses some of the difficulties with symbolic reasoning and reviews some of the newer methods. He answers "Is reasoning too hard?" with a hopeful "Maybe not."

As mentioned here and elsewhere, the representation of knowledge plays a crucial role in choosing a reasoning system. The collection [2] of readings in knowledge representation is somewhat dated but still quite useful. The special issue [14] of *Artificial Intelligence* contains newer papers.

Biographical Sketch

Lotfi A. Zadeh (1921–)

Born Lotfi Aliaskerzadeh in Baku, Azerbaijan, he received a bachelor's degree in electrical engineering from the American College in Teheran, Iran. In 1944 he moved to the United States where he received masters and doctorates from MIT and Columbia. In 1959 he joined UC Berkeley, where he is now professor emeritus of computer science.

In a 1954 paper Zadeh coined the term "systems theory" for his area of research. He introduced fuzzy sets in his seminal 1965 paper [27] and related them to systems theory in the 1970s. According to legend, the vagueness of beauty led Zadeh to develop fuzzy set theory: Unable to resolve an argument with a friend over who had the more beautiful wife, he became interested in finding a way of expressing such ideas. Since then, he and other researchers have used fuzzy methods in systems theory.

References

1. E. B. Baum (ed.), *Computational Learning and Cognition. Proceedings of the Third NEC Research Symposium*, SIAM, Philadelphia (1993).

2. R. J. Brachman and H. J. Levesque (eds.) *Readings in Knowledge Representation*, Morgan Kaufmann, San Mateo, CA (1985).

3. D. Dubois, P. Garbolino, H. E. Kyburg, H. Prade, and P. Smets with P. Gärdenfors, G. Paass, and E. Ruspini, Quantified uncertainty, *Journal of Applied Non-Classical Logics*, **1** (1991) 105–197.

4. D. Dubois and H. Prade, *Possibility Theory. An Approach to Computerized Processing of Uncertainty*, Plenum Press, New York (1988). This is a translation with revisions of the French edition (1985).

5. D. Dubois, H. Prade, and R. R. Yager (eds.), *Readings in Fuzzy Sets for Intelligent Systems*, Morgan Kaufmann, San Mateo, CA (1993).

6. E. R. Dougherty and C. R. Giardina, *Mathematical Methods for Artificial Intelligence and Autonomous Systems*, Prentice Hall, Englewood Cliffs, NJ (1988).

7. *Fuzzy Sets and Systems*, **8** No. 3 (1988).

8. M. Ginsberg, *Essentials of Artificial Intelligence*, Morgan Kaufmann, San Mateo, CA (1993).

9. A. Kandel (ed.), *Fuzzy Expert Systems*, CRC Press, Boca Raton, FL (1992).

10. G. J. Klir, Developments in uncertainty-based information. In [26] 255–332.

11. B. Kosko, *Neural Networks and Fuzzy Systems: A Dynamical Approach to Machine Intelligence*, Prentice Hall, Englewood Cliffs, NJ (1991).

12. T. Langari and M. Tomizuka, Stability of fuzzy linguistic control systems, *Proceedings of the 29th Conference on Decision and Control, Honolulu, Hawaii*, IEEE, New York, NY (1990) 2185–2190.

13. H. J. Levesque, Is reasoning too hard? In [1] 163–176.

14. H. J. Levesque and R. Reiter (eds.), *Aritificial Intelligence* **49** nos. 1–3 (1991). Reprinted as *Knowledge Representation*, MIT Press, Cambridge, MA (1992).

15. R. López de Mántaras, *Approximate Reasoning Models*, Ellis Horwood Ltd., Chichester (1990)

16. D. McNeill and P. Freiberger, *Fuzzy Logic. The Discovery of a Revolutionary Technology—and How It Is Changing Our World*, Simon and Schuster, New York (1993).

17. J. B. Paris, *The Uncertain Reasoner's Companion: A Mathematical Perspective*, Cambridge University Press, Cambridge, Great Britain (1994).

18. G. Shafer and J. Pearl (eds.), *Readings in Uncertain Reasoning*, Morgan Kaufmann, San Mateo, CA (1990).

19. L. Shastri (ed.), A fuzzy logic symposium, *IEEE Expert*, **9**, no. 4 (Aug. 1994) 3–49. This symposium consists of a paper by Elkan and reactions to the issues it raises. The original version of the paper appeared as C. P. Elkan, The paradoxical success of fuzzy logic, *Proceedings of The Eleventh National Conference on Artificial Intelligence*, MIT Press, Cambridge, MA (1993) 698–703.

20. J. Skilling (ed.) *Maximum Entropy and Bayesian Methods. Cambridge, England, 1988*, Kluwer, Dordrecht (1989).

21. C. R. Smith and G. Erickson, From rationality and consistency to Bayesian probability. In [20] 29–44.

22. T. Terano, K. Asai, and M. Sugeno, *Fuzzy Systems Theory and Its Applications*, Academic Press, San Diego (1992).

23. B. E. Tonn and R. T. Goeltz, Psychological validity of uncertainty combining rules in expert systems, *Expert Systems* **7** (1990) 94–100.

24. R. R. Yager, J. Kacprzyk, and M. Fedrizzi (eds.), *Advances in the Dempster-Shafer Theory of Evidence*, John Wiley and Sons, New York (1994).

25. R. R. Yager and L. A. Zadeh (eds.), *An Introduction to Fuzzy Logic Applications in Intelligent Systems*, Kluwer, Boston (1992).

26. M. C. Yovits (ed.), *Advances in Computers*, Vol. 36, Academic Press, San Diego (1993).

27. L. A. Zadeh, Fuzzy sets, *Information and Control* **8** (1965) 338–353.

28. L. A. Zadeh and J. Kacprzyk (eds.), *Fuzzy Logic for the Mangagement of Uncertainty*, John Wiley and Sons, New York (1992).

10

What Is It?

[R]easoning is a specialization of pattern-recognition applied to language, and logic is a further specialization of reasoning ...
—Howard Margolis (1987)

The classification of the constituents of a chaos, nothing less here is essayed.
—Herman Melville (1819–1891)

Introduction

Where there's life, there's classification: organisms constantly classify sensory input patterns. People do it consciously, classifying objects by using nouns, such as "chair," but all living things do it—"food," "enemy," etc. Expert systems are classifiers, too. We often call their classifications conclusions. In fact, pattern classification pervades AI. So far you've only studied classifiers that have their classifications "hardwired" in by means of a knowledge base. This chapter begins the study of classifiers that learn.

In *learning*, the organism or program is developing classifications based on input patterns. The learning may be *supervised*—patterns and their classifications are supplied and should be learned. The learning may be *unsupervised*—input patterns are supplied and significant features should be found in them. In either case, after the learning phase, the organism or program should be able to assign novel patterns to the most appropriate classification category. This is called *recognition*. The entire process is called *automatic pattern classification*.

There are various types of classifiers. In AI, the simpler classification problems are often referred to as *pattern classification*, and the more complex as *machine learning*. The simpler classifiers focus purely on the simplest aspects of the structure (syntax) of the patterns, ignoring more abstract patterns and meaning (semantics). In the next few chapters we'll deal with pattern classification, especially "neural nets."

But aren't systems based on the methods in the previous chapters better than those based on simple automatic pattern classifiers? After all, aren't the former carefully crafted using the knowledge of experts while the latter rely on some crude general-purpose methods that ignore the higher level symbolic structure of the knowledge domain? Not necessarily:

- Converting expert knowledge into an expert system is difficult. There's plenty of room for errors and oversights.

- An expert may be unable to (a) articulate how he/she makes decisions or (b) estimate important numerical values.

- Finally, there may be no expert to provide information. This is typical for subconscious decisions such as interpretation of visual input.

Prerequisites: The definitions of a directed graph and of a decision tree from Section 2.1 are needed, but I'll remind you of them in case you've forgotten.

Used in: This chapter prepares you for the next three chapters by giving an overview and by introducing some important concepts and issues.

10.1   Types of Automatic Classifiers

> *Classification is a tricky business. Yet it is difficult to think of any area of human enterprise in which this is not a fundamental activity. Classification is the way we sort and order, a means to identify and understand. How we classify reflects our biases and priorities.*
>
> —Doris Schattschneider (1991)

Before we begin, we'd better be clear about what a pattern classifier is. The following definition is based on the ideas in the introduction.

Definition 10.1   Pattern Classification

Pattern classification (or *recognition*) is the process of mapping a large space (or set) into a smaller one. An input to the procedure is a *pattern* and an output is a *classification*. A pattern classifier is *automatic* if it develops its own algorithm when it is given a *training set*, which consists of a collection of inputs, possibly paired with desired outputs. In *supervised learning*, desired outputs are present and the goal is to reproduce the associated output as closely as possible. In *unsupervised learning*, there are no desired outputs and the goal is to extract significant features from the input so that the input can be divided into sets that differ significantly.

We'll focus on supervised learning.

Here's a block diagram for training a generic classifier.

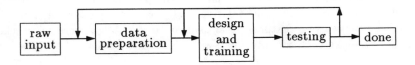

After feedback from a testing failure, the designer changes either the data preparation, the nature of the classifier, or the training algorithm. Once trained, the classifier simply takes input, prepares the data, and classifies the pattern. This sequence of steps is implicit in the testing block. It's also used in a functioning deployed classifier.

Since pattern classification is simply a function from one set (or space) to another, why not just look at all possible functions? Even in simple cases, the number of functions is *huge*. Let's calculate this number in a simple case. Suppose we're looking at functions f from n-long binary vectors to k-long binary vectors. The domain contains $d = 2^n$ points and the range contains $r = 2^k$ points. Since there are r choices for $f(x)$ at each x in the domain, there are

$$r \times r \times \cdots \times r = r^d = 2^{k2^n}$$

possible functions. Even for moderate values such as $n = 10$ and $k = 1$, the number of possible functions greatly exceeds the estimated number of atoms in the universe!

Obviously we need a better approach than simply trying to look at all possible functions.

Some approaches to classification were originally developed in statistics. These often involve fitting some simple function to the data. *Decision trees* fit more complicated functions to the data. This is accomplished by composing rather simple functions—the individual decisions. In another direction, AI researchers have developed *neural nets* as another method of fitting complicated functions to data. The complexity of decision trees and neural nets allows greater flexibility. The price for this flexibility is the need for many more calculations and less understanding of the process. Nevertheless, the flexibility appears worth the cost.

Statistically Based Functions

Typically, statistical classifiers assume that the pattern can be repesented as a point in space, say in $\mathbb{R}^n$. The relative positions of the patterns in space are taken to be the relevant features for classification. Although there are methods that do not make such assumptions, we'll limit this discussion to those that do.

Unfortunately, we need more background in probability theory to discuss probabilities in $\mathbb{R}^n$. For the present discussion, it suffices to know that there is a concept called the "probability density function" which is essentially a generalization of the function Pr introduced in Definition 7.1 (p. 260).

Bayes' Theorem provides an ideal pattern classifier. Let I denote the training set together with any other information we have about the material we must classify. Let X be a random variable whose range is the set of possible classifications for patterns. Suppose that, given a pattern P, there is not enough information in the pattern to decide on a definite classification. For example, symptoms may be inadequate to determine a disease with certainty, or information extracted from pixels may be inadequate to determine unambiguously the boundaries of objects in a scene. The error rate will be minimized if x is chosen so as to maximize $\Pr(X = x \mid P \wedge I)$. Bayes' Theorem enters because it's frequently easier to calculate $\Pr\big(P \mid (X = x) \wedge I\big)$.

Estimating the necessary probabilities usually requires drastic simplifying assumptions. One common assumption in statistics is that a probability density function can be approximated by a "normal density." This leads to two classical statistical approaches to pattern classification:

- Discriminant analysis: Typically, but not necessarily, there are two categories. Discriminant analysis attempts to select parameters in a simple function (usually linear or quadratic) of the inputs so that the sign of the function is highly correlated with the category the input belongs to.

  Ideally, when constructing a linear discriminant for two categories, we would find a vector $\vec{w}$ of "weights" and a parameter θ such that the class of any vector $\vec{x}$ is determined by the sign of the linear function $f(\vec{x}) = \theta + \sum w_i x_i$. If $\vec{x}$ has length n, $f(\vec{x}) = 0$ describes a hyperplane in n-dimensional space such that the points in one category lie on one side and the points in the other category lie the other side. Real life is seldom so neat. Usually we must be satisfied with a high degree of correlation. Sometimes even this is impossible. For example, let the class of a 0/1-vector be the parity of the number of ones it contains. The case $\vec{x} = (x_1, x_2)$ is referred to as the XOR (exclusive or) problem, since the exclusive or of x_1 and x_2 is 0 when $\vec{o}$ contains an even number of ones and is 1 if $\vec{x}$ contains an odd number of ones.

- Regression analysis: The categories are a subset of the real numbers; this subset is typically either discrete or an interval. Regression analysis attempts to select parameters in a simple function (often linear) of the inputs so that the value of the output is close to the category the input belongs to. Thus, it could be viewed as a continuous version of discriminant analysis.

These approaches are called *parametric* by statisticians because they assume that the underlying probability density functions are known, except for some parameters that must be estimated. There are also *nonparametric* methods. These have the advantage of assuming less about the underlying structure and the disadvantage of requiring much more data. Here are two popular nonparametric methods.

- Nearest neighbor: The idea here is to look at the nearest neighbor, or k nearest neighbors, to a given input and classify it according to how its neighbors have been classified.
- Parzen estimators: The idea here is that knowing the classification of some input $\vec{p} \in \mathbb{R}^n$ to be x tells us that the probability density function for $X = x$ should be nonzero in a neighborhood of $\vec{p}$. All these effects are added up to obtain an estimate for the probability density function for patterns conditioned on classification.

Finally, there are unsupervised learning techniques, the most important statistical one being

- Cluster analysis: The various techniques in this category attempt to group patterns into geometrically meaningful groups. The most obvious possibility is that each group should consist of patterns that are near one another and far from other patterns. Hence the term "cluster" analysis.

There's one other statistical method that deserves some mention although it is not a method of pattern classification:

- Principal-component analysis: Imagine a collection of vectors in $\mathbb{R}^n$. There may tend to be linear relationships among the components of the vectors. For example, if each vector measures meteorological information, the components might be barometric pressure, humidity, and so on. Meteorologists would expect lower barometric pressure to often be associated with higher humidity. Thus, there is an approximate dependence among the variables. Geometrically, this means that the vectors tend to lie in a lower dimensional subspace of $\mathbb{R}^n$. The goal of *principal-component analysis* is to find a new coordinate system in $\mathbb{R}^n$ such that, in the new coordinate system, our collection of vectors is nearly constant in as many coordinates as possible. We can then ignore those coordinates with very little loss of information.

Neural Networks

> **Network**: *Anything reticulated or decussated, at equal distances,*
> *with interstices between the intersections.*
> —Samuel Johnson (1755)

As the name suggests, *neural networks* have their origin in the idea of designing a system that behaves like the neurons in a brain. Neural networks are also called *connectionist systems*. The expectation of many researchers is that neural networks will eventually be implemented in hardware by a very large number of simple, highly interconnected processors with rather limited memories.

Given the biological inspiration for neural nets, it's natural to expect a split—those researchers who attempt to use neural nets to understand brains better (cognitive science) versus those researchers who regard human neurons as the model for designing a tool (AI). Our approach is entirely from the tool side.

It's hoped that, when properly designed, such a system will possess some properties of human brains. These include (a) the abilities we associate with "intelligence," (b) the use of massively parallel computation, and (c) the ability to compensate considerably for damage to hardware and software.

A connectionist system can be thought of as a (large) interconnected network of simple processing units operating in parallel and exchanging a limited amount of numerical information. Adjustable "weights" determine how each processor uses the information it receives. The concept can be described in purely mathematical terms, which is essential for the design and analysis of algorithms. There are many different designs for neural nets; most—but not all—are encompassed by the following definition.

Definition 10.2   Neural Net

Recall that a directed graph (V, E) consists of a set V called vertices and a set $E \subseteq V \times V$ called edges and that the edge (u, v) is represented by an arrow from u to v. A *neural net* consists of the following:

- a directed graph (V, E),
- two sets $I, O \subseteq V$, called the inputs and outputs,
- a weight function $w : E \to \mathbb{R}$, usually written w_e instead of $w(e)$, and
- real-valued functions f_v for the vertices $v \in V$.

If $e = (i, j)$, we'll write $w_{i,j}$ instead of $w_{(i,j)}$. The vertices not in $I \cup O$ are called *hidden vertices*.

Time-dependent values o_v are associated with the vertices as follows. The initial values are supplied by the user for $v \in I$, and these values may or may not vary after initialization. For $v \notin I$, o_v is initialized somehow.

Vertices are selected in some manner, and for each selected vertex v, the value of o_v is a function f_v of its previous value and of those o_x and $w_{x,v}$ for which $(x, v) \in E$. The functions f_v often have the form

$$f_v = f\left(\theta_v + \sum_{x \in \text{In}(v)} o_x w_{x,v}\right), \text{ where } \text{In}(v) = \{\, x \mid (x, v) \in E \,\} \quad (10.1)$$

and $f : \mathbb{R} \to \mathbb{R}$. After some number of repeated selections of vertices, the process is terminated. The outputs of the net are the values o_v for $v \in O$. You can think of o_v as the "excitation" of v, which is propagated as a signal along all edges of the form (v, y). How the excitation affects the "neuron" y is determined by its strength and the factor $w_{v,y}$. We "train" the network by adjusting the weights w_e and the parameters such as θ_v in the functions f_v.

A linear discriminant can be viewed as a very simple neural net called a *perceptron*: Let

$$O = \{z\}, \quad V = I \cup O, \quad E = I \times O, \quad f(x) = \text{sign}(x).$$

The values o_i, $i \in I$, are held fixed at the values of the pattern to be classified and the computed value of o_y is the classification. Thus,

$$o_z = f_z = \text{sign}\left(\theta_z + \sum_{i \in I} o_i w_{i,z}\right).$$

This is the linear discriminant described earlier in slightly different notation. In particular, $\vec{x}$ has become $\vec{o}$. As we saw earlier, linear discriminants are too simple for real life. However, we could build a more complicated neural net by sending the input into many perceptrons and then sending the outputs of these perceptrons into another perceptron which produced the output of the net.

It's difficult to find adequate algorithms for training nets and for using trained nets. There is an algorithm for training the perceptron. Unfortunately, it doesn't generalize to the net consisting of layers of perceptrons described at the end of the preceding paragraph.

In the next chapter, we'll discuss two types of neural nets. The focus will be on algorithms for choosing neural net parameters (called "training") and for using trained neural nets. The study of nets continues briefly in Chapter 13 after some probabilistic tools have been developed.

Decision Trees

Decision—or classification—trees were developed by statisticians and, independently, by AI researchers. They're among the most widely used learning methods. Let's begin by quickly reviewing the definition given in Section 2.1.

Definition 10.3   Decision (Classification) Tree

A *rooted tree* is a directed graph that can be built as follows:

- Select a vertex, say r, for the root of the tree.
- Select some number (possibly zero) of rooted trees, $T_1, \ldots, T_k$.
- Add the edges (r, r_i) for $1 \leq i \leq k$, where i is the root of T_i.

A vertex with no edges leading out is called a *leaf* or *terminal vertex*. (See Figure 10.1.) A *decision tree* is a rooted tree such that

- Associated with each nonterminal vertex v is a decision procedure (also called a *rule*) for selecting an edge (v, x) based on the input pattern.
- Associated with each terminal vertex is classification information, either deterministic ("Hot tomorrow") or probabilistic ("30% chance of rain tomorrow").

Since a decision procedure contains, at least implicitly, a list of the possible decisions at each vertex, we can order the edges leading out of a vertex if we wish. The definition of a decision tree in Section 2.1 included such an ordering.

Part of an imaginary decision tree for weather prediction is shown in Figure 10.2. We use a decision tree by starting at the root r and using the input pattern and the root's decision procedure to select an edge (r, x). (A decision procedure might direct us to one of three vertices depending on whether the pattern had a rising, falling, or steady barometer.) Having selected x, we move to it and repeat the procedure there, and so on until we reach a leaf, which gives us a classification.

Since we can think of a decision procedure as a rule, we may think of a decision tree as a type of rule-based expert system. Using automatic tree classification algorithms, people have constructed such expert systems containing thousands of rules.

There's nothing in the definition of a decision tree about how it should be created. Instead of creating one automatically, we might create it by hand. (To do so, we must decide on its structure, rules, and classifications.) In fact, many expert systems based on rules obtained from experts can be viewed as decision trees. A simple example of such a system is a procedure for determining the species of an organism by means of a series of questions.

By changing the nature of the leaves, we can use a decision tree in a hybrid expert system. The decision tree can be used to divide the problem

Figure 10.1   Building rooted trees recursively. We start on the left with a single-vertex rooted tree. This lets us build more rooted trees like those in the middle, which let us build still more, and so on. By convention, the trees are drawn upside down, so the root is the topmost vertex. Open circles are leaves.

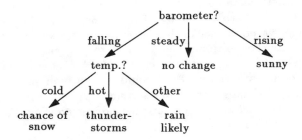

Figure 10.2   An imaginary decision tree for weather prediction. The input pattern consists of "temperature" and "barometer change." A decision procedure consists of a question at a vertex with possible answers on the outgoing edges. The classifications are shown at the leaves. A real decision tree would be larger, probably have more complicated decisions, and might quantify "chance" and "likely."

space into smaller groups amenable to some other expert system approach. Thus the tree is an expert system all of whose leaves are other expert systems. Since this more modular approach allows the construction of larger systems, it is likely to become increasingly important.

To construct a decision tree automatically requires (a) choosing decisions and (b) deciding when the tree is big enough. We'll discuss the construction of decision trees in Chapter 13.

Structural and Syntactic Methods

The methods discussed so far largely ignore structural and syntactic features except for those that can be extracted by simple numerical manipulation of the patterns. The definition of structural and syntactic pattern recognition is somewhat hazy. The basic underlying theme is that it looks for some underlying patterns in the data other than relatively unstructured numerical relationships. At one extreme (the highly symbolic), this could be taken to include inductive learning (p. 596). At the other extreme it could be taken to include purely numerical approaches that involve more structure than usual, such as "hidden Markov models" (p. 593) and some approaches to large neural nets (p. 567). Since the structure of a decision tree is determined by the data, this method could also be called structural pattern recognition. The central area of structural methods lies somewhere between these extremes.

In neural nets, some underlying family of functions is specified parametrically. (The parameters are the weights of the net.) We then work within this domain: for example, by adjusting parameters to obtain a good fit to the training set. Structural and syntactic pattern recognition are similar. Instead of specifying some underlying family of functions, we might specify certain types of graphs or grammars. We then attempt to find a good fit to the data within this family. For example, our data might consist of strings of symbols that we want to classify into two categories. A syntactic classifier would attempt to produce a grammar that would generate all the strings in first category and none of the strings in the second.

Research on structural and syntactic methods is growing. For more information see the References.

Further Remarks

The complexity of reality usually means that compromises must be made in designing pattern classifiers. In statistically based classifiers, these assumptions are clear as we see, for example, in the independence of some random variables or in the particular forms for some probabilities. The assumptions behind other classifiers are usually less clear; for example, neural net classifiers are based on the assumption that certain sets of complicated functions contain functions that fit the data (both observed and unobserved) fairly well.

This suggests that we should use statistical methods. Indeed, all else being equal, we should. Unfortunately, statistical methods exact a price—the necessary simplifying assumptions may be quite unrealistic. The gain from the clarity of the assumptions may not be worth the price.

It can be shown that, given certain assumptions about how decisions should be made, it follows that probability theory—in particular Bayes' Theorem—should be used. In technical terms, the Bayesian approach is *normative*—it tells us how a rational person should act. As a result, some researchers are using Bayesian methods in constructing decision trees and neural nets. We'll come back to this in Chapter 12.

Exercises

10.1.A. What is pattern classification? Give an example of pattern classification by humans.

10.1.B. What is the difference between pattern classification and machine learning?

10.1.C. What are the input and output of a classifier called?

10.1.D. What is automatic pattern classification? How does automatic classification differ from an expert system written in Prolog to classify organisms by species?

10.1.E. In the context of pattern classification, what is a training set?

10.1.F. Distinguish between supervised and unsupervised learning in terms of the information presented and the goals.

10.1.G. What is a neural network? How is it used for classification?

10.1.H. What is a perceptron in statistical terms? in neural net terms?

10.1.I. What is a decision tree? How is it used for classification?

10.1.1. By drawing a picture in two dimensions, show that the exclusive or (XOR) problem cannot be solved by a linear discriminant. Repeat for $\vec{x} = (x_1, x_2, x_3)$.

*10.1.2. Let's return to the XOR problem. Imagine a neural net in which the two inputs i_1 and i_2 both feed into two perceptrons v_1 and v_2 which feed their outputs to a third perceptron o. Find θ's and w's so that this net solves the XOR problem; that is, o has one output when (i_1, i_2) contains a single 1 and a different output otherwise.

10.2   Applications

> *If reason determined us, it would proceed upon the principle, that*
> *instances, of which we have had no experience, must resemble*
> *those, of which we have had experience, and that the course of*
> *nature continues always uniformly the same.*
>
> —David Hume (1748)

Statistical classifiers are used in a variety of disciplines, resulting in the formation of such subdisciplines as econometrics, which is devoted to the analysis
of (= search for patterns in) economic data. In AI, statistical classifiers are
used primarily in vision research and secondarily in speech understanding.

Statisticians developed decision trees and applied them to the design of
expert systems, most notably in medical diagnosis situations in which experts
are unclear about the connections between the patients' medical data and
the underlying diseases. AI researchers also developed decision trees and used
them in other situations where rules relating data to classification were crude
or unknown. Some expert system shells construct decision trees.

Neural networks are by far the largest research area in the automatic
classifier field. They've found a wide range of applications as expert systems.
Medicine, finance, chemistry, and game playing are just a few areas of successful use outside AI. In AI, neural nets have been applied in a variety of fields
including vision, language, robot motion, and more traditional expert systems.
So far, the size of these systems has been small compared to larger rule-based
systems. (More on the problems of building large neural nets in Chapter 13.)
Expert system shells based on neural nets are commercially available.

Researchers are currently studying ways of combining an automatic pattern classifier with another system, either symbolic or automatic.

- Automatic classifiers are used to organize the data so that another system
  can be constructed. For example, an automatic classifier may discover a
  way to cluster the data that leads to rules for a fuzzy controller. (An
  example was mentioned on p. 381.)

- A variant on this idea is to extract information from raw data using an
  automatic classifier. This information is then used by another system,
  either symbolic or subsymbolic.

- A different approach is to divide up the problem domain using one system.
  Other systems then handle the different pieces. The dividing up could be
  based on clustering discovered in the raw data by an unsupervised learning
  algorithm, or it could be based on a clustering of problems (= patterns)
  based on an expert's opinion as to what represents a meaningful division.
  More interestingly, the partitioning could be developed as the system
  is being trained. For example, first train a classifier A_1. Then train a

classifier B_1 to separate data into two sets, one in which A_1 does well and one in which A_1 does poorly. Finally train a classifier A_2 on the latter set. Now we can use B_1 as a decision maker that chooses between the classifiers A_1 and A_2. This could be iterated.

10.3 Data Preparation

> *Errors using inadequate data are much less than those using no*
> *data at all.*
> —Charles Babbage (1792–1871)

You might think that raw data requires little if any preparation before it is fed to an automatic pattern classifier. This is seldom true. Generally, the more effort spent in data preparation, the better a classifier will work—there's no "free lunch." Unfortunately it's tempting to ignore data preparation and get on with the more exciting aspects of the work.

Data Friendliness

Pattern classifiers find some data more friendly than others. Obviously, any classifier will find garbage unfriendly, but there are other considerations. For example:

- Because of the functions they use, some classifiers have to be trained longer if an important input variable ranges from 1,000 to 1,001 than if it ranges from 0 to 1 or from −0.5 to 0.5, even though the difference between minimum and maximum is the same in all cases.

- Since most pattern classifiers perform arithmetic operations on numerical data, they may have difficulty with information like telephone area codes—so called "nominative" data.

How can data be made more friendly for a pattern classifier? That depends on the nature of the classifier and the nature of the input. Suppose the input is in the form of real numbers. The meaning of the numbers can be lumped into three general categories as follows:

- **Nominative:** A nominative value simply *names* something. It need not even be a number. In fact, if it is a number, you should think of the number as simply a name. Generally, such data should be encoded as m-bit binary vectors. Each bit of the vector is interpreted as a real number. For example, suppose we wish to encode the days of the week. Since there are seven days and $2^3 = 8 \geq 7$, we could use 3-bit binary vectors. Seven vectors are assigned to the days of the week in some fashion. The eighth vector is never used.

- Ordinal: Ordinal numbers simply indicate an ordering of the data. For example, say someone is in the 80th percentile on an exam. It makes no sense to say that person is twice as knowledgeable as someone in the 40th percentile. Using such numbers is usually okay, even though the classifier performs arithmetic operations on them.

- Arithmetic: These are numbers for which it makes sense to perform numerical operations. For example, the annual income of a loan applicant or values of probabilities.

It may be best to rescale ordinal and arithmetic data in some manner. For some classifiers such as neural nets, it is usually best if all data is reduced to a similar size. For example, if x is one of the variables in the data, we might choose a and b so that the values of $y = ax + b$ range from -1 to $+1$. The value of y is then used in place of x. If x can be any real number, we might use $y = \tanh(ax)$ to convert to values in the interval $(-1, 1)$. More complicated functions of x may be better, but discussing that takes us too far afield.

Sometimes we can make input data more friendly by doing calculations on the input. For example, it may be better to input a date as "day of year" and "year" rather than the more usual "month," "day of month" and "year." More generally, if we expect that certain operations will need to be performed on the input, it is often better to do them and include the results as part of the input data.

Garbage in the Training Data

> *GIGO: Garbage in; garbage out.*
> —Anonymous programmer

> *The art of being wise is the art of knowing what to overlook.*
> —William James (1890)

What sorts of things in training data should be regarded as garbage? Obvious possibilities are

- Incorrect data.
- Incorrect expert opinions: They may be hard to detect.

Here are three subtler sources of garbage.

- Bias: Training data can be a biased sample of the patterns we plan to classify. Obviously, if you omit a pattern whose classification cannot be inferred from similar patterns, the classifier will not be able to classify it. More subtle is the fact that classifiers are usually less accurate on rarer training patterns. Thus, a classifier may do poorly if a rare training situation arises frequently in applications.

- **Redundancy**: Some of the information in a pattern may be redundant. As noted in data friendliness, redundancy may be good if it makes the input more friendly. Other redundancy is usually bad.

- **Irrelevancy**: Some of the input information may be irrelevant to the classification of the pattern. For example, Social Security number, name, and eye color are probably irrelevant data if we are trying to teach an expert system to assess a person's creditworthiness.

Irrelevancy needs some explanation. Won't the classifier simply ignore irrelevant data? Not always. Given enough irrelevant data, spurious correlations between some irrelevant data and the desired classification are likely to arise. There is no reason to expect future patterns to exhibit the same correlation. Such correlations can be quite strong. In one case, for example, a classifier was trained to detect camouflaged tanks. It so happened that the pictures with tanks were taken on a sunny day and those without, on a cloudy day. The classifier learned to detect sunlight, not tanks.

How can we guard against irrelevance in the data? In the case of tank recognition, it's simply a matter of adding more pictures or replacing the cloudy-day pictures. Sometimes it's not just a matter of adding or throwing out data. For example, a person's scores on individual problems on an exam contain irrelevant information if we want to predict the exam grade—only the sum of the scores is relevant. In this case, it's clear that several numbers should be replaced by a single number. Usually the situation is *much less* clear-cut. Obtaining relevant information in a form the classifier can use is known as *feature extraction*.

Removing irrelevancies can be quite difficult even when we're aware of them. For example, imagine that we want to analyze a picture that's been presented as a set of black and white pixels. Our goal is detecting squares. In this case, only the position of one pixel relative to another matters—absolute position is irrelevant. You may be able to devise a way of arranging the data to eliminate this irrelevancy. A much bigger challenge is the fact that size and rotation are also irrelevant. Since your eyes and brain can eliminate these irrelevancies, there must be a way to do so. How? See Exercise 10.3.1.

What can be done in general?

Statistical tools can provide some guidance and support in rooting out irrelevancy and redundancy. For example, cluster analysis might reveal a sharp grouping of some components of the data. By labeling the groups and using the labels in place of the data that determines the groups, we may eliminate some irrelevancy. As another example, suppose that principal-component analysis provided new coordinate axes in which a few coordinates contained much of the variation in the data. By reparameterizing the input and using these few new coordinates, we eliminate redundant and/or irrelevant information. More generally, a pattern classifier might be used to look for relevant aspects of the data. Using these various mathematical methods is a complex subject, which we lack the time and tools to pursue.

Of course, instead of focusing on data we can focus on *understanding* the particular classification problem. Through such understanding, we may be able to extract useful features or identify irrelevant ones.

Incomplete Data

Data is often incomplete. It may be missing as in questionnaires that are only partially filled out, or it may lack the required accuracy as sometimes happens in dating fossils. What we can do with incomplete data depends on the classification method and the nature of the data.

- If there's plenty of complete data, we can simply ignore the incomplete data. Let's suppose that's not the case.

- Some methods allow the use of incomplete data; however, most methods demand complete data. How can we make complete data out of incomplete data?

- In some cases, we could introduce a new data category—"misssing." In most cases, this is unsatisfactory.

- We could create many "complete" data items from each incomplete one. For example, if we have an estimate of the probabilities of the various possible values for a missing item, we could construct a representative collection of values. Naturally, each of these completed items should be given less weight than an item that was complete to begin with. This method has two drawbacks: It can increase the amount of data tremendously and we may be unable to estimate the probabilities that are required.

There is a more insidious form of incomplete data: We may lack data about a part of reality that the classifier will be called upon to deal with. A weaker form of this is when the data contains more representatives of some areas of reality than others. For example, data on causes of death in one region, A, of the world may differ from that in another region, B. If we construct a classifier based on data from A and attempt to apply it in region B, we are likely to get poor results. The only advice I have here is "Be careful!"

Exercises

10.3.A. What are the differences between nominative, ordinal, and arithmetic data? Give an example of each.

10.3.B. Irrelevant data is an insidious form of garbage. How can irrelevant data cause problems? How can irrelevant data be detected?

10.3.1. This exercise requires some familiarity with complex numbers. The *Fourier transform* of a function is another function. We follow the convention of using a lowercase letter for the original function and the corresponding uppercase letter for its Fourier transform. The definition of the Fourier transform of f is

$$F(y) = \int_{-\infty}^{\infty} f(t)e^{-ity}\,dt. \tag{10.2}$$

Ignore questions of convergence and similar issues.

(a) Let $g(x) = f(x + h)$. Show that $|F(y)| = |G(y)|$, where $|z|$ is the absolute value of the complex number z. If f is the description of a one-dimensional object, then $|F|$ provides information about the object that does not depend on its absolute position.

(b) Extend (10.2) to a function of two variables, $f(x_1, x_2)$ by using a double integral. With $g(x_1, x_2) = f(x_1 + h_1, x_2 + h_2)$, show that $|F(y_1, y_2)| = |G(y_1, y_2)|$. Thus the absolute value of the Fourier transform provides translation-independent information about a two-dimensional object, too.

*(c) While the previous approach removes translation effects, it does not remove rotation or scaling effects. We do this now. Define a new function $h(\rho, \theta) = |F(y_1, y_2)|$, where ρ, θ, y_1, and y_2 are related by $\exp(\rho + i\theta) = y_1 + iy_2$. Show that $|H(*, *)/H(0,0)|$ has rotation and scaling removed. How can we remove scaling but not rotation? rotation but not scaling?

10.4 Evaluation

There are various aspects to evaluating a classifier. The most obvious is accuracy. Complexity and interpretation are also important.

Measuring Accuracy

Let's assume that the classifier makes simple yes/no type classifications—nothing like "The probability of rain tomorrow is $P\%$." The simplest measure of accuracy is the percentage of incorrect classifications the classifier makes. This is called its *error rate*. There are two types of error rate:

- The *training error rate* is error rate on the training set.

- The *generalization error rate* is error rate on all possible data. It's called the generalization error rate since it measures how poorly the classifier generalizes to cases it hasn't seen.

(The word "rate" is often dropped.) Computing the training error is straightforward. We usually want the generalization error. It's almost always greater than the training error, often significantly.

In this discussion, I've slipped in the notion of generalization error as if it were an obvious concept. It's not. What does "all possible data" mean? We often don't know. It frequently refers to an infinite set that is hard to manage or is inherently unknown. For example, "all possible data" for a chess-move classifier is a fantastically large, hard-to-manage set. On the other hand, "all possible data" for a weather predictor contains future weather and so is inherently unknown.

If we don't know what our classifier must generalize to, how can we possibly measure the generalization error rate? We can't! All we can do is hope that the given data is somehow typical and use it to make an estimate. Since generalization error is based on data not used in training, only some of the given data can be used for the training set. How do we go about choosing it? It should be "representative" and so be as large as possible. Often the available data is limited, so how much should be used for training and how much reserved for computing the generalization error? Are there ways of estimating generalization error that don't require reserving much data? In Section 12.3, we'll discuss some ways of approaching these problems.

Classifier Complexity

As we increase the complexity of a classifier (e.g., larger neural nets or bigger decision trees), the added complexity allows the classifier to fit the data better. Thus a good training algorithm will produce classifiers with decreasing training error rates as shown in Figure 10.3. Unfortunately, beyond a certain complexity, this improvement is often an artifact—the generalization error rate often begins to increase. *This behavior is common but not universal.* There's a close connection between classifier complexity and irrelevancy.

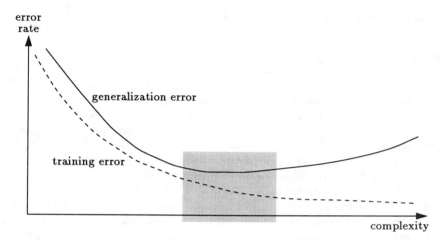

Figure 10.3 The horizontal axis measures classifier complexity in some manner and the vertical axis measures error rate. The dashed curve shows how training error rates behave—they decrease as complexity increases. The solid curve shows how generalization error rates often behave. The rapid decrease in generalization error rate followed by a slow increase is typical. The shaded region indicates a reasonable range of complexity.

In effect, a classifier that is too complex for the problem at hand contains irrelevant parts.

Obviously, we should choose a classifier with the best level of complexity. Unfortunately, no one knows a simple method for making such a choice. The best level of complexity depends in some manner on the classifier, the training set, and "all possible data." Fortunately, there's usually a fairly broad range of acceptable complexity. We'll return to this problem later: See Example 12.4 (p. 502), which discusses the "bias–variance dilemma," and Example 11.10 (p. 485), which discusses the interaction between complexity and parameter estimation in the context of neural nets.

In summary, here are the typical situations for the three intervals of complextity in Figure 10.3.

- **Low** (left of gray region): Insufficient complexity leads to poor fitting of the training data.

- **High** (right of gray region): Excess complexity leads to good fitting of the training data and results in poor generalization.

- **Correct** (in gray region): The correct level of complexity leads to good fitting of the training data and to good generalization. Unfortunately, this level is hard to find and, even if it's known, it may be difficult to train without the freedom excess parameters provide.

These situations aren't universal—just typical.

What Does It Mean?

With some types of classifiers, such as decision trees, it's often easy to identify "rules" that the classifier has developed for its own use. With others, such as neural nets, it's usually difficult to do so. However, obtaining rules may be important if the classifier is being used as a first stage in a multilevel process. Researchers have made progress in extracting rules from classifiers.

When we find such rules, it's very tempting to impute an explanatory significance to them. Unfortunately, they merely describe *correlations* between patterns and their classifications, not causality. Since people usually expect explanations to be rooted in causality, thinking of such rules as explanations can be misleading. You may enjoy reading Armstrong's discussion of this issue [1].

Exercises

10.4.A. What is training error? generalization error? How are they related to the complexity of the classifier?

10.4.B. If we accept the tempting equation

rule found by classifier = explanation of classification,

we can be misled. Explain this statement.

Notes

I've sketched a variety of methods for pattern recognition. Which is best? That depends on both the problem and the expertise of the researcher. Some comparisons have been done for particular problems. For example, Ripley [21] and Kim, Weistroffer, and Redmond [16] compared systems based on neural nets, on statistical methods, and on decision trees. The latter paper deals with bond rating and the former with three examples. An important problem is determining if differences are significant. See Section 12.3 for some discussion.

Part I of the handbook [6] discusses a variety of pattern recognition methods. Later parts discuss applications to robotics and vision. Weiss and Kulikowski [23] have written a readable, practical introduction to a variety of pattern classifiers. In addition to discussing them in some detail, the authors also give experimental results and make suggestions about how and when they should be used. Pao [18] discusses pattern recognition from an AI perspective and provides critical assessments of a variety of methods. A much more abstract approach is taken by Pavel [19], who draws on category theory and topology. It remains to be seen how fruitful such abstractions will be.

There are a variety of books that deal with statistical pattern classifiers. Stork is working on a new edition [9] of Duda and Hart's classic book [8]. It will be greatly expanded to include material on methods that have been developed since the original edition, including nonstatistical topics such as neural nets and learning theory. In addition to the chapter in Weiss and Kulikowski [23], you might look at James's readable book [14], which contains algorithms in BASIC. The books by Devijver and Kittler [7] and by Fukunaga [11] are more mathematically oriented. Jain [13] gives a brief survey of the state of statistical pattern recognition research in 1986. A good introduction to principal-component analysis is Jolliffe's text [15]. In the first part of his book, Bow [2] discusses a variety of methods.

Syntactic pattern recognition was largely founded by K. S. Fu. An introduction to the subject and related mathematics can be found in the books by Fu [10], Miclet [17], and Pavlidis [20] and in the chapter by Bunke [4]. The text by Schalkoff [22] is a readable account of statistical, syntactic, and neural net methods of pattern classification.

The papers reprinted in Chapter 4 of [3] discuss some aspects of pattern recognition in an AI context. The papers in [12] are written for a rather general audience and discuss some applications of probability and statistics in AI and vice versa.

A variety of classification methods, including Bayesian nets, have been unified by Buntine [5] using a graphical framework. This paper uses terminology you may not be familiar with. Most important is the concept of a *maximal clique*. Given a graph with vertices V and edges E, a *clique* is a subset $C \subseteq V$ such that $\{c, c'\} \in E$ for every pair $c, c' \in C$. If there is no clique containing C as a proper subset, then C is a *maximal clique*.

References

1. J. S. Armstrong, Derivation of theory by means of factor analysis or Tom Swift and his electric factor machine, *American Statistician* **21** (1967) 17–21. Reprinted in R. L. Day and L. J. Parsons (eds.), *Marketing Models. Quantitative Applications*, Intext Educational Publishers, Scranton, PA (1971), 413–421.

2. S.-T. Bow, *Pattern Recognition and Image Preprocessing*, Marcel Dekker, New York (1992).

3. B. G. Buchanan and D. C. Wilkins (eds.) *Readings in Knowledge Acquisition and Learning*, Morgan Kaufmann, San Mateo, CA (1993).

4. H. Bunke, Structural and Syntactic Pattern Recognition. Chapter 1.5 in [6], pp. 163–209.

5. W. L. Buntine, Operations for learning with graphical models, *Journal of Artificial Intelligence Research* **2** (1994) 159–225. *JAIR* is available electronically. See the newsgroup `comp.ai.jair.announce` for more information.

6. C. H. Chen, L. F. Pau, and P. S. P. Wang (eds.), *Handbook of Pattern Recognition and Computer Vision*, World Scientific, Singapore (1993).

7. P. A. Devijver and J. Kittler, *Pattern Recognition: A Statistical Approach*, Prentice-Hall International, London (1982).

8. R. O. Duda and P. E. Hart, *Pattern Classification and Scene Analysis*, John Wiley and Sons, New York (1973).

9. R. O. Duda, P. E. Hart, and D. G. Stork, *Pattern Classification and Scene Analysis*, John Wiley and Sons, New York (planned for early 1996).

10. K. S. Fu, *Syntactic Pattern Recognition and Applications*, Prentice Hall, Englewood Cliffs, NJ (1982).

11. K. Fukunaga, *Introduction to Statistical Pattern Recognition*, 2d ed., Academic Press, Boston (1990).

12. W. A. Gale (ed.), *Artificial Intelligence and Statistics*, Addison-Wesley, Reading, MA (1986).

13. A. K. Jain, Advances in statistical pattern recognition. In P. A. Devijver and J. Kittler (eds.), *Pattern Recognition Theory and Applications*, Springer-Verlag, Berlin (1987) 1–19.

14. M. James, *Classification Algorithms*, Collins, London (1985).

15. I. T. Jolliffe, *Principal Component Analysis*, Springer-Verlag, Berlin (1986).

16. J. W. Kim, H. R. Weistroffer, and R. T. Redmond, Expert systems for bond rating: A comparative analysis of statistical, rule-based and neural network systems, *Expert Systems* **10** (1993) 167–171.

17. L. Miclet, *Structural Methods in Pattern Recognition*, North Oxford Academic, London (1986).

18. Y.-H. Pao, *Adaptive Pattern Recognition and Neural Networks*, Addison-Wesley, Reading, MA (1989).

19. M. Pavel, *Fundamentals of Pattern Recognition*, 2d ed., Marcel Dekker, New York (1993).

20. T. Pavlidis, *Structural Pattern Recognition*, Springer-Verlag, Berlin (1977).

21. B. D. Ripley, Flexible non-linear approaches to classification. In V. Cherkassky et al. (eds.) *From Statistics to Neural Networks: Theory and Pattern Recognition Applications*, Springer-Verlag, Berlin (to appear).

22. R. J. Schalkoff, *Pattern Recognition: Statistical, Structural and Neural Approaches*, John Wiley and Sons, New York (1992).

23. S. M. Weiss and C. A. Kulikowski, *Computer Systems That Learn. Classification and Prediction Methods from Statistics, Neural Nets, Machine Learning and Expert Systems*, Morgan Kaufmann, San Mateo, CA (1991).

11

Neural Networks
and
Minimization

The real hope of von Neumann and his coworkers, the story goes,
was to construct parallel processors capable of manipulating large
blocks of information at the same time. ...
It is an exquisite irony that parallel processors and neural
architectures, which he is said to have pioneered, are continually
compared to the "von Neumann approach" to computing.

—Maureen Caudill and Charles Butler (1990)

Introduction

You've probably heard "the brain is a computer" so often that you no longer
think about it. AI attempts to emulate some of the abilities of the human
brain using ordinary computers. Some researchers believe that this is the
wrong approach because brains have some desirable features that computers
lack. These include:

- "associative" memory,
- an architecture that apparently facilitates learning,
- a self-programming ability,
- an ability to obtain results after relatively few machine cycles,
- an ability to compensate for processor and storage degradation.

People researching neural nets hope to develop artificial systems with such
capabilities by emulating, to some extent, the structural features of brains.
These structural features include

- massive parallelism (much more than we can currently achieve) of

- highly connected, simple, imprecise processing units with

- information stored in self-adjusting interconnection weights.

The progress that has been made on connectionist systems (neural nets) is fairly modest when compared with the capabilities of even fairly simple brains. Researchers disagree on the ultimate usefulness of neural nets. At one extreme are those who believe neural nets will eventually supplant other AI techniques. At the other, there are those who believe the entire approach is doomed.

It can be difficult to sort through the wide variety of neural nets that are currently being studied. They can be roughly classified according to the following properties:

- **Computing elements:** This refers to the functions f_v in Definition 10.2 (p. 408). There are two main types of allowed outputs—either all real numbers or just two values, usually denoted by either 0-1 or ± 1. In the real-valued case, a variety of continuous functions have been proposed (see p. 478). Examples of nets with two-valued outputs appear in Sections 11.1 and 11.3.

- **Network architecture:** A major architectural feature in supervised training is the presence or absence of *feedback*. Nets with feedback are called *recurrent* and those without are called *nonrecurrent*. A feedback loop is simply a directed cycle in the digraph of the network. Thus, a net is nonrecurrent if and only if its digraph is acyclic.

  Another aspect of architecture is whether the digraph is predetermined or is developed as part of the training process.

- **Training:** There are many algorithms for training nets. The most important division is between supervised and unsupervised learning. A supervised net is given the responses it should produce while an unsupervised net should find "interesting" features in the data.

The previous discussion provides eight major categories for classifying nets,

discrete vs. continuous
recurrent vs. nonrecurrent
supervised vs. unsupervised,

and additional refinements of them. Here are some other ways of classifying nets.

- **Purpose:** One purpose is *associative memory*: When given partial information about a stored pattern, the net retrieves the complete pattern. Another purpose might be termed *information processing*. Actually there is less to this difference than meets the eye. In both cases, the net is performing a pattern classification. The distinction is primarily in how

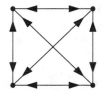 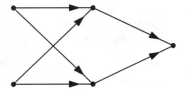

Figure 11.1 In the small Hopfield net on the left, the weights on a given edge are the same in both directions. In the small feedforward net shown on the right, all paths lead rightward.

---

the net output describes the classification (a complicated pattern for associative memory versus probably one or two numbers for information processing) and how we think about the result.

- **Time:** Except for the timing of internal computations, time is irrelevant for some neural nets. An example is an expert system that suggests promising pathways for chemical synthesis. For many recurrent nets, timing of inputs is important. For example, an expert system for converting from spoken to written language would rely heavily on input from the recent past.

Any significant study of the various types of nets would require at least an entire book, so let's limit our attention to two types of nets—Hopfield and feedforward. Both of them use supervised learning; however, they differ from each other in architecture, in computing elements, in algorithms, and in relevant mathematics.

In a Hopfield net, if $(i, j) \in E$, then (j, i) is also an edge and $w_{i,j} = w_{j,i}$. Frequently, E consists of all (i, j) with $i \neq j$. See Figure 11.1. Given (V, E) and a collection of patterns, the values of $w_{i,j}$ can be computed quickly. Given a new pattern to be analyzed, the determination of the outputs for that pattern is time-consuming. The mathematical formulation is akin to that for energy functions in statistical mechanics—we attempt to minimize an "energy."

Feedforward nets are nonrecurrent; that is, every way of moving through the graph along directed edges, (i, j) to (j, k) to (k, l) and so on, eventually leads to some vertex m such that there is no edge of the form (m, x); that is, m has no edges directed "out." In such a net, these vertices form the set O, and the set I consists of vertices v with no edges "in." For $v \notin I$, o_v is initialized to "unknown." Then o_v is set to the value of (10.1) (p. 409) when o_x is known for all x for which $(x, v) \in E$. Hence, unlike the Hopfield nets, computation of outputs is straightforward. On the other hand, given (V, E) and a collection of patterns, the determination of the weights is a difficult minimization problem. The goal is usually to minimize some measure of the error in the net's attempt to map input patterns to desired output patterns.

Since Definition 10.2 (p. 408) allows such a wide variety of neural nets, how do researchers select designs? There are several important points to keep in mind. You should refer back to them when reading material on neural nets.

- Any neural net design involves algorithms.
    - The algorithms must work. (This is not facetious—some neural net algorithms behave poorly.)
    - The amount of calculation required by the algorithms must be reasonable.
- Since a major goal is to implement neural nets with parallel processors, perhaps using very large-scale integration (VLSI), we have the following conditions.
    - The algorithms should involve simple calculations that can be performed in a distributed manner at the vertices and perhaps the edges.
    - The structure of the digraph (V, E) of the neural net must be feasible from a hardware design viewpoint.
- Since neural nets should be competitive with other methods, we have the following conditions.
    - The network should be flexible enough to handle a variety of problems. (This is seldom a problem.)
    - The network size and running time needed to handle a problem should be reasonable. (This is often a problem.)

We'll give the most emphasis to the first category—algorithms—and the least to the last—usage.

Algorithms for neural nets involve finding the minimum of some function. As already mentioned, for Hopfield nets, training is straightforward and the minima must be found when using a trained net. In contrast, for feedforward nets, the minima must be found during the training and usage is straightforward. Unless we are lucky, the problem of minimizing a function can be quite difficult. Here's an outline of the chapter and an indication of the role minimization plays. Since the chapter is so long, you may want to refer back to this outline from time to time.

- We'll begin with an introduction to Hopfield nets since the minimization there is fairly simple.
- In Section 11.2 we'll review some mathematical tools—matrices, linear algebra, and multivariate calculus—most of which should be familiar to you. If it's not review, you may want to do some supplementary reading in introductory texts on linear algebra (or matrix theory) and multivariate calculus (found in most standard calculus texts).
- Before pursuing more complicated aspects of minimization, we'll briefly examine a classic example of feedforward nets in Section 11.3—perceptrons

with no hidden units. As with Hopfield nets, there's a straightforward minimization algorithm.

- Next, we'll tackle the thorny topic of minimization itself in Section 11.4. The treatment differs markedly from that found in numerical analysis texts because it's oriented toward minimization for feedforward nets.

- Section 11.5 on feedforward nets is brief because minimization techniques were already discussed.

- In Sections 11.6 and 7, we'll explore some important issues in algorithm design. (I've marked those two sections with asterisks because they are important in actual design work but not in conceptual understanding.)

- The chapter concludes with a brief history.

Some aspects of neural nets, especially those requiring additional background in probability and statistics, are discussed in Chapters 12 and 13.

Prerequisites: You should have read the preceding chapter which provides some background for the present chapter. In a couple of places, I refer to basic probability theory concepts, but these can be skipped.

Used in: Chapter 12 relies only slightly on the material here, so you can read that chapter without first reading this one. The neural net discussions in Chapter 13 refer back to some concepts in this chapter. Since I'll provide specific references in Chapters 12 and 13, you needn't read this entire chapter beforehand.

Exercises

11.A. What is associative memory?

11.B. What is the difference between a recurrent and a nonrecurrent net?

11.1 Hopfield Networks

Suppose that we have a questionnaire with a variety questions whose answers we believe tend to be interrelated. This is the basis of such things as personality tests, career aptitude tests, and so forth: Some answers are expected to be able to predict what other answers (personality types, aptitudes, etc.) are likely to be. Of course, some answers are simple to obtain—you just ask the person—while others are much more difficult, requiring such methods as in-depth psychological interviews. The hope is that the pattern of easily obtained answers can be used to predict the other answers. Various statistical methods can be used to help locate such patterns. Hopfield nets provide another approach.

More generally, we can think of the answers a subject gives to easy questions together with the (unknown) answers to the remaining questions as a pattern. All we have is a degraded pattern—the answers to the easy questions. We want to recover the entire pattern.

Suppose the questions are all of the true/false type. The idea behind a Hopfield net is to assign one question to each vertex V of the digraph. Thinking in terms of bit patterns, we let each vertex correspond to one bit in the pattern. The edges are $E = V \times V$ and an edge (u, v) measures how alike we expect the answers to questions u and v to be.

Definition 11.1   Hopfield Net

A *Hopfield net* is a particular type of neural net. See Definition 10.2 (p. 408) for neural net notation. Let $V = \{1, 2, \ldots, n\}$ and let E consist of all (i, j) with $i \neq j$ and $i, j \in V$. Frequently, $I = O = V$, but this need not be the case. Let $\vec{o}$, $\vec{P}(1)$, and so on denote vectors. Define

$$f_v(\vec{o}) = \text{sign}\left(\theta_v + \sum_x w_{x,v} o_x\right) \quad \text{and} \quad \text{sign}(a) = \begin{cases} +1, & \text{if } a > 0, \\ -1, & \text{if } a < 0, \\ \pm 1, & \text{if } a = 0, \end{cases} \quad (11.1)$$

where the sign is either chosen arbitrarily at 0 or is chosen to equal the sign of the present value of o_v.

A *pattern* is an n-long sequence of $+1$'s and -1's. Suppose we're given a collection of patterns $\vec{P}(1), \vec{P}(2), \ldots$. Define

$$w_{i,i} = 0 \quad \text{and} \quad w_{i,j} = \sum_k P_i(k)P_j(k) \text{ for } i \neq j. \quad (11.2)$$

The formula for $w_{i,j}$ is often called *Hebb's rule*. When we have computed $\vec{w}$, we say that the patterns have been stored in the Hopfield net.

What does this definition say?

- It defines the vertices to correspond to the bits in the pattern.

- The presence of θ_v in the definition of f_v tends to bias f_v toward $\text{sign}(\theta_v)$. The larger $|\theta_v|$, the larger the bias.

- The definition of $w_{i,j}$ measures how much o_i and o_j tend to agree in the stored patterns. If they always agree, all terms in the sum for $w_{i,j}$ are $+1$ and so the sum is s, the number of patterns. If they always disagree, all terms in the sum for $w_{i,j}$ are -1 and so the sum is $-s$. If the agreements and disagreements are "random," about half the terms are $+1$ and half are -1, and so the absolute value of the sum is much smaller than s.

- The sign of $o_x w_{x,v}$ is a prediction of the sign of o_v given o_x, and $|w_{x,v}|$ is some measure of the confidence of the prediction.

$$* \qquad * \qquad * \qquad \text{Stop and think about this!} \qquad * \qquad * \qquad *$$

Consequently, f_v can be thought of as a prediction of o_v based on the observation of o_x for all vertices $x \in V$. In other words,

> An input pattern $\vec{o}$ is likely to be a solution to the set of $|V|$ linear equations $o_v = f_v(\vec{o})$ for all $v \in V$. $\qquad$ (11.3)

This is the key to recovering stored patterns—start at some $\vec{o}$ and change it by replacing o_v with $f_v(\vec{o})$. Of course, we may need to repeat the changes many times.

The previous discussion leads to some natural questions about Hopfield nets and algorithms for recovering stored patterns. Here are some together with their answers.

(a) Question: Does this procedure stop or can it get stuck going in circles?
    Answer: Yes to both, depending on the algorithm.

(b) Question: If it stops, is (11.3) right; that is, does it do so at a stored pattern?
    Answer: If there aren't too many stored patterns, almost certainly yes.

(c) Question: As a function of $|V|$, how many patterns can we expect to be able to store and recover in this way?
    Answer: Approximately $Cn/\log n$.

Question (a) will be dealt with in this section. The others require additional material on probability theory, which is introduced in the next chapter.

Assume the results in the preceding paragraph. Also assume an algorithm for moving from a starting point $\vec{o}$ to a local minimum of $\Phi(\vec{o})$ that is close to the start. Then we can use a Hopfield net to solve our problems:

- **Degraded Patterns:** Store the given patterns in the net. When a degraded pattern is received, use it as a starting point for $\vec{o}$, set $\vec{i} = \vec{0}$, and find a local minimum. This should be the original pattern corresponding to the degraded image.

- **Questionnaire**: Again, store the given complete patterns in the net. When partial answers are received, use them as starting points $\vec{o}$ for the algorithm. Those components of $\vec{o}$ for which no answers were received could be started randomly at ± 1 or they could be set to 0. Set $\theta_v = 0$ if o_v is unknown. If o_v is known, set θ_v to some multiple of it, thus biasing new values of o_v toward the original value. (Show that if the multiple exceeds $\sum_x |w_{x,v}|$, then o_v will never change.) When the algorithm stops, read off the answers it gives for the missing responses.

The concept of the *energy* of $\vec{o}$ is quite useful in studying Hopfield nets. It's called "energy" because of the close analogy with the energy of what physicists call "spin systems." Researchers have exploited this by converting theorems concerning spin systems to theorems concerning Hopfield nets. That's beyond the level of sophistication of this text, but the energy concept is still helpful.

Definition 11.2   Energy of a Hopfield Net

The energy $\Phi(\vec{o})$ of a particular vector $\vec{o}$ in a Hopfield net with biases θ_v and weights $w_{x,v}$ is

$$\Phi(\vec{o}) = -\sum_{v \in V} \theta_v o_v - \tfrac{1}{2} \sum_{x,v \in V} w_{x,v} o_x o_v. \tag{11.4}$$

We'll see that the local minima of $\Phi(\vec{o})$ are solutions of $o_v = f_v(\vec{o})$. You should be familiar with the concept of a local minimum from calculus. Unfortunately, calculus deals with functions whose arguments are real numbers and the domain of Φ is the set $\{-1, +1\}^n$ of 2^n points, not the n-dimensional interval $[-1, +1]^n$. Thus, our first task is to define the notion of a local minimum.

Definition 11.3   Neighbors and Local Minima

We say that $\vec{o}$ and $\vec{o}'$ from $D = \{-1, +1\}^n$ are *neighbors* if they differ in exactly one component and if that component is allowed to vary. (In the questionnaire example, those components corresponding to known responses are not allowed to vary.)

Let g be a function from D to the real numbers $\mathbb{R}$. We call $\vec{o}$ a *local minimum* of g if $g(\vec{o}) \leq g(\vec{o}')$ for all neighbors $\vec{o}'$ of $\vec{o}$.

The following two theorems show the close relationship between local minima of Φ, the functions f_v, and stored patterns. The first is stated rather vaguely since stating and proving it require concepts and tools we haven't yet discussed. A crude heuristic proof is given in Example 11.1 (p. 436). The theorem is stated more precisely in the discussion beginning on page 568, and a better heuristic proof is given there. In contrast, the second theorem is easily proved.

Theorem 11.1   Minimum Energy at Stored Patterns

Suppose that $\theta_v = 0$ for all v and that $w_{i,j}$ is given by (11.2) (p. 430). Under suitable assumptions, the values $\vec{o} = \vec{P}(k)$ and $\vec{o} = -\vec{P}(k)$ for $k = 1, 2, \ldots$ will almost certainly be local minima of $\Phi(\vec{o})$ and Algorithm 11.1 below will converge to such a state if it starts nearby.

Theorem 11.2   Minimum Energy at Fixed Points of f_v

The vector $\vec{o}$ is a local minimum of $\Phi(\vec{o})$ if and only if $o_v = f_v(\vec{o})$ for all $v \in V$, where the value of sign(0) is the value of o_v. (This does not assume $\theta_v = 0$.)

Proof (of Theorem 11.2):   Let $\vec{o}\,'$ equal $\vec{o}$ in all components except the vth. With a little calculation,

$$\Phi(\vec{o}\,') - \Phi(\vec{o}) = -\theta_v(o_v' - o_v) - \tfrac{1}{2}\sum_{\substack{i,j:\\ i\neq j}} w_{i,j}(o_i' o_j' - o_i o_j), \qquad \text{since } w_{i,i} = 0,$$

$$= -\theta_v(o_v' - o_v) - \tfrac{1}{2}\sum_{i:\, i\neq v} w_{i,v}(o_i o_v' - o_i o_v)$$

$$\qquad\qquad - \tfrac{1}{2}\sum_{j:\, j\neq v} w_{v,j}(o_v' o_j - o_v o_j),$$

$$\qquad\qquad \text{since } o_i' o_j' = o_i o_j \text{ unless } j = v \text{ or } i = v,$$

$$= -(o_v' - o_v)\left(\theta_v + \sum_{i:\, i\neq v} w_{i,v} o_i + \sum_{j:\, j\neq v} w_{v,j} o_j\right)$$

$$= -(o_v' - o_v)\left(\theta_v + \sum_{i:\, i\neq v} w_{i,v} o_i\right), \qquad \text{since } w_{v,i} = w_{i,v},$$

$$= 2o_v\left(\theta_v + \sum_{i:\, i\neq v} w_{i,v} o_i\right), \qquad \text{since } o_v' = -o_v.$$

Since $w_{i,i} = 0$, we've just shown that

$$\Phi(\vec{o}\,') - \Phi(\vec{o}) = 2o_v\left(\theta_v + \sum_x w_{x,v} o_x\right). \qquad (11.5)$$

It follows from the definition of f_v in (11.1) that the right side of (11.5) will be nonnegative if and only if $o_v f_v(\vec{o})$ is nonnegative. This happens if and only if $o_v = f_v(\vec{o})$ since they are both ± 1. ∎

The criterion $o_v = f_v(\vec{o})$ for a local minimum given in Theorem 11.2 suggests an algorithm for finding such minima, while Theorem 11.1 tells us that these local minima probably correspond to input patterns $\vec{P}(k)$. Here's the algorithm.

Algorithm 11.1   Hopfield's Algorithm

Given a Hopfield net as described in the text, the following algorithm converges to a local minimum:

- **Step 1**: Initialize $\vec{o}$ based on the given values.

- **Step 2**: For each $v \in V$, replace o_v by $f_v(\vec{o})$ as given by (11.1), choosing sign(0) to be the old value of o_v. These computations are to be done sequentially, not in parallel. That is, when some o_i has been replaced, its new value is used in all future computations of f_v's.

- **Step 3**: If no components of $\vec{o}$ were changed in Step 2, stop; otherwise, repeat Step 2.

Note that the algorithm can be implemented in a local fashion: Each vertex o_v simply looks at the weights on the edges (x, v) to which it belongs and the values of o_x at at the other ends of the edges.

It's fairly easy to show that the algorithm stops after a finite number of steps—see Exercise 11.1.1. When it stops, Theorem 11.2 guarantees that it has reached a local minimum. In fact, more can be shown: The function $\Phi(\vec{x})$ is a nonincreasing function of time. More generally, suppose that $\vec{x}$ is any collection of variables that change with time according to some algorithm. If $\Phi(\vec{x})$ is a nonincreasing function of time, it is called a *Lyapunov function* for the algorithm.

Aside. Lyapunov functions are important in the study of differential equations. A differential equation for $\vec{x}(t)$ is an algorithm for changing $\vec{x}$. If we can find a Lyapunov function which is bounded below, we can often use it to prove that the solution converges. As a simple example, consider a pendulum whose motion is governed by $\theta' = \omega$ and $\omega' = -a\sin\theta + F(\omega)$ where F is a frictional force whose sign is opposite that of the angular velocity ω. Let $\Phi(\theta, \omega) = \omega^2 - 2a\cos\theta$. Then

$$d\Phi/dt = 2\omega\omega' + 2a(\sin\theta)\theta' = 2\omega(\omega' + a\sin\theta) = 2\omega F(\omega),$$

which is nonpositive by assumption. This can be used to show that the pendulum gradually slows toward a stopped position.

Let's stop and review what a Hopfield net is and how it works. This is just a rough description to help you organize your thoughts on the important points.

- The values at a vertex must be ± 1.

- Every vertex receives input from every other vertex and the weight for an edge (i, j) reflects agreement minus disagreement of vertices i and j in the training patterns—the patterns "stored" in the net. As a result, $w_{i,j} = w_{j,i}$.

- The goal of classification is to recover a stored pattern that is close to the input pattern.

- Since the weights are simply sums, training is very fast. In contrast, pattern recovery is more complicated and time-consuming.

- Pattern recovery can be thought of as choosing each o_v so that its sign agrees with the sign predicted with the weighted sum of the outputs of the other vertices. It can also be thought of as minimizing the energy function Φ.

Many variations on the Hopfield algorithm are possible. Here are a few of them.

- Since the algorithm can be implemented in a local fashion, we could have the vertices simultaneously compute new values in Step 2. While this usually gives more rapid results than the serial computations in the algorithm, it opens the door to the possiblity that the algorithm will not converge—see Exercise 11.1.2.

- Instead of the synchronous (or "lock-step") parallelism we just proposed, we can use a random approach: Imagine each vertex updating its o_l using (11.1), but selecting the times for doing so in a random manner. It can be shown that such a procedure converges *almost surely*—a technical term in probability theory which means that convergence is practically guaranteed. This illustrates an important point:

> If an algorithm can exhibit some sort of undesirable behavior, it may be helpful to introduce some randomness into the algorithm.   (11.6)

The ramifications of this idea extend far beyond computer algorithms. It is "known" to evolution which has used it often in "designing" organs; for example, heart muscles are stimulated in a slightly irregular manner and this helps prevent fibrillation.

- The order in which the bits of o are adjusted to ± 1 could have an important effect on which local minimum is found. One way to overcome this problem is with multiple starts with different (perhaps random) orders for replacing the components in Step 2.

- We can introduce a continuous version by allowing o_v to be in the interval $[-1, +1]$ rather than just ± 1. See Exercise 11.1.3.

Let's conclude with a rough explanation of why we can expect local minima at $\pm \vec{P}(k)$ from Hopfield nets with $\vec{\theta} = \vec{0}$, as claimed in Theorem 11.1 (p. 433). The calculations are used in the discussion starting on page 568, which improves on the conclusions reached in this example.

Example 11.1   Why We Expect Local Minima at $\pm \vec{P}(k)$

In this example, we'll assume that $\vec{\theta} = \vec{0}$ in (11.1) (p. 430) and (11.4) (p. 432). You should be able to easily show that $\Phi(-\vec{o}) = \Phi(\vec{o})$ for any $\vec{o}$. Thus it suffices to consider $\vec{P}(k)$. The idea is to use Theorem 11.2 with $\vec{o} = \vec{P}(k)$ and replace the weights with their definitions (11.2) (p. 430). Let $\vec{P}'(k)$ equal $\vec{P}(k)$ except in the vth component. By (11.5) (p. 433)

$$\Phi(\vec{P}'(k)) - \Phi(\vec{P}(k)) = 2P_v(k) \sum_x w_{x,v} P_x(k), \qquad \text{by (11.5) (p. 433)},$$

$$= 2P_v(k) \sum_{x:\, x \neq v} \left(\sum_t P_x(t) P_v(t) \right) P_x(k)$$

$$= 2P_v(k) \sum_{x:\, x \neq v} \left(P_x(k) P_v(k) + \sum_{t:\, t \neq k} P_x(t) P_v(t) \right) P_x(k)$$

$$= 2 \sum_{x:\, x \neq v} \left(P_x(k) P_v(k) \right)^2 + 2 \sum_{\substack{x:\, x \neq v \\ t:\, t \neq k}} P_v(k) P_x(t) P_v(t) P_x(k).$$

Letting $n = |V|$, the number of vertices in the Hopfield net, and recalling that the components of the $\vec{P}$'s are ± 1, this becomes

$$\Phi(\vec{P}'(k)) - \Phi(\vec{P}(k)) = 2(n-1) + 2 \sum_{\substack{x:\, x \neq v \\ t:\, t \neq k}} P_v(k) P_x(t) P_v(t) P_x(k). \qquad (11.7)$$

Suppose that the $\vec{P}(l)$'s are, in some sense, random. In that case, the term $P_v(k) P_x(t) P_v(t) P_x(k)$ will be randomly ± 1. Since the sum in (11.7) involves terms that are randomly ± 1, there should be a considerable amount of cancellation in that sum. Thus the sum will be much smaller in magnitude than the total number of terms, which is $(n-1)(s-1)$ when s is the number of patterns. How much smaller will be discussed in the next chapter in connection with the Central Limit Theorem. For now, we can simply conclude that if the number of patterns s is not too large—how large depends on n—then the sum in (11.7) should be smaller than $(n-1)$ in magnitude. If this holds for all v, then $\vec{P}(k)$ is a local minimum of Φ. ∎

Exercises

11.1.A. What does classification mean for a Hopfield net?

11.1.B. How is a Hopfield net trained?

11.1.C. What do we do when recovering a pattern from a Hopfield net? (Not an algorithm—just the broad view.)

11.1.D. Give an example of how randomness may be useful in designing an algorithm.

11.1.1. The purpose of this exercise is to prove the convergence of Algorithm 11.1. To aid in the proof, I'll define a number Δ. Let

$$\Delta(\vec{\epsilon}) = \min \left| \sum_x w_{x,v} \epsilon_x \right|,$$

where the minimum is over all v such that the sum is not zero. (If there are no sums satisfying this condition, $\Delta(\vec{\epsilon}) = +\infty$.) Let Δ be the minimum of $\Delta(\vec{\epsilon})$ over all choices of ± 1 for the components of $\vec{\epsilon}$.

(a) Prove that any change made in Step 2 after the first pass causes $\Phi(\vec{o})$ to decrease by at least 2Δ.

(b) Let $M = \frac{1}{2} \sum_{i,j} |w_{i,j}|$. Prove that $M \geq \Phi(\vec{o}) \geq -M$.

(c) Prove that the algorithm terminates in at most $2 + M/\Delta$ iterations of Step 2.
Hint. To obtain the 2, remember that the last pass and, possibly, the first pass of Step 2 are different from the others.

11.1.2. Alter Algorithm 11.1 so that the changes in Step 2 are all computed simultaneously. Suppose that $m = 2$, $n = 4$, $o_3 = +1$, $o_4 = -1$, and $w_{i,j} = +1$ for all i, j. Show that it is possible to have $o_1 = +1$ and $o_2 = -1$ after the first execution of Step 2. Show that at each succeeding application, both o_1 and o_2 change.

11.1.3. In this exercise, you'll briefly look at a continuous version of the Hopfield net and a differential equation algorithm for it. We could imagine that the differential equation might be implemented by some sort of analog circuitry. Let $w_{i,j}$ and Φ be given by (11.2) and (11.4). Suppose that $o_i(0)$ is given for all i.

(a) Let $o_i'(t) = \sum_{j \neq i} w_{i,j} o_j(t)$ determine the values of the o_i's. Prove that $d\Phi(\vec{o})/dt \leq 0$ with equality if and only if $o_i'(t) = 0$ for all t.

(b) Unfortunately, the above algorithm may lead to very large values of o_i. To keep them in the interval $[-1, +1]$, let $o_i = \tanh x_i$ and determine the values of the x_i's by the initial values of the o_i's at $t = 0$ and then by $x_i'(t) = \sum_{j \neq i} w_{i,j} o_j(t)$ for $t > 0$. Again, prove that $d\Phi(\vec{o})/dt \leq 0$ with equality if and only if $o_i'(t) = 0$ for all t.

11.1.4. A *heteroassociative memory* is one in which each pattern $\vec{P}(k)$ is a pair $\vec{x}(k), \vec{y}(k)$ such that when we input $\vec{x}(k)$ the memory recovers $\vec{y}(k)$.

(a) Why can this be done using a Hopfield net in which V is partitioned into I and O where I is associated with $\vec{x}$, O is associatied with $\vec{y}$, and the outputs of $v \in I$ are held fixed to the input $\vec{x}$?

(b) Suppose we modify the preceding net by setting $w_{u,v} = 0$ whenever $u, v \in O$. Show that Algorithm 11.1 converges after one iteration of Step 2. Why is this a bad way to implement a heteroassociative memory?

11.2   Some Mathematics: Mostly a Review

We'll begin with a review of the basic properties of vectors and matrices. Except for the fact that indices need not be $\{1, 2, \ldots, n\}$, this is standard, calculus-level material. Next we'll discuss vector spaces, including some material that is probably new. Since that material will not be used extensively, the presentation is brief and the proofs sketchy. Your instructor may decide to provide more details. Finally we'll discuss some basic tools from multivariate calculus. Again, this should be review with the possible exception of the concept of the Hessian and results about approximations.

Vectors and Matrices

The classical vector is a 3-tuple of real numbers that gives the position of a point in space. It is often denoted (x_1, x_2, x_3). There's nothing special about the number three and the set $\{1, 2, 3\}$ of indices—unless you insist on a simple physical interpretation, which is not relevant here.

Definition 11.4   Vector Notation

Let $\mathbb{R}$ be the real numbers. A *vector* indexed by the set I is a function $\vec{x} : I \to \mathbb{R}$. The value of $\vec{x}$ at $i \in I$ is denoted by x_i (instead of by $\vec{x}(i)$) and is called the *i*th component of $\vec{x}$. The set of all such vectors is denoted by $\mathbb{R}^I$. When $I = \{1, 2, \ldots, n\}$, the set is denoted by $\mathbb{R}^n$ rather than the cumbersome $\mathbb{R}^{\{1,2,\ldots,n\}}$. In all our work, the set I will be finite. The *dot product* of two vectors $\vec{x}, \vec{y} \in \mathbb{R}^I$ is given by

$$\vec{x} \cdot \vec{y} = \sum_{i \in I} x_i y_i.$$

The *length* of $\vec{x}$ is given by $|\vec{x}| = \sqrt{\vec{x} \cdot \vec{x}}$.

We're using a more general notation for the index set of a vector so that we can use such things as $\mathcal{X}$ and the edges of a graph as index sets.

You've probably encountered matrices in connection with linear algebra where they arise from linear transformations of vector spaces. Our use of matrices is much less sophisticated—they simply serve as a convenient device for manipulating numerical data. In particular, we'll be multiplying and adding matrices and vectors.

Definition 11.5   Matrix Terminology

A matrix with index set $I \times J$ is a function $A : (I \times J) \to \mathbb{R}$. The value at (i,j) is denoted by $A_{i,j}$, $(A)_{i,j}$, or, sometimes, $a_{i,j}$. We call $A_{i,j}$ the (i,j)th entry of A and say that it is in row i and column j. We refer to A as an $I \times J$ matrix, or simply an $m \times n$ matrix if $I = \{1, \ldots, m\}$ and $J = \{1, \ldots, n\}$.

Let $|I|$ be the number of elements in I. When $|I| = 1$ we may think of A as a vector and call it a *row vector*. Similarly, if $|J| = 1$, we speak of a *column vector*. If $|I| = |J| = 1$, we may think of A as a real number.

The *transpose* of a matrix $A : (I \times J) \to \mathbb{R}$ is the matrix $A^{\mathrm{t}} : (J \times I) \to \mathbb{R}$ given by $A^{\mathrm{t}}_{j,i} = A_{i,j}$.

The identification of a matrix having $|I| = |J| = 1$ with a real number could sometimes lead to problems. That will never be the case here, so

We'll always interpret a matrix with $|I| = |J| = 1$ as a real number.

Finally, we come to the arithmetic operations on matrices. Since vectors are a special case of matrices, these definitions apply to them as well.

Definition 11.6   Arithmetic of Matrices

Let A and B be matrices with index sets $I \times J$ and $K \times L$, respectively. Let c be a real number. The matrix cA has index set $I \times J$ and is defined by $(cA)_{i,j} = c(A_{i,j})$. The matrix sum $A + B$ is defined if and only if $I = K$ and $J = L$. In this case, $(A + B)_{i,j} = A_{i,j} + B_{i,j}$. The matrix product AB is defined if and only if $J = K$. In this case, $AB : (I \times L) \to \mathbb{R}$ and

$$(AB)_{i,l} = \sum_{j \in J} a_{i,j} b_{j,l}.$$

When a vector appears in matrix calculations, it should be thought of as a column vector. In particular, if $\vec{x}, \vec{y} \in \mathbb{R}^I$, then $\vec{x} \cdot \vec{y} = \vec{x}^{\mathrm{t}} \vec{y}$.

The essential feature of the arithmetic operations on matrices is that they behave in the same way that arithmetic operations on real numbers behave, except that $AB \neq BA$ and division hasn't been defined. Theorem 11.3 lists these properties, as well as some results about transposes of matrices and lengths of vectors.

Theorem 11.3   Properties of Matrices and Vectors

Let A be a matrix with index set $I \times J$; let B, C, and D be matrices with index sets $J \times K$; let E be a matrix with index set $K \times L$; and let r and s be real numbers. We have associative and distributive laws and *some* commutative laws:

(a) $B + (C + D) = (B + C) + D$, $(rs)A = r(sA)$, and $A(BE) = (AB)E$;

(b) $r(B+C) = rB+rC$, $A(B+C) = AB+AC$, and $(B+C)E = BE+CE$;

(c) $B + C = C + B$ and $A(rB) = (rA)B = r(AB)$.

Transposition "passes through" some operations:

(d) $(B+C)^t = B^t + C^t$, $(rA)^t = r(A^t)$, and $(AB)^t = (B^t)(A^t)$, reversing the order of multiplication.

There are matrices that behave like 1 in multiplication:

(e) Let I_n be the $n \times n$ matrix with $(I_n)_{i,j}$ equal to 1 if $i = j$ and equal to 0 otherwise. If A and B are matrices such that AI_n and $I_n B$ are defined, then $AI_n = A$ and $I_n B = B$. We call I_n the $n \times n$ *identity matrix*.

Some properties of the lengths of vectors $\vec{x}, \vec{y} \in \mathbb{R}^I$ are

(f) $|r \cdot \vec{x}| = |r| \, |\vec{x}|$, where $|r|$ is the absolute value of r;

(g) $\vec{x} \cdot \vec{y} \le |\vec{x}| \, |\vec{y}|$, with equality if and only if $\vec{x} = \vec{0}$ or $\vec{y} = p\vec{x}$ for some real number $p \ge 0$;

(h) (the *triangle inequality*) $|\vec{x} \pm \vec{y}| \le |\vec{x}| + |\vec{y}|$ or, an equivalent form which is sometimes used, $|\vec{x} \pm \vec{y}| \ge |\vec{x}| - |\vec{y}|$.

Since $\vec{x} \cdot \vec{y} = \vec{x}^t \vec{y}$, properties of matrix products hold for dot products, too. Let A be $n \times n$. If there is a matrix B such that $AB = BA = I_n$, we call B the inverse of A and write $B = A^{-1}$. This is the analog of the inverse (reciprocal) of a real number and so provides a means for defining division. (I won't pursue this or discuss how to compute A^{-1} from A.)

Proof (of (g) and the triangle inequality):   The rest of the parts and some of the details in this proof are left as an exercise. Squaring both sides of the desired inequality in (g) and rearranging, we see that it would be implied by $0 \leq |\vec{x}|^2|\vec{y}|^2 - (\vec{x} \cdot \vec{y})^2$. By simple algebra,

$$|\vec{x}|^2|\vec{y}|^2 - (\vec{x} \cdot \vec{y})^2 = \sum_{i \in I}\sum_{j \in I} x_i^2 y_j^2 - \sum_{i \in I}\sum_{i \in I} x_i y_i x_j y_j$$

$$= \tfrac{1}{2}\sum_{i \in I}\sum_{j \in I}\left((x_i y_j)^2 + (x_j y_i)^2 - 2(x_i y_j)(x_j y_i)\right)$$

$$= \tfrac{1}{2}\sum_{i \in I}\sum_{j \in J}(x_i y_j - x_j y_i)^2.$$

Since this is a sum of squares, the inequality follows. The inequality is strict unless $x_i y_j = x_j y_i$ for all i and j. Suppose that $\vec{x} \neq \vec{0}$. Let i be such that $x_i \neq 0$ and let $p = y_i/x_i$. It follows that

$$y_j = \frac{y_i x_j}{x_i} = px_j; \quad \text{that is,} \quad \vec{y} = p\vec{x}.$$

Hence $\vec{x} \cdot \vec{y} = p(\vec{x} \cdot \vec{x}) = p|\vec{x}|^2$. It follows that we must have $p \geq 0$.

We now prove the first version of the triangle inequality. It will be true if it is true for the square of both sides. We have

$$\left(|\vec{x}| + |\vec{y}|\right)^2 = \vec{x} \cdot \vec{x} + 2|\vec{x}|\,|\vec{y}| + \vec{y} \cdot \vec{y}$$

$$|\vec{x} \pm \vec{y}|^2 = (\vec{x} \pm \vec{y}) \cdot (\vec{x} \pm \vec{y}) = \vec{x} \cdot \vec{x} \pm 2\vec{x} \cdot \vec{y} + \vec{y} \cdot \vec{y}.$$

By canceling common terms, we see that it suffices to prove that

$$2|\vec{x}|\,|\vec{y}| \geq \pm 2\vec{x} \cdot \vec{y}.$$

This follows from (g). ■

Exercises

11.2.A. Define vectors, dot product, and vector length.

11.2.B. Define matrices, transpose, matrix addition, and matrix multiplication.

11.2.C. Express the dot product of two vectors in terms of the matrix operations defined in the previous exercise.

11.2.1. Prove the rest of Theorem 11.3.

*11.2.2. Suppose that A and B are $n \times n$ matrices such that $AB = I_n$. Prove that $B = A^{-1}$ and $A = B^{-1}$. In other words, prove that $BA = I_n$.

11.2.3. Using vector and matrix notation, rewrite the material on Hopfield nets; that is, regard $\vec{\imath}$, $\vec{P}(k)$, $w_{i,j}$, et cetera as vectors and matrices with appropriate index sets and then rewrite (11.1), (11.2), (11.4), and the derivation of (11.5). In doing this, let n be the dimension of the vectors and let s be the number of patterns $\vec{P}(k)$.

Hint. It may be helpful to define $\vec{e}_v$ to be a vector that consists entirely of zeros except for a $2o_v$ in its vth component because then $\vec{o} - \vec{e}_v$ differs from $\vec{o}$ in just the vth component.

11.2.4. Given a list of pairs $(\vec{x}_1, \vec{y}_1), \ldots, (\vec{x}_s, \vec{y}_s)$, where $\vec{x}_i \in \{-1, +1\}^m$ and $\vec{y}_i \in \{-1, +1\}^n$, we want to create a bidirectional *associative memory*; that is, given $\vec{x}_i$ retrieve $\vec{y}_i$, and given $\vec{y}_i$ retrieve x_i. Define an $m \times n$ matrix by $M = \sum_{i=1}^{s} \vec{x}_i \vec{y}_i{}^{\mathrm{t}}$.

(a) Assuming that the components of the $\vec{x}_i$'s and the $\vec{y}_i$'s are ± 1 and are random, show that we can expect to have

$$M\vec{y}_k \approx n\vec{x}_k \quad \text{and} \quad \vec{x}_k{}^{\mathrm{t}} M \approx m\vec{y}_k{}^{\mathrm{t}}$$

provided s is not too large.

Hint. See Example 11.1 (p. 436).

(b) Explain how you could use the previous observation to design an associative memory.

Linear Algebra

We'll begin with the notion of a vector space and bases. Next we introduce some concepts and results concerning eigenvalues and eigenvectors. Our definition of a vector space won't be as general as the one used by mathematicians. Since we won't use this material much, you may want to skim it and then refer back to it as needed.

Definition 11.7   Vector Spaces and Bases

Let F stand for either $\mathbb{R}$ or $\mathbb{C}$, the real or complex numbers, and let I be a finite set of indices. A subset V of F^I is called a *vector space* if

- $\vec{u} + \vec{v} \in V$ for every $\vec{u}, \vec{v} \in V$ and

- $f\vec{v} \in V$ for every $\vec{v} \in V$ and every $f \in F$.

Let V be a vector space containing $\vec{v}_1, \ldots, \vec{v}_n$ such that, for every $\vec{v} \in V$ there exist *unique* $f_1, \ldots, f_n \in F$ for which $\vec{v} = f_1\vec{v}_1 + \cdots + f_n\vec{v}_n$. We call $\vec{v}_1, \ldots, \vec{v}_n$ a *basis* for V, call n the *dimension* of V, and call $\vec{f} \in F^n$ the *coordinates* of $\vec{v}$ in terms of the basis $\vec{v}_1, \ldots, \vec{v}_n$.

A simple example of a vector space is all of F^I. We can take $\vec{v}_i$ to be the vector that is 0 in all components except the ith, where it is 1. The dimension of F^I is $|I|$.

Given a basis for a vector space, the map $\vec{v} \to (f_1, \ldots, f_n)$ shows that we can think of the vector space simply as F^n. This is *very important and convenient* because it allows us to forget about the abstract notion and work with matrices and vectors as defined earlier. That's what we'll usually do. In order to do that, we need to know that bases always exist. That's what the following theorem says.

Theorem 11.4   Existence of Basis and Dimension

Every vector space has a basis. If $\vec{v}_1, \ldots, \vec{v}_n$ is a basis and $\vec{w}_1, \ldots, \vec{w}_m$ is another basis, then $n = m$.

Given two bases $\vec{v}_1, \ldots, \vec{v}_n$ and $\vec{w}_1, \ldots, \vec{w}_n$ for a vector space and some vector $\vec{u}$ in the space, let $\vec{f}$ and $\vec{g}$ be the coordinates of $\vec{u}$ in terms of the two bases; that is, $\vec{u} = \sum_{i=1}^{n} f_i \vec{v}_i = \sum_{i=1}^{n} g_i \vec{w}_i$. How are $\vec{f}$ and $\vec{g}$ related?

Define B by

$$\vec{w}_j = b_{1,j}\vec{v}_1 + \cdots + b_{n,j}\vec{v}_n.$$

Since

$$\vec{u} = \sum_{j=1}^{n} g_j \vec{w}_j = \sum_{j=1}^{n} g_j \left(\sum_{i=1}^{n} b_{i,j} \vec{v}_i \right) = \sum_{i=1}^{n} \left(\sum_{j=1}^{n} b_{i,j} g_j \right) \vec{v}_i,$$

we see that $\vec{f} = B\vec{g}$.

Definition 11.8   Eigenvalues and Eigenvectors

Let A be an $n \times n$ matrix of real or complex numbers. If $\lambda \in \mathbb{C}$ and $\vec{v} \in \mathbb{C}^n$ are such that $A\vec{v} = \lambda\vec{v}$ and $\vec{v} \neq \vec{0}$, we call λ an *eigenvalue* of A and $\vec{v}$ an *eigenvector* of A. The terms *characteristic value* and *characteristic vector* are also used. Note that if $\vec{v}$ is an eigenvector, then so is $c\vec{v}$ for all nonzero $c \in \mathbb{C}$.

It can be shown that A has at most n eigenvalues. By allowing appropriate multiplicities, it turns out that A has exactly A eigenvalues. This is not an ad hoc adjustment—it is closely related to the fact that an nth-degree equation over the complex numbers has exactly n roots when multiplicities are handled appropriately. Explaining how this works would take us too far afield, so you'll have to take it on faith or look in a linear algebra text.

Since many of our matrices are symmetric, the following result will be useful.

Theorem 11.5 Spectral Theorem

Suppose that A is an $n \times n$ symmetric matrix with eigenvalues $\lambda_1, \ldots, \lambda_n$. Then the following are true:

- All of the eigenvalues are real numbers.

- If $\vec{x}^{\mathrm{t}} A \vec{x} \geq 0$ for all $\vec{x} \neq \vec{0}$, then $\lambda_i \geq 0$ for all i. Furthermore, one inequality is always strict if and only if the other is.

- Let $\vec{x} \in \mathbb{R}^n$. There is a "rigid motion" of $\mathbb{R}^n$ that fixes the origin and is such that, if the components of $\vec{y}$ are the coordinates of $\vec{x}$ in the new coordinate system, then

$$\vec{x}^{\mathrm{t}} A \vec{x} = \lambda_1 y_1^2 + \cdots + \lambda_n y_n^2.$$

A *rigid motion* is the sort of motion we allow in geometry—one that preserves straight lines, angles, and distances.

In two and three dimensions, the only rigid motions fixing the origin are rotations and reflections about the origin. In higher dimensions, the situation is more complicated. It turns out that rigid motion in any number of dimensions is equivalent to the statement that the dot product of the coordinates of any two vectors is the same in both coordinate systems. (See Exercise 11.2.5.)

We can describe a rigid motion in terms of our change-of-coordinates equation $\vec{f} = B\vec{g}$ derived earlier. The dot product of two vectors *in the first coordinate system* has the form $\vec{f} \cdot \vec{f}' = \vec{f}^{\mathrm{t}} \vec{f}'$ and, in the second coordinate system, $\vec{g}^{\mathrm{t}} \vec{g}'$. Since $\vec{f} = B\vec{g}$ and $\vec{f}' = B\vec{g}'$, we have

$$\vec{f} \cdot \vec{f}' = \vec{f}^{\mathrm{t}} \vec{f}' = (B\vec{g})^{\mathrm{t}} (B\vec{g}') = \vec{g}^{\mathrm{t}} (B^{\mathrm{t}} B) \vec{g}',$$

which must equal $\vec{g}^{\mathrm{t}} \vec{g}'$ for all $\vec{g}$ and $\vec{g}'$ if values of the dot product are unchanged from one coordinate system to the other. Thus $\vec{g}^{\mathrm{t}} (B^{\mathrm{t}} B) \vec{g}' = \vec{g}^{\mathrm{t}} \vec{g}'$ for all $\vec{g}$ and $\vec{g}'$. This implies $B^{\mathrm{t}} B = I_n$, the $n \times n$ identity matrix—just take $\vec{g}$ and $\vec{g}'$ to be all 0's except for a 1 in the ith and jth coordinates, respectively. This explanation of eigenvalues and rigid motions is quite sketchy. A fuller explanation would take longer than the applications merit. If you want more information, see a good linear algebra text.

Proof (of Theorem 11.5):   For those of you with the background, I'll indicate how most of the theorem is proved when the eigenvalues are distinct. Let $\vec{v}_k$ be an eigenvector corresponding to λ_k.

Suppose that $i \neq j$. We have

$$\begin{aligned} \lambda_j \vec{v}_i \cdot \vec{v}_j &= \vec{v}_i \cdot (A\vec{v}_j) = \vec{v}_i{}^{\mathrm{t}} A \vec{v}_j \\ &= (A\vec{v}_i)^{\mathrm{t}} \vec{v}_j, \qquad \text{since } A^{\mathrm{t}} = A \text{ by symmetry,} \\ &= \lambda_i \vec{v}_i \cdot \vec{v}_j. \end{aligned}$$

Since $\lambda_i \neq \lambda_j$, it follows that $\vec{v}_i \cdot \vec{v}_j = 0$.

We can prove by contradiction that all eigenvalues are real. Suppose that λ is not real. Then its complex conjugate λ^* is also an eigenvalue and, by the conclusion in the last paragraph, $\vec{v} \cdot \vec{v}^* = 0$. This is impossible since $\vec{v} \cdot \vec{v}^* = 0$ is a sum of squares.

Since the eigenvalue λ is real, $\vec{v}$ is determined by a set of linear equations with real coefficients, namely $A\vec{v} = \lambda\vec{v}$. Consequently, we may assume that $\vec{v}$ is real. Since an eigenvector can always be multiplied by a constant, we may also assume that $|\vec{v}| = 1$.

We now have a set of eigenvectors such that $\vec{v}_i \cdot \vec{v}_j$ is 1 or 0 according to whether $i = j$ or not. These provide the unit vectors for our new coordinate system. In terms of these coordinates, $\vec{x} = y_1\vec{v}_1 + \cdots + y_n\vec{v}_n$. A little matrix algebra shows that $\vec{x}^t A\vec{x}$ equals $\lambda_1 y_1^2 + \cdots + \lambda_n y_n^2$. ∎

*Example 11.2 Principal-Component Analysis

Suppose we're given a sequence of vectors $\vec{v}_1, \ldots, \vec{v}_t \in \mathbb{R}^n$. We want to extract some new coordinates that contain most of the information.

To simplify our discussion, we assume that $\sum_i \vec{v}_i = \vec{0}$. This is not really a restriction because we could simply replace $\vec{v}_i$ with $\vec{v}_i - \vec{m}$ where $\vec{m} = \frac{1}{t} \sum_i \vec{v}_i$.

Define an $n \times n$ matrix $A = \sum_i \vec{v}_i \vec{v}_i^t$. Note that A is symmetric. By the Spectral Theorem, all the eigenvalues of A are real, say $\lambda_1 \geq \lambda_2 \geq \cdots \geq \lambda_n$. We claim that the eigenvalues are nonnegative. To see this, let λ and $\vec{w} \in \mathbb{R}^n$ be an eigenvalue and corresponding eigenvector. Since $|\vec{w}| > 0$ and

$$\lambda |\vec{w}|^2 = \vec{w}^t(\lambda\vec{w}) = \vec{w}^t A\vec{w} = \sum_i (\vec{w}^t \vec{v}_i)(\vec{v}_i^t \vec{w}) = \sum_i (\vec{w} \cdot \vec{v}_i)^2 \geq 0,$$

it follows that $\lambda \geq 0$.

By the Spectral Theorem, there is a rigid-motion change of coordinates B such that $\vec{x} = B\vec{y}$ changes $\vec{x}^t A\vec{x}$ to $\sum \lambda_i y_i^2$. Thus

$$B^t A B = \begin{pmatrix} \lambda_1 & 0 & \cdots & 0 \\ 0 & \lambda_2 & \cdots & 0 \\ \vdots & & \ddots & 0 \\ 0 & \cdots & 0 & \lambda_n \end{pmatrix}$$

This is just the change of coordinates needed to extract the principal components of the $\vec{v}_i$'s, but we have to unwind things a bit. Define $\vec{z}_i = B^t \vec{v}_i$. Since

$$B^t A B = \sum_i B^t \vec{v}_i \vec{v}_i^t B = \sum_i B^t \vec{v}_i (B^t \vec{v}_i)^t.$$

In summary, there is a rigid motion B such that

$$\sum_i \vec{z}_i \vec{z}_i^t = \begin{pmatrix} \lambda_1 & 0 & \cdots & 0 \\ 0 & \lambda_2 & \cdots & 0 \\ \vdots & & \ddots & 0 \\ 0 & \cdots & 0 & \lambda_n \end{pmatrix} = B^t A B, \text{ where } \vec{z}_i = B^t \vec{v}_i. \quad (11.8)$$

The transformation from $\vec{v}_i$ to $\vec{z}_i$ is the transformation to "principal-component" form.

From (11.8), the sum of the squares of the jth components of the $\vec{z}_i$'s is $\sum_i \left((\vec{z}_i)_j \right)^2 = \lambda_j$. Thus, if the eigenvalues beyond λ_k are small, we can ignore all but the first k components of the $\vec{z}_i$'s with little loss of information. ∎

Exercises

11.2.D. Define vector space, basis, and coordinates.

11.2.E. Why are bases important?

11.2.F. Explain the following statement—what does it mean and how is it done? "The idea behind principal-component analysis is to change coordinates so that some of the coordinates become unimportant and can be ignored."

*11.2.5. Show that $4\vec{f} \cdot \vec{f}' = |\vec{f} + \vec{f}'|^2 - |\vec{f} - \vec{f}'|^2$. Conclude that if lengths are unchanged, then so are dot products.

11.2.6. Let $\vec{x}$ and $\vec{y}$ be vectors in $\mathbb{R}^n$ and let $\vec{v}_1, \ldots, \vec{v}_n$ be a basis for a $\mathbb{R}^n$. Suppose that $\vec{f}$ and $\vec{g}$ are the coordinates of the vectors $\vec{x}$ and $\vec{y}$ in terms of that basis; that is, $\vec{x} = \sum_i f_i \vec{v}_i$ and $\vec{y} = \sum_i g_i \vec{v}_i$.

    (a) Let V be a matrix whose columns are the vectors $\vec{v}_1, \ldots \vec{v}_n$. Show that $\vec{x} = V\vec{f}$.

    (b) Show that $\vec{x} \cdot \vec{y} = \vec{f}^{\,t}(V^t V)\vec{g} = \sum_{i,j} f_i(\vec{v}_i \cdot \vec{v}_j)g_j$.

11.2.7. This exercise requires familiarity with linear independence and bases. You are to prove the claims after Definition 11.8. Suppose that A is an $n \times n$ real or complex matrix with n distinct eigenvalues $\lambda_1, \ldots, \lambda_n$. Let $\vec{v}_i$ be an eigenvector associated with λ_i.

    (a) Suppose that $c_1 \vec{v}_1 + \cdots + c_k \vec{v}_k = \vec{0}$ and not all of $c_1, \ldots, c_k$ are zero. Prove that we can assume that $c_k = 0$.
    Hint. Let $\vec{w} = c_1 \vec{v}_1 + \cdots + c_k \vec{v}_k$ and look at $A\vec{w} - \lambda_k \vec{w}$.

    (b) Conclude that $v_1, \ldots, v_n$ form a basis.

11.2.8. Let $\vec{f} = B\vec{g}$ convert coordinates in the basis $\vec{w}_1, \ldots, \vec{w}_n$ to those in the basis $\vec{v}_1, \ldots, \vec{v}_n$. Show that B^{-1} exists and that $\vec{g} = B^{-1}\vec{f}$.
Hint. Show that there must be some matrix A for converting from $\vec{v}_1, \ldots, \vec{v}_n$ coordinates to $\vec{w}_1, \ldots, \vec{w}_n$ coordinates, and use uniqueness to conclude that $AB = I_n$.

Multivariate Calculus

Modern scientific thought has been formed from the concepts
of calculus and is meaningless outside this context. ...
We teach calculus because it is important for an understanding
of who we are as a society.

—David M. Bressoud (1992)

The main concepts we need from multivariate calculus are partial derivatives, the chain rule, and approximations by first and second derivatives.

Definition 11.9   Multivariate Calculus Terminology

Suppose g is a real-valued function defined on (some subset of) $\mathbb{R}^I$.

- The *partial derivative* of g with respect to x_i is given by

$$\frac{\partial g}{\partial x_i}(\vec{x}) = \lim_{h \to 0} \frac{g(\vec{x} + h\vec{e_i}) - g(\vec{x})}{h}$$

if the limit exists. Otherwise, $\partial g / \partial x_i$ is said not to exist at $\vec{x}$.

- The *gradient* of g, is a vector with index set I and

$$(\nabla g)_i = \frac{\partial g}{\partial x_i}.$$

Note that the components of the gradient are *functions*, so we have to expand our notion of vectors and allow components of vectors to be functions.

If each of the functions in the components of ∇g is evaluated at $\vec{a}$, we write $\nabla g(\vec{a})$ for the resulting vector in $\mathbb{R}^I$.

- The *Hessian* of g is the $I \times I$ matrix of functions given by

$$(H(g))_{i,j} = \frac{\partial}{\partial x_i}\left(\frac{\partial g}{\partial x_j}\right).$$

The results in the following theorems are incomplete because they require "well-behaved" functions—a vague concept we won't try to make precise. The functions that arise in practice are "well-behaved," so the condition can be ignored. For a precise statement of the theorems and for complete proofs, consult any good multivariate calculus text.

Theorem 11.6

If g is well-behaved, the order of differentiation is irrelevant; that is,

$$\frac{\partial}{\partial x_i}\left(\frac{\partial g}{\partial x_j}\right) = \frac{\partial}{\partial x_j}\left(\frac{\partial g}{\partial x_i}\right), \quad \text{and we write these as} \quad \frac{\partial^2 g}{\partial x_i \, \partial x_j}.$$

In particular $(H(g))_{i,j} = (H(g))_{j,i}$ and so $(H(g))^t = H(g)$.

Proof: Let $\vec{e}_k$ be a vector in which all the components are 0 except the kth, which equals 1. We have

$$\frac{\partial g}{\partial x_j} = \lim_{h \to 0} \frac{g(\vec{x} + h\vec{e}_j) - g(\vec{x})}{h}$$

and so

$$\frac{\partial}{\partial x_i} \left(\frac{\partial g}{\partial x_j} \right)$$

$$= \lim_{h' \to 0} \left(\frac{1}{h'} \left(\lim_{h \to 0} \frac{g(\vec{x} + h'\vec{e}_i + h\vec{e}_j) - g(\vec{x} + \vec{h}'e_i)}{h} \right. \right.$$

$$\left. \left. - \lim_{h \to 0} \frac{g(\vec{x} + h\vec{e}_j) - g(\vec{x})}{h} \right) \right)$$

$$= \lim_{h' \to 0} \left(\lim_{h \to 0} \left(\frac{g(\vec{x} + h'\vec{e}_i + h\vec{e}_j) - g(\vec{x} + \vec{h}'e_i) - g(\vec{x} + h\vec{e}_j) + g(\vec{x})}{h'h} \right) \right).$$

If g is well-behaved, we can interchange the order of the limits in the last expression. Then the steps can be reversed and we end up with $\vec{e}_i$ and h' interchanged with $\vec{e}_j$ and h. ∎

Partial derivatives behave like ordinary derivatives except that the generalization of $(f(g(x)))' = f'(g(x))g'(x)$ is not obvious. This generalization is called the *chain rule*:

Theorem 11.7   Chain Rule

Suppose that $f \colon \mathbb{R}^I \to \mathbb{R}$, and for all $i \in I$, $g_i \colon \mathbb{R}^J \to \mathbb{R}$ are well-behaved functions. Let $h(\vec{x})$ be $f(\vec{y})$ with y_i replaced by $g_i(\vec{x})$. Then the *chain rule* is

$$\frac{\partial h(\vec{x})}{\partial x_j} = \sum_{i \in I} \frac{\partial f(\vec{y})}{\partial y_i} \frac{\partial g_i(\vec{x})}{\partial x_j}. \tag{11.9}$$

We often use the same letter to denote f and h, distinguishing between the two functions by the letters used for the arguments. We also often use y_i to denote the function g_i since we are setting $y_i = g_i(\vec{x})$ in f. Rewriting the chain rule in this form, we have

$$\frac{\partial f}{\partial x_j} = \sum_i \frac{\partial f}{\partial y_i} \frac{\partial y_i}{\partial x_j}.$$

Example 11.3   The Chain Rule

Suppose that $F : \mathbb{R}^n \to \mathbb{R}$ is given by $F(\vec{x}) = G(\vec{a} \cdot \vec{x})$. What is $\partial F/\partial x_i$? Apply the chain rule with

$f = G$ and $I = \{1\}$, a function of a single variable,

$g_1(\vec{x}) = \vec{a} \cdot \vec{x}$, and $J = \{1, \ldots, n\}$.

The result is $\partial F/\partial x_i = G'(\vec{a} \cdot \vec{x})a_i$ since $\partial g_1/\partial x_i = a_i$.

Now suppose that $x_i = H_i(\vec{y})$ and we want $\partial F/\partial y_j$. We again use the chain rule:

$$\frac{\partial F}{\partial y_j} = \sum_{i=1}^{n} \frac{\partial F}{\partial x_i}\frac{\partial H_i}{y_j} = \sum_{i=1}^{n} G'(\vec{a} \cdot \vec{x})a_i \frac{\partial F}{\partial x_i}\frac{\partial H_i}{y_j}.$$

The chain rule can be conveniently written in matrix notation. Let $\partial \vec{y}/\partial \vec{x}$ be the matrix whose (i, j)th entry is $\partial y_i/\partial x_j$. Similarly, for a single function $\partial f/\partial \vec{x}$ is a $1 \times n$ matrix—a single row, not a column vector. The chain rule is then

$$\frac{\partial f}{\partial \vec{x}} = \frac{\partial f}{\partial \vec{y}}\frac{\partial \vec{y}}{\partial \vec{x}}.$$

This has the form of the chain rule for functions of a single variable. We can then easily state results for functions of functions of functions analogous to $f'(u) = f'(y)y'(x)x'(u)$:

$$\frac{\partial f}{\partial \vec{u}} = \frac{\partial f}{\partial \vec{y}}\frac{\partial \vec{y}}{\partial \vec{x}}\frac{\partial \vec{x}}{\partial \vec{u}}.$$

This shows you some of the convenience and power of matrix notation in multivariate calculus. ■

For this text, the essential property of the gradient and the Hessian is that they allow us to estimate nearby values of the function. These estimates correspond to the approximations of functions of a single variable by parabolas and tangent lines, respectively.

Theorem 11.8   Quadratic Approximation

Suppose that $g : \mathbb{R}^I \to \mathbb{R}$ is a well-behaved function. The following is the analog to approximating a function with a Taylor series using terms through the second derivative. Suppose that $\epsilon > 0$. Let H be the Hessian of g at $\vec{a}$. There is a $\delta > 0$ depending on ϵ, g, and $\vec{x}$ such that

$$\left|\left(g(\vec{a}) + (\nabla g(\vec{a})) \cdot \vec{h} + \tfrac{1}{2}\vec{h}^t H \vec{h}\right) - g(\vec{a} + \vec{h})\right| < \epsilon|\vec{h}|^2 \qquad (11.10)$$

whenever $|\vec{h}| < \delta$.

In practice, we usually won't be able to obtain the Hessian of a function, so we'll have to make compromises. One possibility is to use the following weaker result instead of (11.10).

Theorem 11.9   Linear Approximation

Suppose that $g : \mathbb{R}^I \to \mathbb{R}$ is well behaved.

(a) If $\vec{a}$ is a local maximum or minimum of g, then $\nabla g(\vec{a}) = \vec{0}$.

(b) If $\epsilon > 0$, then there is a $\delta > 0$ depending on ϵ, g, and $\vec{a}$ such that

$$\left| \left(g(\vec{a}) + (\nabla g(\vec{a})) \cdot \vec{x} \right) - g(\vec{x} + \vec{a}) \right| < \epsilon |\vec{x}| \tag{11.11}$$

whenever $|\vec{x}| < \delta$.

(c) If $\vec{x} \cdot (\nabla g(\vec{a})) < 0$, then there is a $\delta > 0$ depending of g, $\vec{a}$, and $\vec{x}$ such that

$$g(\vec{a} + c\vec{x}) < g(\vec{a}) < g(\vec{a} - c\vec{x}) \tag{11.12}$$

whenever $0 < c < \delta$.

Exercises

11.2.G. Define partial derivative, gradient, and Hessian.

11.2.H. State the chain rule as a summation and in matrix notation.

11.2.I. What is the multivariate statement corresponding to the single-variable statement "$|f(x + h) + hf'(x) - f(x)|$ is small compared to $|x|$ when $|h|$ is small"?

11.2.J. Deduce Theorem 11.9(a) from the standard calculus result for extrema of a function of a single variable. Do this by holding all components of $\vec{x}$ fixed and thinking of $g(\vec{x})$ as a function of just x_i.

11.2.9. Suppose that $g : \mathbb{R}^n \to \mathbb{R}$ and that B is an $m \times n$ matrix. Let $h : \mathbb{R}^m \to \mathbb{R}$ be given by $h(\vec{x}) = g(B\vec{x})$. Prove that $\nabla h = B^t \nabla g$.

11.2.10. Deduce Theorem 11.9(b) from (11.10).

11.2.11. Deduce Theorem 11.9(c) from (b).

11.2.12. Using your knowledge of multivariate calculus, show that

$$f(\vec{x} + \vec{a}) = g(\vec{a}) + (\nabla g(\vec{a})) \cdot \vec{x}$$

is a plane tangent to g at $\vec{a}$ when $i = \{1, 2\}$. What does (11.11) say about approximating g by a tangent plane?

11.3   A Brief Introduction to Perceptrons

The purpose of this brief section is to familiarize you with some of the ideas and problems associated with feedforward nets and to introduce the concept of gradient descent, which plays a central role in the next section.

A perceptron is a particularly simple feedforward neural net. Instead of describing it in those terms, let's use terminology more in keeping with statistics. We are given a collection of vectors $\vec{y}$, each of which is classified as being either "in" or "out." We seek a real number θ and a vector $\vec{w}$ such that $\theta + \vec{w} \cdot \vec{y}$ is positive when $\vec{y}$ is in and is negative when $\vec{y}$ is out. It is, of course, entirely possible that no such parameters θ and $\vec{w}$ exist. For example, suppose that $\vec{v}_1 = (1,1)$ and $\vec{v}_2 = (3,3)$ are in and $\vec{v}_2 = (2,2)$ is out. Using the requirements and a bit of algebra, we have the contradiction

$$0 > 2(\theta + \vec{w} \cdot \vec{v}_2) = (\theta + \vec{w} \cdot \vec{v}_1) + (\theta + \vec{w} \cdot \vec{v}_1) > 0.$$

If it is possible to find the parameters, statisticians say that the in and out sets are *linearly separable*.

Before discussing how to solve the problem of finding parameters, let's simplify it a bit. First, we'll add an additional component to all the vectors. For the given vectors $\vec{y}$, this component is always 1. For $\vec{w}$, this component is θ. The expression $\theta + \vec{w} \cdot \vec{y}$ in the old notation becomes simply $\vec{w} \cdot \vec{y}$ in the new. Second, if a vector $\vec{y}$ is out, we replace it by $-\vec{y}$ and call the new vector by the old name. These two operations (adding a component and then changing signs) have changed the problem to the following:

Given a collection of vectors $\vec{y}$, we wish to find a vector $\vec{w}$ such that $\vec{w} \cdot \vec{y} > 0$ for all $\vec{y}$ in the collection.

The following theorem provides an algorithm for solving the problem whenever a solution exists. It can be thought of as a minimization algorithm since we're trying to minimize the number of errors. However, the algorithm doesn't work in general—it won't stop unless there's a $\vec{w}$ that produces no errors.

Theorem 11.10   Perceptron Algorithm

Let $\mathcal{Y}$ be a finite set of vectors such that for some $\vec{w}$ we have $\vec{w} \cdot \vec{y} > 0$ for all $\vec{y} \in \mathcal{Y}$. Let $\vec{w}(1)$ be arbitrary and let $c > 0$ be arbitrary. Define

$$\mathcal{Y}(t) = \left\{ \vec{y} \in \mathcal{Y} \mid \vec{w}(t) \cdot \vec{y} \le 0 \right\} \quad \text{and} \quad \vec{w}(t+1) = \vec{w}(t) + c \sum_{\vec{y} \in \mathcal{Y}(t)} \vec{y}$$

for $t \ge 1$. Then, for all sufficiently large t, $\mathcal{Y}(t) = \emptyset$. This is equivalent to the statement that for all sufficiently large t, $\vec{w}(t) \cdot \vec{y} > 0$ for all $\vec{y} \in \mathcal{Y}$.

Proof: Here is a brief but unenlightening proof that $\mathcal{Y}(t)$ is eventually zero. The equivalence of the last two statements is obvious from the definition of $\mathcal{Y}(t)$. In fact, note that if $\mathcal{Y}(t) = \emptyset$, then $\mathcal{Y}(t') = \emptyset$ for all $t' \geq t$.

Let δ be the minimum of $\vec{w} \cdot \vec{y}$ over all $\vec{y} \in \mathcal{Y}$, let $\vec{s}(\mathcal{V})$ be the sum of all $\vec{y}$ in $\mathcal{V}$, and let S be the maximum of $|\vec{s}(\mathcal{V})|$ over all subsets $\mathcal{V} \subseteq \mathcal{Y}$. Choose a constant $r > 0$ and let α be such that

$$2\alpha\delta - cS^2 > r.$$

Since $\delta > 0$, this can be done. We claim that

$$\mathcal{Y}(t) \neq \emptyset \text{ implies that } \left|\vec{w}(t+1) - \alpha\vec{w}\right|^2 < \left|\vec{w}(0) - \alpha\vec{w}\right|^2 - crt.$$

Since the left side of the inequality is positive and the right side decreases without limit, the procedure must stop.

Here's the proof of the claim.

$$\begin{aligned}
\left|\vec{w}(t+1) - \alpha\vec{w}\right|^2 &= \left|\vec{w}(t) + c\vec{s}(\mathcal{Y}(t)) - \alpha\vec{w}\right|^2 \\
&= \left|\left(\vec{w}(t) - \alpha\vec{w}\right) + c\vec{s}(\mathcal{Y}(t))\right|^2 \\
&= \left|\vec{w}(t) - \alpha\vec{w}\right|^2 + c^2\left|\vec{s}(\mathcal{Y}(t))\right|^2 \\
&\quad + 2c\left(\vec{w}(t) \cdot \vec{s}(\mathcal{Y}(t)) - \alpha\vec{w} \cdot \vec{s}(\mathcal{Y}(t))\right).
\end{aligned}$$

Since $\vec{w}(t) \cdot \vec{y} \leq 0$ for all $y \in \mathcal{Y}(t)$ and since $|\vec{s}(\mathcal{Y}(t))| \leq S$, we have

$$\begin{aligned}
\left|\vec{w}(t+1) - \alpha\vec{w}\right|^2 &\leq \left|\vec{w}(t) - \alpha\vec{w}\right|^2 + c^2S^2 - 2c\alpha\vec{w} \cdot \vec{s}(\mathcal{Y}(t)) \\
&\leq \left|\vec{w}(t) - \alpha\vec{w}\right|^2 + c^2S^2 - 2c\alpha\delta|\mathcal{Y}(t)| \\
&< \left|\vec{w}(t) - \alpha\vec{w}\right|^2 + cr,
\end{aligned}$$

since $|\mathcal{Y}(t)| \geq 1$ because the set is not empty. The claim now follows easily by induction on t. ∎

There are a variety of variants on the algorithm. A significant one involves cycling through the vectors $\vec{y} \in \mathcal{Y}$ one by one, and each time $\vec{w}(t) \cdot \vec{y} < 0$, setting $\vec{w}(t+1) = \vec{w}(t) + c\vec{y}$. This holds the promise of quicker convergence since we don't examine the entire set $\mathcal{Y}$ before changing $\vec{w}(t)$.

Let's assess what's been accomplished and what hasn't been accomplished. On the positive side, we've stated an algorithm and proved that it converges. This is a significant accomplishment since such results are rare in this field. On the negative side, we've fallen down in three important areas:

- **Failure Detection:** The theorem gives no indication of what happens when no linear separator $\vec{w}$ exists. This can be dealt with but isn't worth the time since we'll abandon perceptrons shortly.

- **Time Estimates**: It's impossible to extract from the proof an estimate of how long convergence will take without having δ, which depends on already having a solution. This is typical—useful time estimates for minimization algorithms are usually impossible to obtain.

- **Motivation**: The apparently ad hoc definition of $\vec{w}(t+1)$ in Theorem 11.10 leaves us in the dark about how the algorithm might be improved upon or how it might be extended to more general situations.

How can changing the weight vector by $c\vec{s}(\mathcal{Y}(t))$ be made less ad hoc? Let's change notation slightly: $\vec{w}$ will now be the current vector $\vec{w}(t)$. There are various ways of looking at the problem of changing $\vec{w}$. One fruitful approach is to think in terms trying to increase

$$f(\vec{w}) = \sum_{\vec{y} \in \mathcal{Y}(t)} \vec{w} \cdot \vec{y}$$

since it's a sum of terms, each of which we want to make positive.

You should be able to show that the gradient is given by

$$\nabla f(\vec{w}) = \sum_{\vec{y} \in \mathcal{Y}(t)} \vec{y} = \vec{s}(\mathcal{Y}(t)).$$

Hence the algorithm changes $\vec{w}$ by $c\nabla f(\vec{w})$, a procedure known as *gradient descent*—or, more properly in this case, gradient *ascent* since we're trying to increase $f(\vec{w})$. Gradient descent is the first algorithm we'll look at in the next section.

Exercises

11.3.1. Prove that a solution exists for classification by a simple perceptron if and only if a linear separator exists. (This involves little beyond translating from one definition to another.)

11.3.2. Show that a linear separator for two inputs can be interpreted as a line in the plane such that all inputs on one side are classified 0 and all on the other are classified 1. What is the interpretation for three inputs?

11.4   Finding Minima

> *[Minimum methods] are the first refuge of the computational scoundrel, and one feels at times that the world would be a better place if they were quietly abandoned. But even if these techniques are frequently misused, it is equally true that there are problems for which no alternative solution method is known.*
>
> —Forman S. Acton (1970)

> *In practice there are serious convergence issues for such algorithms, with, unfortunately, very few analytical tools to address them.*
>
> —Stuart Geman, Elie Bienenstock, and René Doursat (1992)

Since we're moving through the domain of a function seeking values that make the function small, we can think of minimization as a form of search. Of course, the methods are rather different from those we studied for "combinatorial" type searches in Chapter 2. Since the minimization problems are in Euclidean space, we usually use tools from calculus. In this section, we introduce some of the methods for minimizing functions of many variables that are particularly suited for use in neural networks. We'll also discuss some of the difficulties that arise.

Types of Algorithms

Let's suppose we want to find the minimum of some complicated function $g : \mathbb{R}^n \to \mathbb{R}$. From Theorem 11.9(a), it would suffice to find the solutions $\vec{a}$ of $\nabla g(\vec{a}) = \vec{0}$. It's usually unrealistic to expect to solve such equations directly. Instead, minimization problems are solved by search techniques called *descent methods*. Maximization techniques are called *hill-climbing methods*. Of course, replacing $g(\vec{x})$ by $-g(\vec{x})$ converts a maximization problem into a minimization problem and vice versa. We can describe many max/min algorithms in general terms by the following algorithm.

Algorithm 11.2   Generic Hill Climbing/Descent

Given a fairly smooth function $g : \mathbb{R}^n \to \mathbb{R}$ and a termination criterion, proceed as follows:

1. **Start:** Choose a value $\vec{x}_0$ and set $k = 0$.

2. **New Guess:** Using information at $\vec{x}_k$—namely some or all of $g(\vec{x}_k)$, $\nabla g(\vec{x}_k)$, and $H(g(\vec{x}_k))$—and perhaps previous information, choose

    - a direction to move and
    - a distance to move,

    which together determine a vector $\vec{h}_k$. Set $\vec{x}_{k+1} = \vec{x}_k + \vec{h}_k$ and increment k. We call $\vec{h}_k$ the kth *step*.

3. **Terminate or Iterate:** If the termination criterion is satisfied, stop; otherwise, go to Step 2.

Depending on the information used in Step 2 of the algorithm, we can broadly classify the methods as follows:

- **Simple Search:** In this case, only the function values are used—derivatives and previous function values are ignored. The amount of calculation required by such methods is generally too large.

- **Linear Approximation:** These methods make use of the gradient ∇g and the approximation (11.11). The most naive approach is steepest descent, also called gradient descent. Because it has severe problems, several modifications have been proposed.

- **Quadratic Approximation:** The most direct way to obtain a quadratic approximation is to use both the gradient and the Hessian $H(g)$. Such methods are potentially much better than methods based only on linear approximation, and numerical analysts advise using them whenever possible.

    Some methods use only first derivatives but make use of information gained on previous steps to approximate the Hessian. This circumvents the need to compute second-order partial derivatives but still provides a quadratic approximation similar to (11.10).

    Unfortunately, calculating the Hessian, or even just storing any matrix approximating it, violates the requirement that neural net calculations be carried out in a distributed manner. There are some methods that circumvent this problem for special situations, either by avoiding the need for second derivatives or by computing just the vector $H\vec{v}$ instead of the Hessian matrix H.

Now we'll look at some algorithms. As always, the goal is to indicate tools and ways of thinking—not to provide a mere catalog of methods. As a result, some details will be omitted.

While on the subject of thinking, it's worth noting that comparing algorithms is often difficult and is particularly so for neural nets. People often compare algorithms by collecting data from many runs of the algorithms. As in practically all experiments, pitfalls await the unwary:

- Comparisons may vary from problem to problem.
- Comparisons may be sensitive to the choice of $\vec{x}_0$.
- Comparisons may be sensitive to the criteria for stopping the algorithms. (Instead of reaching a local minimum, we normally just get close.)

Newton's Method

Before discussing methods that are actually used, let's see how the Hessian can be used to find local minima.

For $g : \mathbb{R} \to \mathbb{R}$, a function of one variable, we have $\nabla g = g'$ and $H = g''$ in Theorem 11.8 (p. 449). Thus (11.10) yields the approximation

$$g(a + h) \approx g(a) + g'(a)h + \tfrac{1}{2}g''(a)h^2 \quad \text{for } |h| \text{ small.} \tag{11.13}$$

This approximation is simply a parabola and its extreme point is found by setting the derivative with respect to h equal to zero. Solving for h, we obtain $h = -g'(a)/g''(a)$. This suggests that we take $h_k = -g'(x_k)/g''(x_k)$ in Step 2 of Algorithm 11.2 (p. 455). This is known as *Newton's method*.

The idea is too naive. In the first place, it is attempting to find a point where $g'(x) = 0$, which could be a maximum or point of inflection instead of a minimum. In the second place, we must be careful with division since $|g''(x_k)|$ might be nearly zero at times. Various methods exist for circumventing such problems, but we needn't go into them because they are primarily ad hoc and would contribute little to our overall understanding of minimization methods.

Now suppose $g : \mathbb{R}^n \to \mathbb{R}$. How can Newton's method be extended to a function of several variables? Replace (11.13) by

$$g(\vec{a} + \vec{h}) \approx g(\vec{a}) + (\nabla g(\vec{a})) \cdot \vec{h} + \tfrac{1}{2}\vec{h}^t H(g(\vec{a})\vec{h}. \tag{11.14}$$

In view of Theorem 11.9(a), we set the gradient of the right side (as a function of $\vec{h}$) equal to zero. After a bit of calculation (which you should do) we obtain

$$\nabla g(\vec{a}) + H(g(\vec{a}))\vec{h} = \vec{0}.$$

(Remember that $H_{i,j} = H_{j,i}$ when you do this.) Solving this equation for the step $\vec{h}_k$, we get

$$\vec{h}_k = -\Big(H(g(\vec{x}_k))\Big)^{-1}\nabla g(\vec{x}_k), \tag{11.15}$$

which leaves us facing the same problems as in the single-variable case plus the fact that finding H and $H^{-1}\nabla$ may involve considerable computation.

This computation is nonlocal in nature—it can't be carried out by simple computing units on the edges and vertices of the neural net. As a result, let's abandon this naive use of the Hessian.

Linear Methods

Using the gradient directly gives us only the linear approximation

$$g(\vec{x} + \vec{h}) \approx g(\vec{x}) + (\nabla g(\vec{x})) \cdot \vec{h}. \tag{11.16}$$

We cannot attempt to find the minimum of this approximation since the values of the approximation are not bounded below. Indeed, if you're familiar with basic n-dimensional analytic geometry, you may have recognized that the right side of (11.16) is the equation for a hyperplane. (The one- and two-variable cases are straight lines and planes, respectively.) Such a function has neither a minimum nor a maximum.

What can we do?

Suppose we choose $\vec{h}$ so that $(\nabla g(\vec{x})) \cdot \vec{h} < 0$. Then (11.16) suggests that $g(\vec{x} + \vec{h}) < g(\vec{x})$ for small enough $\vec{h}$. In fact, this is just what is stated in (11.12). How should $\vec{h}$ be chosen? How big should it be? What direction should it point in?

The obvious choice for direction is that opposite the gradient because this provides the most rapid rate of decrease for small $|\vec{h}|$. This choice of direction for $\vec{h}$ is called *gradient descent*. Although the obvious choice, it is often not the best. Thus, newcomers to minimization often try gradient descent while full-fledged initiates look for other methods. In the next example, we explain why gradient descent has problems and also estimate the rate of convergence for gradient descent.

Example 11.4   A Difficulty with Gradient Descent

Consider the function of two variables

$$g(x, y) = \delta x^2 + y^2, \quad \text{where } \delta > 0 \text{ is small.} \tag{11.17}$$

You should be able to quickly see that the minimum of $g(x, y)$ occurs at $x = y = 0$.

The gradient at (x, y) is $2(\delta x, y)$, so gradient descent gives us $\vec{h} = -2\eta(\delta x, y)$ for some $\eta > 0$. Then

$$(x, y) + \vec{h} = ((1 - 2\delta\eta)x, (1 - \eta)y).$$

If we start at (x_0, y_0) and repeat gradient descent k times, with the same η throughout, we reach the point

$$((1 - 2\delta\eta)^k x_0, (1 - 2\eta)^k y_0). \tag{11.18}$$

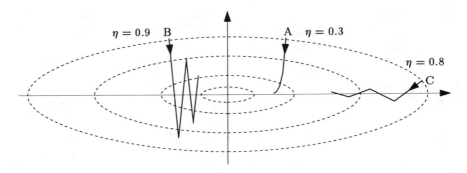

Figure 11.2   The dashed ellipses are level curves of $g(x, y) = 0.09x^2 + y^2$: i.e., curves along which $g(x, y)$ is constant. From multivariate calculus, the gradient at a point is perpendicular to the level curve through the point—a very different direction than toward the minimum at the origin. The solid lines show three possible gradient descent paths, each with five steps. The stepping rules are given by $\vec{h} = -\eta\nabla g$, where η is the learning rate.

How fast could this converge? We must have $|1 - 2\eta| < 1$ to keep the second component from increasing. That is, $0 < \eta < 1$. The Taylor series for log truncated at the linear term gives us $\log_e(1 - z) \approx -z$. Thus

$$(1 - 2\delta\eta)^k = \exp\big(k\log_e(1 - 2\delta\eta)\big) \approx \exp(-2k\delta\eta) > e^{-2k\delta} \tag{11.19}$$

since $\eta < 1$. When δ is very small, the convergence of $(1 - 2\delta\eta)^k$ to zero will be very slow, even when $\eta \approx 1$. (How large must k be to reduce x_0 by a factor of 10 when $\eta \approx 1$ and $\delta = 0.01$?—Get out your calculator!) Figure 11.2 illustrates (11.17). The value of δ is fairly large so that the figure won't be too squashed in a vertical direction to be easily seen.

The factor η in $\vec{h} = \eta\nabla g$ is called the *learning rate*. This may be somewhat of a misnomer since it is a measure of step size. On the other hand, rate of convergence can be thought of as the learning rate and (11.18) and (11.19) connect η with the rate of convergence when $0 < \eta < 1$.

The reasoning for (11.17) can be extended to a general function g near a local minimum $\vec{a}$. The role played by δ will be taken by the ratio of eigenvalues of the Hessian. We do this now.

Using (11.14), we get $g(\vec{a} + \vec{h}) \approx g(\vec{a}) + \frac{1}{2}\vec{h}^t H\vec{h}$, where H is the Hessian of g at $\vec{a}$ and there is no gradient term because the gradient vanishes at a local minimum.

Imagine applying Theorem 11.5 (p. 444) with $A = H$ and $\vec{x} = \vec{h}$. In the resulting new coordinate system, $\vec{h}^t H\vec{h} = \lambda_1 y_1^2 + \cdots + \lambda_n y_n^2$. Since we are supposed to be near a minimum, $\vec{h}^t H\vec{h}$ should be positive for all $\vec{h} \neq \vec{0}$. The only way that this can happen for all choices of $\vec{y}$ is to have $\lambda_k > 0$ for all k.

Using Exercise 11.2.9 (p. 450), we can show that gradient descent with learning rate η in the new coordinates and gradient descent with learning rate η in the old coordinates lead to the same points. Our method for studying (11.17) can be applied in the new coordinates. Let $\lambda_{\max}$ be the largest of the λ_i's. It can be shown that after k steps the ith coordinate has decreased by about a factor of $\exp(-2\mu k \lambda_i / \lambda_{\max})$, where $0 < \mu < 1$. We omit the details. Thus the rate of convergence of gradient descent depends on the ratio $\lambda_{\min}/\lambda_{\max}$ of the smallest to the largest eigenvalue of the Hessian. Often this ratio is quite small and so convergence is slow. Phrased another way, the number of steps needed to at least halve all coordinates is proportional to $\lambda_{\max}/\lambda_{\min}$. Thus, we say that the *time constant* for gradient descent is proportional to $\lambda_{\max}/\lambda_{\min}$. When the time constant is large, convergence is slow. ∎

How common is the sort of behavior in the previous example? Empirical evidence indicates that $\lambda_{\max}/\lambda_{\min}$ is usually quite large for neural nets. What can we do to overcome the problem?

One approach is to carefully choose the *distance* that we go in the gradient direction. This is a tricky problem: We try to find the least positive t_k that minimizes $g(\vec{x}_k - t_k \nabla g(\vec{x}_k))$.

At first, this appears to be no gain at all—we've introduced yet another minimization problem. But the new problem is *one-dimensional* instead of n-dimensional because the variable t_k is a real number, not a vector. For large n, the new problem is much simpler than minimizing $g : \mathbb{R}^n \to \mathbb{R}$.

Finding t_k is called *line search*. This is a popular method in numerical analysis texts. Unfortunately it may be impractical for neural nets because computing g is usually costly. The following example shows how the idea might be adapted and indicates a major difficulty.

Example 11.5   Approximating Line Search

In gradient descent, we have $\vec{h}_k = -\eta_k \nabla g(\vec{x}_k)$ and want to choose the *learning rate* η_k to be close to the t_k that gives a minimum for line search. The following observation suggests a method.

Let $h(t) = g(\vec{x}_k - t \nabla g(\vec{x}_k))$ and let t_k be the least positive t such that $h(t)$ is at a local minimum. Let $\vec{x}_k(t) = \vec{x}_k - t \nabla g(\vec{x}_k)$ and let $\vec{x}_{k+1} = \vec{x}_k(\tau)$

for some τ. Then there is a $T > t_k$ such that

$$\left(\nabla g(x_{k+1})\right) \cdot \left(\nabla g(x_k)\right) \begin{cases} \geq 0, & \text{if } 0 < \tau < t_k, \\ = 0, & \text{if } \tau = t_k, \\ \leq 0, & \text{if } t_k < \tau < T. \end{cases} \qquad (11.20)$$

The proof is left as an exercise.

We might adjust our value of η_k from step to step based on (11.20): Adjust it up or down, depending on whether the dot product of the two previous gradients was positive or negative. This does not tell us *how much* of an adjustment to make, but at least it tells us whether to increase or decrease η_k. In making such adjustments, we must be careful that η_k not get too large because it might exceed T.

This proposal is not a purely local condition since the computation of the dot product $(\nabla g(\vec{x}_k)) \cdot (\nabla g(\vec{x}_{k-1}))$ requires that all the terms be summed up at a central location. On the other hand, this is a relatively simple calculation.

This seems like a good way to approximate line search. Unfortunately, the step size that makes $(\nabla g(\vec{x}_k)) \cdot (\nabla g(\vec{x}_{k-1}))$ nearly zero will probably be quite different from the one that makes $(\nabla g(\vec{x}_{k+1})) \cdot (\nabla g(\vec{x}_k))$ nearly zero. Thus, adapting η_k may be impractical since the best value now and the best value next can differ quite a bit.

There's another approach based on the observation that even though computing H is impractical, computing $H\vec{v}$ is practical [28]. Let's drop the subscript k that indicates the step and let $\vec{v} = \nabla g(\vec{x})$. The line search problem is then to choose t to minimize $g(\vec{x} + t\vec{v})$. By the quadratic approximation formula in Theorem 11.8 (p. 449) with $\vec{a} = \vec{x}$ and $\vec{h} = t\vec{v}$, this is similar to minimizing the quadratic

$$g(\vec{x}) + (\nabla g) \cdot (t\vec{v}) + \tfrac{1}{2}(t\vec{v})^{\text{t}} H(t\vec{v}) = g(\vec{x}) + \left((\nabla g)^{\text{t}}\vec{v}\right) t + \tfrac{1}{2}\left(\vec{v}^{\text{t}} H \vec{v}\right) t^2,$$

where ∇g and H are evaluated at $\vec{x}$. The critical point of this quadratic is at

$$t = -\left(\frac{(\nabla g)^{\text{t}}\vec{v}}{\vec{v}^{\text{t}} H \vec{v}}\right),$$

which is a minimum if $\vec{v}^{\text{t}} H \vec{v} > 0$. Since $H\vec{v}$ can be computed locally and in about the same amount of time it takes to compute ∇g, this method is practical. ∎

Another approach to the gradient descent problem is to move in a direction other than the gradient direction. One idea is to learn from previous movement, which is the idea behind "momentum." Another approach is to use some quadratic method.

Example 11.6 Momentum

Look back at Figure 11.2. Notice that as we use gradient descent we either take steps that are too small (path A) or tend to reverse our direction with each step (paths B and C). If we learned from our past mistakes, so to speak, we would take less timid steps than in path A while avoiding the excessive motion in the y direction in paths B and C. How can this be done?

To learn from past mistakes, we need some sort of simple memory. In physical systems, there is a simple memory that tends to damp out radical changes in direction—momentum. The minimization method based on this idea is also referred to as momentum. It sets

$$\vec{h}_k = \eta_k \left(-\nabla g(\vec{x}_k) + \mu_k \vec{h}_{k-1} \right) \tag{11.21}$$

where η_k and μ_k are some positive numbers that must be chosen. As in gradient descent, η_k is called the learning rate. We call μ_k the *momentum*. When the momentum equals zero, (11.21) reduces to gradient descent.

Suppose $\eta_k = \eta$ is fixed and $\mu_k = \mu(\eta)$ is fixed at an optimum value that depends on η. (The optimum μ cannot be determined *a priori*, so algorithms are designed to adjust the value of μ based on previous steps.) It can be shown that, under reasonable assumptions, the time constant for the momentum method is then proportional to $\sqrt{\lambda_{\max}/\lambda_{\min}}$, where $\lambda_{\max}$ and $\lambda_{\min}$ are the largest and smallest eigenvalues of H. Since this is the square root of the result for gradient descent, momentum should outperform gradient descent when the learning rate is fixed. It does.

There's an interesting contrast between momentum learning and the previous example. The previous example suggests that we should try to adjust the learning rate so that the dot product of successive gradients vanishes; that is, they are perpendicular. On the other hand, momentum tries to straighten out the changes in direction. Studying conflicting ideas such as these two is often rewarded by a deeper insight into the situation. What can we say here? As noted at the end of Example 11.5, adjusting the learning rate as desired could be difficult. It seems easier to adjust the learning rate and momentum in a good manner. However, I'm not aware of any empirical or theoretical confirmation of this.

How good is the momentum method? It varies considerably, depending on the nature of the surface. The method is an ad hoc approach to the problem of descending "ravines"—regions with steep sides but a slow overall descent rate. In other situations, it sometimes performs poorly. ■

Quadratic Methods

Quadratic methods are usually significantly better than linear ones. Unfortunately, they often involve computing a matrix that measures changes in g with respect to pairs of values x_i, x_j. This appears to cause both storage and computational problems since both grow at least quadratically in the number of parameters rather than linearly. Furthermore, such computations cannot possibly be done locally since local storage is limited and the number of local computation sites is essentially equal to the number of parameters. On the other hand:

- If the number of parameters is not large and you're not interested in doing calculations in a local manner, the problems are nonexistent.

- Total work equals number of steps times the work per step. Since quadratic methods usually converge significantly faster, the need for fewer steps could easily mean less total work for a quadratic method than for a linear one. This doesn't address the problem of local computation.

- There's a practical use of the Hessian, which we saw while discussing line search on page 460. The idea is that $H\vec{v}$ doesn't require a quadratic amount of storage. Pearlmutter [28] discusses the efficient computation of $H\vec{v}$.

- Finally, a quadratic method need not use the Hessian or a quadratically growing amount of storage. That idea is discussed here.

While linear methods have dominated network minimization algorithms, it's likely that quadratic methods will dominate in the near future.

Example 11.7 Sums of Many Squares

When a problem has some special structure, we should try to exploit that structure. Anticipating Section 11.5, let's note that in feedforward nets we're often trying to minimize a function of the form

$$g(\vec{x}) = \tfrac{1}{2} \sum_{s=1}^{S} \left(f_s(\vec{x}) - t_s \right)^2. \tag{11.22}$$

(The value $|f_s - t_s|$ is the error in an output unit for a particular pattern.) A look at books on minimization reveals that minimizing sums of squares is a common problem and that special methods have been developed. Can we use any of them here?

Of course, an expression for the gradient can easily be written down:

$$\nabla g = \sum_{s=1}^{S} (f_s - t_s) \nabla f_s. \tag{11.23}$$

What's more interesting is that an approximation to $H^{-1}\nabla$ can be found, thereby opening up the possibility of using (11.15). Differentiating (11.23), we have

$$H_{i,j} = \sum_{s=1}^{S} (f_s - t_s) \frac{\partial^2 f_s}{\partial x_i \, \partial x_j} + \sum_{s=1}^{S} \frac{\partial f_s}{\partial x_i} \frac{\partial f_s}{\partial x_j}. \tag{11.24}$$

We can hope that the terms in the sums will be randomly positive and negative, and so tend to cancel out—except in the second sum when $i = j$ since all the terms are then squares. Thus, we have the approximation

$$H_{i,j} \approx \begin{cases} \sum_{s=1}^{S} (\partial f_s / \partial x_i)^2, & \text{if } i = j, \\ 0, & \text{otherwise.} \end{cases}$$

Hence

$$(H^{-1})_{i,j} \approx \begin{cases} \dfrac{1}{\sum_{s=1}^{S} (\partial f_s / \partial x_i)^2}, & \text{if } i = j, \\ 0, & \text{otherwise,} \end{cases}$$

and so

$$-(H^{-1}\nabla)_i \approx \frac{-\sum_{s=1}^{S} (f_s - t_s) \, \partial f_s / \partial x_i}{\sum_{s=1}^{S} (\partial f_s / \partial x_i)^2}.$$

From (11.15), the n-variable Newton's method, it seems reasonable to define

$$(\vec{h})_i = \frac{-\sum_{s=1}^{S} (f_s - t_s) \, \partial f_s / \partial x_i}{\sum_{s=1}^{S} (\partial f_s / \partial x_i)^2}. \tag{11.25}$$

This approximation does not have the computational problems of the general formula (11.15).

The approach in this example is closely related to an algorithm of Becker and Le Cun [5]. Their method does not rely on having a sum of squares and, even for the sum of squares (11.22), gives a step that differs from (11.25). See Exercise 11.5.4 (p. 476). ∎

In the previous example, we approximated the Hessian H, obtaining the quadratic approximation

$$g(\vec{a} + \vec{h}) \approx g(\vec{a}) + (\nabla g) \cdot \vec{h} + \tfrac{1}{2} \sum_i c_i h_i^2,$$

whose right side we then minimized. This approximation has an interesting feature: There are no terms of the form $h_i h_j$ present. What this means is that interaction between different components of $\vec{h}$ has been neglected in estimating $g(\vec{x} + \vec{h})$. Perhaps we should look for other methods that ignore such interaction. That idea provides the basis for the next example.

Example 11.8   Quickprop

We'll look at minimizing a general function $g(\vec{x})$ and neglect the interaction between different components to obtain a quadratic approximation. In contrast to the previous example, we won't attempt to approximate the Hessian. Instead, we'll use only the function and the gradient.

Suppose we have a function of one variable and want to approximate it with a parabola using just function values and derivatives. Since a parabola is an arbitrary quadratic, $y = Ax^2 + Bx + C$, there are three unknown coefficients. Since we want to determine three numbers A, B, and C, we need three numerical pieces of information about $g(x)$.

Since we're not using the Hessian, only the function g and its derivative are available. Evaluating them at x_k gives only two data points, which is not enough. Hence, values of g and g' at previous points must also be used. How should we choose among g and g' at the current point x_k and the previous points, x_{k-1}, x_{k-2}, ... ? Since we have three constants to determine—A, B, and C for the parabola—it suffices to choose three of these values. Which ones? A good rule of thumb is

> In iterative algorithms, give the most emphasis to the most recent information.

This tells us that we should use $g(x_k)$ and $g'(x_k)$ together with either $g(x_{k-1})$ or $g'(x_{k-1})$.

As we'll see, $g(x_k)$, $g'(x_k)$, and $g'(x_{k-1})$ are the easiest to use because of a nice property of parabolas:

If $p(x) = Ax^2 + Bx + C$, then $\dfrac{p'(b) - p'(a)}{b - a} = 2A$ whenever $b \neq a$.   (11.26)

Suppose that $p'(c) = 0$ so that c is a maximum (if $A < 0$) or minimum (if $A > 0$) of $p(x)$. Put $a = c$ in (11.26) to obtain

$$\frac{p'(b) - 0}{b - c} = 2A = \frac{p'(b) - p'(a)}{b - a} \quad \text{and so} \quad c - b = \frac{p'(b)(b - a)}{p'(a) - p'(b)}.$$

Approximating $p(x)$ with $g(x)$, setting $a = x_{k-1}$, $b = x_k$, and $c = x_{k+1}$, leads to

$$h_k = \begin{cases} \dfrac{(x_k - x_{k-1})g'(x_k)}{g'(x_{k-1}) - g'(x_k)}, & \text{if } \dfrac{g'(x_k) - g'(x_{k-1})}{x_k - x_{k-1}} > 0, \\ \text{??? (undecided)}, & \text{otherwise.} \end{cases} \quad (11.27)$$

The undecided case corresponds to a parabola whose critical point is a maximum. If g is reasonably well-behaved, this should occur rarely if ever. Nevertheless, we must decide what to do. About all that can be done is to head down the slope; that is, somehow choose a value for h_k that is opposite in sign from $g'(x_k)$. If this is done so that $|h_k|$ is small enough, then $g(x_{k+1}) < g(x_k)$. (We also need to be careful of small denominators in the first case.)

You may have noticed that (11.27) uses only two values, namely $g'(x_k)$ and $g'(x_{k-1})$, but we pointed out that three values are needed to determine A, B, and C. What happened?

$*$ $*$ $*$ Stop and think about this! $*$ $*$ $*$

We computed only one number, the minimum point of the parabola. The x coordinate of the minimum depends only on slope information. Since a single y coordinate provides no slope information, $p(x_k)$ does not enter the calculations. If we'd used $p'(x_k)$, $p(x_k)$, and $p(x_{k-1})$, the answer would have involved $p'(x_k)$ and $\big(p(x_k) - p(x_{k-1})\big) \big/ \big(x_k - x_{k-1}\big)$.

The above calculations were done for a function of a single variable. Since we're ignoring the effect of interactions between different components of $\vec{h}$, we can use (11.27) for $g(\vec{x})$. All we need do is (a) replace $g'(x)$ with $\partial g(\vec{x})/\partial x_i$ and (b) replace h_k and $x_k - x_{k-1} = h_{k-1}$ with the ith components of the vectors $\vec{h}_k$ and $\vec{h}_{k-1}$. This is the core of Fahlman's Quickprop algorithm [8]. ■

Exercises

11.4.A. Describe a generic hill-climbing/descent algorithm.

11.4.B. Distinguish between simple search, linear approximation methods, and quadratic approximation methods for minimization. Which are usually infeasible because of the amount of calculation? the amount of storage?

11.4.C. Give three reasons why it is often difficult to compare minimization algorithms by testing them on a variety of inputs.

11.4.D. Why are quadratic methods good? Why are they bad?

11.4.E. What is gradient descent? Why is it bad? What is the momentum method and how does it attempt to overcome the problem with gradient descent?

11.4.F. What basic principle of iterative algorithm design does Quickprop use?

11.4.G. What is the idea (assumption) behind Quickprop? In other words: Although Quickprop requires only first derivatives, it gives a quadratic approximation if a certain assumption holds. What is the assumption?

11.4.1. Show that, of all vectors with $|\vec{h}| = \delta$, the one that makes $\nabla g(\vec{x}) \cdot \vec{h}$ a maximum is given by

$$\vec{h} = \left(\frac{\delta}{|\nabla g(\vec{x})|} \right) \nabla g(\vec{x}).$$

Show that the vector for a minimum is given by $-\vec{h}$.

11.4.2. Fill in the details of the proof at the end of Example 11.4 about the rate of convergence in the general case. Don't forget to show that using gradient descent in the new coordinates is the same as using it in the original coordinates.

11.4.3. The goal is to prove (11.20).

(a) Show that $h'(t) = -\big(\nabla g(\vec{x}_k)\big) \cdot \big(\nabla g(\vec{x}_k(t))\big)$.

(b) Use the previous part and single-variable calculus to deduce (11.20).

11.4.4. In this exercise, you'll verify the comment at the end of Example 11.5 in a particular case. Let $g(\vec{x}) = \frac{1}{2}(10x_1^2 + 3x_2^2 + x_3^2)$.

(a) Show that to have $\nabla g(\vec{x}_{k+1})$ and $\nabla g(\vec{x}_k)$ perpendicular requires that

$$\eta_k = \frac{10^2 x_{k,1}^2 + 3^2 x_{k,2}^2 + x_{k,3}^2}{10^3 x_{k,1}^2 + 3^3 x_{k,2}^2 + x_{k,3}^2}.$$

(b) Suppose $\vec{x}_0 = (10, 10, 10)$. Show that $\eta_0 = 0.107$, $\vec{x}_1 = (-0.7, 6.8, 9.9)$, and $\eta_1 = 0.19$.

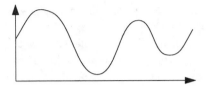

Figure 11.3   A textbook example with few relative minima and maxima versus the sort of picture that may be closer to what occurs in real-world problems.

11.4.5. In the momentum method, show that, for sufficiently small η_k, the step goes down whenever

$$\left|\nabla g(\vec{x}_k)\right|^2 + \lambda_k\left(\nabla g(\vec{x}_k)\right) \cdot \vec{h}_{k-1} > 0.$$

Also, show that it goes up if the inequality is reversed.

11.4.6. In the momentum method, it is important that η_k and λ_k not be too large. To begin analyzing the situation, let's make the totally false assumption that $\nabla g(\vec{x})$ is independent of $\vec{x}$. This is a sort of worst-case assumption because all the vectors will point in the same direction and so there will be no cancellation. Also, set $\eta_k = \eta$ and $\lambda_k = \lambda$ for all k. To start the momentum algorithm, set $\vec{h}_0 = \vec{0}$.

 (a) Show that

$$\vec{h}_k = -\eta\frac{1 - (\eta\lambda)^k}{1 - \eta\lambda}\nabla g.$$

 (b) Show that $|\eta\lambda| < 1$ is necessary and sufficient for $\vec{h}_k$ to converge.

11.4.7. Fill in the details of the derivation of (11.27).

11.4.8. What do we obtain in place of (11.27) for the step h_k if we use $g(x_k)$, $g'(x_k)$, and $g(x_{k-1})$?

*Domains of Attraction

Real-world functions are often more complicated than those in calculus texts. For functions of a single variable, Figure 11.3 shows what you encounter in textbooks as opposed to what you may encounter in the real world. Both of the local minima in the interior of the left-hand curve might be acceptable, but many of the local minima of the right-hand curve probably would not be. It then becomes important to reach an acceptable local minimum. In AI, we usually have *many* variables—so the situation is even more complicated.

A useful concept is "domain of attraction." Suppose that $g : \mathbb{R}^2 \to \mathbb{R}$. Then we can construct a physical model of g, with a vertical axis for $g(x, y)$ and horizontal axes for x and y. Imagine a drop of water released on this surface at $(x, y, g(x, y))$. If it rolls without inertia, it will reach a local minimum $\vec{v}$. We say that (x, y) is in the domain of attraction of $\vec{v}$. More generally:

Definition 11.10   Domain of Attraction

Let $\vec{v}$ be a local minimum of a continuous function $g : \mathbb{R}^n \to \mathbb{R}$. We say that $\vec{x}$ is in the *domain of attraction*, or *attraction basin*, of $\vec{v}$ if there is a "downhill path" from $\vec{x}$ to $\vec{v}$.

Aside. For those who want precision, "downhill path" can be made precise as follows. We assume that g is differentiable. If $p : [0, 1] \to \mathbb{R}^n$ is a differentiable function such that $p(0) = \vec{x}$, $p(1) = \vec{v}$ and $(g(p(t))' \leq 0$ for $t \in (0, 1)$, then p is a *downhill path* from $\vec{x}$ to $\vec{v}$.

Suppose that some fraction r of all possible starting points $\vec{x}_0$ belong to the basins of unacceptable local minima. For example, they may be unacceptable because the function values are too large. If we choose $\vec{x}_0$ at random, our probability of reaching an acceptable local minimum is roughly $1 - r$. (We said "roughly" because a minimization algorithm may sometimes take a step from one basin to another.) Here are some ways we might improve the $1 - r$ probability.

- We could change the function g. This may be impractical.

- We could try several random starts and choose the best answer. With k independent random starts, the probability that all of them lie in a bad basin is roughly r^k and so the probability of success is roughly $1 - r^k$. For example, if $r = 0.3$, quadrupling the amount of work by setting $k = 4$ changes the chances of success from 70% to more than 99%.

- Unacceptable local minima often tend to have smaller basins of attraction. We can we try to escape from them by sometimes taking large steps or by sometimes going uphill. This is more likely to get us out of a bad basin than out of a good one. It's common to try to eliminate uphill tendencies. Ironically, this might make an algorithm worse. A systematic approach to uphill movement is provided by "simulated annealing."

Example 11.9   Simulated Annealing for Hopfield Nets

When a Hopfield net is storing many patterns, spurious local minima arise. These typically have larger function values than minima associated with patterns stored in the net. Our definition for basins of attraction does not apply since vertex outputs are limited to the discrete values ± 1 (and sometimes 0), while our definition requires real values. Nevertheless, the problem of unacceptable local minima still exists. In the previous section, we gave an algorithm for the Hopfield net which had the property that each step took it downhill. As a result, it might stop at a spurious local minimum. We'll indicate how the algorithm can be modified to escape from bad minima by using *simulated annealing*.

Simulated annealing receives its name from the physical process of annealing in which a metal is tempered by gradually lowering the temperature. At any time, an atom can move to a new state. The probability of choosing a state depends on the energy difference and the temperature. With slow cooling the crystalline state of the metal achieves a lower overall potential energy because the probability of escaping from local potential energy minima is greater at higher temperatures.

Call the quantity in (11.5) (p. 433) $\Delta_l(\vec{o})$. Since it equals $\Phi(\vec{o}') - \Phi(\vec{o})$, it measures how much the energy will increase if the lth component of $\vec{o}$ is changed. If the absolute "temperature" is T, then simulated annealing would change the lth component of $\vec{o}$ with probability

$$\Pr(\text{change } o_l) = \frac{1}{1 + \exp(\Delta_l(\vec{o})/T)}.$$

Whether o_l changes or not is decided at random using this probabilty. (Many versions of simulated annealing set $\Pr(\text{change } o_l) = 1$ when $\Delta_l(\vec{o}) < 0$.)

To complete this discussion, we must specify T as a function of time t. This is referred as an *annealing schedule*. It is known that if $T(t) = C/\log t$ for some constant C that depends on the problem, then the simulated annealing will converge to a global minimum. Unfortunately, the time required for this leads to excessive computation, so some compromises are needed. Choosing effective annealing schedules is a difficult problem.

The algorithm I've just described brings us close to the concept of a *Boltzmann net*, which I won't discuss. ∎

Unfortunately, understanding the basins of attraction of functions that arise in practice is quite difficult. Success at the theoretical level is hampered by the complexity of the situation. Success at the experimental (numerical) level is hampered by the considerable number of calculations needed to obtain a reasonable amount of information.

Exercises

11.4.H. What is a domain of attraction?

11.4.I. How can repeated random starts of a algorithm help overcome the problem of bad local minima?

11.4.J. How can simulated annealing help overcome the problem of bad local minima?

11.4.9. The following method is proposed for finding domains of attraction. Select an algorithm that finds a local minimum given a starting point $\vec{x}_0$. Select a set of points so that the space of possible $\vec{x}$'s is well covered. For each $\vec{x}$ in the set, run the algorithm with $\vec{x}_0 = \vec{x}$ and note the minimum $\vec{m}(\vec{x})$ to which it converges. We then claim that $\vec{x}$ belongs to the domain of attraction of $\vec{m}(\vec{x})$, with due allowance for the fact that $\vec{m}(\vec{x})$ may be sightly off from a minimum because the algorithm will probably not *exactly* reach a minimum.

(a) Explain why this method will probably not give an accurate picture of the domains of attraction.

(b) Defend or attack the thesis that the result obtained by the above algorithm is more relevant than the actual domains of attraction. (There are arguments for both sides.)

11.4.10. Using the ideas in Example 11.9, write out an algorithm like Algorithm 11.1 (p. 434).

*11.4.11. Let $\vec{x}_1, \ldots, \vec{x}_P \in \mathbb{R}^n$ be such that every vector in $\mathbb{R}^n$ can be written in the form $\sum_p c_p \vec{x}_p$ for some $c_p \in \mathbb{R}$. (The c_p's may not be unique.) Let $y_1, \ldots, y_P \in \mathbb{R}$. Define $g : \mathbb{R}^n \to \mathbb{R}$ by

$$g(\vec{w}) = \frac{1}{2} \sum_{p=1}^{P} (\vec{w} \cdot \vec{x}_p - y_p)^2,$$

Prove that g has a unique local minimum and that its basin of attraction is the entire space $\mathbb{R}^n$.

11.5   Backpropagation for Feedforward Nets

Backpropagation is an application of the chain rule to the computation of partial derivatives in feedforward nets. A *feedforward net* is a net whose directed graph has no cycles. We call such a graph a DAG—directed acyclic graph. A feedforward net is a DAG with

$$I = \left\{ v \in V \mid \text{there is no } x \in V \text{ with } (x, v) \in E \right\}$$

and

$$O = \left\{ v \in V \mid \text{there is no } x \in V \text{ with } (v, x) \in E \right\}$$

The simplest sort of feedforward net is one with no hidden units. Unless $|I|$ is large, such a net is not very useful. The next step up is a feedforward net with

$$V = I \cup H \cup O \quad \text{and}$$
$$E \subseteq (I \times H) \cup (H \times O).$$

We call H the hidden layer. More generally, we can define

$$V = I \cup H_1 \cup \cdots \cup H_k \cup O \quad \text{and}$$
$$E \subseteq (I \times H_1) \cup (H_1 \times H_2) \cup \cdots \cup (H_{k-1} \times H_k) \cup (H_k \times O)$$

to obtain a feedforward net with k hidden layers. Of course, a feedforward net need not have the hidden vertices in layers. These ideas are illustrated in Figure 11.4 on the following page.

For Hopfield nets, training is simple—just set $w_{i,j} = \sum P_i(k)P_j(k)$. On the other hand, getting an answer (i.e., recovering a pattern) is relatively time-consuming. This may cause problems in applications: A net may be trained once and then used repeatedly. Using feedforward nets is simple because a pattern is simply propagated forward from input vertices to the output vertices. In contrast, training is a time-consuming minimization process based on the methods of the previous section. You've already had a taste of this with the perceptron training algorithm in Theorem 11.10 (p. 451). The algorithms for general feedforward nets are much slower and offer no guarantees that they will find good local minima.

Feedforward nets have been studied more than any other neural net architecture. This is probably due to their ability to fit training sets and the concept of "backpropagation," which forms the basis for a variety of learning algorithms. In this section, we'll limit ourselves to discussing feedforward nets with just one hidden layer for the following reasons:

- The behavior of these nets is sufficiently rich to be useful.

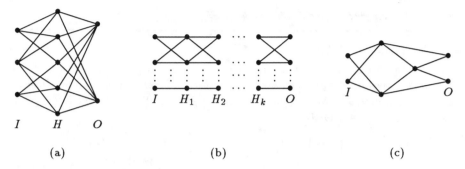

$I \qquad H \qquad O$

$I \quad H_1 \quad H_2 \qquad H_k \quad O$

$I \qquad\qquad O$

(a)   (b)   (c)

Figure 11.4 Feedforward nets that have (a) one hidden layer, (b) k hidden layers, (c) hidden units not in layers. All edges point toward the right. (Arrowheads have been omitted to avoid clutter.)

- The limitation helps us keep discussions simple, especially in equations.
- The computational ideas and the observations apply to general feedforward nets.

After introducing notation, we'll state the backpropagation algorithm for localized computation of the partial derivatives. Some concerns associated with implementation are discussed in the following section.

Assumptions and Notation

The special form we're assuming in this section is a feedforward net with one hidden layer, where

- all possible edges are present so $E = (I \times H) \cup (H \times O)$,
- all vertex functions have the form in (10.1) (p. 409), and
- the parameters should be chosen to minimize the sum of the squares of the output errors over the training set.

The top part of Figure 11.5 shows how calculations move through the net.

The following notation will be used throughout

$$|I| = a, \text{ elements of } I \text{ are denoted by } \alpha\text{'s},$$
$$|H| = b, \text{ elements of } H \text{ are denoted by } \beta\text{'s},$$
$$|O| = c, \text{ elements of } O \text{ are denoted by } \gamma\text{'s},$$

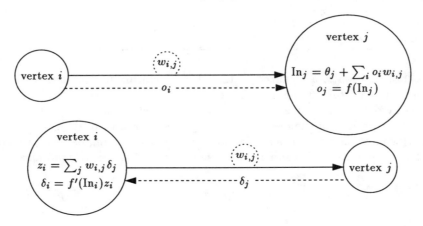

Figure 11.5   The top diagram shows information flow from i to j during output calculation. The bottom diagram shows information flow from j to i during backpropagation. Dashed arrows indicate information flow. (The variable z_i was introduced just for the figure.)
Warning: It's common in the neural net literature to write $w_{j,i}$ for the weight of edge (i, j), so you may need to reverse the order of indices when reading the literature.

$$P = \text{number of patterns in the training set,}$$
$$o_\alpha(p) = \text{output of } \alpha \text{ given in the } p\text{th training pattern, and}$$
$$o_\gamma^*(p) = \text{desired output of } \gamma \text{ for } p\text{th training pattern.}$$

Exercises

11.5.A. What is a feedforward net? a hidden layer?

11.5.B. Explain the following statements. Training times differ markedly between Hopfield nets and feedforward nets. The same is true for the times required to classify a pattern after training.

11.5.1. Here we consider the possibility of designing a simple neural net with no hidden states. We also assume that the outputs are all ± 1. Define

$$w_{\alpha,\gamma} = \sum_{p=1}^{P} o_\alpha(p) o_\gamma^*(p)$$

and let $f_v = \pm 1$ according to the sign of $\sum_x o_x w_{x,v}$. Compute the argument of f_γ when pattern $o_\alpha(l)$ is fed into the network and use it to show that for small enough P we can expect to get output $o_\gamma^*(l)$.

Backpropagation

Backpropagation is a localized method for computing partial derivatives by using the chain rule. The calculations are localized in the sense that computations are done at the edges and vertices, each of which has a severely limited storage capacity, and information is propagated backward along the edges. This localization makes it feasible to consider implementing backpropagation in a massively parallel manner. In our notation, the goal is to minimize

$$g(\vec{w},\vec{\theta}) = \sum_{p=1}^{P} g(p,\vec{w},\vec{\theta}), \quad \text{where} \quad g(p,\vec{w},\vec{\theta}) = \tfrac{1}{2}\sum_{\gamma \in O}\left(o_\gamma(p) - o_\gamma^*(p)\right)^2, \quad (11.28)$$

$\vec{w}$ is the vector of all weights $w_{i,j}$, $\vec{\theta}$ is the vector of all θ's,

$$o_\gamma(p) = f\left(\theta_\gamma + \sum_{\beta \in H} o_\beta(p)w_{\beta,\gamma}\right), \quad\quad (11.29)$$

$$o_\beta(p) = f\left(\theta_\beta + \sum_{\alpha \in I} o_\alpha(p)w_{\alpha,\beta}\right), \quad\quad (11.30)$$

and the values of $o_\alpha(p)$ are given by the training pattern.

Since the values of $o_\gamma(p)$ are built up by functional compositions, this is a natural situation for the chain rule, Theorem 11.7 (p. 448). Let In_γ and In_β denote the arguments of f in (11.29) and (11.30), respectively. We use this notation because the argument of f at a vertex v is thought of as the signal coming *into* the vertex, which then sends *out* the signal $f(\text{In})$. For $(i,j) \in E$, the following partial derivatives are useful:

$$\frac{\partial g(p,\vec{w},\vec{\theta})}{\partial \text{In}_\gamma} = \left(o_\gamma(p) - o_\gamma^*(p)\right)f'(\text{In}_\gamma), \qquad \text{for starting,}$$

$$\frac{\partial \text{In}_j}{\partial \text{In}_i} = w_{i,j}\,f'(\text{In}_i), \qquad\qquad \text{for backpropagating,} \quad (11.31)$$

$$\frac{\partial \text{In}_j}{\partial w_{i,j}} = o_i \quad \text{and} \quad \frac{\partial \text{In}_j}{\partial \theta_j} = 1, \qquad \text{for computing the gradient.}$$

You should derive these formulas.

Using (11.31) gives rise to *backpropagation*, so called because information is transmitted along edges of the DAG (V,E) in the reverse direction of the edges, starting at O and moving back to I. An important observation follows from the facts that (11.28) is a sum over patterns and the derivative of a sum is the sum of the derivatives:

> Derivative information can be collected at each vertex by backpropagating information for each pattern separately and then adding up the results.

Figure 11.5 may help you picture what is happening in the backpropagation algorithm that follows.

Algorithm 11.3   Backpropagation

Suppose that a feedforward net with a single hidden layer is given. Suppose that a training pattern of matched input ($o_\alpha(p)$ for $\alpha \in I$) and desired output ($o_\gamma^*(p)$ for $\gamma \in O$) are given. First carry out the forward computation in the usual manner:

$$\text{In}_\beta = \theta_\beta + \sum o_\alpha w_{\alpha,\beta}, \quad o_\beta = f(\text{In}_\beta)$$

and then similarly for In_γ and o_γ. After this, information for the pth pattern is backpropagated as follows.

1. **Send to γ:** Send to all $\gamma \in O$ the value $(o_\gamma(p) - o_\gamma^*(p))$. Each γ multiplies what it receives by $f'(\text{In}_\gamma)$ and calls the result δ_γ.

2. **Send to β and (β, γ):** Each $\gamma \in O$ sends to all $\beta \in H$ and to each edge (β, γ) the value δ_γ. Each (β, γ) multiplies what it receives by o_β and calls the result $\delta_{\beta,\gamma}$. Each β multiplies what it receives by $w_{\beta,\gamma} f'(\text{In}_\beta)$, sums over γ, and calls the result δ_β.

3. **Send to (α, β):** Each $\beta \in H$ sends to all edges (α, β) for $\alpha \in I$ the value δ_β. Each (α, β) multiplies what it receives by o_α and calls the result $\delta_{\alpha,\beta}$.

When the algorithm is done, δ_i is the partial derivative of the pth term of (11.28) with respect to In_i, which is also the partial with respect to θ_i. Also, $\delta_{i,j}$ is the partial derivative of the pth term of (11.28) with respect to $w_{i,j}$.

The proof of the algorithm's correctness uses (11.31) and is left as an exercise.

As a result of backpropagation, each edge and vertex has stored information for the pth pattern in a value called δ. Let each vertex and each edge sum its δ's over all training patterns. The result is the gradient of $g(\vec{w}, \vec{\theta})$, where the partial with respect to $\theta_v \in V$ is the sum at vertex v and the partial with respect to $w_e \in E$ is the sum at edge e.

The gradient is sufficient for the momentum and Quickprop equations, (11.21) (p. 461) and (11.27) (p. 465), respectively.

In the Newton's method approximation (11.25) (p. 463), more information is needed: We must also compute the sum of squares in the denominator of (11.25). To do this, we backpropagate to compute the values of $C_v = \partial o_\gamma / \partial \theta_v$ and $\partial o_\gamma / \partial w_e$. They begin $C_\gamma = f'(\text{In}_\gamma)$ and are propagated backward just as the δ_v's are. Each edge and vertex then takes what it has, squares it, and sums the squares over all input patterns.

Exercises

11.5.C. Explain the meaning and importance of the statement that backpropagation involves localized calculations.

11.5.2. Prove the claims in Algorithm 11.3 by judicious use of the chain rule.

11.5.3. Write out explicitly the propogation rules for the C_v's for using (11.25) (p. 463) and verify that they give the correct partials.

*11.5.4. Becker and Le Cun [5] have proposed an alternative method to that in Example 11.7 (p. 463) for neglecting diagonal terms. It doesn't depend on g's being a sum of squares. You'll study it here.

(a) Derive a formula for $\partial^2 g(p,\ldots)/(\partial \ln_\gamma)^2$ where $g(p,\ldots)$ is given by (11.28).

(b) Becker and Le Cun approximate $\partial^2 g(p,\ldots)/(\partial \ln_\beta)^2$ by

$$\sum_\gamma \frac{\partial^2 g(p,\ldots)}{(\partial \ln_\gamma)^2} \left(w_{\beta,\gamma} f'(\ln_\beta)\right)^2 + \sum_\gamma \frac{\partial g(p,\ldots)}{\partial \ln_\gamma} w_{\beta,\gamma} f''(\ln_\beta).$$

What terms are being neglected? Does this seem reasonable?

(c) Becker and Le Cun approximate $\partial^2 g(p,\ldots)/(\partial w_{i,j})^2$ by

$$\left(\partial^2 g(p,\ldots)/(\partial \ln_j)^2\right)\left(f(\ln_i)\right)^2.$$

What terms are being neglected? Does this seeem reasonable?

11.5.5. This exercise deals with the backpropagation algorithm for a general DAG.

(a) Describe the numerical calculations needed to compute the information in backpropagation.

(b) Describe how to time the propagation of information backward. You may find it useful to assign two states to vertices and/or edges to indicate whether or not they have received (or sent?) backpropagated information.

11.5.6. Algorithm 11.3 focuses on partials with respect to $\ln_v$. Can the algorithm be modified to focus on o_v instead? If yes, do so. If no, explain what the problem is.

*11.5.7. While it is obviously not feasible to compute all second-order partials by a local algorithm, it might be feasible to compute $\partial^2 g/(\partial w_{i,j})^2$ and $\partial^2 g/(\partial \theta_j)^2$ by a modification of backpropagation.

(a) Explain how to do this in a local manner for θ_γ and $w_{\beta,\gamma}$.

(b) Can it be done in a local manner for θ_β and $w_{\alpha,\beta}$? Justify your answer.

*11.5.8. Training a neural net is often time-consuming. It has been suggested that this could be sped up by taking steps more frequently. Here are some suggestions. Let $\delta_j(p)$ be δ_j for the pth pattern as computed by the backpropagation algorithm.

(i) Sum $\delta_j(p)$ over all p and use it to take a step. Repeat this. This is the method discussed in the text.

(ii) Sum $\delta_j(p)$ over m patterns and use it to take a step. Repeat this using another set of m patterns, either the next m repeatedly cycling through the patterns, or a randomly chosen set of m. At one extreme, $m = 1$; at the other, m is the number of patterns in the training set and we have (i) again.

(iii) Start with $\delta_j = 0$. As in (ii) sum m values of $\delta_j(p)$ but now define a new δ_j equal to this sum plus r times the old δ_j where $0 \leq r < 1$. When $r = 0$, we have (ii) again.

Can you suggest other methods? What do you think might be the pros and cons of these various methods? Some issues you might address are:

• Will it converge? If so, would it be faster or slower?

• Will it help in the escape from local minima?

• How might these ideas help if a net must be constantly retrained because the environment (typical patterns encountered) is changing (slowly?) with time.

This exercise has no one right answer and might best be used for class discussion.

11.6   Parameter Issues in Feedforward Nets

Various issues arise in designing networks and their algorithms. Ignoring these issues lessens your chances of writing a successful program, but an awareness of them is not central to a theoretical understanding of neural net algorithms. Hence, unless you are planning some programming or some research on algorithms, you may wish to skim the next two sections.

In this section, we focus on parameters; in the next, data. The discussion of some issues is largely independent of network and algorithm specifics, while other discussions are highly dependent on specifics. In the context of feedforward nets, here's a list of parameter issues that progresses, more or less, from very independent to very dependent.

• **Network geometry:** What should the size and structure of (V, E) be?

• **Activation functions:** What should we use for the activation functions f_v?

• **Initialization:** How should the starting values of the parameters be selected?

- **Excessive parameter growth**: What should be done if a parameter is moving outside the range that's considered acceptable?—and what *is* an acceptable range anyway?
- **Stopping**: How should we decide that the algorithm has converged sufficiently?
- **Internal parameters**: How should the algorithm's own parameters be set?

Discussing an algorithm's internal parameters requires focusing on a particular algorithm, which we won't do. Network geometry for feedforward nets is discussed to some extent in Chapter 13. The remaining issues are discussed here.

Activation Functions

How should we choose the functions f_v? Since we plan to differentiate them, we should choose functions whose derivatives are easily computed. Here are some possibilities.

- **Linear**: Linear functions are obvious candidates. This is a *very bad choice* because the output of the network is simply a linear combination of the inputs regardless of the number and arrangement of the hidden units. (See Exercise 11.6.2.) In other words, the net reduces to a perceptron, which was discussed in Section 11.3.
- **Logistic**: The output of a biological neuron is a nonnegative increasing function of its input. A simple function of this form, which is also easily differentiated, is

$$f(s) = \frac{1}{1 + e^{-s}} \quad \text{since} \quad f'(s) = f(s)\big(1 - f(s)\big). \tag{11.32}$$

  This is called the *logistic function* and is shown in Figure 11.6. We then set $f_v = f\left(\theta_v + \sum_x o_x w_{x,v}\right)$.
- **Radial basis**: A different approach is to design functions that look at the form of the vector of inputs instead of just reacting to total strength. Thus f_v looks for o_x to be close to some number, say $w_{x,v}$, for each $x \in \text{In}(v)$. We cannot use $\sum o_x w_{x,v}$ to measure this closeness. A reasonable alternative is $\sum (o_x - w_{x,v})^2$ since it is small if and only if the inputs o_x are close to the parameters $w_{x,v}$. The *radial basis functions* are based on this idea and are largest when $o_x = w_{x,v}$ for all $x \in \text{In}(v)$. We take

$$f_v = \exp\left(-\sigma_v^{-2} \sum_{x \in \text{In}(v)} (o_x - w_{x,v})^2\right). \tag{11.33}$$

There are two reasons for using σ_v^2 in the denominator instead of just a multiplicative factor of θ_v. First, θ_v could become negative but a square

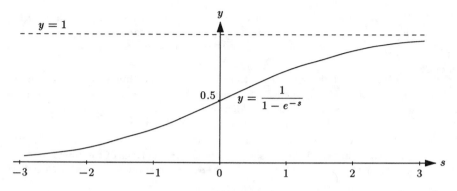

Figure 11.6 A graph of the logistic function (11.32).

cannot. Second, (11.33) looks more like a "normal density function," which we'll study in the next chapter.

- **More general:** Let $\vec{o}$ and $\vec{w}_v$ be vectors indexed by In(v). The linear and logistic functions have the form $f_v(\vec{o} \cdot \vec{w}_v)$ while the radial basis functions have the form $f_v(|\vec{o} - \vec{w}_v|)$. Obviously, we could use other maps $(\vec{o}, \vec{w}_v) \to \mathbb{R}$. Some researchers have done so.

Aside. It can be shown that a variety of nonlinear functions, including logistic and radial basis, have a nice property: Suppose we are given a training set and a $G > 0$. There is a feedforward net using the selected function such that $g(\vec{w}, \vec{\theta}) < G$; that is, the net is a good approximator to the training set. Such a net can even be built with only one hidden layer. While this sort of theoretical result is comforting, its practical implications are somewhat limited.

First, it says nothing useful about the number of hidden units that are needed. Too many units can lead to poor generalization. See Figure 10.3 (p. 421).

Second, it says nothing about the shape of $g(\vec{w}, \vec{\theta})$. It may be so bad that minimization algorithms are unable to find the values of $\vec{w}$ and $\vec{\theta}$ that make g small. Experimenters have reported situations in which g from nets having one hidden layer was difficult to minimize but g from nets having two hidden layers was not.

Exercises

11.6.A. List several parameter issues for feedforward nets.

11.6.B. Why are linear activation functions a bad choice?

11.6.C. How do logistic functions and radial basis functions differ in what they look for in inputs: that is, in what produces large output versus what produces small output?

11.6.1. Show that we can convert $\vec{\theta}$ into edge weights by adding another vertex that has no inputs, has constant output 1, and is connected to all noninput vertices. Why can't this be done with $\vec{\sigma}$ in radial basis functions?

11.6.2. The purpose of this exercise is to show that a linear function is a bad choice for f_v. Let $z_v = \sum_x o_x w_{x,v}$. We assume that $f_v = \lambda_v z_v + \theta_v$, which is more general than a linear function of the form $f_v = f(z_v + \theta_v)$ since λ_v also depends on v. We will also work with an arbitrary DAG.

   (a) Prove that if all the inputs to v are linear functions of the input pattern, then so is o_v. In other words, if there are constants $a_{x,i}$ and b_x such that

$$o_x = \sum_{i \in I} a_{x,i} o_i + b_x, \text{ for all } x \text{ such that } (x, v) \text{ is an edge,}$$

   then o_v also has this form.

   (b) Use the previous result and induction to prove that o_v is a linear combination of the inputs for all $v \in V$.

11.6.3. Show that the logistic function always lies between 0 and 1 and that it is symmetric about $(0, \frac{1}{2})$. The latter means that $-(f(s) - \frac{1}{2}) = f(-s) - \frac{1}{2}$; in other words, $f(-s) = 1 - f(s)$.

11.6.4. Develop a backpropagation algorithm for computing the gradient when radial basis functions (11.33) are used.

11.6.5. There are those who believe the logistic function is a poor choice because the weight change algorithm should react equally to the extremes of the input. We explore that here.

   (a) Show that the $w_{i,j}$ component of the gradient does not behave symmetrically with regard to o_i; that is, the values differ radically for $o_i = \epsilon$ and $o_i = 1 - \epsilon$.

   (b) Suppose the logistic function $f(s)$ is replaced by the hyperbolic tangent $\frac{1}{2} \tanh(s/2) = f(s) - \frac{1}{2}$. Show that the $w_{i,j}$ component of the gradient now has a symmetry in the two extreme values $o_i = -1 + \epsilon$ and $o_i = 1 - \epsilon$.

   (c) Let $\vec{w}$ and $\vec{\theta}$ be a set of parameters for a feedforward net using the logistic function. Determine parameters $\vec{w}^*$ and $\vec{\theta}^*$ such that the same feedforward net with the function $f(s) - \frac{1}{2}$ will behave in the "same manner." The "same manner" means that the inputs and outputs must be translated from the interval $[0, 1]$ to $[-\frac{1}{2}, \frac{1}{2}]$ by subtracting $\frac{1}{2}$.

*11.6.6. In this exercise, you look at an alternative to the sum of squares definition (11.28) for $g(\vec{w}, \vec{\theta})$.

(a) There are those who point out that momentum methods using the logistic function—or any monotonic function to $(0, 1)$—have problems near 0 and 1. Explain why this is so by looking at how $f'(s)$ contributes to ∇g when s is near 0 or 1.

(b) Let's replace (11.28) with

$$g(p, \vec{w}, \vec{\theta}) = \sum_{\gamma \in \mathcal{O}} \left(o_\gamma^*(p) \log o_\gamma(p) + \left(1 - o_\gamma^*(p)\right) \log\left(1 - o_\gamma(p)\right)\right).$$

Show that $a \log x + (1 - a) \log(1 - x)$ has a maximum at $x = a$ and use that to argue that minimizing g is a reasonable goal. The new g is related to "cross entropy" and has been proposed by various people. See Exercise 12.4.7 (p. 541).

(c) What changes must be made in (11.31)?

(d) How does the new value of g change $\partial g / \partial w_{\beta, \gamma}$. Does this help the problem discussed in (a)?

(e) How does the new value of g change $\partial g / \partial w_{\alpha, \beta}$. Does this help the problem discussed in (a)?

Initialization

A natural choice for starting values would be $\vec{w} = \vec{0}$ and $\vec{\theta} = \vec{0}$, but this is a very bad choice. The major problem with this choice is *symmetry*. Since all the nodes in H are connected to all the nodes in I and in O, permuting the nodes in H gives the same DAG. To be more precise suppose that π is a permutation of H; that is, $\pi : H \rightarrow H$ is one-to-one. You should be able to easily see that the following are true. (Drawing yourself a picture may be useful.)

- The outputs computed by a network with parameters $\vec{w}$ and $\vec{\theta}$ will be the same as those computed by a network with parameters given by

$$w'_{\alpha, \beta} = w_{\alpha, \pi(\beta)}, \quad w'_{\beta, \gamma} = w_{\pi(\beta), \gamma}, \quad \theta'_\beta = \theta_{\pi(\beta)}, \quad \theta'_\gamma = \theta_\gamma. \qquad (11.34)$$

- Furthermore, an algorithm that adjusts $\vec{w}$ and $\vec{\theta}$ will adjust $\vec{w}'$ and $\vec{\theta}'$ so that the equalities (11.34) will continue to hold. In particular, if $\vec{w} = \vec{w}'$ and $\vec{\theta} = \vec{\theta}'$ at some step, they will remain equal from then on.

In other words, an algorithm does not destroy symmetries. Although it cannot destroy them, it can create them. Thus, we have a basic principle for algorithms:

> Always initialize parameters so that *no symmetries* are present. If possible, start the parameters so that they are not even near values with symmetries.

In our case, this means that we want to choose $\vec{w}$ and $\vec{\theta}$ so that if

$$w_{\alpha,\beta} = w_{\alpha,\pi(\beta)}, \quad w_{\beta,\gamma} = w_{\pi(\beta),\gamma}, \quad \theta_\beta = \theta_{\pi(\beta)} \tag{11.35}$$

for all α, β, and γ, then π is the trivial function $\pi(\beta) = \beta$.

One method for doing this is to choose the starting values at random. If this is done, we can be practically certain that there is no symmetry. To see this, notice that a nontrivial π in (11.35) will require several nontrivial equalities between parameters, and notice that two randomly chosen numbers are not likely to be equal. Random initialization is just an application of (11.6) (p. 435).

There are two natural questions at this point: Is random choice the best we can do? and How should the random numbers be chosen? The answers aren't known. Some researchers believe it's good practice to choose "normally distributed" (= "gaussian") random numbers. These are described on page 512.

Excessive Parameter Growth

A false balance is an abomination to the Lord, but a just weight is God's delight.

—Proverbs

Experimenters have observed that when the parameters of a neural net are large, the net tends to perform poorly. Apparently the descent algorithm finds a poor local minimum. They have also observed that if the algorithm is going to go off the track, it usually starts to do so fairly early.

Given these empirical observations, what should we do? Here are three possible ways to deal with the weights. The biases $\vec{\theta}$ can be handled similarly.

- If some weights are getting too large, restart the algorithm. Assuming initialization is "random," a new start may do better. Restarting is not as drastic as it sounds because we may want to run an algorithm several times anyway, and then select the best result.

- Insist that $|w_{i,j}| \leq w_{\max}$ for all i and j. The value of $w_{\max}$ might depend on the number of steps taken. We can force $|w_{i,j}| \leq w_{\max}$ by simply setting a value back to $\pm w_{\max}$ when it becomes too large. Exercise 11.6.8 discusses another method.

- Change the function we are minimizing to include a penalty for larger parameters. For example, we could minimize

$$\hat{g}(\vec{w}, \vec{\theta}) = g(\vec{w}, \vec{\theta}) + \frac{\lambda}{2}\left(\sum_{i,j} w_{i,j}^2 + \sum_i \theta_i^2\right), \qquad (11.36)$$

  where g is given by (11.28). This procedure is often referred to as *weight decay*. The value of $\lambda > 0$ must be chosen somehow and might depend on the number of steps taken. Some aspects of (11.36) are discussed further in Examples 11.10 (p. 485) and 13.5 (p. 563).

- Insist that the algorithm converge after some number of steps. If it has not, discard the result. Choosing the number of steps correctly is critical and may be difficult to do. It is probably best to set the number large and combine this with some other method.

Exercises

11.6.D. Why should symmetries be avoided when initializing parameters?

11.6.E. Describe some methods for preventing excessive parameter growth.

11.6.7. Suppose that we are trying to constrain the weights by using $\hat{g}$ from (11.36). We'll look at gradient descent.

   (a) Compute $\nabla \hat{g}$.

   (b) Explain how the difference between $\nabla \hat{g}$ and ∇g makes it harder for $|w_{i,j}|$ to get large. (Assume that the learning rates are the same for $\hat{g}$ and g.)

   (c) Explain how to modify (11.25) to use $\hat{g}$ instead of g.

11.6.8. Suppose we want $|w_{i,j}| \le w_{\max}$.

   (a) Show that this will be the case if

$$w_{i,j} = w_{\max}\left(2f(u_{i,j}) - 1\right),$$

   where f is given by (11.32). (Incidentally, $2f(s) - 1 = \tanh^{-1}(s/2)$.)

   (b) Modify the backpropagation algorithm to use $u_{i,j}$ instead of $w_{i,j}$ as the parameters.

11.6.9. We consider the effect of weight decay on some parameter x. Show that whenever $|(\nabla g)_x| < \lambda|x|$, the magnitude of x is decreased by a gradient step based on $\hat{g}$. In other words, the larger $|x|$ is, the bigger the gradient component must be to keep x from moving toward 0.

Stopping

Stopping encompasses three problems because there's more than one situation in which we should stop.

- How can we tell that we've done training sessions with enough different initial values so that at least one of the trained nets is acceptable?

- How can we tell that the parameters are close enough to a local minimum so that we can stop the training session in success?

- How can we tell that the parameters have gone astray so that we can stop the training session in failure?

These are all difficult problems. No good solutions are known.

How many starts should we make? If we knew what the domains of attraction were like, we might be able to answer this. Of course, we could simply continue making starts until we get a network that is good enough for the problem at hand—but what if there are *no values* of $\vec{w}$ and $\vec{\theta}$ that give a good enough network? Or what if we set our sights too low and could have gotten a much better network?

"Close enough to a local minimum" is a hard issue—particularly since we don't know where the local minima are. In an ideal run of an algorithm, there will be a period of time when the parameters are changing relatively rapidly followed by a slowing down as a local minimum is approached. This could be used as a sign to stop. Unfortunately, other things such as a saddle point can cause the rate of parameter change to slow down and then increase again. Various heuristics for stopping have been developed.

It would seem that the only harm in running an algorithm past the time when we've gotten close enough to a local minimum is that we would waste time getting closer than necessary. The situation is often more serious than that: Stopping training early may result in better generalization ability. A graph like Figure 10.3 (p. 421) is often obtained, where the horizontal axis now measures time spent in training rather than network complexity. This phenomenon is referred to as *overtraining*. Although overtraining, complexity and parameter growth seem quite different, the next example shows heuristically that they may be closely related.

Example 11.10   Complexity: Overtraining Gives Bad Parameters

The argument in this example is based almost entirely on handwaving and proof by intimidation. Nevertheless, it seems somehow to capture the essence of what happens.

We'll use the number of edges as a rough measure of complexity of the net. Our first step is to estimate how the size of the set of good weights varies with complexity. Using this result, we reason that excessive training is likely to produce unreasonably large weights.

For any set of classification data C and a net with edges E, let $W(C, E)$ be the set of weights $\vec{w} \in \mathbb{R}^{|E|}$ for which the net classifies the data in C well. In some sense we won't make precise, W is a space of some dimension, say $d(W)$. As we increase the number of parameters by increasing E, we can expect $d(W)$ to increase. We might even expect that the dimensions of both spaces increase by about the same amount; that is,

$$d\big(W(C, E_2)\big) - d\big(W(C, E_1)\big) \approx |E_2| - |E_1|. \tag{11.37}$$

How d varies with C is unclear, but it should tend to decrease as C gets larger.

Now imagine that we have a classification problem for a feedforward neural net. The training set will be T and the set of all the data—training data plus generalization testing data—will be U. If the problem is reasonable, it's reasonable to expect a solution with weights of a reasonable size. Suppose we need at least edges $|E_1|$ to achieve this.

Imagine training a net with $|E_2|$ edges where $|E_2|$ is rather larger than $|E_1|$—unnecessary complexity. From (11.37),

$$d\big(W(T, E_2)\big) \geq d\big(W(T, E_2)\big) - d\big(W(T, E_1)\big) \approx |E_2| - |E_1|,$$

which is fairly large. In this large-dimensional space, we hope we have found a point in the smaller dimensional space $W(U, E_2)$—"smaller" because $U \supset T$.

Large-dimensional objects tend to have an interesting property: A lot of the object tends to be far from the origin. This tendency increases rapidly with dimension. (See Exercise 11.6.10 below.) So what? A large part of $W(T, E_2)$ will probably be far from the origin. A rather smaller part of $W(U, E_2)$ will be far from the origin. With a considerable amount of handwaving and some dubious assumptions we've just shown that

$$\Pr\big(\vec{x} \in W(U, E_2) \mid \vec{x} \in W(T, E_2)\big)$$

decreases as $|\vec{x}|$ increases, where we're using the notion of probability loosely. In words:

> Heuristic arguments suggest that the more unnecessary complexity a network possesses, the more likely it is that parameters which fit the training data well will provide poor generalization. These poor generalizers tend to contain large parameter values. (11.38)

What does this tell us? Since an algorithm is usually not started with large parameter values, it begins by approaching points in $W(T, E_2)$ fairly near the origin. The longer it runs, the more likely it will approach points in $W(\mathcal{T}, E_2)$ far from the origin since most of the set is there. These points are more likely to give poor generalizations than those close to the origin.

What can be done about this? We might try to choose E_2 close to E_1, but we don't have any way of estimating what E_1 must be except by experimenting with $\mathcal{U}$. Is there another solution? Essentially, (11.38) tells us that we should try to keep the weights near the central values; that is, avoid excessive parameter growth, a subject recently discussed.

This discussion is merely suggestive, not a proof. Since researchers have constructed examples in which the conclusions are wrong, any theorem about overtraining would have to contain some restrictive hypotheses. I'm not aware of any useful theorem of this sort. ■

Exercises

11.6.F. How are complexity, poor generalization, and overtraining related?

11.6.10. In the example just completed, it was claimed that points in high-dimensional regions are likely to be far from the center. This exercise provides some support for the claim.

(a) What fraction of points in the $2n$-dimensional cube given by $|x_i| \le 1$ lie further than distance 1 from the origin?
Hint. The volume of a sphere of radius r in $2n$ dimensions is $\pi^n r^{2n}/n!$.

(b) It can be shown that $n! > (n/e)^n$. Let f be the fraction of points in our cube whose distance from the origin is at most $n^{1/3}$. Show that $f \to 0$ as $n \to \infty$.

*11.7   Data Issues

Various aspects of data preparation were discussed in Section 10.3 and some training sets—another data preparation issue—were discussed briefly in the succeeding section. Additional aspects closely related to (feedforward) neural nets are discussed here.

Training Sets

Sometimes a training set may be quite large. In that case it seems wasteful to go through the entire set before our algorithm can take a step in parameter space. Wouldn't it be faster to climb on a random subset of the patterns at the beginning? Going to extremes, why not let the algorithm take a step after each pattern?

Our algorithms are based on trying to minimize a sum that is taken over the entire training set. Actually, this is not what we really want to do. It would be preferable to sum over all possible patterns—not just training patterns. Thus, our function $g(\vec{w}, \vec{\theta})$ is simply an approximation to what we want. There is nothing magic about this particular approximation. We could take a sum over some smaller set of patterns. We could even change the set we are summing over from step to step. After all, why should we prefer one approximating sum to another?

If we follow the suggestion in the last paragraph, we're shooting at a moving target because the function g keeps changing. This should cause no problems as long as our approximations are good enough: that is, as long as we sum over a large enough set of patterns. This gives us a way to modify our algorithm:

1. Select some number K of patterns at random from our training set. Since accuracy is less important initially, we may want K to increase somehow with time rather than stay fixed.

2. Backpropagate errors and cumulate them to compute the needed derivatives.

3. Take the step dictated by whatever minimization algorithm we're using.

We've now introduced K, yet another parameter that we have no idea how to choose. There's another problem. Our surface changes at each step because we change g. Thus, Quickprop (11.27) and momentum (11.21) may be in difficulty because they use information about the surface from previous steps.

We could go even further with the previous idea and take a step after each backpropagation provided we use previous information. Simply replace the first two steps with

1. Select a pattern at random from the training set.

2. Backpropagate the error and add it to α times the "cumulated error" to produce a new "cumulated error." We should have $0 \leq \alpha < 1$.

If our algorithm takes fairly small steps, this approach may give good results. We won't pursue these ideas.

Data Preparation

> *For real-life applications, how we represent data is at least as important as what neural network we choose.*
>
> —Vladimir Cherkassky and Hossein Lari-Najafi (1992)

The shape of the surface $g(\vec{w}, \vec{\theta})$ can make the difference between success and failure. Improved data preparation leads to improved surface shape. That's practically a tautology—a definition of good data preparation.

Feature Detection

In Section 10.3 we saw that eliminating irrelevancy is desirable. Using a pre-processing net for feature detection is a way of trying to do this. The raw data are the preprocessor's inputs and the features are somehow read from the net. These features form the inputs for the classifier we're training. Some possibilities for a feature detection net are

- A net that performs principal-component analysis (p. 407). The principal components form the inputs of the net we wish to train.
- A multilayer feedforward net with a "bottleneck" hidden layer consisting of relatively few hidden units. The net is trained with desired output equal to input. After training, the output of the bottleneck layer is used.
- An unsupervised net consisting of feedforward layers and inhibitory connections *within* each layer. The outputs of any layer represent features based on some sort of clustering of the data.

Clustering has also been used to extract rules for fuzzy controllers. In this case, a net input pattern consists of both the sensor inputs and the desired controller action.

Nonlinear Calculations

Suppose the net is using logistic functions.

Consider what happens at a vertex: A linear combination of the inputs is formed and fed into the logistic function. When its input is small, the output of a logistic function is nearly a constant times its input. Thus, there's a lot of linearity and near linearity in a such a network. This suggests a principle:

> Reduce the need for the network to do nonlinear computations.

How can this be done?

As an example, imagine a network-based expert system that is supposed to evaluate an application for a home mortgage. Annual income I, amount of mortgage M, and appraised value of the home H are important, so we might use them as inputs. The ratios M/I and M/H are more important than the actual numbers M, and H. As a result, the network will learn to approximate the ratios. This requires learning time and additional hidden units. We could have chosen I, M/I, and M/H as inputs instead of I, M, and H. From the network's viewpoint, this elimination of nonlinear operations (division) has simplified the problem. To summarize:

> When you believe that certain nonlinear calculations will need to be done on the input data, precompute the values and use them as part of the input data. This will probably give a better net.

11.8 Some History

The development of neural nets began in the 1940s with two biologically inspired models, one by McCulloch and Pitts [24], the other by Hebb [13].

In our terminology, the McCulloch-Pitts model can be described as follows:

$$w_{i,j} = +1 \text{ or } -\infty,$$

$$f_v = m(s) = \begin{cases} +1, & \text{if } s > 0, \\ 0, & \text{otherwise,} \end{cases} \quad \text{where} \quad s = \sum_{x \in \text{In}(v)} o_x w_{x,v}. \quad (11.39)$$

In other words, o_v is nonzero if and only if it receives at least one positively weighted input and no negatively weighted inputs. The inputs to v at time t determine its output at time $t + 1$. They proved that such units could be interconnected to produce any two-valued function of the inputs and that stable cycles of neuron firings could be created to store information.

The McCulloch-Pitts model depends entirely on structural differences to produce different results because the weights are fixed. In contrast, Hebb proposed that long-term memory was based on the modification of weights.

McColloch-Pitts neurons can be connected to simulate a Turing machine, hence any computer and, presumably, the cognitive abilities of human beings. Von Neumann pointed out that it may happen that a full description of a mechanism for simulating human cognitive abilities might be more complex than the human brain itself. This has important implications for designing AI with human-like abilities. It would be too complex to build piece by piece like a rule-based expert system. Instead, it must be allowed to grow on its own. A first step in this direction is the modification of network weights using parallel algorithms.

In 1957, Rosenblatt introduced his perceptrons, which we discussed earlier. A *simple* perceptron consists of an input layer, a single output vertex, and no hidden vertices. It looks for a linear separator, but as we noted, many cases of interest lack linear separators. It's claimed that Minsky and Papert's 1969 book [25] on the limitations of simple perceptrons sounded the death knell for neural nets in AI. Actually, most researchers had already turned away from neural nets. They knew that more general perceptrons were needed, but no learning algorithm could be found. This is hardly their fault, since even the supercomputers of the 1960s would require lengthy runs using present-day training algorithms. In other words, the loss of interest was due at least as much to lack of computing power as it was to Minsky and Papert's book.

Interest in neural nets revived in the early 1980s, thanks in part to funding by DARPA (Defense Advanced Research Projects Agency) and the existence of more powerful computers. The PDP (Parallel Distributed Processing) group's publication of their results in 1986 [30] triggered an explosive growth in research. Several journals and annual meetings are now devoted to neural network research and new books are appearing at a rapid rate. In 1992, the IEEE Standards Board approved a committee to work on standardizing nomenclature, tools, benchmarks, and software and hardware interfaces.

Since backpropagation is simply a means of computing the gradient efficiently, it has been rediscovered several times. The first discovery may have been by Werbos who described it in 1974 in his thesis at Harvard. However, it wasn't until a decade later that backpropagation found its way permanently into AI lore after the PDP group publicized the method and their applications of it.

This isn't intended to be a comprehensive history. We've omitted important researchers such as Hopfield and Grossberg as well as many network architectures. For further information, see the books mentioned in the next section.

Applications of neural nets are emerging at an ever greater rate. A few of these are

- traditional types of expert systems such as loan applications and diagnosis problems,
- quality prediction in industrial production,
- speech recognition and synthesis,
- character recognition, and
- a backgammon champion.

Thanks to increasing computer power, researchers are building larger and larger nets. This has brought them face to face with a new problem: It's normally impossible to use descent methods to estimate parameters for large nets because the surface is so poorly behaved. Consequently, interest has been increasing in finding methods for building up large neural nets. The importance of this problem will continue to grow since it must be overcome if connectionists are going to avoid a perennial AI problem—computational requirements grow much more rapidly than problem size.

Notes

> *The hardest part of a controlled experiment, it turns out, is design; a poorly designed experiment will almost never yield meaningful and reliable data. One must understand clearly, from the outset, precisely what one wants to learn.*
>
> —Nathaniel S. Borenstein (1991)

Neural nets can be very seductive in two ways that our earlier AI topics were not. First, they seem to offer something for nothing—an issue partially discussed in Section 10.3 (p. 415). Second, because there are so many things that could be tweaked in minimization routines and in net structure, there is a temptation to replace thinking with programming. To counteract this I've devoted most of this chapter to *thinking* about aspects of minimization that might tempt you to (nearly) thoughtless programming. I applied these ideas to two types of neural nets—Hopfield and feedforward—and ignored other areas of neural net research completely. Perhaps the most important neglected topics are unsupervised learning, training recurrent nets, and building large networks. The last appears briefly in Chapter 13. Unsupervised learning refers to training in which no desired outputs are given with the training set. Such nets are often used to detect potentially interesting features in the data. They can look for principal components (see below) or clustering. Training occurs by having units "compete." (One approach to this is adaptive resonance theory [6].) Algorithms for training recurrent nets exist, but better methods are needed and research is being done. See some of the texts mentioned below for an introduction to these as well as other topics.

Are neural networks a better choice than statistical methods of classification? Yes and no. Neural nets provide a richer set of models than statistics and hence are more likely than statistical methods to give good results (low generalization error). If they can be used, traditional statistical methods are usually a better choice because they're quicker and often produce better results. On the other hand, neural nets can sometimes be viewed as estimating posterior probabilities in the sense of Bayes' Theorem. Hampshire and Pearlmutter [12] and Richard and Lippman [29] discuss this further.

Isn't a fuzzy controller like a feedforward net? If so, why bother to think about rules when we can just use a net? Yes, a fuzzy controller is like a net, but there's no free lunch. If you understand the situation well enough to write rules, you'd be foolish to throw the knowledge away. If you don't understand the situation, then you're certainly better off with a net since you can't begin to use fuzzy logic. There's another caveat with using a net: As discussed in Sections 10.3 and 11.7, a net is no better than the training data.

For more information on standard minimization methods in numerical analysis, see [9]. Adaptations of and variations on these methods appear in many papers in the neural net literature. Fahlman [8] discusses some of the problems encountered in designing computer experiments to compare different methods. He then compares Quickprop with the momentum method. Van der Smagt [32, Sec. 4] discusses some common methods.

Simulated annealing was introduced in by Kirkpartrick, Gelatt, and Vecchi in 1983 [18]. See Azencott's article [3] for a quick introduction and the book by Aarts and Korst [1] for an extensive discussion of the connection between simulated annealing and Boltzmann machines. Freeman and Skapura's text [10, Ch. 5] contains a discussion of simulated annealing in the context of Boltzmann machines—a type of neural net.

Another optimization method having a probabilistic basis is genetic algorithms. These were introduced in the 1970s by John Holland. While simulated annealing was inspired by the statistical mechanics of cooling, genetic algorithms were inspired by the processes of evolution—selection, recombination, mutation, and crossover. See Section 14.1 (p. 589) for further discussion.

Researchers have been exploring a variety of other methods for solving the minimization problem. These methods are generally found only in the research literature and, I think, haven't yet raised enough interest to warrant inclusion here.

Introductory books on neural nets are continually appearing. The texts by Hecht-Nielsen [14], Freeman and Skapura [10], Kosko [19] and Zurada [36] have exercises and broad coverage. Over half of Zeidenberg's book [35] is devoted to applications. Wasserman [33] presents a collection of relatively self-contained and elementary chapters on a variety of topics. Gallant [11] also presents a variety of neural net algorithms and discusses their application in expert systems. A thorough mathematical discussion of many aspects of neural nets is provided by Hertz, Krogh, and Palmer [15], including the physical aspects of Hopfield nets. These physical aspects are also discussed in Müller and Reinhardt's book [26], which is based on a course given in a physics department.

Chauvin and Rumelhart [ChauvinR] have edited a collection that discusses many aspects of feedforward nets.

Various proofs that feedforward nets are good approximators have been developed. For example, see [4] or [22].

Various authors have discussed weight decay. See for example, [21].

My discussion of principal-component analysis in Chapter 10 was extremely brief. Jolliffe has written a book on the subject [17]. Leen, Rudnik, and Hammerstrom [23] indicate how to construct a net for extracting principal components and discuss its usefulness. Kramer [20] discusses the bottleneck approach to feature extraction.

Neural net texts often contain some history. In this category, see Hecht-Nielsen [14]. Johnson and Brown [16] have written a historically oriented account of neural nets for the layman. Olazaran [27] has written a lengthy article.

Biographical Sketches

John J. Hopfield (1933–)

Born in Chicago, he received his bachelor's degree from Swarthmore and his doctorate in physics from Cornell. He was a MacArthur Prize fellow from 1983 to 1988. (Not well known outside of academic circles, the MacArthur prize is a grant awarded to particularly creative individuals—not necessarily academics—to allow them to pursue their ideas.)

After establishing a reputation in physics and biophysics, he became a professor at Caltech and also became interested in neural networks. His research and lectures in the first half of the 1980s contributed greatly to the rebirth of research in neural networks. It's said that, in 1986 about one third of the researchers were involved because of Hopfield.

Gottfried Wilhelm Leibnitz (1646–1716)

Born in Leipzig, Saxony, he studied at Leipzig and Nuremberg. He worked in philosophy and mathematics; however, his mathematical work did not really commence until 1672 when he took up the study of geometry. In 1676, he was appointed librarian to the Duke of Hanover, a post that allowed ample free time for his studies. He developed calculus in 1677 and published it in 1684. His work grew out of the geometric problems of finding areas and finding tangents to curves. We owe our notation, such as dy/dx and $\int f(x)dx$, to him.

Considerable dispute raged over who developed calculus first—Leibnitz or Newton. Today it is common practice to publish discoveries as soon as they are made, but 300 years ago that was often not the case. The issue is further confused by the fact that Leibnitz visited England and corresponded with Newton. It is unlikely that the issue will ever be entirely resolved.

See the biographical sketch of Newton for a discussion of foundational problems.

Isaac Newton (1642–1727)

Born in Lincolnshire, England, he studied at Trinity College, Cambridge, where he remained until 1696. At Trinity College he worked on mathematics and natural philosophy, as physics was then called. Building on the work of others, Newton developed both calculus and physics, using calculus as a tool for physics. His major publications in physics were the *Principia* (1687), in which he developed Newtonian mechanics and his law of gravitation, and *Optics* (1704). In the former, he used his method of "fluxions" (derivatives) to derive results which were then proved geometrically. In the latter, he had appendices on quadrature (integration), infinite series, and the method of fluxions. Newton denoted derivatives by dots as in $\dot{x}$ and $\ddot{x}$, which today is often used in physics, but is seldom used in mathematics.

Lacking an adequate notion of limit, Newton computed the derivative of $f(x)$ by rearranging $\frac{f(x+h)-f(x)}{h}$ so that h was absent from the denominator and then deleting all terms containing h. Deleting the terms was justified by regarding h as equal to zero. On the other hand, the initial division by h was justified by regarding h as not really zero. Bishop Berkeley wrote a scathing criticism of this approach—h is a quantity and so must be either zero or nonzero, but not both. It took over a century for mathematicians to develop an alternative, consistent foundation for calculus—the theory of limits. The concept of limit leads to the dreaded ϵ-δ arguments found in calculus. After another century, logicians finally created a firm foundation for the more intuitive 17th-century approach—nonstandard analysis. In nonstandard analysis, infinitesimals like h and dx exist in an extension of the real numbers and the need for ϵ-δ arguments disappears.

References

1. E. Aarts and J. Korst, *Simulated Annealing and Boltzmann Machines; A Stochastic Approach to Combinatorial Optimization and Neural Computing*, John Wiley and Sons, New York (1989).

2. J. A. Anderson and E. Rosenfeld (eds.), *Neurocomputing. Foundations of Research*, MIT Press, Cambridge, MA (1988).

3. R. Azencott, Sequential simulated annealing: Speed of convergence and acceleration techinques. In R. Azencott (ed.), *Simulated Annealing: Parallelization Techniques*, John Wiley and Sons, New York (1992) 1–10.

4. P. Baldi, Computing with arrays of bell-shaped and sigmoid functions, *NIPS* **3**, Morgan Kaufmann, San Mateo, CA (1991), 735–742. ("NIPS" stands for *Advances in Neural Information Processing Systems*.)

5. S. Becker and Y. Le Cun, Improving the convergence of back-propagation learning with second order methods. In [31] 29–37.

6. G. A. Carpenter and S. Grossberg, The ART of adaptive pattern recognition by a self-organizing neural network, *Computer* **21** No. 3 (March 1993) 77–88.

7. R. Chauvin and D.E. Rumelhart (eds.), *Back-propagation: Theory, Architectures and Applications*, Lawrence Erlbaum Associates, Hillsdale, NJ (1994).

8. S. E. Fahlman, Faster-learning variations on back-propagation: An empirical study. In [31] 38–51.

9. R. Fletcher, *Practical Methods of Optimization*, 2d ed., John Wiley and Sons, New York (1987).

10. J. A. Freeman and D. M. Skapura, *Neural Networks: Algorithms, Applications, and Programming Techniques*, Addison-Wesley, Reading, MA (1991).

11. S. I. Gallant, *Neural Network Learning and Expert Systems*, MIT Press, Cambridge, MA (1993).

12. J. B. Hampshire II and B. A. Pearlmutter, Equivalence proofs for multilayer perceptron classifiers and the Bayesian discriminant function *Proceedings of the 1990 Connectionist Models Summer School*, Morgan Kaufmann, San Mateo, CA (1990) 159–172.

13. D. O. Hebb, *The Organization of Behavior*, John Wiley and Sons, New York (1949). Partially reprinted in [2] 45–56.

14. R. Hecht-Nielsen, *Neurocomputing*, Addison-Wesley, Reading, MA (1990).

15. J. A. Hertz, A. Krogh, and R. G. Palmer, *Introduction to the Theory of Neural Computation*, Addison-Wesley, Reading, MA (1991).

16. R. C. Johnson and C. Brown, *Cognizers: Neural Networks and Machines That Think*, John Wiley and Sons, New York (1988).

17. I. T. Jolliffe, *Principal Component Analysis*, Springer-Verlag, Berlin (1986).

18. S. Kirkpatrick, C. D. Gelatt, Jr., and M. P. Vecchi, Otimization by simulated annealing, *Science* **220** (1983) 671–680. Reprinted in [2] 554–568.

19. B. Kosko, *Neural Networks and Fuzzy Systems: A Dynamical Approach to Machine Intelligence*, Prentice Hall, Englewood Cliffs, NJ (1991).

20. M. A. Kramer, Nonlinear principal component analysis using autoassociative neural networks, *J. American Inst. of Chemical Engineers* **37** (1991) 233–243.

21. A. Krogh and J. A. Hertz, A simple weight decay can improve generalization, *NIPS* **4**, Morgan Kaufmann, San Mateo, CA (1992) 950–957. ("NIPS" stands for *Advances in Neural Information Processing Systems*.)

22. V. Kůrková, Kolmogorov's theorem is relevant, *Neural Computation* **3** (1991) 617–622.

23. T. Leen, M. Rudnik, and D. Hammerstrom, Hebbian feature discovery improves classifier efficiency, *IJCNN* 1990, IEEE, New York, I.51–56. ("IJCNN" stands for *International Joint Conference on Neural Networks*.)

24. W. S. McCulloch and W. Pitts, A logical calculus of the ideas immanent in nervous activity, *Bull. Math. Biophysics* **5** (1943) 115–133. Reprinted in [2] 18–27.

25. M. L. Minsky and S. Papert, *Perceptrons: An Introduction to Computational Geometry*, MIT Press, Cambridge, MA (1969). It has been reprinted with some additional discussion as *Perceptrons: An Introduction to Computational Geometry, Expanded Edition* (1988).

26. B. Müller and J. Reinhardt, *Neural Networks. An Introduction*, Springer-Verlag, Berlin (1991).

27. M. Olazaran, A sociological history of the neural network controversy. In [34] 335–425.

28. B. A. Pearlmutter, Fast exact multiplication by the Hessian, *Neural Computation* **6** (1994) 147–160.

29. M. D. Richard and R. P. Lippmann, Neural network classifiers estimate Baysian *a posteriori* probabilities, *Neural Computation* **3** (1991) 461–483.

30. D. E. Rumelhart, J. L. McClelland, and the PDP Research Group, *Parallel Distributed Processing. Explorations in the Microstructure of Cognition*. Vols. 1 and 2, MIT Press, Cambridge, MA (1986).

31. D. S. Touretzky, G. E. Hinton and T. J. Sejnowski (eds.), *Proceedings of the 1988 Connectionist Models Summer School*, Morgan Kaufmann, San Mateo, CA (1988).

32. P. P. van der Smagt, Minimization methods for training feedforward neural networks, *Neural Networks* **7** (1994) 1–11.

33. P. D. Wasserman, *Neural Computing: Theory and Practice*, Van Nostrand Reinhold, New York (1989).

34. M. C. Yovits (ed.), *Advances in Computers*, Vol. 37, Academic Press, San Diego (1993).

35. M. Zeidenberg, *Neural Network Models in Artificial Intelligence*, Ellis Horwood, New York (1990).

36. J. M. Zurada, *Introduction to Artificial Neural Systems*, West Publ. Co., St. Paul (1992).

12

Probability
Statistics
and
Information

Mathematics is not the art of computation but the art of minimal computation.

—Anonymous

Probability theory would be effective and useful even if not a single numerical value were accessible.

—William Feller (1957)

Introduction

In the following sections we'll discuss mean and variance, continuous random variables, some statistical methods, and information theory—all concepts based on probability theory. In contrast to Chapter 7, the range of random variables will be limited:

> Unless noted otherwise, the ranges of random variables in this chapter will be contained in $\mathbb{R}$ or, for vectors, $\mathbb{R}^n$ for some $n > 0$.

I'll begin by introducing mean and variance. Although it's traditional to introduce them early in the study of probability theory, I haven't done so because Chapter 7 already had quite a bit of material and I was able to avoid using them. So far, the probability space $\mathcal{E}$ has been finite. This is too great a limitation. To simplify the presentation in Section 12.2, I'll limit attention to the most important infinite situation and focus on random variables.

Since statistics could easily take up a year-long course, a single section can give only a skewed and limited view of the field. My discussion of statistics begins with the problem of estimating generalization error. While this is not a major area of statistics, it plays a major role in classifier testing. Next I'll add a little bit of a major statistical discipline—hypothesis testing.

The twin concepts of information and entropy play important roles in communications theory. Since statistics is concerned with extracting information from data and since information is central to AI, it's natural that some statisticians and AI researchers have made use of concepts and tools from information theory. At present, the subject is still somewhat peripheral, so that section is starred.

The first section of this chapter is essential for the remainder; however, the other sections are largely independent of one another so you can skip whichever you choose.

Prerequisites: Basic probability, Chapter 7, is essential. Some reference is made to material in Chapter 11, but it's not essential.

Used in: This material is used in Chapter 13. See that chapter for more detailed information.

12.1   Mean and Variance

We first encountered random variables in Definition 7.4 (p. 266) and the notion of their independence in Definition 7.10 (p. 292). You may want to review those ideas now.

Definition 12.1   Expectation and Variance

Let $X : \mathcal{E} \to \mathbb{R}$ be a random variable on a probability space $(\mathcal{E}, \mathrm{Pr})$ and let $B \subseteq \mathcal{E}$. The *conditional expectation* or *expected value* of X given B is

$$\mathrm{E}(X \mid B) = \sum_{e \in \mathcal{E}} \mathrm{Pr}(e \mid B) X(e) = \frac{1}{\mathrm{Pr}(B)} \sum_{e \in B} \mathrm{Pr}(e) X(e), \qquad (12.1)$$

where the latter equality follows from the definition of $\mathrm{Pr}(e \mid B)$. The *variance* of X is

$$\mathrm{var}(X \mid B) = \mathrm{E}\Big(\big(X - \mathrm{E}(X \mid B) \big)^2 \,\Big|\, B \Big).$$

Expectation and variance conditioned on some set $\mathcal{Y}$ of random variables is defined in a similar manner. If there is no conditioning (i.e., $B = \mathcal{E}$), we simply drop "$|B$."

The expected value of X is simply its average value. The variance is more complicated; however, it can be seen from the definition that the variance is a measure of how much X deviates from its expected value. Why not define $\operatorname{var}(X)$ to be $\operatorname{E}(|X - \operatorname{E}(X)|)$, where the vertical bars denote absolute value? The answer is twofold. First, an absolute value is often harder to work with than a square. (Remember the formulas for $d(t^2)/dt$ and $d(|t|)/dt$. Which is simpler?) Second, the given definition turns out to be the "right" one because of its appearance in a variety of theoretical results.

Example 12.1 Computing Mean and Variance of Dice

A fair die is tossed. What are the mean and variance of the number of pips that appear? The probability space should be obvious—the six die faces, each with probability $\frac{1}{6}$. Let $X(e)$ be the number of pips. Then $\Pr(X = i) = \frac{1}{6}$ for $1 \le i \le 6$ and so

$$\operatorname{E}(X) = \tfrac{1}{6} \times 1 + \tfrac{1}{6} \times 2 + \tfrac{1}{6} \times 3 + \tfrac{1}{6} \times 4 + \tfrac{1}{6} \times 5 + \tfrac{1}{6} \times 6 = \tfrac{21}{6} = \tfrac{7}{2}$$

$$\operatorname{var}(X) = \sum_{i=1}^{6} \tfrac{1}{6} \times \left(i - \tfrac{7}{2}\right)^2$$

$$= \tfrac{1}{6}\left(\tfrac{-5}{2}\right)^2 + \tfrac{1}{6}\left(\tfrac{-3}{2}\right)^2 + \tfrac{1}{6}\left(\tfrac{-1}{2}\right)^2 + \tfrac{1}{6}\left(\tfrac{1}{2}\right)^2 + \tfrac{1}{6}\left(\tfrac{3}{2}\right)^2 + \tfrac{1}{6}\left(\tfrac{5}{2}\right)^2$$

$$= \tfrac{1}{24}\left(25 + 9 + 1 + 1 + 9 + 25\right) = \tfrac{35}{12}.$$

Now suppose we toss two fair dice and look at the sum of the pips. The computations are much more complicated, but the answers turn out to be 7 and $\frac{35}{6}$, simply twice the previous answers. There's a way we can pretend to toss two dice: Toss just one and look at the top and bottom. It turns out that dice are made so that the sum of the pips on opposite faces is always 7. It follows immediately from the definition that the mean is still 7 but the variance is now 0. It appears that the independence of the two values is important for the variance but not for the mean. This, and other results, appear in the next theorem. ∎

Theorem 12.1 Properties of Expectation and Variance

Let $X, X_1, \ldots, X_n$ be random variables and let $c_1, \ldots, c_n$ be real numbers.

(a) $\operatorname{var}(X \mid B) = \operatorname{E}(X^2 \mid B) - \operatorname{E}(X \mid B)^2$.

(b) If X has mean μ and variance σ^2, then $aX + b$ has mean $a\mu + b$ and variance $a^2\sigma^2$.

(c) For any c_i's and X_i's,

$$\operatorname{E}(c_1 X_1 + \cdots + c_n X_n \mid B) = c_1 \operatorname{E}(X_1 \mid B) + \cdots + c_n \operatorname{E}(X_n \mid B). \quad (12.2)$$

(d) If, for all $i \ne j$, X_i and X_j are independent given B, then

$$\operatorname{var}(c_1 X_1 + \cdots + c_n X_n \mid B) = c_1^2 \operatorname{var}(X_1 \mid B) + \cdots + c_n^2 \operatorname{var}(X_n \mid B). \quad (12.3)$$

Note that the expectation is always linear, but *the linearity result for the variance requires pairwise independence*. This requirement arises because we need $E(X_iX_j \mid B) = E(X_i \mid B) E(X_j \mid B)$. Here's how this is used. (I left off the $\mid B$ to make the formulas more readable.)

$$\operatorname{var}(X_1 + \cdots + X_n) = E\left(\left(\sum_{i=1}^{n}(X_i - E(X_i))\right)^2\right)$$

$$= E\left(\sum_{i,j=1}^{n}(X_i - E(X_i))(X_j - E(X_j))\right)$$

$$= \sum_{i,j=1}^{n} E\Big((X_i - E(X_i))(X_j - E(X_j))\Big), \quad \text{by (12.2)}.$$

The terms with $i = j$ in the sum give $\operatorname{var}(X_i)$. If $i \neq j$, we have

$$E\Big((X_i - E(X_i))(X_j - E(X_j))\Big)$$

$$= E\Big(X_iX_j - E(X_i)X_j - E(X_j)X_i + E(X_i)E(X_j)\Big)$$

$$= E(X_iX_j) - E(X_i)E(X_j)$$

$$\quad - E(X_j)E(X_i) + E(E(X_i)E(X_j)), \qquad \text{by (12.2)},$$

$$= E(X_iX_j) - E(X_i)E(X_j), \qquad \text{since } E(\text{constant}) = \text{constant}.$$

The remainder of the proof of the theorem is left as an exercise.

Example 12.2   Computing Means and Variances of Coins

We return to our friend, the probability space $(\mathcal{E}_k, \operatorname{Pr}_k)$ of k-long sequences of heads and tails, all of equal probability 2^{-k}. Let X_i keep track of heads on the ith toss; that is,

$$X_i(\vec{e}) = \begin{cases} 1, & \text{if } e_i \text{ (the } i\text{th toss) is heads,} \\ 0, & \text{if } e_i \text{ is tails.} \end{cases}$$

We've seen that $\operatorname{Pr}(X_i = 0) = \operatorname{Pr}(X_i = 1) = \frac{1}{2}$ and that the X_i's are mutually independent. You should be able to show that $E(X_i) = \frac{1}{2}$ and $\operatorname{var}(X_i) = \frac{1}{4}$.

Let X be the total number of heads in the entire sequence. Obviously, $X = X_1 + \cdots + X_k$. It follows from Theorem 12.1 that $E(X) = k/2$. This is intuitively obvious since it simply says we expect half the tosses to be heads. The theorem also gives the less obvious result that $\operatorname{var}(X) = k/4$. Try to derive these results directly from the definition of the expectation and variance of X without using the formula $X = X_1 + \cdots + X_k$ or Theorem 12.1. It's not so easy! ∎

Example 12.3   Estimating Probabilities by Sampling

Suppose we have a weighted die and want to estimate the probability of each of its six faces' coming up. This could be done by the frequency approach. For concreteness, let's estimate the probability of rolling ⚅. Suppose we make n rolls and make the reasonable assumption that they're independent. Let X_i be 1 or 0 according to whether ⚅ does or does not appear on the ith roll. Let $X = X_1 + \cdots + X_n$. By the frequency approach, the estimate is $\Pr(⚅) = X/n$, which we'll call Y.

We refer to the n rolls as a *sample* and this method of estimating probabilities as *sampling*.

You should be able to show that

$$E(Y) = \frac{E(X)}{n} = \frac{n\,E(X_1)}{n} = \Pr(⚅)$$

and

$$\mathrm{var}(Y) = \frac{\mathrm{var}(X)}{n^2} = \frac{n\,\mathrm{var}(X_1)}{n^2} = \frac{\mathrm{var}(X_1)}{n}.$$

In other words, our estimator Y—the average number of ⚅s in n rolls—has an average value equal to the probability we want to estimate, $\Pr(⚅)$. When the expected value of an estimator equals what we are trying to estimate, statisticians say we have an *unbiased estimator*.

What sense can we make of $\mathrm{var}(Y)$? Since variance measures how much the square of Y deviates from its expectation, the square root of the variance should measure how much Y deviates from $\Pr(⚅)$. In other words $\sqrt{\mathrm{var}(Y)}$ should give us some idea of how large an error to expect when we use Y as an estimate for $\Pr(⚅)$. This is indeed the case, in a sense that will be made more precise in Example 12.7 (p. 515). The difference between Y and $\Pr(⚅)$ is called the *sampling error*.

From the previous paragraph, we can expect the estimate $Y \approx \Pr(⚅)$ to have an error that is proportional to $\sqrt{\mathrm{var}(Y)} = \sqrt{\mathrm{var}(X_1)}/\sqrt{n}$. Thus we can expect the error to be proportional to $1/\sqrt{n}$. Notice that, if n increases by a factor of 100, this decreases only by a factor of 10. In other words, we would have to do one hundred times as many rolls for one more digit of accuracy in our estimate of $\Pr(⚅)$. This is disappointing since it says the frequency method of estimating probabilities requires large samples. Unfortunately, sampling and outright guessing are often the only available options.

When a simulation program uses the frequency method to estimate probabilities, we call it *Monte Carlo Simulation*. One of the original applications of this method was the determination of the critical mass of ^{235}U for an A-bomb. The probabilities associated with the behavior of individual atoms and particles were fairly well understood, but the overall rate of neutron capture by a large mass of atoms was not. ∎

*Example 12.4   The Bias–Variance Dilemma for Classifiers

As we saw some time ago, creating a pattern classifier for a problem can often be viewed as choosing parameters in some specified function, such as choosing $\vec{w}$ and $\vec{\theta}$ for a neural net. This example uses mean and variance to explore the errors such functions make.

Imagine a probability space $(\mathcal{E}, \text{Pr})$ in which we associate with each elementary event two random variables, $\vec{X}$ and Y, the input pattern and the desired output (classification). In general, $\vec{X}$ and Y are vectors of real numbers, but we'll assume that $Y \in \mathbb{R}$ for simplicity. For simplicity, we'll also make the often unrealistic assumption that Y is a function of $\vec{X}$; that is, $\vec{X}$ is sufficient to determine Y. The argument for the general case is similar to that for the special case.

We want to choose f from a specified set so that $\text{E}((f(\vec{X}) - Y)^2)$ is minimized. (This expectation is over the probability space $(\mathcal{E}, \text{Pr})$ of the previous paragraph.) To do this, we select a training set $\mathcal{T} = (\vec{X}_1, Y_1), \ldots, (\vec{X}_t, Y_t)$ and try to choose f so as to minimize $\sum_{i=1}^{t} \text{E}((f(\vec{X}_i) - Y_i)^2)$. Thus, the training set leads to a function $f(\vec{X}; \mathcal{T})$ that depends on $\mathcal{T}$.

Now imagine that $\mathcal{T}$ is chosen randomly using $(\mathcal{E}, \text{Pr})$. We'll do this by choosing t independent samples from $\mathcal{E}$. This can be described by a new probability space $(\mathcal{E}^{(t)}, \text{Pr}^{(t)})$ which is the t-fold product of $(\mathcal{E}, \text{Pr})$ with itself. Let $\text{E}_{\mathcal{T}}$ denote expectation over this space (so the $\vec{X}_i$'s and Y_i's in the training set vary). Let E with no subscript denote the expectation over $(\mathcal{E}, \text{Pr})$ (so $\vec{X}$ and Y to which the selected function f is applied vary). Since computing $\text{E}_{\mathcal{T}}$ involves selecting a training function, f differs from term to term in the sum for $\text{E}_{\mathcal{T}}$. The average square error of $f(\vec{X}; \mathcal{T})$ is $\text{E}((f(\vec{X}; \mathcal{T}) - Y)^2)$. We would like to choose our set of allowed f's so that the expected value of the average square error is minimized. In other words, we want to minimize

$$\text{E}_{\mathcal{T}} \left(\text{E}((f(\vec{X}; \mathcal{T}) - Y)^2) \right). \tag{12.4}$$

It can be shown that $\text{E}_{\mathcal{T}}(\text{E}(\ldots)) = \text{E}(\text{E}_{\mathcal{T}}(\ldots))$. (Do it! This requires a clear understanding of what sums are involved.) Let's begin by focusing on $\text{E}_{\mathcal{T}}$. Note that $\vec{X}$ and Y are fixed when computing $\text{E}_{\mathcal{T}}$. Using Theorem 12.1 repeatedly, we have

$$\text{E}_{\mathcal{T}} \left((f(\vec{X}; \mathcal{T}) - Y)^2 \right) = \text{E}_{\mathcal{T}}(f(\vec{X}; \mathcal{T})^2) - 2Y \, \text{E}_{\mathcal{T}}(f(\vec{X}; \mathcal{T})) + Y^2$$

$$= \left(\text{E}_{\mathcal{T}}(f(\vec{X}; \mathcal{T})) - Y \right)^2$$

$$+ \text{E}_{\mathcal{T}}(f(\vec{X}; \mathcal{T})^2) - \left(\text{E}_{\mathcal{T}}(f(\vec{X}; \mathcal{T})) \right)^2$$

$$= \left(\text{E}_{\mathcal{T}}(f(\vec{X}; \mathcal{T})) - Y \right)^2 \qquad \text{bias} \tag{12.5}$$

$$+ \text{var}_{\mathcal{T}}((f(\vec{X}; \mathcal{T})) \qquad \text{variance.} \tag{12.6}$$

The bias measures how much the average function value at $\vec{X}$ deviates from Y. The variance measures how much the function values at $\vec{X}$ vary from one training set to another. Note that neither of these is negative, so there can be no cancellation between them. To average over all $(\vec{X}, Y)$, we simply take the expectations E of (12.5) and (12.6) over $(\mathcal{E}, \text{Pr})$.

It follows that, to make the expected mean square error small, we ought to reduce both the bias and the variance. How can this be done?

If the allowed functions f are too simple, they are unlikely to be able to capture some of the aspects of the data. In particular, for a particular pair $(\vec{X}, Y)$, there may be a general tendency to overestimate or a general tendency to underestimate—a bias that will make (12.5) large. The obvious solution is to choose $f(\vec{X}; \mathcal{T})$ from a larger class of functions. In other words, increasing the set of possible functions should help reduce bias.

On the other hand, this increased complexity causes problems. We may find a function that fits the training set well but tends to behave wildly off the training set. For example, given n points (x_i, y_i) with distinct x-coordinates, we can always fit them exactly with a polynomial of degree $n - 1$ or greater, but the polynomial often has large coefficients and thus behaves poorly. In other words, for any particular pair $(\vec{X}, Y)$, we may obtain a wide range of values for $f(\vec{X}; \mathcal{T})$ as we vary the training set $\mathcal{T}$.

This is the bias–variance dilemma: Attempts to decrease bias by introducing more parameters often tend to increase variance. Attempts to reduce variance by reducing parameters often tend to increase bias. Figure 10.3 (p. 421) illustrates this problem. When bias is the dominant effect, increasing complexity reduces generalization error; but when variance is the dominant effect, increasing complexity increases the generalization error. Example 11.10 (p. 485) discusses generalization error from a different approach. ∎

Don't be misled by the preceding example into thinking that the bias–variance dilemma is well understood. Experts still debate how to handle the tradeoff between bias and variance. Many aspects of classifiers relate to the problem. For example, choosing different classifiers obviously leads to different possible bias–variance tradeoffs. How the classifier is trained is important: Example 11.10 (p. 485) suggested one approach for neural nets—don't train too long. Less obvious is the fact that even data preparation influences the possible tradeoffs.

Exercises

12.1.A. Define mean and variance.

12.1.B. State the formulas for computing the mean and variance of a linear combination of $X_1, \ldots, X_n$ in terms of the individual means and variances. Be sure to explain the role of independence.

12.1.C. Explain the concept of estimation by sampling. Why does it often give a rather poor estimate?

12.1.D. What is the bias–variance dilemma?

12.1.1. A penalty is often given on multiple choice exams to counteract guessing. What is the effect of guessing?

Suppose a question has k answers and someone guesses at random. Construct a probability space in which e_i, $1 \le i \le k$ are the possible answers, e_1 is the correct answer, e_0 is no answer and $\Pr(e_i)$ is the probability that e_i is chosen by the student. The value of $\Pr$ will be determined by how a person guesses.

(a) Let X be a random variable equal to the number of points received where $X(e_0) = 0$, $X(e_1) = 1$, and $X(e_i) = -w$ otherwise. Describe $\Pr$ when the person guesses at random. Show that $w = (k-1)^{-1}$ makes $E(X) = 0$ when a person guesses randomly.

(b) Determine the variance of X for random guessing

(c) Suppose a person is able to somehow decide (correctly) that j of the possible answers are wrong and then guesses randomly. What is $\Pr$? What are the mean and variance of X now?

12.1.2. Tweedledum and Tweedledee have decided to go their separate ways.

(a) Alice is carefully wrapping their collection of 60 distinctive matched cups and saucers which she passes to the Mad Hatter. He distributes them randomly between the twins' boxes. Since each cup and each saucer is wrapped separately, there's no telling which goes with which. How many matched cups and saucers can Tweedledee expect to receive?
Hint. Introduce a random variable X_i that is 1 if he receives the ith cup and ith saucer and is 0 otherwise.

(b) The White Rabbit decides this is unfair and rearranges things so that each twin gets 30 cups and 30 saucers. Now how many matched cups and saucers can Tweedledee expect to receive? (He can tell a wrapped saucer from a wrapped cup, but he can't match them up.)

(c) Alice and the Mad Hatter repeat the process with the collection of 60 pairs of salt and pepper shakers. How many matched salt and pepper shakers can Tweedledee expect to receive?

(d) Again the White Rabbit intervenes, but all he can do is make sure each twin receives 60 wrapped shakers. Now how many matched salt and pepper shakers can Tweedledee expect to receive?
Hint. Be careful on this one!

12.1.3. Complete the proof of Theorem 12.1 as follows:

(a) Prove the first two parts of the theorem.

(b) Prove (12.2).

(c) Show that $\operatorname{var}(cX \mid B) = c^2 \operatorname{var}(X \mid B)$ and that $c_i X_i$ and $c_j X_j$ are independent given B whenever X_i and X_j are. Conclude that it suffices to prove (12.3) when all the c_i's equal 1.

(d) Prove the claim in the text that $\operatorname{E}(X_i X_j \mid B) = \operatorname{E}(X_i \mid B)\operatorname{E}(X_j \mid B)$ when X_i and X_j are independent given B.

(e) Complete the proof of the theorem.

12.1.4. The parts of this exercise are not stated in the most general form because that would make the statements and proofs more complex. More general forms can be obtained by repeatedly applying the results given here. Suppose that $X_1, \ldots, X_n$ are independent given B.

(a) Let S be a subset of the real numbers. Define

$$Y(e) = \begin{cases} 1, & \text{if } X_1(e) \in S, \\ 0, & \text{if } X_1(e) \notin S. \end{cases}$$

Prove that $Y, X_2, \ldots, X_n$ are independent given B.

(b) Let a and b be real numbers. Prove that $(aX_1 + bX_2), X_3, \ldots, X_n$ are independent given B.

(c) Let $a_{t+1}, \ldots, a_n$ be real numbers. Prove that $X_1, \ldots, X_t$ are independent given $C = (X_{t+1} = a_{t+1} \wedge \cdots \wedge X_n = a_n) \wedge B$.

*12.1.5. Let X and Y be random variables. Define a new random variable Z by

$$Z(e) = \begin{cases} \operatorname{E}\left(X(e) \,\middle|\, Y = Y(e)) \wedge B\right), & \text{if } e \in B, \\ \text{anything}, & \text{if } e \notin B. \end{cases}$$

More briefly, we simply write $Z = \operatorname{E}(X \mid Y \wedge B)$.

(a) Show that $\operatorname{E}(Z \mid B) = \operatorname{E}(X \mid B)$; that is,

$$\operatorname{E}\left(\operatorname{E}(X \mid Y \wedge B) \,\middle|\, B\right) = \operatorname{E}(X \mid B).$$

(b) It would be nice to extend the idea to make some reasonable sense out of $\operatorname{E}\left(\operatorname{E}(X \mid A \wedge B) \,\middle|\, B\right) = \operatorname{E}(X \mid B)$. Do so.

12.1.6. This exercise returns to the Hopfield net. See Section 11.1 (p. 430) for notation.

(a) Show that, except for a constant factor, we may think of $w_{i,j}$ as being $\operatorname{E}(P_i P_j)$, where the training data are regarded as observations of random variables $\vec{P}$.

*(b) In explaining local minima at $\pm \vec{P}(k)$ in Example 11.1 (p. 436), I assumed that the patterns were randomly ± 1. Show that my argument breaks down if the patterns are biased in certain components; that is, if some $\operatorname{E}(P_i)$ are not zero.

12.2   Probability Spaces and Density Functions

Since computers are finite-state machines, our probability space can always be considered finite. Nevertheless, there are times when the fiction of an infinite space is useful because the underlying mathematics is simpler. For example, we talk about the uniform distribution on $[0, 1]$, by which we mean that all real numbers on $[0, 1]$ are "equally likely." We can't make sense of this in terms of the elementary event model: Since the value of $\Pr(e)$ would be some constant c, the sum of $\Pr(e)$ over $e \in \mathcal{E} = [0, 1]$ would be zero or infinity depending on whether $c = 0$ or not. Unfortunately, this sum is supposed to $\Pr(\mathcal{E}) = 1$.

Calculus suggests a way out—replace the sum by an integral. Then $\Pr(e)$ will be replaced by something like $f(x)dx$, which is an infinitesimal, not a number. Rather than trying to base a theory on infinitesimals, we abandon simple events in favor of compound events. The interpretation of $f(x)dx$ as the probability of an elementary event is then a convenient heuristic aid, like thinking of dy/dx as a fraction.

A solid, general, mathematical foundation of probability requires fancy tools, namely sigma algebras and measure theory. This section relies only on basic calculus. The resulting theory is not the most general, but it's all we need.

Probability Spaces

The following definition is *not a generalization* of the definition of a finite probability space given in Chapter 7. The more general definition given in mathematical probability theory subsumes both definitions and requires more mathematical background. Since we don't need the general definition, let's agree to use the simpler one.

Definition 12.2   Probability Space and Random Variable

Let $\mathcal{E} = \mathbb{R}^n$ for some n. If $f : \mathcal{E} \to \mathbb{R}$ is such that $f(\vec{x}) \geq 0$ for all $\vec{x} \in \mathbb{R}^n$ and such that

$$\int_{-\infty}^{\infty} \cdots \int_{-\infty}^{\infty} f(\vec{x})dx_1 \cdots dx_n = 1,$$

then the pair $(\mathcal{E}, f)$ is called a *probability space* and f is called a *probability density function*, or *pdf*. When $\mathcal{E} = \mathbb{R}$, the function $F(x) = \int_{-\infty}^{x} f(x)dx$ is called the (cumulative) *distribution function*. By the Fundamental Theorem of Calculus, $f = F'$.

12.1.3. Complete the proof of Theorem 12.1 as follows:

   (a) Prove the first two parts of the theorem.

   (b) Prove (12.2).

   (c) Show that $\text{var}(cX \mid B) = c^2 \, \text{var}(X \mid B)$ and that $c_i X_i$ and $c_j X_j$ are independent given B whenever X_i and X_j are. Conclude that it suffices to prove (12.3) when all the c_i's equal 1.

   (d) Prove the claim in the text that $\text{E}(X_i X_j \mid B) = \text{E}(X_i \mid B)\,\text{E}(X_j \mid B)$ when X_i and X_j are independent given B.

   (e) Complete the proof of the theorem.

12.1.4. The parts of this exercise are not stated in the most general form because that would make the statements and proofs more complex. More general forms can be obtained by repeatedly applying the results given here. Suppose that $X_1, \ldots, X_n$ are independent given B.

   (a) Let S be a subset of the real numbers. Define

$$Y(e) = \begin{cases} 1, & \text{if } X_1(e) \in S, \\ 0, & \text{if } X_1(e) \notin S. \end{cases}$$

   Prove that $Y, X_2, \ldots, X_n$ are independent given B.

   (b) Let a and b be real numbers. Prove that $(aX_1 + bX_2), X_3, \ldots, X_n$ are independent given B.

   (c) Let $a_{t+1}, \ldots, a_n$ be real numbers. Prove that $X_1, \ldots, X_t$ are independent given $C = (X_{t+1} = a_{t+1} \wedge \cdots \wedge X_n = a_n) \wedge B$.

*12.1.5. Let X and Y be random variables. Define a new random variable Z by

$$Z(e) = \begin{cases} \text{E}\Big(X(e) \,\Big|\, Y = Y(e)) \wedge B\Big), & \text{if } e \in B, \\ \text{anything}, & \text{if } e \notin B. \end{cases}$$

More briefly, we simply write $Z = \text{E}(X \mid Y \wedge B)$.

   (a) Show that $\text{E}(Z \mid B) = \text{E}(X \mid B)$; that is,

$$\text{E}\Big(\text{E}(X \mid Y \wedge B) \,\Big|\, B\Big) = \text{E}(X \mid B).$$

   (b) It would be nice to extend the idea to make some reasonable sense out of $\text{E}\Big(\text{E}(X \mid A \wedge B) \,\Big|\, B\Big) = \text{E}(X \mid B)$. Do so.

12.1.6. This exercise returns to the Hopfield net. See Section 11.1 (p. 430) for notation.

   (a) Show that, except for a constant factor, we may think of $w_{i,j}$ as being $\text{E}(P_i P_j)$, where the training data are regarded as observations of random variables $\vec{P}$.

   *(b) In explaining local minima at $\pm \vec{P}(k)$ in Example 11.1 (p. 436), I assumed that the patterns were randomly ± 1. Show that my argument breaks down if the patterns are biased in certain components; that is, if some $\text{E}(P_i)$ are not zero.

12.2   Probability Spaces and Density Functions

Since computers are finite-state machines, our probability space can always be considered finite. Nevertheless, there are times when the fiction of an infinite space is useful because the underlying mathematics is simpler. For example, we talk about the uniform distribution on $[0, 1]$, by which we mean that all real numbers on $[0, 1]$ are "equally likely." We can't make sense of this in terms of the elementary event model: Since the value of $\Pr(e)$ would be some constant c, the sum of $\Pr(e)$ over $e \in \mathcal{E} = [0, 1]$ would be zero or infinity depending on whether $c = 0$ or not. Unfortunately, this sum is supposed to $\Pr(\mathcal{E}) = 1$.

Calculus suggests a way out—replace the sum by an integral. Then $\Pr(e)$ will be replaced by something like $f(x)dx$, which is an infinitesimal, not a number. Rather than trying to base a theory on infinitesimals, we abandon simple events in favor of compound events. The interpretation of $f(x)dx$ as the probability of an elementary event is then a convenient heuristic aid, like thinking of dy/dx as a fraction.

A solid, general, mathematical foundation of probability requires fancy tools, namely sigma algebras and measure theory. This section relies only on basic calculus. The resulting theory is not the most general, but it's all we need.

Probability Spaces

The following definition is *not a generalization* of the definition of a finite probability space given in Chapter 7. The more general definition given in mathematical probability theory subsumes both definitions and requires more mathematical background. Since we don't need the general definition, let's agree to use the simpler one.

Definition 12.2   Probability Space and Random Variable

Let $\mathcal{E} = \mathbb{R}^n$ for some n. If $f : \mathcal{E} \to \mathbb{R}$ is such that $f(\vec{x}) \geq 0$ for all $\vec{x} \in \mathbb{R}^n$ and such that

$$\int_{-\infty}^{\infty} \cdots \int_{-\infty}^{\infty} f(\vec{x})dx_1 \cdots dx_n = 1,$$

then the pair $(\mathcal{E}, f)$ is called a *probability space* and f is called a *probability density function*, or *pdf*. When $\mathcal{E} = \mathbb{R}$, the function $F(x) = \int_{-\infty}^{x} f(x)dx$ is called the (cumulative) *distribution function*. By the Fundamental Theorem of Calculus, $f = F'$.

If $A \subseteq \mathbb{R}^n$ is a set over which it is possible to integrate, we define $\Pr(A)$ to be the integral of $f(\vec{x})$ over the set A.

A function $X : \mathcal{E} \to \mathbb{R}$ is called a (real-valued) *random variable*. The distribution function of X is $G(x) = \Pr(X \leq x)$ and its density function is G'.

We often talk about the pdf g of a random variable without mentioning a probability space. In this case, we can usually take $(\mathbb{R}, g)$ to be the probability space and take $X(x) = x$. To see that this works out correctly, simply note that

$$\Pr(X \leq x) = \int_{-\infty}^{x} g(x)dx \quad \text{and so} \quad \frac{d\Pr(X \leq x)}{dx} = g(x).$$

Example 12.5   Generating Random Numbers

We saw in the previous chapter that we use random numbers in connection with neural nets. How are such numbers generated?

The support software for most computer languages includes a procedure that generates a "uniform random variable" on $[0, 1)$. Let $\mathcal{E} = \mathbb{R}$. The uniform distribution on $[0, 1)$ is given by the density function

$$f(x) = \begin{cases} 1, & \text{if } 0 \leq x < 1, \\ 0, & \text{otherwise.} \end{cases}$$

Obviously $f(x) \geq 0$ and $\int_{-\infty}^{\infty} f(x)dx = 1$, so f is a pdf. A random variable X with pdf f is called uniform on $[0, 1)$. More generally, a uniform random variable on $A \subset \mathbb{R}^n$ is a random variable whose pdf is constant on A and 0 otherwise. In order to have such a constant and have the integral of the pdf equal 1, the "measure" of A must be finite. When $n = 1$, measure is length; when $n = 2$, measure is area, and so forth.

A computer installation frequently lacks procedures for generating other random variables. Here's a way to do it. Suppose $G(x) = \Pr(X \leq x)$. To generate a random variable Y based on this distribution, first generate U uniformly on $[0, 1)$ and then let $Y = G^{-1}(U)$. Here's a proof that this works. Since G is a nondecreasing function, $Y \leq y$ if and only if $G(Y) \leq G(y)$. Since $G(Y) = U$, we have

$$\Pr(Y \leq y) = \Pr(U \leq G(y)) = \int_{0}^{G(y)} 1\, dy = G(y).$$

This idea can also be used for finite probability spaces. For example, suppose we want to randomly choose an elementary event from $\mathcal{E} = \{e_1, \ldots, e_n\}$. Let $p_i = \Pr(e_i)$, let $P_0 = 0$, and let $P_i = p_1 + \cdots + p_i$. To choose an e_k, generate U uniformly at random on $[0, 1)$ and let k be such that $P_{k-1} \leq U < P_k$. The proof that this works is left as an exercise. ∎

The ideas and results from finite probability spaces carry over to the new framework. The differential $f(\vec{x})dx_1 \cdots dx_n$ and integration play the same roles as $\Pr(e)$ and summation did for finite probability spaces. In working out results, it's sometimes convenient to think in terms of elementary events, regarding $f(\vec{x})dx_1 \cdots dx_n$ as the "probability" of the event $\vec{x}$. For example,

$$\mathrm{E}(X) = \int_{-\infty}^{\infty} \cdots \int_{-\infty}^{\infty} X(\vec{x})f(\vec{x})dx_1 \cdots dx_n.$$

In fact, this correspondence is so strong that people (including me) sometimes write $\Pr(\vec{x})$ for the pdf. From the previous discussion, the correct analogy is $\Pr(\vec{x})$ for $f(\vec{x}))d\vec{x}$, not for $f(\vec{x})$; however, the $d\vec{x}$'s make equations messier and it's easy to put them in integrals as needed.

Computing something like $\Pr(X \leq a)$ can be tricky—we need to integrate $f(\vec{x})$ over $\{\,\vec{x} \mid X(\vec{x}) \leq a\,\}$. A full study of this problem requires measure theory. To avoid that, our random variables will usually be of the form $X_i(\vec{x}) = x_i$. Since $\{\,\vec{x} \mid X_i(\vec{x}) \leq a\,\} = \{\,\vec{x} \mid x_i \leq a\,\}$, $\Pr(X_i \leq a)$ is computed by integrating x_i from $-\infty$ to a and all the other x_j's from $-\infty$ to ∞. When $n = 1$, this becomes simply $\Pr(X \leq a) = F(a)$, the distribution function.

Definition 12.3   Marginal Density and Independence

If $I \subseteq \{1, \ldots, n\}$, define $f_I : \mathbb{R}^I \to \mathbb{R}$ to be the multiple integral of $f(\vec{x})$ over all x_j with $j \notin I$. For example, when $n = 4$

$$f_{\{2,4\}}(x_2, x_4) = \int_{-\infty}^{\infty} \int_{-\infty}^{\infty} f(x_1, x_2, x_3, x_4)dx_1 dx_3.$$

We call f_I the *marginal density function* with respect to I. Write f_i for $f_{\{i\}}$.

Let $X_i : \vec{x} \to x_i$ be random variables and let $\mathcal{X} = \{\,X_i \mid i \in I\,\}$. We say that the variables $\mathcal{X}$ are *independent random variables* if $f_I = \prod_{i \in I} f_i$.

This definition of independence is quite limited because of the restricted form of the X_i. There's a general definition, but we don't need it.

The use of the definition is often more straightforward than its statement. Typically, we have random variables, say X_1 and X_2, with distribution functions $\Pr(X_i \leq x) = F_i(x)$. To create a space in which they are independent, we set $\mathcal{E} = \mathbb{R} \times \mathbb{R}$ and $f(x_1, x_2) = F_1'(x_1)F_2'(x_2)$. (We constructed finite probability spaces of independent random variables in the same manner.) The proof that the definition of independence is satisfied follows simply from the

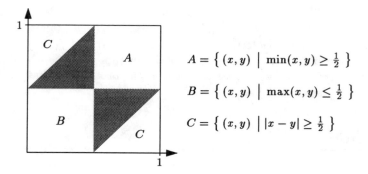

$$A = \big\{ \, (x,y) \;\big|\; \min(x,y) \geq \tfrac{1}{2} \, \big\}$$

$$B = \big\{ \, (x,y) \;\big|\; \max(x,y) \leq \tfrac{1}{2} \, \big\}$$

$$C = \big\{ \, (x,y) \;\big|\; |x-y| \geq \tfrac{1}{2} \, \big\}$$

Figure 12.1 The shaded region is the part of $[0,1]^2$ that gives rise to triangles, as discussed in Example 12.6.

statements that

$$f = F_1' F_2', \quad \int_{-\infty}^{\infty} f(\vec{x}) dx_2 = F_1'(x_1), \quad \int_{-\infty}^{\infty} f(\vec{x}) dx_1 = F_2'(x_2).$$

The next example uses this idea.

Example 12.6   Making Triangles

Suppose we choose two numbers independently at random according to the uniform distribution on the interval $[0,1]$ and then divide $[0,1]$ at those two points. (You can think of this as breaking a stick at two random points.) What is the probability that the three pieces of the interval $[0,1]$ can be assembled to form a triangle? You should convince yourself that the pieces will form a triangle if and only if the length of the longest piece is less than $\tfrac{1}{2}$. To do this, remember that any two sides of a triangle add up to more than the third side.

We create a new probability space equal to the product of the uniform distribution space with itself: $\mathcal{E} = \mathbb{R}^2$ and

$$f(x,y) = \begin{cases} 1, & \text{for } (x,y) \in [0,1]^2, \\ 0, & \text{otherwise.} \end{cases}$$

Let's look at the choices for (x,y) that fail to give a triangle. To do this, we check the length of each of the three pieces into which $[0,1]$ is divided. If $\min(x,y) \geq \tfrac{1}{2}$, the left piece $[0, \min(x,y)]$ shows that we do not have a triangle. Similarly, $\max(x,y) \leq \tfrac{1}{2}$ fails to give a triangle because the right piece is too long. Finally, if $|x-y| \geq \tfrac{1}{2}$, the middle piece (between x and y) is too long. These three conditions determine regions in $[0,1]^2$, labeled A, B, and C in Figure 12.1. Each of A, B, and C has area $\tfrac{1}{4}$.

Since $f = 1$, integrating it over a region simply gives the area of the region. We need not be concerned about whether to include or exclude boundary points or boundary lines because they are regions in $\mathbb{R}^2$ of zero area and so they contribute nothing to the integral of $f(x, y)$. Thus, the probability that the three pieces form a triangle is the integral of $f(x, y)$ over the shaded regions in Figure 12.1. Since $f(x, y) = 1$ in those regions, the integral equals the area of the regions. This proves that the probability that the three pieces form a triangle is $\frac{1}{4}$. ∎

Exercises

12.2.A. Define a probability density function.

12.2.B. Explain the connection between marginal density functions and independent random variables of the form $X_i(\vec{x}) = x_i$.

12.2.C. What is the uniform distribution?

12.2.1. Suppose that $(\mathbb{R}, f)$ is a probability space. For any set A of real numbers, define $\chi_A(x)$ to be 1 when $x \in A$ and 0 otherwise. Prove that, if $\int_{-\infty}^{\infty} \chi_A(x) f(x) dx$ exists, then it equals $\Pr(A)$.

12.2.2. Let $(\mathbb{R}, f_i)$ be probability spaces for $1 \leq i \leq n$. Define

$$f(\vec{x}) = f_1(x_1) \cdots f_n(x_n) \text{ for } \vec{x} \in \mathbb{R}^n.$$

(a) Prove that $(\mathbb{R}^n, f)$ is a probability space.

(b) Let $X_i(\vec{x}) = x_i$. Prove that the X_i are independent random variables.

12.2.3. Prove that the method described in the last paragraph of Example 12.5 works.

12.2.4. Prove Theorem 12.1 (p. 499), without conditioning on B, for infinite probability spaces.

12.2.5. Two random numbers are chosen uniformly and independently on $[1, 3]$. What is the probability that their sum exceeds their product?

12.2.6. Let X be a random number (variable) with probability distribution function $F(x)$ and density function $f(x) = F'(x)$. Show the the distribution and density functions of $Y = -X$ are given by $G(x) = 1 - F(-x)$ and $g(x) = f(-x)$.

12.2.7. Let $X_1, \ldots, X_n$ be independent random variables and let $A_1, \ldots, A_n \subseteq (-\infty, \infty)$. Assuming all the integrals involved exist, prove that

$$\Pr\big((X_1 \in A_1) \wedge \cdots \wedge (X_n \in A_n)\big) = \Pr(X_1 \in A_1) \times \cdots \times \Pr(X_n \in A_n)$$

12.2.8. Suppose that n independent random numbers $X_1, \ldots, X_n$ are chosen and that each has the probability distribution function $G(x)$.

   (a) Let Y be the maximum of $X_1, \ldots, X_n$. Show that the distribution function for Y is $(G(x))^n$ and that the probability density function for Y is $nG'(x)(G(x))^{n-1}$.

   (b) Let Z be the minimum of $X_1, \ldots, X_n$. What is the distribution function for Z?
   Hint. $\min(X_1, \ldots, X_n) = -\max(-X_1, \ldots, -X_n)$

The Normal Distribution and the Central Limit Theorem

> *There must be something mysterious about the normal law since mathematicians think it is a law of nature whereas physicists are convinced that it is a mathematical theorem.*
>
> —Henri Poincaré (1854–1912)

Recall that a matrix B is called *symmetric* if $b_{i,j} = b_{j,i}$ for all i, j. The following defines one of the most important pdfs.

Definition 12.4   Multivariate Normal (or Gaussian) Distribution

Let $\mathcal{E} = \mathbb{R}^n$, let $\vec{\mu} \in \mathbb{R}^n$, and let B be an $n \times n$ real symmetric matrix with positive eigenvalues $\lambda_1, \ldots, \lambda_n$. We call

$$f(\vec{x}) = \frac{\exp\left(\frac{1}{2}(\vec{x} - \vec{\mu})^t B^{-1}(\vec{x} - \vec{\mu})\right)}{\sqrt{(2\pi)^n \lambda_1 \cdots \lambda_n}} \tag{12.7}$$

a *normal* or *gaussian density function*.

Of course, we have to *prove* that B^{-1} exists and that f satisfies the conditions for a density function, namely that $f \geq 0$ and that the integral of f over $\mathcal{E}$ equals 1. It's obvious that $f \geq 0$. The other conditions are beyond the mathematics we've developed here.

Aside. For those familiar with determinants, the product of the eigenvalues is simply the determinant of the matrix. Also, a square matrix has an inverse if and only if its determinant is nonzero, which is equivalent to having no zero eigenvalues.

   For those familiar with changing coordinates in n-dimensional integrals, use the rigid motion in Theorem 11.5 (p. 444) to convert the integrand to a simple form. The Jacobian of a change of basis just involves the change of basis matrix, which has determinant ± 1 for a rigid motion.

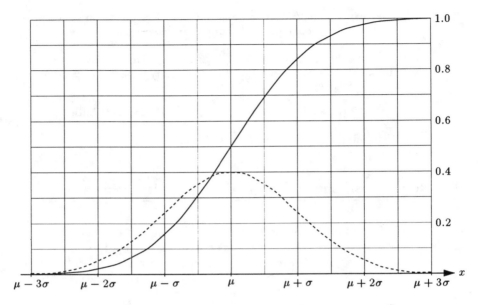

Figure 12.2   The normal distribution with mean μ and variance σ^2. The dotted curve is the associated density function. The vertical lines are at multiples of 0.5σ. For the distribution function, the horizontal lines are at multiples of 0.1. For the density function, the horizontal lines are at multiples of $0.1/\sigma$. Thus the vertical scales differ for the two curves unless $\sigma = 1$.

The case $n = 1$ is particularly important, and we'll limit our attention to it. If $n = 1$, then $\vec{\mu}$ and B are just numbers and $B = \lambda_1$. Set $\sigma = \sqrt{B}$. Then (12.7) becomes

$$\text{(one-dimensional) normal pdf:}\quad f(x) = \frac{\exp\left((x-\mu)^2/2\sigma^2\right)}{\sqrt{2\pi}\,\sigma}. \qquad (12.8)$$

The one-dimensional normal distribution and density functions are shown in Figure 12.2. A random variable with pdf (12.8) is said to be normally distributed. It's left as an exercise to show that the random variable's mean and variance are μ and σ^2.

The integral of (12.8) is often evaluated in calculus classes when integration in polar coordinates is discussed. First, set $x = \sigma t + \mu$:

$$\frac{1}{\sqrt{2\pi}\,\sigma} \int_{-\infty}^{\infty} f(x)dx = \frac{1}{\sqrt{2\pi}} \int_{-\infty}^{\infty} e^{-t^2/2}\,dt,$$

Multiply the latter integral by itself after replacing t with x and y. Change to polar coordinates and evaluate. In symbols:

$$\left(\int_{-\infty}^{\infty} e^{-t^2/2}\, dt\right)^2 = \int_{-\infty}^{\infty}\int_{-\infty}^{\infty} e^{-(x^2+y^2)/2}\, dx\, dy = \int_{0}^{2\pi}\int_{0}^{\infty} e^{-r^2/2}\, r\, dr\, d\theta$$

$$= \left(\int_{0}^{2\pi} d\theta\right)\left(\int_{0}^{\infty} e^{-r^2/2}\, r\, dr\right) = (2\pi)\left(-e^{-r^2/2}\Big|_{r=0}^{\infty}\right) = 2\pi.$$

It's left as an exercise for you to show that the mean and variance of the random variable given by $\Pr(X \leq x) = \int_{-\infty}^{x} f(x)dx$ are μ and σ^2, respectively.

Important properties of normally distributed random variables are given in the next two theorems. The first says that "normality" is preserved by addition. The second says that addition creates random variables that are close to normal. We'll prove the first theorem shortly, but the proof of the second is beyond the mathematical level of this text. See Feller [10; vol. 2, pp. 259–262] for a proof of the Central Limit Theorem and a discussion of other versions of it.

Theorem 12.2   Sums of Independent Normal Variables

The sum of mutually independent, normally distributed random variables is again normally distributed.

Theorem 12.3   Central Limit Theorem

Let $X_1, X_2, \ldots$ be an infinite sequence of random variables with means and variances μ_i and σ_i^2. Assume that the σ_i^2's are bounded and that $X_1, \ldots, X_n$ are mutually independent for all n. Define

$$\mu(n) = \mu_1 + \cdots + \mu_n, \quad \sigma(n)^2 = \sigma_1^2 + \cdots + \sigma_n^2,$$

and

$$Y_n = \frac{(X_1 + \cdots + X_n) - \mu(n)}{\sigma(n)}.$$

Then, for any fixed y,

$$\lim_{n\to\infty} \Pr(Y_n \leq y) = (2\pi)^{-1/2} \int_{-\infty}^{y} e^{-t^2/2}\, dt; \tag{12.9}$$

that is, the distribution of Y_n approaches a normal distribution with mean 0 and variance 1.

Although the Central Limit Theorem talks about a limit, relatively small values of n often give a good approximation to a normal distribution. (See Exercise 12.2.11.) The practical importance of the Central Limit Theorem lies primarily in the point of view it leads to:

> If a variety of primarily additive and rather independent random effects are contributing to a measurement, we expect the measurement to have a distribution that is close to normal.

As a result, we frequently approximate an unknown distribution by a normal distribution, even when we have no way of knowing whether the conditions for near normality are satisfied. In fact, the normal distribution is often used when it's known that the approximation is poor! This may be done out of ignorance ("Isn't it always done?"), desperation ("I can't find the true distribution"), or hope ("It'll probably be good enough"). In fact a normal distribution, or any other reasonable distribution, is often good enough. It's most likely to fail when you're using the tail of the distribution, that is, estimating the probability of a random variable's being far from the mean.

Proof (of Theorem 12.2):   We begin with a sum of two normally distributed independent random variables X and Y. If $Z = X + Y$, what is the distribution of Z? Before attempting to answer this, we need to sort out what it means. Since these separate functions are normal and since the variables are independent, their joint pdf is

$$f(x,y) = \frac{1}{2\pi\sigma_x\sigma_y} \exp\left((x - \mu_x)^2/2\sigma_x^2 + (y - \mu_y)^2/2\sigma_y^2\right).$$

The probability of $Z = z$ in the finite case would be the sum over all x and y such that $x + y = z$. The analogous statement here is that the probability density function for $Z = z$ is given by the corresponding integral. Since $x + y = z$, we substitute $y = z - x$ and then integrate over all x. This will be a somewhat messy calculation. At its heart will be the exponential of

$$\frac{-(x - \mu_x)^2}{2\sigma_x^2} + \frac{-(z - x - \mu_y)^2}{2\sigma_y^2}.$$

When this is rearranged, it has the form $-a(x - l(z))^2 - bz^2 - cz - d$, for some constants a, b, c, and d and some linear function $l(z)$. We can move all but $\exp\left(-a(x - l(z))^2\right)$ outside the integral over x. When the integral is evaluated, we obtain something of the form $A\exp(-b(z + c/2b)^2)$. Thus the result is a normal distribution.

The proof of the theorem will be complete if we show that the independence of $X_1, \ldots, X_n$ implies the independence of $Y_1, \ldots, Y_{n-1}$, where $Y_i = X_i$ for $i \leq n-2$ and $Y_{n-1} = X_{n-1} + X_n$. Let f_i be the density function for X_i and let f be the density function for $\vec{X}$. Then the density function for $\vec{Y}$ is given by

$$\int_{-\infty}^{\infty} f(y_1,\ldots,y_{n-2},y_{n-1}-t,t)\,dt$$

$$= \int_{-\infty}^{\infty} f_1(y_1)\cdots f_{n-2}(y_{n-2})f_{n-1}(y_{n-1}-t)f_n(t)\,dt$$

$$= f_1(y_1)\cdots f_{n-2}(y_{n-2})\int_{-\infty}^{\infty} f_{n-1}(y_{n-1}-t)f_n(t)\,dt,$$

which is just what we need since the density function for Y_{n-1} is the right-hand integral. ∎

Example 12.7   Sampling Revisited: Confidence Intervals

In Example 12.3 (p. 501) we discussed the estimation of parameters by sampling. Specifically, we looked at estimating $\Pr\left(\boxed{\cdot\cdot}\right)$ by rolling a die n times and using the frequency of $\boxed{\cdot\cdot}$ as an estimate for the probability. This estimate was given by $Y = (X_1 + \cdots + X_n)/n$ where the X_i were independent random variables. Since the X_i's are identically distributed random variables, $\mathrm{E}(Y) = \mathrm{E}(X_i)$. Our data give us one sample of the random variable Y. How close is this sample likely to be to the number we want, that is, the expected value of Y? We can phrase this more precisely:

What is $\Pr(|Y - \mathrm{E}(Y)| < \delta)$?

The Central Limit Theorem tells us that the density function f of Y is approximately a normal with mean μ and variance σ^2/n where μ and σ^2 are the mean and variance of X_i. (This requires a bit of algebra using Theorem 12.1 (p. 499), which you can do in Exercise 12.2.12.) Using this approximation and setting $t = (y-\mu)/\sigma\sqrt{n}$, we have

$$\Pr(|Y - \mathrm{E}(Y)| < \delta) = \int_{\mu-\delta}^{\mu+\delta} f(y)\,dy$$

$$\approx \frac{1}{\sqrt{2\pi}\,\sigma}\int_{\mu-\delta}^{\mu+\delta} \exp\left(-(y-\mu)^2 n/2\sigma^2\right)dy$$

$$= \frac{1}{\sqrt{2\pi}}\int_{-\delta\sqrt{n}/\sigma}^{\delta\sqrt{n}/\sigma} e^{-t^2/2}\,dt. \tag{12.10}$$

The value of the integral can be estimated from Figure 12.2, read from tables, or obtained from a commonly available computer subroutine. Here's a table.

Values of $y = (2\pi)^{-1/2}\int_{-x}^{x} e^{-t^2/2}\,dt$							
x	0.5	1.0	1.5	2.0	2.5	3.0	3.5
y	0.38	0.68	0.87	0.955	0.988	0.997	0.9998

$$(12.11)$$

In order to use our formula, we need an estimate for σ. Since the range of X_i is $\{0,1\}$, we have

$$\mathrm{var}(X_i) = \mathrm{E}(X_i^2) - \mathrm{E}(X_i)^2 = \Pr(X_i = 1) - \Pr(X_i = 1)^2 = \mu - \mu^2 = \mu(1-\mu).$$

The maximum possible value of this expression is $\frac{1}{2}(1 - \frac{1}{2}) = \frac{1}{4}$. If $\mu(X_i) = \frac{1}{6}$, the variance is $\frac{5}{36}$. Hence, σ is probably about 0.37 and certainly no larger than 0.5.

Now that we have our numbers, we can use (12.10) and (12.11). For example, suppose we wanted an estimate that will be within ± 0.01 of the true value with probability 0.95—that's one chance in 20 of our estimate's being off by more than ± 0.01. According to the table, we want $x \approx 2.0$. Thus $n \approx x^2 \sigma^2 / \delta^2 \approx 2.0^2 \left(\frac{5}{36}\right) / 0.01^2 \approx 60$. This number of rolls is feasible, but if we reduce the error δ by a factor of ten to 0.001, the number of rolls needed is an unreasonably large 6,000. (If, as in this paragraph, an estimate has a 95% probability of being within ± 0.01 of an estimate E, statisticians call $[E - 0.01, E + 0.01]$ the 95% *confidence interval.*) ∎

Exercises

12.2.D. Write down the formula for the pdf of a one-dimensional normal distribution. Identify the mean and variance.

12.2.E. What important property do sums of independent normal variables have?

12.2.F. What does the Central Limit Theorem say?

12.2.G. What is the practical importance of the Central Limit Theorem?

12.2.H. Suppose you're trying to estimate the probability of an event by sampling. How does the expected accuracy of your estimate vary with the number of samples? Why is this a disappointment?

12.2.9. Let X be a normally distributed random variable with (12.8) as its pdf. Show that the mean and variance of X are μ and σ^2.

12.2.10. This exercise uses the results and notation of Exercise 12.1.1 (p. 504), which deals with guessing on multiple-choice exams.

  (a) Suppose the student guesses randomly on N problems, each of which has 4 choices. Use the Central Limit Theorem to estimate the probability distribution of the student's total score.

  (b) Do the same as in the previous part when the student is able to correctly eliminate one choice in each of the N problems. When $N = 5$, approximately what is the probability that the student will obtain a negative score? When $N = 30$? (Use Figure 12.2.)

12.2.11. In the Central Limit Theorem, let each X_i be a random variable that has $\Pr(X_i = 1) = \Pr(X_i = -1) = \frac{1}{2}$.

  (a) Show that the mean and variance of X_i are 0 and 1, respectively.

  (b) On a graph of the normal distribution with mean 0 and variance 1, superimpose a graph of $\Pr(Y_5 \leq x)$. (You can simply trace Figure 12.2.)

12.2.12. Prove the claim about the mean and variance for the Central Limit Theorem application in Example 12.7.

12.2.13. Let $X_1, \ldots, X_n$ be independent identically distributed random variables with mean μ and variance σ^2. We want to use the X_i's to estimate μ and σ^2. Let $Y = (X_1 + \cdots + X_n)/n$. An observation of Y is an estimate of μ. Since $\sigma^2 = \mathrm{E}((X_i - \mu)^2)$, it's tempting to estimate σ^2 by computing $S^2(Y) = \frac{1}{n} \sum_{i=1}^{n} (X_i - Y)^2$.

(a) Show that $\mathrm{E}(s^2(Y)) = \frac{n-1}{n} \sigma^2$.

(b) Use this result to describe a better method for estimating σ^2.

(c) Show that the variance of Y is σ^2/n and conclude that an estimate of the variance of Y is

$$\hat{\sigma}^2(Y) = \frac{1}{n(n-1)} \sum_{i=1}^{n} (X_i - Y)^2.$$

(d) Use the Central Limit Theorem to show that Y is roughly normally distributed with mean μ and variance $\hat{\sigma}^2(Y)$.

12.3   Some Statistics

The great fly in the ointment of controlled experiments is always variation. ... Error, bias, and unexplained variance may creep into experiments in any number of ways.
—Nathaniel S. Borenstein (1991)

There are three kinds of lies: lies, damned lies, and statistics.
—attributed to Benjamin Disraeli (1804–1881)

Statistics is the branch of mathematics that treats the analysis of data. As a consequence, it is also concerned with how to design experiments so that the data will be as useful as possible. You'll need to know a bit about two areas of statistics: estimation of parameters and hypothesis testing.

We discussed simple Monte Carlo methods for estimating parameters in Example 12.7 (p. 515). The estimation problem in this section is more specific: How can we estimate the generalization error of a classifier?

There are two types of hypothesis testing. One is the sure thing. For example, I can easily test the hypothesis "This coin is two-headed." I need only look at it. Suppose I'm not allowed to look at the coin but am given only the fact that you tossed it ten times and got heads every time. How confident can I be that the coin is two-headed? This is an example of the question "Given the data, what sort of evidence do we have concerning the hypothesis?" This is dealt with in the theory of hypothesis testing.

Estimating Generalization Error

> *[I]t is impossible to justify a correlation between reproduction of a training set and generalization error off of the training set using only a priori reasoning. As a result, the use in the real world of any generalizer that fits a hypothesis function to a training set (e.g., the use of back-propagation) is implicitly predicated on an assumption about the physical universe.*
>
> —David H. Wolpert (1992)

A pattern classifier might assign a definite class to each input pattern, or it might assign a probability distribution over the classes (for example, 0.7 probability that the input pattern represented a "B," 0.2 probability a "D," and so on). We'll study just the nonprobabilistic classifiers.

We're interested in the error rate of the classifier. More specifically, the *generalization error rate*. Suppose there are only a finite number of possible patterns $e_1, \ldots, e_n$ and that the probability of seeing e_i is p_i. (We've just described a probability space.) Define a random variable X by

$$X(e) = \begin{cases} 1, & \text{if } e \text{ is misclassified,} \\ 0, & \text{if } e \text{ is correctly classified.} \end{cases}$$

The generalization error is $\mathrm{E}(X)$, the expected value of X. This definition can be extended to an infinite probability space of patterns, but we won't bother to do so. Unfortunately, we normally don't know the probability space and so must estimate the generalization error rate $\mathrm{E}(X)$ rather that computing it.

How can we estimate the generalization error rate? As Wolpert points out in the opening quote, such estimates are based on either explicit or implicit assumptions about reality. Since the classifier is expected to be able to generalize beyond the training data, we'll need to give it a training data set that is representative of the sorts of patterns it will encounter when it's being used. Suppose we've done this.

The simplest way to estimate $\mathrm{E}(X)$ is by testing the classifier on the training data. The average error per input is called the *resubstitution error rate* by statisticians and the *training error rate* by connectionists. This rate underestimates the true classifier error rate, sometimes drastically. To see why this is so, let's imagine a *thought experiment*. A good thought experiment can be a useful method for improving understanding. To conduct a thought experiment, we create a simple situation and then reason out what will happen. For resubstitution error, imagine a training set that contains only one pattern which the classifier easily learns. Thus the resubstitution error rate is 0. It's likely that other input will be classified almost randomly and so have a high error rate.

An estimator like the training error rate is called *biased* because it underestimates the true value. In general, an estimator whose expected value is the true value is called *unbiased* while an estimator whose expected value differs from the true value is called *biased*. Obviously, it's desirable to have unbiased estimators.

Bias isn't the whole story. Variance is also important. For example, you may be interested in the probability that a coin will land heads. You can obtain an unbiased estimator for each coin by simply tossing the coin once and reporting either that the probability of heads is 1 or that it is 0 depending on how the coin lands. This estimator is unbiased—its expected value is the probability of heads. However, you wouldn't use it because of it's high variance. To achieve low variance, we usually need a large amount of data. Example 12.7 (p. 515) discusses this.

What conclusions does the previous discussion lead to? First, here are three important points in the discussion.

- To be sure we have a representative collection of examples, our training data set must be fairly large.
- To obtain a low variance estimate of generalization error, the testing data set must be fairly large.
- To avoid bias in estimating the generalization error, we apparently need separate testing and training data sets.

Thus we apparently need two rather large sets of data. Often data is difficult or expensive to obtain. Can we get by with less? That's what this section is about.

Setting the Stage

Let $\mathrm{err}(T, D)$ denote the error rate on D of a classifier trained on T and let U be the (perhaps infinite) set of all possible inputs and desired classifications. We want $\mathrm{err}(T, U)$. You know that the resubstitution error rate $\mathrm{err}(T, T)$ may underestimate $\mathrm{err}(T, U)$, perhaps severely. So you should be able to see that $\mathrm{err}(T, D)$ could also be a bad estimate for $\mathrm{err}(T, U)$. *We must have U to compute the generalization error rate.* With less than U, we can only obtain estimates. Statisticians have devised various methods for estimating the generalization error rate. Three of their methods are called

- (a) test-sample estimation,
- (b) cross-validation, and
- (c) bootstrapping.

Statisticians have shown that these methods provide unbiased or nearly unbiased estimators of the generalization error rate—provided certain conditions are satisfied. ("Conditions" are also called assumptions or hypotheses.) We won't state these hypotheses, so our discussion of these methods will be purely descriptive. Why not state the assumptions? Some assumptions are satisfied only by some simple statistical classifiers and the others can't be verified in practice. What good are results based on such assumptions? Hypotheses in theorems and mathematical theories are seldom satisfied (e.g., geometric lines don't exist in the real world, coin tossing is never entirely random) and are sometimes unverifiable (e.g., these dice are fair, these events are independent). Nevertheless, we apply mathematics. Perhaps the assumptions hold, perhaps the real world is a close enough approximation, or perhaps weaker assumptions are enough; whatever the case, mathematicians haven't managed to state and prove such a result. People are seldom so careful: They often apply theorems either in blissful ignorance of the unverified assumptions, or with their fingers crossed, or simply wishing that things will work out okay. With generalization error, you have to cross your fingers *and* wish hard.

Test-Sample Estimation

Test-sample estimation of error is based on an obvious idea for correcting the resubstitution problem. Divide the data into two parts, the training data and the testing data. After the classifier has been trained on the training data, compute its error rate on the testing data. Unfortunately, we're seldom so rich in data that we can afford to set aside a large portion of it for testing. As a result, we're pushed to make the testing data set small. This leads to error estimates that have a large variance, a problem we met in Example 12.7 (p. 515). Because of these problems, this method is seldom used in AI.

Cross-Validation

Cross-validation largely avoids the data-wasting problem of test-sample estimation. It pays for this improvement with greatly increased training time.

First divide the training data T into V sets T_i of nearly equal size. Next, for $1 \leq i \leq V$, train a classifier C_i on $T - T_i$ and test it on T_i. Finally, average the V different classifier error rates to get an overall error rate estimate. We now have an estimate for the generalization error, but we don't have a classifier to use! Rather than selecting one of those that was trained on only part of the data, train a classifier C on all the data and use it.

The extreme case in which $V = |T|$ and so $|T_i| = 1$ is often referred to as the *leave-one-out method*.

What can be said pro (+) and con (−) regarding cross-validation?

− The amount of work is an obvious issue: We must train $V + 1$ classifiers.

+ We're in the fortunate position of using all the data to train our final classifier C and using all the data to estimate the generalization error rate.

+ The fraction of the original data in $T - T_i$ is about $\frac{V-1}{V}$. Thus, choosing V large allows us to use nearly all the data for training the ith classifier.

± Since each classifier is tested on only a small fraction of the data, its error rate estimate will have a large variance. However, averaging all V error rate estimates will give a number with much smaller variance.

— The error rate estimate we obtain is not based on any one classifier. In fact, the classifier C that we finally use has not even been tested! We need to look at this more closely.

Implicit in the entire procedure is the assumption that the classifiers are, in some sense, reasonably similar. If the C_i differ greatly from one another, it hardly makes sense to average their error rates to obtain an estimate for yet another different classifier C. There is a possible way around this. The difference between generalization and resubstitution error rate may be more stable than the generalization error rate. Thus, we could estimate

$$\Delta(T, U) := \text{err}(T, U) - \text{err}(T, T) \qquad (12.12)$$

and add it to the resubstitution error rate $\text{err}(T, T)$ for C to obtain an estimate of C's generalization error rate. From C_i we have the estimate

$$\Delta(T, U) \approx \text{err}(\mathcal{R}_i, T_i) - \text{err}(\mathcal{R}_i, \mathcal{R}_i),$$

where $\mathcal{R}_i = T - T_i$ is the set C_i is trained on. I'm not aware of any theoretical or empirical studies that would let us decide when estimating $\Delta(T, U)$ is better than estimating $\text{err}(T, U)$.

Bootstrapping

Bootstrapping also requires the construction of more than one classifier, but there are major differences between it and cross-validation. Perhaps the most notable is that bootstrapping might be described as an approach rather than as a specific method. The basic theme behind the method is "sampling with replacement." We'll describe the simplest bootstrap method here. See the literature for variations. The most popular variant is probably the .632 estimator.

Let $n = |T|$. Imagine a bag containing T, out of which we draw an item, note what it is, and return it to the bag. This is repeated for $i = 1, \ldots, n$. Now we have some collection T^*, which may not be a set since it can contain multiple copies of the same element. We say that T^* was chosen from T by *sampling with replacement* because an element of T may appear more than once in T^*. The bootstrap estimate, which we won't justify, is given by

$$\Delta(T, U) \approx \mathrm{E}_{T^*}\Big(\text{err}(T^*, T) - \text{err}(T^*, T^*)\Big). \qquad (12.13)$$

The right-hand side contains an expectation that can't be computed in a reasonable manner. Instead, it's estimated by sampling: Repeatedly choose T^*'s and average the results to estimate the expectations. The variance of $\text{err}(T^*, T) - \text{err}(T^*, T^*)$ is often low so that not many T^*'s are needed to obtain a reasonable estimate.

Unlike cross-validation, bootstrapping has the virtue of using all the data for testing each classifier. What are its drawbacks? It uses only about 63% of T

for each classifier it trains. To see this, suppose that $X_i \in \mathcal{T}$. The probability that $X_i \notin \mathcal{T}^*$ is $(1 - 1/n)^n$. It's a standard calculus exercise to show that this approaches $1/e$ as $n \to \infty$. Thus, the expected number of elements in $\mathcal{T}$ that are not in $\mathcal{T}^*$ is

$$n(1 - 1/n)^n \approx n/e \approx 0.368n$$

Using only 63% of the training set might cause problems if the data is very limited.

Summary

The various error estimates you've seen are

$$\Delta(\mathcal{T}, \mathcal{U}) \approx \begin{cases} 0, & \text{resubstitution} \\ \operatorname{err}(\mathcal{T} - \mathcal{D}, \mathcal{D}) - \operatorname{err}(\mathcal{T} - \mathcal{D}, \mathcal{T} - \mathcal{D}), & \text{test sample} \\ \frac{1}{V} \sum_{i=1}^{V} \Big(\operatorname{err}(\mathcal{T} - \mathcal{T}_i, \mathcal{T}_i) - \operatorname{err}(\mathcal{T} - \mathcal{T}_i, \mathcal{T} - \mathcal{T}_i) \Big), & \text{cross-validation} \\ \mathrm{E}_{\mathcal{T}^*} \Big(\operatorname{err}(\mathcal{T}^*, \mathcal{T}) - \operatorname{err}(\mathcal{T}^*, \mathcal{T}^*) \Big), & \text{bootstrap} \end{cases}$$

The first three of these are also used to provide a direct estimate of $\operatorname{err}(\mathcal{T}, \mathcal{U})$:

$$\operatorname{err}(\mathcal{T}, \mathcal{U}) \approx \begin{cases} \operatorname{err}(\mathcal{T}, \mathcal{T}), & \text{resubstitution} \\ \operatorname{err}(\mathcal{T} - \mathcal{D}, \mathcal{D}), & \text{test sample} \\ \frac{1}{V} \sum_{i=1}^{V} \operatorname{err}(\mathcal{T} - \mathcal{T}_i, \mathcal{T}_i), & \text{cross-validation} \end{cases}$$

What's the method of choice? Resubstitution is a clear loser and data limitations often preclude test-sample estimation. Can we choose between cross-validation and bootstrapping? Not easily, if at all. Applying bootstrapping blindly can lead to unfortunate results—see Exercise 12.3.3. On the other hand, there's evidence that bootstrapping methods give superior results in some situations. Whether any particular case is such a situation is often best determined by hindsight. Young's remark [22, p. 411] concerning bootstrapping applies to the other methods, too:

> The bootstrap is no surrogate for careful thought on a statistical problem . . . . Applied blindly, the bootstrap often cannot be trusted, and it is always necessary to formulate in precise terms the problem being tackled.

Whatever method you use, it's important to avoid reading too much into the results. Here are some common errors.

- **Ignoring Bias in Data:** It's difficult—perhaps impossible—to know if $\mathcal{T}$ is a biased sample of $\mathcal{U}$. Regardless of the testing done on $\mathcal{T}$, it only tells us about a classifier's behavior on $\mathcal{T}$, never about its behavior on $\mathcal{U}$. To make any predictions about generalization to $\mathcal{U}$, we must assume that $\mathcal{T}$

is "typical" of $\mathcal{U}$. Remember the classifier that detected tanks by sunlight (page 417)? No amount of testing with the biased $\mathcal{T}$ would have predicted err$(\mathcal{T},\mathcal{U})$ correctly.

- **Ignoring Bias in Estimates**: It can be shown that, under suitable assumptions, the estimates of generalization are unbiased. Such assumptions are usually not satisfied in practice, but we still hope that the estimates will be unbiased whenever the data is. This is usually the case; however, some researchers have found that bias was significant in problems they were looking at, so be careful.

- **Ignoring Monte Carlo Variance**: Since $\mathcal{T}$ is a more or less random sample of $\mathcal{U}$, we're dealing with a Monte Carlo type of situation. Consequently, the estimate of generalization error is expected to be no closer to the true value than about $C/\sqrt{|\mathcal{T}|}$. Unfortunately, C is not known.

- **Ignoring Classifier Variation**: Cross-validation and bootstrapping do not test the classifier C that is finally produced. Instead, they combine data from other classifiers and expect it to predict the behavior of C. If there's considerable variation between classifiers, it may be unwise to have confidence in such estimates.

Some research is being done on these problems, but it's unclear if much progress can be expected.

Exercises

12.3.A. What is test sample estimation and why is it seldom used?

12.3.B. What is cross-validation and what are some of its strengths and weaknesses?

12.3.C. What is sampling with replacement? How does bootstrapping use it?

12.3.1. In cross-validation, it's been suggested that, instead of training a new classifier, we might use the C_i for which err$(\mathcal{R}_i, \mathcal{T})$ is smallest. Explain how you might use an estimate of $\Delta(\mathcal{T},\mathcal{U})$ to estimate the generalization error rate for C_i.

12.3.2. Describe sampling with replacement by constructing a probability space on $\mathcal{E} = \mathcal{T}$ using the uniform distribution $\Pr(e) = 1/n$ and regarding $\mathcal{T}^*$ as the instantiation of n independent random variables each of which has $X(e) = e$.

12.3.3. Suppose we have a classifier that ignores the number of times a sample is repeated in the test data. Let T^* be the sample drawn for a bootstrap training of the classifier and let T' be T^* with duplications eliminated.

  (a) Show that $\text{err}(T^*, T) = \text{err}(T', T)$.

  (b) Conclude that $\text{err}(T^*, T) \approx \frac{e-1}{e}\text{err}(T', T') + \frac{1}{e}\text{err}(T', T - T')$.

  (c) Argue that $\text{err}(T^*, T^*) \approx \text{err}(T', T')$.

  (d) Conclude that the bootstrap estimate is about 37% of the cross-validation estimate, namely

$$\text{bootstrap} \quad \Delta(T, \mathcal{U}) \approx \frac{1}{e}\, \text{E}\Big(\text{err}(T', T - T') - \text{err}(T', T')\Big).$$

*12.3.4. The discussion in the text assumed that the classification algorithm provides a definite assignment to a class. Suggest a method for measuring the error rate when the classifier provides only probability distributions. Your method should agree with the measurement for definite classifiers when the probability distributions all collapse to certainty—one probability 1 and the rest 0. (This is an open question intended to provoke thought and discussion—there's no "right" answer.)

Is It Significant?—Hypothesis Testing

> *Statistical inference is a kind of inverse to probability theory in which we start with data and then reason backwards to some initially unknown information about an underlying random process*
>
> —Fred Kochman (1993)

Probability theory tells us how to compute the probability of a sequence, $\vec{x}$, say, of heads and tails given that $\Pr(\text{heads}) = p$. In contrast, statistics tells us how to decide between competing hypotheses concerning p when we are given $\vec{x}$.

A *hypothesis* can be regarded as a statement about parameters; for example, "the probability of heads on coin A differs from that on coin B." Given a hypothesis, we always have the complementary hypothesis—in this case, the two probabilities of heads are equal. Statisticians refer to the one hypothesis as the *null hypothesis* H_0 and the other as the *alternative hypothesis* H_1. In the coin example, we might choose

H_0 is "Coins A and B have the same probability of heads."

H_1 is "Coins A and B have differing probabilities of heads."

Of course, we could also reverse the roles of H_0 and H_1. This is important because statisticians view the null hypothesis and the alternative hypothesis asymmetrically—we want to find support for the null hypothesis. We might decide either that

1. The data makes it likely that H_0 is true. (We "accept" H_0.)

2. The data's evidence is inadequate to support H_0. (We "reject" H_0.)

The terminology in the latter case is somewhat unfortunate since "rejecting" H_0 does not imply "accepting" its negation H_1—the evidence for H_1 may be inadequate, too, so we remain undecided. If you read much on statistics, you will run into the expressions *Type I error* and *Type II error*. A Type I error is a rejection of H_0 when it's true and a Type II error is an acceptance of H_0 when it's false. In a situation that requires us to accumulate information, a Type I error results from a failure to decide (or act) and a Type II, from acting too soon. Thus, Type I is associated with hesitation or inertia and Type II with rashness.

Although there's a natural symmetry between H_0 and H_1 (each is the complement of the other), acceptance and rejection are asymmetric. Thus, it can be important to decide which is H_0 and which is H_1. Various methods have been advanced for making this decision. Often the null hypothesis is the simpler one—it says that nothing interesting is going on.

The theory of hypothesis testing is extensive enough to provide material for an entire course. We'll discuss only two topics:

- Given estimates of two numbers, what can we say about the probability that one number (the actual number, not the estimate) is larger than the other? For example, suppose we estimate the generalization error rate for two classifiers. What can we say about the chances of one classifier's being better than the other?

- The other topic is essentially the coin problem we began with: deciding if two samples come from the same probability distribution. For example, a vertex in a decision tree splits some training data into two sets. Do the sets differ significantly; that is, do the outputs appear to be sampled from different distributions?

Is This Number Really Larger Than That One?

We'll phrase this discussion in terms of generalization error to make things more concrete; however, it applies to many other situations. Suppose that two classifiers C_1 and C_2 have generalization error rates r_1 and r_2, both of which are unknown. Using test-sample estimation, we obtain estimates $\hat{r}_1$ and $\hat{r}_2$ of r_1 and r_2. What can we say about $\Pr(r_1 > r_2)$? Comparing $\hat{r}_1$ and $\hat{r}_2$ is not enough because they're estimates—not true values. We need some idea of how good the estimates are.

Let's look at a simple general problem: Let $x_1, x_2, \ldots, x_n$ be independent samples from a distribution having unknown mean and variance r and σ^2. Estimate the probability that $r > 0$. Assume as little as possible about the nature of the distribution. (The statement that the x_i are independent samples from a distribution with pdf f means that each x_i was selected "at random" using the pdf f.)

Let $\hat{r}$ denote the average of $x_1, x_2, \ldots, x_n$. According to Exercise 12.2.13 (p. 517), $\hat{r}$ is nearly normally distributed with mean equal to the mean r of the underlying distribution and with variance σ^2, which is approximately $\hat{\sigma}^2 = \hat{\sigma}^2(\hat{r})$ in the notation of Exercise 12.2.13. In other words, the pdf for $\hat{r}$ is approximately

$$f(\hat{r} \mid r) = \frac{\exp\left(-(\hat{r} - r)^2/2\hat{\sigma}^2\right)}{\sqrt{2\pi}\,\hat{\sigma}}. \tag{12.14}$$

We can't immediately extract information about r from (12.14) because it's the pdf of $\hat{r}$ given r. This suggests taking a Bayesian approach: Multiply by the prior pdf of r and divide by the pdf of $\hat{r}$. As we found in Bayes' Theorem, we don't need the latter since it can be found by summing (integrating in the continuous case) as was done in (7.19) (p. 281). We must have the pdf of r and we have no idea what it is. A rather questionable but common approach is to assume a uniform distribution. In fact, we can't have a uniform distribution on $(-\infty, \infty)$. We can have one on $[-L, L]$ and let $L \to \infty$. I'll skip the details and simply tell you the result: (12.14) is an estimate of the pdf for r given $\hat{r}$. In other words, if

$$\hat{r} = \frac{1}{n}\sum_{i=1}^{n} x_i \quad \text{and} \quad \hat{\sigma}^2 = \frac{1}{n(n-1)}\sum_{i=1}^{n}(x_i - \hat{r})^2, \tag{12.15}$$

then (12.14) is approximately the pdf for the mean r of the distribution from which the values x_i were obtained.

We're almost done. We've now estimated the pdf for the unknown mean r based on values $\hat{r}$ and $\hat{\sigma}$ calculated from the data by (12.15). Thus

$$\Pr(r > 0) = \frac{1}{\sqrt{2\pi}\,\hat{\sigma}} \int_0^\infty \exp\left(\frac{-(t - \hat{r})^2}{2\hat{\sigma}^2}\right) dt.$$

The substitution $u = \frac{(t - \hat{r})}{\hat{\sigma}}$ converts this into

$$\Pr(r > 0) = \frac{1}{\sqrt{2\pi}} \int_{-\hat{r}/\hat{\sigma}}^{\infty} e^{-u^2/2} du \qquad (12.16)$$

Numerical values of this integral can be looked up in tables. Since (12.16) exceeds 0.95 when $\hat{r}/\hat{\sigma} > 2$, statisticians refer to $\hat{r}/\hat{\sigma} = 2$ as the 95% *confidence level*. It means that $\Pr(r \leq 0) < 1/20$ when $\hat{r}/\hat{\sigma}$ is 2 or larger. In much of scientific research, the 95% confidence level has become the accepted standard for a significant result. It means that less than 5% of the conclusions drawn from such tests are wrong—provided the data has been correctly collected and the tests have been used correctly.

Let's apply (12.16) to the original problem: Let $y_{i,j} = 1$ if the ith classifier is wrong on pattern j and let it be 0 otherwise. Let $\vec{x} = \vec{y}_1 - \vec{y}_2$ so that $\vec{x}$ and hence $\hat{r}$ provides information about how the two classifiers differ. Let e_i be the number of times the ith classifier is wrong and the other is right. We'll show that

$$\hat{r} = \frac{e_1 - e_2}{n} \quad \text{and} \quad \hat{\sigma}^2 < \frac{e_1 + e_2}{n(n - 1)} \qquad (12.17)$$

and so

$$\frac{\hat{r}}{\hat{\sigma}} > \frac{e_1 - e_2}{\sqrt{e_1 + e_2}} \sqrt{\frac{n - 1}{n}} \approx \frac{e_1 - e_2}{\sqrt{e_1 + e_2}}.$$

As noted after (12.16), this should be at least 2 for the statement that the first error rate exceeds the second to be considered significant. The $\hat{r}$ value is straightforward and the $\hat{\sigma}^2$ inequality is obtained from

$$\hat{\sigma}^2 = \frac{1}{n(n-1)} \sum_{i=1}^{n} (x_i - \hat{r})^2 = \frac{1}{n(n-1)} \sum_{i=1}^{n} \left(x_i^2 - 2x_i\hat{r} + \hat{r}^2 \right)$$

$$= \frac{1}{n(n-1)} \left((e_1 + e_2) - 2(e_1 - e_2)\hat{r} + n\hat{r}^2 \right)$$

$$= \frac{e_1 + e_2}{n(n-1)} - \frac{\hat{r}^2}{n-1}. \qquad (12.18)$$

There are some objections to this approach. Here are two important ones

- The result was based on the assumption of a uniform prior for the mean of the ith distribution, an assumption some people would take exception to. Bayesians claim that priors are needed if we want to answer questions like "what is $\Pr(r_1 > r_2)$?" I believe they're correct. Given that priors must be used and we don't know what they are, what should we do? One favorite choice of Bayesians is the "conjugate prior." Various textbooks contain discussions of it. See for example [5]. Non-Bayesians reject the notion of creating a prior and look for other approaches. In the remainder of this section we'll discuss such a method for testing if two samples come from the same distribution.

- The result was based on test-sample estimation where it's clear how the $y_{i,j}$'s were obtained. In reality, we're likely to use cross-validation or bootstrapping. In these cases, we can certainly obtain estimates of the error rate, but it's not clear what corresponds to the $y_{i,j}$'s. See Exercise 12.3.6 for a partial resolution of this problem.

Exercises

12.3.D. Explain Kochman's statement "Statistical inference is a kind of inverse to probability theory."

12.3.E. How do prior probabilities enter when we want the probability that a parameter r is positive and we have an estimate $\hat{r}$ of r?

12.3.5. In the notation of (12.17), suppose that $e_2 = 0$. What does this mean? How large must e_1 be for $r > 0$ to be accepted as significant? What does all this mean?

12.3.6. Let's return to the problem of $\Pr(r_1 > r_2)$. Now assume that $\hat{r}_i$ and $\hat{\sigma}_i$ have been obtained somehow for r_1 and r_2. Thus we have (12.14) with subscripts $i = 1, 2$ on everything.

   (a) Show that $f(-r_2 \mid \hat{r}_2)$ is roughly normal and determine the mean and variance.

   (b) Show that if r_1 and r_2 are independent, then $f(r_1 - r_2 \mid \hat{r}_1 \wedge \hat{r}_2)$ is roughly normal with mean $\hat{r}_1 - \hat{r}_2$ and variance $\hat{\sigma}_1^2 + \hat{\sigma}_2^2$.

   (c) Obtain a result similar to (12.16) for $\Pr(r_1 > r_2)$.

12.3.7. This is a continuation of the previous exercise. Suppose that $\hat{r}_i$ is determined from $y_{i,1}, \ldots, y_{i,n}$ where the $y_{i,j}$'s are as in the text. Suppose also that the two classifiers make errors independently of each other.

   (a) Show that $\hat{\sigma}_i^2$ in the modified (12.14) is

$$\hat{\sigma}_i^2 = \frac{\hat{r}_i(1 - \hat{r}_i)}{n - 1}.$$

   (b) In the notation of (12.17), show that $\mathrm{E}(e_1) = r_1(1 - r_2)n$ and similarly for $\mathrm{E}(e_2)$.

   (c) Using the results of this exercise, show that the value of $\hat{\sigma}$ in (12.18) is the same as that in the previous exercise.

   (d) Suppose that the errors of the two classifiers are correlated; that is, if one makes an error then the other is more likely to make an error, too. The equality of the previous step now becomes an inequality. Which is larger and why? Which method makes it more difficult to establish significance and why?

*Fisher's Exact Test

Let X and Y be random variables with the same range. Our data consists of some observations of X and some observations of Y. H_0 is the hypothesis that $\Pr(X = a) = \Pr(Y = a)$ for all a. Since the theory is simpler when X and Y can take only two values, we'll assume that the common range of X and Y is $\{\mathrm{T}, \mathrm{F}\}$.

Fisher's exact test is a method for deciding on acceptance or rejection. Let $N_{\mathrm{T},X}$ be the number of observations of X that equal T, and define $N_{\mathrm{F},X}$, $N_{\mathrm{T},Y}$, and $N_{\mathrm{F},Y}$ similarly. If we sum on a subscript, we simply replace it with an asterisk. According to the null hypothesis, we made $N_{*,*}$ observations, obtaining $N_{\mathrm{T},*}$ values of T and $N_{\mathrm{F},*}$ of F. All the observations were then labeled either X or Y so that a total of N_X were labeled X. According to H_1, there are two different distributions present.

We need to formulate H_0 (and thus H_1) carefully. The values of $N = N_{*,*}$, $N_{\mathrm{T},*}$, $N_{\mathrm{F},*}$, $N_{*,X}$, and $N_{*,Y}$ are taken as given. The null hypothesis asserts that the observed $N_{i,j}$ have arisen simply from a random sampling.

How can we calculate the probability of the $N_{i,j}$ given the null hypothesis? Imagine a bag containing $N_{\mathrm{T},*}$ identical balls labeled T and $N_{\mathrm{F},*}$ identical balls labeled F. Select $N_{*,X}$ balls at random from the bag and label them X. The remainder are labeled Y. We can ask for the probability that this process will lead to the observed values of $N_{i,j}$. Note that once $N_{\mathrm{T},X}$ is determined, so are the other $N_{i,j}$ because

$$N_{\mathrm{F},X} = N_{*,X} - N_{\mathrm{T},X}, \quad N_{\mathrm{T},Y} = N_{\mathrm{T},*} - N_{\mathrm{T},X}, \quad N_{\mathrm{F},Y} = N_{*,Y} - N_{\mathrm{T},Y}.$$

Thus, it suffices to calculate the probability of obtaining $N_{\mathrm{T},X}$. The elementary events in our probability space will be all ways of choosing $N_{*,X}$ balls and each will be equally likely. Let $\binom{n}{k}$ be the number of ways of choosing k things from a set of n. The number of ways to take $N_{*,X}$ balls from the bag is then $\binom{N}{N_{*,X}}$. Since each $e \in \mathcal{E}$ is equally likely, $\Pr(e) = 1 \big/ \binom{N}{N_{*,X}}$. If we are to get the correct values of $N_{i,j}$, we must have chosen $N_{\mathrm{T},X}$ from the balls labeled T and $N_{\mathrm{F},X}$ from the balls labeled F. How many elementary events have this property? We can choose the T balls in $\binom{N_{\mathrm{T},*}}{N_{\mathrm{T},X}}$ ways and the F balls in $\binom{N_{\mathrm{F},*}}{N_{\mathrm{F},X}}$ ways. Their product is the number of elementary events that have $N_{\mathrm{T},X}$ balls labeled T and $N_{\mathrm{F},X}$ labeled F. Thus, the probability that we obtained the observed values of $N_{i,j}$ is

$$\Pr(N_{\mathrm{T},X} | H_0) = \binom{N_{\mathrm{T},*}}{N_{\mathrm{T},X}} \binom{N_{\mathrm{F},*}}{N_{\mathrm{F},X}} \frac{1}{\binom{N}{N_{*,X}}}. \tag{12.19}$$

To complete this calculation we need a formula for $\binom{n}{k}$. If the order of choosing were important, there would be $n(n-1)\cdots(n-k+1)$ total choices because there would be $n-i+1$ ways to choose the ith object after removing the first $i-1$. This gives all *ordered* choices. Given k objects, there are $k(k-1)\cdots 1 = k!$

ways to order them—just choose k objects from the set of k with the order of choosing being important. Thus, for each unordered choice there are $k!$ corresponding ordered choices. Hence

$$\binom{n}{k} = \frac{n(n-1)\cdots(n-k+1)}{k!} = \frac{n!}{k!(n-k)!}.$$

Combining this with (12.19), we have the desired probability.

What should we do with the probability?

A first thought is to accept H_0 if (12.19) is large and reject it otherwise. There's a problem with this: Any *particular* values for the $N_{i,j}$ will not be very likely because there are so many possible choices for (i,j).

Presumably $N_{T,X}$ deviates from its expected value, $N_{*,X} N_{T,*}/N$ (see Exercise 12.3.8 for a derivation). Let's change H_0 to be, "Under the assumption that X and Y are assigned at random, $N_{T,X}$ will deviate from its expected value by at least the amount observed." This is easily computed. Let Z be a random variable defined on the probability space we constructed earlier with $Z(e)$ equal to the number of selected balls that are labeled T. Then

$$\Pr\left(|Z - N_{*,X} N_{T,*}/N| \geq |N_{T,X} - N_{*,X} N_{T,*}/N|\right) = \sum_{k \in S} \frac{\binom{N_{T,*}}{k}\binom{N_{F,*}}{N_{*,X}-k}}{\binom{N}{N_{*,X}}}$$

$$(12.20)$$

where

$$S = \left\{ k \ \Big| \ |k - N_{*,X} N_{T,*}/N| \geq |N_{T,X} - N_{*,X} N_{T,*}/N| \right\}.$$

Some algebra shows that S is the same as the set of k outside the open interval between $N_{T,X}$ and $2N_{*,X} N_{T,*}/N - N_{T,X}$. Fisher's exact test rejects the hypothesis that X and Y have the same probability distribution whenever (12.20) is small. How small? That depends on the user.

Example 12.8   Applying Fisher's Exact Test

Rice [18, p. 434] applies Fisher's test to some data of Rosen and Jerdee. Each of 48 male supervisors was asked to examine an imaginary employee's file and decide whether to promote the employee to a new position or to interview other candidates. The qualifications in all the files were the same; however, 24 of the files identified the candidate as male and 24 as female. The 48 supervisors made decisions that led to the following table:

	$X =$ Female	$Y =$ Male	
T = Promote	14	21	$N_{T,*} = 35$
F = Interview	10	3	$N_{F,*} = 13$
	$N_{*,X} = 24$	$N_{*,Y} = 24$	$N = 48$

We want to know if there is sex bias, that is, a significant difference between the two distributions.

Each supervisor corresponds to an elementary event. The random variable X is measured at some elementary events and the random variable Y at others. The expected value of $N_{T,X}$ is $N_{*,X} N_T/N = 24 \times 35/48 = 17.5$. Here's a table of (12.19) based on the given values of N_i:

$N_{T,X}$	13	14	15	16	17	18	19	20	21	22
Pr from (12.19)	.004	.021	.072	.162	.241	.241	.162	.072	.021	.004

Larger and smaller values of $N_{X,T}$ have probability 0 to three places. The probability that $N_{X,T}$ deviates from its expected value by at least as much as the observed $|14 - 17.5|$ is the sum of the values up to 14 and above 21. This gives us a probability of 0.05 of seeing a deviation as large as or larger than that which was observed. It seems reasonable to reject H_0 and so believe that sex bias is present. ∎

Actually, we might argue that the probability in (12.20) is not quite the appropriate thing to be interested in. It works fine in a situation with certain symmetries as in the example.

Instead, statisticians look at all 2×2 arrays A with the same values of $N_{*,X}$, $N_{*,Y}$, $N_{T,*}$, and $N_{F,*}$. For each array, they compute $\Pr(A)$. They then add the probabilities over all arrays that have a probability not exceeding the probability of the given array; in other words, all arrays that are at least as unlikely as A. In the previous example, this reduces to (12.20).

This idea has the advantage that it easily carries over to $r \times c$ tables. This is important if we have $c > 2$ possible classifications and/or allow $r > 2$ possible splits at a decision. If you want more information on how to compute the probabilities, see [16] or [17].

When the $N_{i,j}$ are large, Fisher's exact test is harder to apply because of the computation involved. In this case, we can use an approximation, the *chi-square test*. In fact, the chi-square test is just as easily stated for $r \times c$ tables. Statisticians usually recommend that all $N_{i,j}$ be at least 5 before using the test. Unfortunately, this is neither necessary nor sufficient to guarantee that the chi-square test is a good approximation. Fortunately, the niceties of test accuracy are often unimportant in the light of other approximations and assumptions that are made in designing expert systems. Hence, it's probably reasonable to use the chi-square test if the average value of $N_{*,j}$ is at least 5.

Here's the chi-square test for c classifications. Define

$$X^2 = \sum_{i=T,F} \sum_{j=1}^{c} \frac{(N_{i,j} - N_{i*}N_{*,j}/N)^2}{N_{i,*}N_{*,j}/N}, \qquad (12.21)$$

where a term $\frac{0}{0}$ is considered to be 0. (Such terms arise when $N_{i,*}N_{*,j} = 0$.) In a table of chi-square, look up $\Pr(\chi^2 \geq X^2)$ for $c - 1$ degrees of freedom.

Pr	degrees of freedom												
	1	2	3	4	5	6	7	8	9	10	15	20	30
.50	0.46	1.39	2.37	3.36	4.35	5.35	6.35	7.34	8.34	9.34	14.3	19.3	29.3
.20	1.64	3.22	4.64	5.99	7.29	8.56	9.80	11.0	12.2	13.4	19.3	25.0	36.3
.05	3.84	5.99	7.82	9.49	11.1	12.6	14.1	15.5	16.9	18.3	25.0	31.4	43.8
.02	5.41	7.82	9.84	11.7	13.4	15.0	16.6	18.2	19.7	21.2	28.3	35.0	48.0
.01	6.63	9.21	11.3	13.3	15.1	16.8	18.5	20.1	21.7	23.2	30.6	37.6	50.9

Figure 12.3   The left column is the chi-square probability $\Pr(\chi^2 \geq X^2)$. Each of the remaining columns gives the value of X^2 needed to obtain that probability. For larger tables, see almost any statistics text. Many computer systems have subroutines for chi-square.

This is the approximation to the probability in the generalization of Fisher's exact test. A table of chi-square appears in Figure 12.3.

Exercises

12.3.F. What is Fisher's exact test used for?

12.3.8. Suppose the assignment of the labels X and T to our balls is done independently. Express the probability that a ball is labeled both X and T in terms of N, $N_{*,X}$ and $N_{T,*}$. Use your result to compute $E(N_{T,X})$.

12.3.9. Suppose the experiment described in the example was repeated, but this time the resulting entries were $\begin{pmatrix} 20 & 15 \\ 4 & 9 \end{pmatrix}$. Is it reasonable to assume that sex bias is present? Why? What if the numbers were $\begin{pmatrix} 16 & 19 \\ 8 & 5 \end{pmatrix}$?

12.3.10. You and I each perform an experiment t times. The possible outcomes are P and Q. All t times that I did it, I obtained P. All t times that you did it, you obtained Q. What is the result of applying Fisher's exact test to the problem of deciding whether or not we were doing experiments with the same probability of having P as an outcome?

12.3.11. Lynd and Lynd [15] gave 415 female and 369 male high school students a list of ten attributes and asked them to choose the two they considered most desirable in a father. The attribute "college graduate" was chosen by 86 males and 55 females. Do males and females rate this attribute differently? Why?

12.3.12. For the 2×2 case, prove that (12.21) simplifies to

$$X^2 = \frac{4(N_{\mathrm{T},X}\, N - N_{*,X}\, N_{\mathrm{T},*})^2}{N_{*,X}\, N_{*,Y}\, N_{\mathrm{T},*}\, N_{\mathrm{F},*}\, N}.$$

12.4 Information Theory

Uncertainty has played an important role in most of this text and we've discussed various approaches to dealing with it. Information theory is one more approach; however, it's not a tool for manipulating uncertain knowledge. Instead, it's a tool for measuring uncertainty. We can think of the amount of our knowledge ("information") and lack of it ("uncertainty") as complementary concepts—more of one means less of the other. In information theory, uncertainty is measured by a quantity called "entropy." It's similar to, but not the same as, the concept of entropy in physics.

Imagine constructing a pattern classifier when the set of possible outputs is finite. For any input A, let d_A be the desired output. An ideal classifier would be a mapping f so that $f(A) = d_A$ for all A. In other words, once we compute $f(A)$ there is no uncertainty about the value of d_A. Since the ideal is usually unattainable, we might look for an f that is nearly correct. Thus, we want to choose f so that, given $f(A)$, the uncertainty about the value of d_A is as small as possible.

In order to do this, we must have some numerical measure of uncertainty. We'll use the axiomatic approach: First we'll list and defend some properties we want a measure of uncertainty (or information) to possess. Next we'll show that the properties uniquely determine the measure.

Suppose we have a probability space with a random variable X whose range is finite, say $\{1, 2, \ldots, n\}$. What characterizes the uncertainty about the value of X? Since the uncertainty arises from not knowing the value of X, all that should matter is $\mathrm{Pr}(X)$. For this reason, we'll assume that the uncertainty of X is a function $H(p_1, \ldots, p_n)$ where $p_k = \mathrm{Pr}(X = k)$. (Sometimes we'll write $H(\vec{p})$.) This gives us

Axiom 1: Uncertainty is measured by a real-valued function $H(\vec{p})$ that is defined for all $\vec{p} \in \mathbb{R}^n$ for which $\sum_{k=1}^n p_k = 1$ and $p_k \geq 0$ for all k.

(If I were being pedantic, I'd say there are an infinite number of H's, one for each value of n.) The next three axioms capture the concepts given in brackets after them.

Axiom 2: $H(\vec{p}) \geq 0$ with equality if and only if some $p_i = 1$. [Uncertainty is nonnegative and there is none if and only if the outcome is certain.]

Axiom 3: If $\vec{q}$ is $\vec{p}$ with all components that are 0 removed, then $H(\vec{p}) = H(\vec{q})$. ⟦Impossible outcomes are irrelevant.⟧

Axiom 4: $H(1/n, \ldots, 1/n)$ is an increasing function of n. ⟦When all outcomes are equally likely, more of them gives more uncertainty.⟧

Axiom 5: $H(\vec{p})$ is a continuous function of $\vec{p}$. ⟦Uncertainty varies smoothly.⟧

These five axioms are straightforward. Unfortunately, they don't uniquely determine H. We need one more axiom. Before stating it, let's explore why we need it and set up some notation.

Suppose $\{1, 2, \ldots, n\}$ is partitioned into subsets $S_1, \ldots, S_m$. We can imagine obtaining the value of X in two steps: First determine the subset containing X and then determine its value given the subset it's in. (The latter sounds very much like a conditional probability. Indeed, conditional probability plays a central role in Axiom 6.) The uncertainty about the result of this two-step process should be the same as the uncertainty in determining Y directly since the end result is the same.

This should give us a relationship between values of H in the first step and the "direct" value of H. To do this we need two assumptions: Uncertainty is additive and, in a random situation, we should take expectations. Exactly what this means will be made clear in the next paragraph. It may sound like assuming additivity is rather ad hoc. For example, why not assume that it's multiplicative? If it were, its logarithm would be additive and there's no reason why we couldn't use the logarithm as the measure of uncertainty.

You should be able to establish the following simple conditional probability formula by simply recalling how S_i was defined:

$$\Pr(X = k \mid X \in S_i) = \begin{cases} p_k/s_i, & \text{if } k \in S_i, \\ 0, & \text{if } k \notin S_i, \end{cases} \quad \text{where} \quad s_i = \sum_{k \in S_i} p_k.$$

Let H_i be H evaluated at $\Pr(X \mid X \in S_i)$; that is, $H_i = H(\vec{q})$ where $q_k = \Pr(X = k \mid X \in S_i)$. Once we know that $X \in S_i$, the remaining uncertainty is given by H_i. Thus, the *expected* uncertainty after determining which S_i contains X is

$$\sum_{i=1}^{m} \Pr(X \in S_i)H_i = \sum_{i=1}^{m} s_i H_i. \tag{12.22}$$

On the other hand, the uncertainty about which set contains X is just $H(\vec{s})$. Thus we have

Axiom 6: In terms of the notation just introduced,

$$H(\vec{p}) = H(\vec{s}) + \sum_{i=1}^{m} s_i H_i.$$

This completes the list of axioms.

The axioms are actually stronger than needed; for example, Axiom 4 is unnecessary. Our goal was to state a reasonable set of axioms from which the unique form of H can be easily deduced. With the weakest known set of axioms, the following theorem is harder to prove.

Theorem 12.4   Uniqueness of H

The only functions that satisfy Axioms 1–6 are

$$H(\vec{p}) = -C \sum_i p_i \log_2 p_i,$$

where the constant C is an arbitrary positive real number. If $p_i = 0$, interpret $p_i \log p_i$ to be 0. By convention, we usually set $C = 1$.

The function H is often referred to as the *entropy*. When $p_i = \Pr(X = a_i)$ and the range of X is $\{a_1, \ldots, a_n\}$, we write $H(X)$ for $H(p_1, \ldots, p_n)$. In other words, $H(X) = \sum_X \Pr(X) \log \Pr(X)$. (Remember that $\sum_X$ means that we sum over all instantiations of X.) We can extend this to a set $\mathcal{X}$ of random variables:

$$H(\mathcal{X}) = \sum_{\mathcal{X}} \Pr(\mathcal{X}) \log(\Pr(\mathcal{X})).$$

Proof (of the theorem):   The proof involves several ideas and steps. Let $f(n) = H(1/n, \ldots, 1/n)$.

- Since we defined $0 \log 0$ to be 0, Axiom 3 allows us to ignore components of $\vec{p}$ that equal 0. We'll do so from now on.

- Because H is continuous (Axiom 5), it's completely determined by its values at those $\vec{p}$ whose components are rational. Thus, the proof focuses on rational $\vec{p}$.

- Other axioms make it possible to express H at any rational $\vec{p}$ in terms of $f(n)$. Thus the proof focuses on $f(n)$.

- Finally, it's shown that f behaves like a logarithm: $f(bc) = f(b) + f(c)$. The monotonicity of f (Axiom 4) is then used to show that f is a logarithm.

Let's get started.

Applying Axiom 6 to $f(n)$ with $|S_i| = a_i$, we have

$$f(n) = H(a_1/n, \ldots, a_m/n) + \sum_{i=1}^{m}(a_i/n)f(a_i). \qquad (12.23)$$

Thus

$$H(a_1/n, \ldots, a_m/n) = f(n) - \sum_{i=1}^{m}(a_i/n)f(a_i). \qquad (12.24)$$

H will be known for rational $\vec{p}$ once f is known. Since H is continuous and we can approximate any real number closer than $1/n$ by a fraction with denominator n, (12.24) allows us to approximate $H(\vec{p})$ for any $\vec{p}$ to any desired degree of accuracy. We'll use this in the last paragraph of the proof.

We claim that there is a $C > 0$ such that, for all positive integers b, we have $f(b) = C \log b$. It will take some time to prove this.

First, apply (12.23) with $n = bc$ and all $a_i = c$ to conclude that

$$f(bc) = f(b) + f(c) \tag{12.25}$$

for all positive integers b and c. By repeated application of this equation,

$$f(b^k) = k f(b). \tag{12.26}$$

We'll use this to show that $f(b)/f(2) = \log_2 b$ and so $f(b) = f(2) \log_2 b$. Choose n so that $2^n \leq b^k < 2^{n+1}$. Taking logarithms and dividing by k, we have

$$\frac{n}{k} \leq \log_2 b < \frac{n}{k} + \frac{1}{k}.$$

Multiplying by $f(2)$ and doing a bit of rearranging, you can show that this implies

$$\frac{f(2)}{k} > f(2) \left| \log_2 b - \frac{n}{k} \right|.$$

From Axiom 4, $f(2^n) \leq f(b^k) \leq f(2^{n+1})$. By (12.26), and some algebra similar to that in the previous few lines, this implies that

$$\frac{f(2)}{k} \geq \left| f(b) - f(2) \frac{n}{k} \right|.$$

Combine the last two displayed inequalities and use the triangle inequality to obtain

$$\frac{f(2)}{k} + \frac{f(2)}{k} > f(2) \left| \log_2 b - \frac{n}{k} \right| + \left| f(b) - \frac{f(2)n}{k} \right| \geq \left| f(2) \log_2 b - f(b) \right|.$$

As $k \to \infty$, the left side goes to 0. Since the right side is independent of k, it must equal 0. This completes the proof that $f(b) = C \log b$ for all b.

You should be able to show by a bit of algebra that (12.24) can be written as

$$H(a_1/n, \ldots, a_m/n) = -C \sum_{i=1}^{n} (a_i/n) \log(a_i/n);$$

that is, $H(\vec{p}) = -C \sum p_i \log(p_i)$ for all rational $\vec{p}$. Both sides of this equation are continuous—the left by Axiom 5 and the right by standard arguments. Since they are equal at all rationals, they are equal everywhere. ∎

Information Reduces Uncertainty

The notion that "information" reduces H, our measure of uncertainty, is important because the goal of an information-consuming process is often to reduce uncertainty. In particular, this is the goal of pattern classifiers: A trained pattern classifier "consumes" input information (the pattern) to reduce the uncertainty about the classification of the input.

Let's explore this concept of reducing entropy by using information. Since Axiom 6 was derived by looking at the consequence of obtaining information, its derivation provides a natural starting point. To begin with, the entropy of X is $H(\vec{p})$. Suppose we ask which S_i contains X. If the answer is S_j, the remaining entropy is just H_j. Thus, the expected value of the entropy after obtaining the answer is given by (12.22). It follows from Axiom 6 that the information on which S_i contains X reduces the entropy on average by $H(\vec{s})$.

If our goal is to reduce uncertainty as much as possible by such questions, we should choose $S_1, \ldots, S_m$ so as to maximize $H(\vec{s})$. It's left as an exercise for you to show that

$$H(s_1, \ldots, s_m) \text{ has a unique maximum at}$$
$$s_1 = \cdots = s_m = 1/m, \text{ namely } \log_2 m. \tag{12.27}$$

This says we should try to choose the S_i's so that the probability is equally partitioned; that is, $\Pr(X \in S_i) = 1/m$ for all i. In practice, we usually can't achieve exact equality.

Example 12.9   Sorting by Comparisons

Suppose we always have $m = 2$, such as a series of yes/no questions. Since $\log_2 m = 1$, we'll have to ask an average of at least H questions to eliminate all uncertainty.

Let's apply this to the problem of sorting a list of n distinct numbers by comparing various numbers in the list with one another. A comparison is a yes/no question of the form "Is $x > y$?"

Construct a probability space $(\mathcal{E}, \Pr)$ whose elementary events consist of all $n!$ possible arrangements of the list of numbers, each of which is equally likely. Let X indicate the correct ordering of the list. Since each $e \in \mathcal{E}$ has a different correct ordering, X takes on $n!$ different values each with probability $1/n!$. Thus the entropy is $H(1/n!, \ldots, 1/n!) = \log(n!)$. From the opening paragraph of this example, it follows that we must make on average at least $\log(n!)$ comparisons to sort the list. This is a fairly tight lower bound.

If certain initial orderings were impossible or highly improbable, the uncertainty of X would be less and so a clever strategy that took advantage of this would use fewer questions on average. This situation is not uncommon. For example, we may have a sorted list of $n - k$ elements to which we've appended k new elements. Even if k is unknown, we might expect to find an algorithm that requires, on average, less that $\log_2(n!)$ comparisons. ∎

Example 12.10   Minimizing the Expected Number of Decisions

Let's explore the idea suggested at the end of the previous example. Imagine that we're given N items and must build a decision tree based on yes/no questions to determine which item is the correct one. We want to ask as few questions as possible. Unlike the previous case, the probability that the ith item is correct is known to be p_i and is not, in general, equal to $1/N$. What can we say about the structure of the tree that minimizes the expected number of questions needed to reach a leaf?

Let d_i be the distance from the root to the leaf associated with the ith item. You should have no trouble proving by induction that $\sum_{i=1}^{N} 2^{-d_i} = 1$. (Remember that each nonleaf vertex—including the root—has exactly two children.)

We claim that the maximum of $\sum p_i \log(q_i)$ over all $q_i \geq 0$ with $\sum q_i = 1$ is achieved at $q_i = p_i$. The proof is left to Exercise 12.4.3.

Combining the results of the two previous paragraphs, we find that the expected number of questions is

$$\sum_i^N p_i d_i = -\sum_i^N p_i \log_2(2^{-d_i}) \geq -\sum_i^N p_i \log_2(p_i),$$

the entropy of $\vec{p}$. Thus we should try to construct the decision tree so that d_i is close to $-\log_2(p_i)$. When $N = n!$ and $p_1 = 1/N$, we obtain the result in the previous example. ∎

I mentioned the use of "conjugate priors" on page 527. Another popular prior is the maximum-entropy prior. Why? Since entropy is the negative of information, assuming the maximum entropy distribution as the prior in some sense involves assuming the least information. The next example illustrates this idea.

Example 12.11   Maximum Entropy Priors

This example requires a knowledge of Lagrange multipliers. Here's what you need to know. Suppose we want to find those $\vec{x} \in \mathbb{R}^n$ that give rise to the critical values (potential local maxima and minima) of $f(\vec{x})$ subject to the equality constraints $g_i(\vec{x}) = 0$ for $1 \leq i \leq k$. Let

$$h(\vec{x}, \vec{\lambda}) = g(\vec{x}) + \lambda_1 g_1(\vec{x}) + \cdots + \lambda_k g_k(\vec{x}).$$

The critical values are found among the solutions $\vec{x}, \vec{\lambda}$ of the $n + k$ equations

$$\frac{\partial h}{\partial x_i} = 0 \text{ for } 1 \leq i \leq n. \tag{12.28}$$

If you're unfamiliar with this result, you could either look in a text on optimization or you could take it on faith.

Suppose someone gives us a die. What should we assume about the probability of the various possible rolls? Let p_i be the probability of rolling i for $1 \leq i \leq 6$. Let $f(\vec{p})$ be the entropy times $\ln 2$. (This means we're using the natural log instead of $\log_2$—it makes computing derivatives easier.) The only constraint is that the probabilities must sum to 1. Thus

$$h(\vec{p}, \lambda_1) = -\sum_{i=1}^{6} p_i \ln p_i - \lambda_1 \left(\sum_{i=1}^{6} p_i - 1\right).$$

The equations corresponding to (12.28) are

$$-\ln p_i - 1 + \lambda_1 = 0 \text{ for } 1 \leq i \leq 6 \quad \text{and} \quad \sum_{i=1}^{6} p_i = 1.$$

From the first six equations, $p_i = e^{\lambda_1 - 1}$. From this and the last equation, $6e^{\lambda_1 - 1} = 1$ and so $p_i = \frac{1}{6}$ for all i. Of course, this isn't at all surprising since we already knew that $H(\vec{p})$ is a maximum when the components of $\vec{p}$ are all equal.

Let's look at something more interesting. Suppose we somehow know (or believe) that the die is loaded in such a way that the average value is 3 instead of $3\frac{1}{2}$. The constraint that the expectation equals 3 adds a new constraint. We obtain

$$h(\vec{p}, \vec{\lambda}) = -\sum_{i=1}^{6} p_i \ln p_i + \lambda_1 \left(\sum_{i=1}^{6} p_i - 1\right) + \lambda_2 \left(\sum_{i=1}^{6} i p_i - 3\right)$$

and the equations

$$-\ln p_i - 1 + \lambda_1 + i\lambda_2 = 0 \text{ for } 1 \leq i \leq 6, \quad \sum_{i=1}^{6} p_i = 1, \quad \text{and} \quad \sum_{i=1}^{6} i p_i = 3.$$

It follows that $p_i = Ab^i$ where $A = e^{\lambda_1 - 1}$ and $b = e^{\lambda_2}$. The values of A and b are determined by

$$A \sum_{i=1}^{6} b^i = 1 \quad \text{and} \quad A \sum_{i=1}^{6} i b^i = 3.$$

We obtained $A = 0.294$ and $b = 0.840$ by solving these equations numerically. Is this prior a reasonable choice? Perhaps. If we knew how dice were loaded and understood the physics of rolling dice, we might be able to work out a better choice.

What about continuous distributions? In this case there's an infinite number of parameters and we'd need the calculus of variations, a subject we won't discuss. ■

Exercises

12.4.A. What is the connection between entropy and information?

12.4.B. What does $H(\vec{p})$ measure and how is it computed?

12.4.1. If you're familiar with Lagrange multipliers, minimize $\sum x_i \log x_i$ subject to the constraints $\sum x_i = 1$ and $x_i \geq 0$. Also, maximize $\sum a_i \log x_i$ subject to the same constraints, given that $a_i \geq 0$ and $\sum a_i = 1$.

12.4.2. In this exercise you'll prove (12.27) without using Lagrange multipliers.

   (a) Show that maximizing H with log equal to $\log_2$ gives the same locations for maxima as with log equal to ln.

   (b) Show that $x \ln x + (a-x) \ln(a-x)$ has a unique minimum at $x = a/2$.

   (c) Conclude that if $H(\vec{s})$ is at a maximum, then $s_i = s_j$ for all i and j. *Hint.* Let $a = s_i + s_j$.

12.4.3. In this exercise, you'll fill in the details of the proof in Example 12.10 without using Lagrange multipliers.

   (a) Prove that $\sum 2^{-d_i} = 1$.

   (b) Suppose that $a, b, c \geq 0$, $c \leq 1$, and $a+b > 0$. Show that $a \ln x + b \ln(c-x)$ has a unique maximum at $x = ac/(a+b)$.

   (c) Suppose $a_i \geq 0$. Conclude that if $\sum a_i \ln(x_i)$ is at a maximum subject to $\sum x_i = 1$, then $\vec{x}$ is a multiple of $\vec{a}$. *Hint.* Using (b), show that $x_i/a_i = x_j/a_j$ and then conclude that this ratio is independent of i.

12.4.4. A bag of n coins contains exactly one counterfeit coin, which is lighter than all the other coins, each of which has weight w. We want to find the counterfeit coin by using some scale.

   (a) Let X be a random variable that indicates which coin is counterfeit. Explain why it is reasonable to assume that $H(X) = \log n$.

   (b) Our scale is the single-pan type—you put the items in a pan and the scale gives you the weight. Show that we'll need at least $\log n$ weighings on average.

   (c) Our scale is the two-pan type—you put some items in one pan and some other items in a second pan and the scale tells you which pan is heavier. Show that we'll need at least $\log_3 n$ weighings on average.

12.4.5. A patient is known to have one of n diseases. Diagnostic tests may be of the yes/no or high/medium/low variety. What can you say about the average number of tests a clever diagnostician would need to make? Be sure to explain your reasoning. *Hint.* This is a rather open question. You should look at various possibilities.

12.4.6. Let $(\text{Pr}, \mathcal{E})$ be a probability space. You want to select n independent elementary events. This can be described by the usual product approach: Define $(\text{Pr}_n, \mathcal{E}^n)$ by $\text{Pr}_n(\vec{e}) = \text{Pr}(e_1)\cdots\text{Pr}(e_n)$. Define the real-valued random variable $X(\vec{e}) = \log(\text{Pr}_n(\vec{e}))$. You want the expectation of X.

(a) Define a random variable $Y_{i,f}(\vec{e})$ to be 1 if the ith component of $\vec{e}$ equals f and 0 otherwise. Show that

$$X(\vec{e}) = \sum_{i=1}^{n}\sum_{f\in\mathcal{E}} Y_{i,f}(\vec{e})\log(\text{Pr}(f)).$$

(b) Show that $E(Y_{i,f}) = \text{Pr}(f)$ and use this to show that

$$E(X) = n\sum_{e\in\mathcal{E}} \text{Pr}(e)\log(\text{Pr}(e)).$$

12.4.7. The *cross entropy* or *Kullback-Leibler statistic* for two random variables with the same range R is

$$D(X\|Y) = \sum_{r\in R} \text{Pr}(X=r)\log\left(\frac{\text{Pr}(X=r)}{\text{Pr}(Y=r)}\right).$$

(a) Show that $D(X\|Y) \geq 0$.
  Hint. Use Exercise 12.4.1 or Exercise 12.4.2.

(b) Look back at the definition of g in Exercise 11.6.6 (p. 481). Explain how, by adding terms depending only on the desired output o^*, the value of g can be made to look like a cross entropy. In spite of this, we should probably not regard g as related to cross entropy because there is no underlying probability space in Exercise 11.6.6.

12.4.8. Let $\mathcal{X}$ and $\mathcal{Y}$ be sets of random variables. We define the *conditional entropy* of $\mathcal{X}$ given $\mathcal{Y}$ by

$$H(\mathcal{X}\mid\mathcal{Y}) = \sum_{\mathcal{Y}} \text{Pr}(\mathcal{Y})H(\mathcal{X}\mid\mathcal{Y}) = -\sum_{\mathcal{X},\mathcal{Y}} \text{Pr}(\mathcal{Y})\text{Pr}(\mathcal{X}\mid\mathcal{Y})\log(\text{Pr}(\mathcal{X}\mid\mathcal{Y})),$$

remembering that summing over a random variable means summing over all instantiations of the variable.

(a) Prove $H(\mathcal{X}\cup\mathcal{Y}) = H(\mathcal{X}) + H(\mathcal{Y}|\mathcal{X})$.

(b) Explain why Axiom 6 is just a special case of the previous formula.

(c) Let $\mathcal{X}_i = \{X_1,\ldots,X_i\}$ and $\mathcal{X}_0 = \emptyset$. Prove that

$$H(\mathcal{X}_n) = \sum_{i=1}^{n} H(X_i\mid\mathcal{X}_{i-1}).$$

Notes

Chapter 6 of the AI text by Dougherty and Giardina [6] discusses some of material in this chapter and also some related subjects.

There are a variety of texts on probability theory. The classic texts by Feller [10] go well beyond the probability in this chapter. Chung [3] has written a shorter and easier introduction to mathematical probability.

See Geman, Bienenstock, and Doursat [11] for an extensive discussion of the bias–variance dilemma. See [20] for another approach. The various approaches to reducing generalization error, such as cross-validation and bootstrapping, all involve assumptions about how the training set and the universe are related to each other. See [21] for a discussion.

I barely touched the surface of statistics. There are a variety of "introductory" texts available, from cookbook texts that require little mathematics through theoretical graduate-level texts. Since you are reading this book, Rice's text [18] is at about the right level. It begins by reviewing and supplementing the probability background you have gained in this text. To study error rate estimation, you'll need to look at more specialized literature such as [7], [13], and [19]. The last is aimed specifically at the interests of the AI research community.

The subject of information theory was fathered by Claude Shannon in 1948. He used it to define "channel capacity" in the theory of communication and proved that increasing the transmission rate need not increase the error rate in a noisy channel, as long as the rate of communication was below the channel capacity. Since then, information theory has been used in many ways. The text by Cover and Thomas [4] covers essentially the entire field of information theory and its applications. Khinchin's articles (translated in [14]) provide an introduction to the basic mathematical theory. Feinstein [9] and Behara [1] derive the form of H using a weaker set of axioms than ours, which follows the approach in [14]. The text [12] by Goldie and Pinch provides an introduction to both the information-theoretic and algebraic sides of coding theory. The expository papers in [8] provide more background on the maximum entropy method and its connection with the Bayesian approach.

Biographical Sketch

Ronald A. Fisher (1890–1962)

Born in London, he attended Cambridge where he pursued statistics and genetics and received a degree in mathematics.

Almost singlehandedly, he pioneered the use of statistical methods for "inductive reasoning" and their applications in biology. This led to his publication of *The Genetical Theory of Natural Selection* in 1930 and the founding of the International Biometric Society in 1947. In this context, inductive reasoning refers to the process of reaching conclusions about hypotheses on the basis of experimental evidence. Statistics is involved both in the design of experiments and in the analysis of the results. This interaction can be seen in our discussions of generalization error and of decision tree algorithms.

When Fisher began his career with a brief talk to fellow undergraduates in 1911 on "Mendelism and biometry," statistics was an undeveloped area that was seldom used in the sciences. Fisher's work over the succeeding two decades changed this: He developed tools and applied them in a variety of scientific fields, particularly genetics.

Two quotes express Fisher's philosophy better than any lengthy discussion here could:

> It is the method of reasoning, and not the subject matter, that is distinctive of mathematical thought. A mathematician, if he is of *any* use, is of use as an expert in the process of reasoning, by which we pass from a theory to its logical consequences, or from an observation to the inferences which must be drawn from it.   [2, p. 240]

And, in response to criticism for his focus on specific applications:

> ... from my point of view, this is a misapprehension, based on the belief that the understanding required can be obtained from the mathematical background rather than, as I think, from the particular peculiarities of the actual body of data to be examined.   [2, p. 244]

This discussion was based on Joan Box's biography [2] of her father.

References

1. M. Behara, *Additive and Nonadditive Measures of Entropy*, John Wiley and Sons, New York (1990).

2. J. F. Box, *R. A. Fisher: The Life of a Scientist*, John Wiley and Sons, New York (1978).

3. K. L. Chung, *Elementary Probability Theory with Stochastic Processes*, Springer-Verlag, Berlin (1979).

4. T. M. Cover and J. A. Thomas, *Elements of Information Theory*, John Wiley and Sons, New York (1991).

5. M. H. DeGroot, *Optimal Statistical Decisions*, McGraw-Hill, New York (1970).

6. E. R. Dougherty and C. R. Giardina, *Mathematical Methods for Artificial Intelligence and Autonomous Systems*, Prentice Hall, Englewood Cliffs, NJ (1988).

7. B. Efron, Estimating the error rate of a prediction rule: Improvement on cross-validation, *J. American Statistical Assn.* **78** (1983) 316–331.

8. G. J. Erickson and C. R. Smith (eds.), *Maximum-Entropy and Bayesian Methods in Science and Engineering. Volume 1: Foundations*, Kluwer, Dordrecht (1988).

9. A. Feinstein, *Foundations of Information Theory*, McGraw-Hill, New York (1958).

10. W. Feller, *An Introduction to Probability Theory and Its Applications*, vol. 1 (3d ed.) and vol. 2 (2d ed.), John Wiley and Sons, New York (1968, 1971).

11. S. Geman, E. Bienenstock, and R. Doursat, Neural networks and the bias/variance dilemma, *Neural Computation* **4** (1992) 1–58.

12. C. M. Goldie and R. G. E. Pinch, *Communication Theory*, London Mathematical Society Student Texts Vol. 20, Cambridge University Press, Cambridge, Great Britain (1991).

13. P. Hall, *The Bootstrap and Edgeworth Expansion*, Springer-Verlag, Berlin (1992).

14. A. I. Khinchin, *Mathematical Foundations of Information Theory*, Dover, New York (1957). Translated by R. A. Silverman and M. D. Friedman from two articles in *Uspekhi Matematicheskikh Nauk.* **8** (1953) 3–20 and **11** (1956) 17–75.

15. R. S. Lynd and H. M. Lynd, *Middletown: A Study in Modern American Culture*, Harcourt Brace, New York (1956).

16. C. R. Mehta and N. R. Patel, A network algorithm for performing Fisher's exact test in $r \times c$ contingency tables, *J. Amer. Statistical Assn.* **78** (1983) 427–434.

17. M. Pagano and K. T. Halvorsen, An algorithm for finding the exact significance levels of $r \times c$ contingency tables, *J. Amer. Statistical Assn.* **76** (1981) 931–934.

18. J. A. Rice, *Mathematical Statistics and Data Analysis*, Wadsworth & Brooks/Cole, Pacific Grove, CA (1988).

19. D. H. Wolpert, On the connection between in-sample testing and generalization error, *Complex Systems* **6** (1992) 47–94.

20. D. H. Wolpert, On overfitting avoidance as bias, Santa Fe Institute paper SFI TR 93-03-016.

21. D. H. Wolpert, Off-training set error and *a priori* distinctions between learning algorithms, Santa Fe Institute paper SFI TR 94-12-123.

22. G. A. Young et al., Bootstrap: More than a stab in the dark? (with commentaries and a rejoinder), *Statistical Science* **9** (1994) 382–415.

13

Decision Trees
Neural Nets
and
Search

*The study of neural networks in recent years has involved
increasingly sophisticated mathematics ... .
A reader unfamiliar with the mathematical tools may find this
more technical literature unapproachable.*
—Stuart Geman, Elie Bienenstock, and René Doursat (1992)

Introduction

Decision trees were introduced in Chapter 2. They appeared again in Chapter 10 as a type of automatic classifier. Until now, we've lacked the mathematics needed to discuss algorithms for the automatic formation of decision trees. Now that we've developed the tools, we can tackle the algorithms. Next we'll discuss neural nets a bit more, including the unfinished business of providing a heuristic proof for the storage capacity of Hopfield nets. Finally we'll return to search—the subject that started our AI investigations. Problem complexity and time constraints often make complete searches undesirable or impossible. This makes heuristics and partial search important. Although these ideas were introduced in Chapter 2, we said very little about them. Now that we have the appropriate mathematical tools, we can discuss them further.

Since the three sections of this chapter are nearly independent of one another, you can pick and choose among them.

Prerequisites: The discussion of decision trees depends on Chapter 10 for terminology and uses the statistics and information theory sections of Chapter 12. The neural net discussion depends on the basic ideas in Chapter 11 and makes some use of the probability theory in Chapter 12. The search tree discussion requires Chapter 2 and the probability theory in Chapter 12.

Used in: This chapter is not used in later chapters.

13.1   Decision Trees

> *The difficulty in life is the choice.*
> —George Moore (1900)

Since decision trees were introduced some time ago, you may want to review pp. 410–411.

Decision tree data can be almost anything from real numbers to purely nominal information. For simplicity, our training data will be of the form $(\vec{Q}, R)$, where $\vec{Q} \in \{T, F\}^k$ and R lies in some finite set $\mathcal{R}$. The input is $\vec{Q}$, a vector of true/false values, and the classification (or desired response) is R. (We're using $\vec{Q}$ to suggest Questions—or, more appropriately, answers to questions.) Given $\vec{Q}$, the ideal decision tree would, after a few decisions based on $\vec{Q}$, yield a leaf that declared the correct classification for $\vec{Q}$. If R is a function of $\vec{Q}$, then such a tree can be constructed for the training data. All it need do is ask enough questions so that each leaf is reached by just one training pair $(\vec{Q}, R)$. From previous discussions of complexity versus generalization error—see Figure 10.3 (p. 421)—it seems likely that such a large tree will be a poor generalizer. This is usually the case: Good trees are not very large and yet classify most of the training data correctly.

It seems we should plan to construct relatively small trees. This means relying on clever questions to divide the data. Consider the following very simple training set:

$$\mathcal{T} = \Big\{ (T, T), 0), \ (T, F), 1), \ (F, T), 1), \ (T, T), 0) \Big\}$$

Neither the first component nor the second component of $\vec{Q}$ gives any information about R; however, the two of them together determine R completely. Thus, if we look only at single components, we'd need a tree like $\wedge\!\!\wedge$, which has one data point at each leaf. The question "Is $Q_1 = Q_2$?" divides the data nicely, so we only need the smaller tree $\wedge$, which has two data points at each leaf. Unfortunately, it's impractical to allow all possible questions. Why? A true/false question can be thought of as a truth table for some function of the

n components of $\vec{Q}$. Such a table has 2^k rows and each row has 2 choices for the value of the function. Hence there are $2 \times \cdots \times 2 = 2^{2^k}$ possible tables. It's not computationally feasible to consider this many questions, so we must limit the number of questions somehow.

We also need a way to measure how good a tree is. With feedforward nets, we used the function $g(\vec{w}, \vec{\theta})$ when training the net and then suggested cross-validation or bootstrapping for estimating the generalization error of a trained net. The situation for decision trees is similar. We need some simple measure of goodness to use while the decision tree classifier is being trained. After training, we need to estimate generalization error.

In summary, regardless of what algorithm is used for constructing a decision, three things are needed:

- a collection of allowable questions that may be asked about the inputs in the training data,

- a way to measure how good a tree is during training, and

- a way to estimate generalization error.

For simplicity, let's limit the allowable questions to the form "Is $Q_i = $ T?" Measuring goodness is more complicated. We'll discuss a method soon, but first we need to set the stage.

As stated above, training data consists of inputs taken from $\{\text{T}, \text{F}\}^k$ and outputs taken from some finite set $\mathcal{R}$. It's useful to think of this in probabilistic terms: There is some unknown underlying probability space $(\mathcal{E}, \text{Pr})$. The training data consists of a collection of pairs of observations of random variables $(\vec{Q}(e), R(e))$. The range of $\vec{Q}$ is the set of vectors of question answers $\{\text{T}, \text{F}\}^k$, and the range of R is $\mathcal{R}$. Note that this viewpoint allows the possibility of repeated inputs' (same $\vec{Q}$'s) having different outputs (R's). In other words, the data $\vec{Q}$ may not uniquely determine the result R. For simplicity, we'll limit questions to simply looking at a component of $\vec{Q}(e)$.

Ideally, we would like to look at all possible decision trees and choose the "best." This is computationally impossible in most situations. The training algorithm we'll discuss consists of three steps:

- **Grow**: We grow a tree one decision at a time by selecting a question at a leaf: that is, by selecting a component Q_i of $\vec{Q}$ to examine. (Recall that a leaf of a decision tree is a vertex at which no decisions are made; that is, no edges lead out from it.) The question is used to "split" the vertex. This operation is repeated a sufficient number of times.

- **Prune**: Since it's difficult to decide when to stop, we grow a tree that is too big and then prune it one decision at a time.

- **Evaluate and Choose**: Since it's difficult to decide when to stop pruning, we consider various levels of pruning and choose the best based on cross-validation.

Growing Decision Trees

The following algorithm describes how to split a vertex and then how to use this process to grow a tree. It does not address the question of when to stop.

Algorithm 13.1   Growing a Decision Tree

We'll store training data at the leaves of the tree. A leaf v can be split using the ith component of $\vec{Q}$ as follows:

- Attach to v the information that we should look at the ith component of $\vec{Q}$.

- Create two edges labeled T and F leading out from v to two new leaves.

- Move all data $(\vec{Q}, R)$ at v with $Q_i = $ T to the leaf that is reached along edge T and the remaining data to the other leaf.

In the process, v ceases to be a leaf. To grow a decision tree, initialize the tree to be a single vertex, called the *root*, containing all the training data. Then, repeatedly split leaves. The question for each split is chosen from among all possible questions at all possible leaves so as to give the greatest increase in the "goodness" of the tree.

This is a *greedy algorithm* because it always chooses the split with the greatest immediate improvement. We'll discuss the meaning of "goodness" soon.

Greediness is a compromise—a clever algorithm might try to look ahead at the effect the current split will have on future splits. A mediocre current split may make a very good future split possible. Such situations can't be detected with the greedy algorithm. The best we can do is split even when the gain is small, hoping for future gains. Later, we can undo poor splits; that is, prune the tree.

Example 13.1   Choosing Questions

The following questions are to be asked of individuals who are living with a "significant other." Imagine that Q has three components corresponding to

1. were children conceived with significant other?

2. is respondent female?

3. is significant other female?

For R, we want to determine if the relationship is heterosexual or homosexual. Algorithm 13.1 would undoubtedly first split the data based on Q_1 since knowing the sex of only one of the partners is useless. The best tree would ignore Q_1 completely and instead split twice using Q_2 and Q_3. Of course, if we'd allowed more general questions concerning $\vec{Q}$, we could simply ask if $Q_2 = Q_3$. ∎

What should the stopping criterion be? That's not very important except that it should be liberal in allowing splits. In the extreme case, we might even split until further splitting is impossible (all $\vec{Q}$'s the same at a vertex) or useless (all R's the same at a vertex).

Measuring Goodness

Without a measure of goodness, Algorithm 13.1 is meaningless because we have no way to decide how a vertex should be split. One measure that's used is based on information theory. The reason for doing so is the idea that a decision tree should provide as much information as possible. The leaves provide the information and entropy provides a measure of information content. Entropy requires a probability space and the obvious one is $(\mathcal{E}, \text{Pr})$ defined earlier. We need to be more precise. At each leaf v we can use the formula for entropy to define the uncertainty at the leaf: namely,

$$H_v = - \sum_{r \in \mathcal{R}} p_r(v) \log p_r(v), \tag{13.1}$$

where

$$p_r(v) = \text{Pr}\big(R = r \mid \vec{Q} \text{ leads to } v\big).$$

If $p(v)$ is the probability that leaf v is reached using a randomly chosen $(\vec{Q}, R) \in \mathcal{E}$, then the expected entropy is simply

$$H = \sum_v p(v) H_v. \tag{13.2}$$

Our goal is to minimize (13.2) and so maximize information.

Since (13.2) is defined for the space $(\mathcal{E}, \text{Pr})$ it's more like generalization error than like the function $g(\vec{w}, \vec{\theta})$ of feedforward nets. As such, it suffers from the same problem that generalization error does—we can't compute it since we don't have $(\mathcal{E}, \text{Pr})$. All we can do is estimate it by using the training set $\mathcal{T}$. There's an obvious approximation. Use frequencies to estimate probabilities: Replace $p(v)$ with $n(v)/n$ and $p_r(v)$ with $n_r(v)/n(v)$, where n is the number of elements in $\mathcal{T}$, $n(v)$ is the number of these at v, and $n_r(v)$ is the number of these with $R = r$. In a little while we'll look at some problems with frequencies.

How is the measure H used? Suppose we're considering splitting a leaf v into two leaves, which we'll denote by the labels of the edges T and F leading to them. The only change in (13.2) caused by the replacement is to remove the H_v term and add H_{T} and H_{F} terms. Since information increases as entropy decreases, this change increases the information by

$$
\begin{aligned}
\frac{n(v)}{n}&H_v - \frac{n(\text{T})}{n}H_{\text{T}} - \frac{n(\text{F})}{n}H_{\text{F}} \\
&= -\sum_{r\in\mathcal{R}} \frac{n_r(v)}{n}\log\left(\frac{n_r(v)}{n(v)}\right) \\
&\quad + \sum_{r\in\mathcal{R}} \frac{n_r(\text{T})}{n}\log\left(\frac{n_r(\text{T})}{n(\text{T})}\right) + \sum_{r\in\mathcal{R}} \frac{n_r(\text{F})}{n}\log\left(\frac{n_r(\text{F})}{n(\text{F})}\right) \qquad (13.3)\\
&= -\frac{1}{n}\left(n(\text{T})\log\left(\frac{n(\text{T})}{n(v)}\right) + n(\text{F})\log\left(\frac{n(\text{F})}{n(v)}\right)\right) \\
&\quad + \frac{1}{n}\sum_{r\in\mathcal{R}}\left(n_r(\text{T})\log\left(\frac{n_r(\text{T})}{n_r(v)}\right) + n_r(\text{F})\log\left(\frac{n_r(\text{F})}{n_r(v)}\right)\right)
\end{aligned}
$$

(The calculations are left as an exercise.)

Except for a factor of $1/n$, this formula for change in goodness depends only on parameters that are local to v. Hence what is done at one leaf has no effect on any other. This leads to a speed up in the tree-growing algorithm because we needn't look at all leaves before deciding on a split.

Algorithm 13.2 Growing a Decision Tree (Revised)

See Algorithm 13.1 (p. 550) for terminology. To grow a decision tree, initialize the tree to be a single vertex, called the *root*, containing all the training data. Mark the vertex splittable. As long as a splittable vertex v exists, mark it as unsplittable and find the question that maximizes (13.3). If the value of (13.3) is large enough, split v and mark the two new vertices as splittable.

The algorithm requires that (13.3) be large enough. If we want to split as long as possible, "large enough" will be "greater than zero."

Example 13.2 A Toy Problem

The example in this problem is unrealistically small. I did that to keep the computations tractable and to highlight the problems caused by small values of n_v.

Representation of digits using subsets of the seven lines 8 should be a familiar sight. Rather than use these to create a toy problem, we'll choose a simpler array using only 4 lines. The possible patterns will be ı⊡ı, ı, ı_, and ı_ı, which we refer to as O, I, L, and U, respectively. Positions will be listed in the order $2^1_4 3$ and 0 (resp. 1) will be used instead of F (resp. T). Thus, our input belongs to $\{1,0\}^4$ and our output to {O, I, L, U} with

$$1111 \to \text{O},\quad 0010 \to \text{I},\quad 0101 \to \text{L},\quad 0111 \to \text{U}.$$

We'll allow each line to "fail" (1 replaced by 0) independently with probability p. The situation can be described analytically; see Exercise 13.1.3. Our goal is to recover O, I, L, or U given the observed pattern.

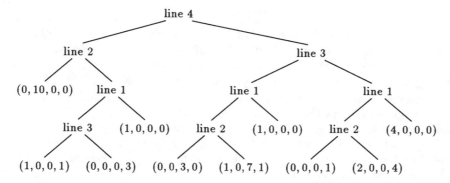

Figure 13.1 See Example 13.2 for a description of the problem. The vector at each leaf indicates the number of garbled training patterns with each of the classifications O, I, L, and U, respectively. A line number indicates the line in the array $2\frac{1}{4}3$ on which the decision is based, with a left son corresponding to absent and a right son to present.

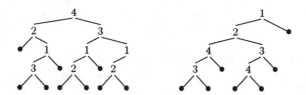

Figure 13.2 The tree on the left is the same as that in Figure 13.1. The other tree is generated from another random training set on the same probability space. A number at a vertex indicates the line that was used for the decision, with a left son corresponding to absent and a right son to present.

By starting with 10 copies of each symbol and a failure rate $p = 0.2$, a program produced a random training set with 40 elements. Our tree-growing algorithm was used to produce Figure 13.1.

Another set of data was generated using the same parameters but a different random start. For comparison, this tree is shown in Figure 13.2 together with the previous result. Note the large difference between the two trees. Such differences arise from random fluctuations in sampling. These are most noticeable when the training set is a small random sample or when the information in $\vec{Q}$ is woefully inadequate for the classification task. This large difference between trees is reduced somewhat by pruning. ∎

*Example 13.3   The Frequency Problem

In Example 12.7 (p. 515) we saw that frequencies based on small samples provide poor estimates for probabilities. The situation is even worse when we're attempting to split data so as to minimize entropy. To see this, let's perform a thought experiment. Suppose that the data at the leaf v is $\{(\vec{Q}, R), (\vec{Q}', R')\}$, where $\vec{Q} \neq \vec{Q}'$ and $R \neq R'$. If we use *any* component i such that $Q_i \neq Q_i'$, to split v, (13.3) tells us that the information will increase by $1/n$. (Do the calculations!) On the other hand, the question chosen is likely to be irrelevant. Thus, small sets may lead to irrelevant splits. More generally, the process of splitting based on observed frequencies tends to accentuate random variations that are present in the training data.

How can we correct the situation? First, let's explore the situation a bit.

What does fixing it gain? Unless the data set is quite small, fixing the problem probably won't alter the numerical values in (13.3) very much. Thus the question that gives the greatest improvement would probably not change. If we needed some absolute assessment of a goodness, the effect would be more pronounced. Fortunately, such assessment is rarely needed.

Since the gain seems to be slight, a time-consuming correction isn't worthwhile. Can we find a quick one? I don't know of any when we are dividing data; however, there is one when we are simply estimating probabilities. The method is referred to as *flattening counts*. A greater degree of flattening might be helpful in constructing decision trees.

In order to explain flattening, we need to carefully explain the problem. In the simplest case, we have n independent observations of a random variable whose range is $\{0, 1\}$. Of these, k have the value 0 and $n - k$ have the value 1. What is the best estimate for $\Pr(X = 0)$?

Let $p = \Pr(X = 0)$. What is the probability of obtaining exactly k observations equal to 0? The probability of k observations of 0 followed by $n - k$ of 1 is $p^k(1 - p)^{n-k}$. In fact, this is the probability regardless of the order of the zeros and ones. Since we don't know that order, we must sum $p^k(1 - p)^{n-k}$ over all possible orders. Let $\binom{n}{k}$ denote the number of ways to form a sequence of k zeros and $n - k$ ones. $\binom{n}{k}$ is called a *binomial coefficient*. You won't need to know how to compute it.

One approach to estimating p is called the *maximum likelihood estimator*. It's obtained by finding the value of p that maximizes the probability $\binom{n}{k}p^k(1 - p)^{n-k}$ of obtaining exactly k zeros. It's a straightforward calculus problem to show that the solution is given by $\hat{p} = k/n$. Do it.

Bayesian methods provide another approach. Let's assume that the prior pdf of the value of p is uniform on $[0, 1]$. Since

$$\Pr(\text{observation} \mid p) = \binom{n}{k}p^k(1 - p)^{n-k},$$

it follows that the posterior pdf for p is

$$f(p \mid \text{observation}) = Cp^k(1 - p)^{n-k} \qquad (13.4)$$

for some constant C. We can determine C from

$$1 = \int_0^1 f(p \mid \text{observation})dp = \int_0^1 Cp^k(1-p)^{n-k}dp.$$

How can we use (13.4) to estimate p? A standard approach is to use the expected value of p:

$$\hat{p} = \int_0^1 pCp^k(1-p)^{n-k}dp = \frac{\int_0^1 p^{k+1}(1-p)^{n-k}dp}{\int_0^1 p^k(1-p)^{n-k}dp}.$$

Repeated use of integration by parts allows us to reduce the power of $(1-p)$ to 0 and gives us an integral we can evaluate. You should carry out these steps to obtain

$$\int_0^1 p^i(1-p)^j\,dp = \frac{j}{i+1}\int_0^1 p^{i+1}(1-p)^{j-1}dp = \frac{j!\,i!}{(i+j)!}\int_0^1 p^{i+j}\,dp$$

$$= \frac{i!\,j!}{(i+j+1)!}.$$

Thus $\hat{p} = \frac{k+1}{n+2}$. This result can be interpreted as follows. Increase the count of zeros by 1 and increase the count of ones by 1. Then use the new counts to produce a frequency estimate. This result is true in general:

> If the range of a random variable X is $\{r_1, \ldots, r_t\}$, if n independent observations have the value r_i exactly k_i times, and if the prior probability distribution for X is uniform, then the Bayesian estimator for $\Pr(X = r_i)$ is $(k_i + 1)/(n + t)$.

This is the method of flattening counts—increase each count by 1 before using frequencies to estimate probabilities. Notice that if we thought the range of X was $\{r_1, \ldots, r_s\}$ with $s < t$, then we'd obtain a different estimate for the probabilities. In practice it may be hard to determine t—are values missing from the observations because they are rare or because they are impossible? Let's not pursue this. ∎

Exercises

13.1.A. Why do we prune decision trees?

13.1.B. Explain how to grow a decision tree?

13.1.C. How is entropy H used to measure goodness?

13.1.D. What is "flattening counts" and why is it used?

13.1.1. Explain how we could think of $\mathcal{E}$ as being $\{\text{T}, \text{F}\}^k \times \mathcal{R}$ and each $(\vec{Q}, R)$ as an elementary event. Now suppose that $\vec{Q}$ determines R. Explain how we can think of $\mathcal{E}$ as being $\{\text{T}, \text{F}\}^k$ in this case and why this doesn't work in general.

13.1.2. Derive (13.3) and interpret the parenthesized terms in the last two lines in terms of information.

13.1.3. This exercise refers to Example 13.2. In the following table, each column gives the probability of the pattern arising from O, I, L, and U. An asterisk stands for either a 0 or a 1, z is the number of asterisks that are 0, and $q = 1 - p$.

	1***	0111	0110 0011	0101	0100 0001	0010	0000
O	$p^z q^{4-z}$	pq^3	$p^2 q^2$	$p^2 q^2$	$p^3 q$	$p^3 q$	p^4
I	0	0	0	0	0	q	p
L	0	0	0	q^2	pq	0	p^2
U	0	q^3	pq^2	pq^2	$p^2 q$	$p^2 q$	p^3

(a) Verify the table.

(b) Construct a similar table whose entries are the probabilities of having started with each of the four possible patterns given the observed pattern. You should need only five columns instead of seven.

(c) Each column of the table gives rise to a rule. State each rule as a probability distribution on O, I, L, and U. State each rule as a hard choice—specify one of O, I, L, and U.

(d) Compute the generalization error for your rules in the hard-choice case.

*(e) Compute the generalization error for your rules in the probability distribution case.

13.1.4. Each leaf in Figure 13.1 gives rise to a rule.

(a) State the ten rules as hard choices—specify one of O, I, L, and U at each leaf. (For one leaf, it is not clear what the choice should be.)

(b) Combine whatever rules you can to obtain a smaller set.

(c) Compute the generalization error for your rules. The table of probabilities in the preceding exercise may be helpful. (Remember that $p = 0.2$ in this case.)

Pruning and Evaluating Decision Trees

> *The riders in a race do not stop short when they reach the goal.*
> *There is a little finishing canter before coming to a standstill.*
>
> —Oliver Wendell Holmes, Jr. (1931)

Before discussing pruning, let's to define our concepts, which will make discussion a bit easier.

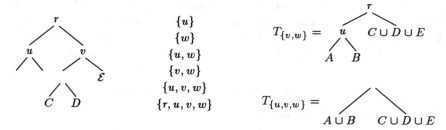

Figure 13.3 A decision tree T is on the left, all prunable sets are in the middle, and two prunings are on the right. Lowercase letters denote decisions and uppercase letters denote data sets.

Definition 13.1   Pruning

Let T be a decision tree. A decision vertex v all of whose sons are leaves is called *prunable*. The *pruning* T_v of v is the decision tree in which the sons of v have been removed and the data they contained has been moved to v. More generally, a set W of nonleaf vertices is called *prunable* if, whenever $w \in W$, every nonleaf vertex below w is in W. The *pruning* T_W of W is the decision tree obtained by sequentially pruning all the decisions in W, starting with the lowest vertices and moving upward. In effect, T_W is T with all of the decisions in W eliminated.

More graphically, a set W is prunable if, when we grab the tree by the root and remove the vertices in W, only leaves fall off. The decisions in W needn't be pruned sequentially. Instead, suppose that $w \in W$ has no vertices in W above it. Remove all descendants of w and move the data at the leaves to W. Figure 13.3 shows a decision tree and some aspects of pruning.

The measure of goodness used for growing a tree didn't take the tree's complexity into account. Complexity is important: Let T be a decision tree grown by Algorithm 13.2 (p. 552) and let W be a prunable set. By the way T was constructed, it will do a better job of classifying the training data than T_W will do. Thus the goodness of T exceeds that of T_W. On the other hand T is more complex than T_W because it contains more vertices.

The previous paragraph shows that, in order to select the prunable set W that gives the best T_W, we must have some way of handicapping trees for their complexity.

Consider the more general setting of an arbitrary classifier. The number of parameters the classifier contains provides a measure of its complexity. For decision trees, the number of parameters is the number of decision vertices. For neural nets, you can see from (10.1) (p. 409) that the number of parameters is the number of edges plus the number of vertices. Statisticians have shown that, in some situations, it's best to minimize a linear combination of the number of parameters and the expected cost due to misclassification [27]. (I won't prove this—you'll have to take it on faith.)

Let's be as simple as possible: The classification at a leaf is just the most frequent classification among the training data at the leaf. The cost of mis-classification is 1. Let $D(T)$ be the number of decisions in the tree T and note that $D(T_W) = D(T) - |W|$. For some $\beta, \gamma > 0$ we want to minimize

$$\beta(\text{error rate}) + \gamma D(T_W) = \beta\Big((\text{error rate}) + (\gamma/\beta)(D(T) - |W|)\Big),$$

which is equivalent to minimizing

$$(\text{error rate}) - \alpha|W|$$

for some $\alpha > 0$. There are four problems with this:

- What is α? For the time being, let's assume that α is known.

- By "error rate" we should mean the generalization error rate, which we don't yet know how to estimate for decision trees. For the time being, we'll use the training error rate.

- For all but the smallest decision trees, we'll have too many prunable sets W to consider. To avoid this, we'll use a greedy algorithm.

- Finally, there's a theoretical issue: The tree was grown using the expected entropy as a measure of goodness but it's being pruned using the error rate. Why? The answer is simple: Although our concern is the error rate, we can't use it while growing a tree since we don't yet have classifications at the leaves.

Algorithm 13.3   Greedy Pruning

Let $\mathcal{T}$ be the training set used to grow the decision tree and let $e(T)$ be $\text{err}(\mathcal{T}, T)$ measured using the decision tree T.

Initialize T to be the decision tree grown using $\mathcal{T}$ and mark all leaves as untested. As long as there is an untested leaf in the tree T, carry out the following sequence of three steps:

- Find a prunable vertex v of T that was split to provide two untested leaves v_T and v_F. Mark v_T and v_F as tested.

- Let T_v be T with the decision v pruned and the new leaf v marked as untested.

- If $e(T) > e(T_v) - \alpha$, replace T with T_v.

It's important, but not immediately obvious, that the two leaves attached to a retractable vertex are either both tested or both untested. (You should prove it.) If this were not true, there could be untested leaves but no prunable vertices both of whose leaves were untested.

We now have a greedy algorithm that uses training error rather than generalization error and contains an unknown parameter α. How can this be fixed? Greediness and training error will stay in the algorithm. What we'll do is describe a method for computing generalization error as a function of α. We then choose the tree with the smallest generalization error.

Here's the idea. The parameter α determines the degree to which complexity is penalized. Larger α means a higher penalty and hence a greater pruning of the tree. For any particular α, we can grow and prune a tree on a subset of $\mathcal{T}$. This makes cross-validation possible since it requires training on a subset of $\mathcal{T}$. In this way, we can associate a generalization error with any value of α.

Since α can be any positive real number, this isn't quite good enough. We need to provide a way of limiting attention to finite number of α's.

Here's how that's done. Forget about cross-validation for the time being—design the tree using all of $\mathcal{T}$. Every value of α determines a tree $T(\alpha)$ based on the growing and pruning algorithms 13.2 (p. 552) and 13.3. Each $T(\alpha)$ is associated with a prunable set W. Since there are only finitely many choices for W, there will be only finitely many distinct trees $T(\alpha)$. Breiman et al. [2] show that $T(\alpha)$ is a nonincreasing function of α and provide a practical method for finding the sequence of trees and the values of α at which $T(\alpha)$ changes values. Thus we now have

$$\text{a sequence } 0 = \alpha_1 < \cdots < \alpha_{n+1} = \infty$$

and

$$\text{a sequence of decreasing trees } T = T_1, \ldots, T_n = \bullet$$

such that

$$T_i \text{ is constructed when } \alpha_i < \alpha < \alpha_{i+1}.$$

(The symbol $\bullet$ stands for the single-vertex tree.) To evaluate T_i:

- Set α to a value intermediate between α_i and α_{i+1}.

- Using cross-validation and this value of α, obtain an estimate of the generalization error.

- Choose that T_i with the lowest estimated generalization error.

See [2] for details.

Exercises

13.1.E. What can often be used as a measure of classifier complexity? What does this measure amount to for decision trees? neural nets?

13.1.F. Describe how to prune a decision tree.

13.1.G. How do we evaluate generalization error and thereby sidestep the problem of directly determining the weight α of complexity in the pruning algorithm 13.3?

13.1.5. Suppose that all the leaves in the decision tree T are at a distance n from the root. For example, the trees for $n = 1$ and $n = 2$ are $\wedge$ and $\wedge\wedge$. Let $P(n)$ be the number of prunable sets (including the empty set). Show that $P(0) = 1$ and $P(n+1) = P(n)^2 + 1$. Use this to prove that $P(k+1) \geq 2^{2^k}$ for $k \geq 0$.

13.1.6. Let $L(T)$ denote the number of leaves of T. Prove that finding the best T_W is equivalent to finding W that minimizes $e(T_W) + \alpha L(T_W)$.

*13.1.7. Discuss how the tests for $2 \times c$ arrays could be used in pruning. (An example of such a test is the chi-square on page 531.)

Extracting Rules from Trees

I've warned against attributing meaning to rules extracted from classifiers (p. 422). So why does this section exist? I didn't say you shouldn't extract rules. I said you shouldn't impute causality to them.

For simplicity, we'll discuss rule extraction in a deterministic framework. Thus we'll assume that the pattern $\vec{Q}$ is sufficient to determine the classification R and that a decision tree has been constructed so that the patterns at each leaf have only one classification. The questions asked on the path from the root to a leaf can be interpreted as a rule: If the questions refer to components i, j, k of $\vec{Q}$, if the answers are $a, b, c \in \{\text{T}, \text{F}\}$, and if the value of R at the leaf is r, then the rule is

$$\Big((Q_i = a) \wedge (Q_j = b) \wedge (Q_k = c)\Big) \to r. \tag{13.5}$$

It may be possible to simplify a set of rules; for example, we can combine (13.5) and

$$\Big((Q_i = \neg a) \wedge (Q_k = c) \wedge (Q_j = b)\Big) \to r \tag{13.6}$$

to obtain $\Big((Q_j = b) \wedge (Q_k = c)\Big) \to r$.

The decision tree approach to extracting rules has some attractive features for the design of expert systems:

- Decision trees can be constructed from whatever information is available.

- Given a large amount of data, a decision tree procedure is likely to produce a better set of rules than a system designer does.

- If the user wants to know why a question is being asked, a response of the form "It will help me decide if the situation is r," can be constructed automatically by choosing the value of r that maximizes information gain. (See Exercise 13.1.8.) If more than one r gives a large gain, they might all be mentioned.

As a result, decision trees are a popular method for extracting rules from data, and expert system shells often contain a decision tree algorithms.

Exercises

13.1.8. In the notation of (13.3) (p. 552), show that the information gain at v attributable to r is given by

$$\left\{ \frac{n_r(\mathrm{T})}{n} \log\left(\frac{n_r(\mathrm{T})}{n_r(v)}\right) + \frac{n_r(\mathrm{F})}{n} \log\left(\frac{n_r(\mathrm{F})}{n_r(v)}\right) \right\}$$

$$- \left\{ \frac{n_r(\mathrm{T})}{n} \log\left(\frac{n(\mathrm{T})}{n(v)}\right) + \frac{n_r(\mathrm{F})}{n} \log\left(\frac{n(\mathrm{F})}{n(v)}\right) \right\}.$$

*13.1.9. The information gain computed in the previous exercise could be large primarily because $n_r(\mathrm{T})/n(\mathrm{T})$ is closer to an end of the interval $[0, 1]$ than $n(\mathrm{T})/n(v)$, or it could be large primarily because $n_r(v)$ is large. Justify this claim and show how the two cases could be used to produce different refinements to the explanation suggested in the text for why a question is being asked.

*13.1.10. Suppose that classification is not always constant at leaves.

   (a) Explain how the rule extraction process can be modified to give probabilistic results.

   *(b) Due to relatively small sample size, such probabilistic results are often rather uncertain. How can confidence information be associated with the probabilities?
   Hint. A starred part of a starred exercise is not easy! It could easily be a research topic.

   *(c) Develop a method for deciding if two probabilistic rules should be combined as (13.5) and (13.6) were.

13.2   Neural Nets Again

Here we'll apply some of the ideas in the previous chapter to neural nets and also discuss issues related to network size. The ideas in the first two parts of this section can be adapted to classifiers other than feedforward nets.

Training and Testing

Random variables arise in a variety of ways in training. As we've already seen, they're often used to initialize network parameters. Experimenters say that independent gaussian random variables appear to work better than uniformly distributed random variables. A good choice for the mean is usually obvious—often it's zero. No one knows how the variance should be chosen, so it's done on the basis of experience or intuition. In this section, we'll look at a couple of other uses of random variables.

Example 13.4   Noisy Data Can Improve Generalization

A subtler use of random variables is made in an attempt to improve the generalization error rate by adding noise to the training data. Suppose the patterns lie in $\mathbb{R}^n$. Each time we want to use a pattern/classification pair from the training set, we use a randomly perturbed version of the pair. Typically, independent gaussian random variables with mean zero are added to the components of the pattern and, perhaps, to the classification. Although it may seem paradoxical at first, adding such "noise" to the data may improve the net. How can this be?

An algorithm attempts to fit a complicated function to the given points, regardless of what happens elsewhere. The training set contains a small collection of isolated points in $\mathbb{R}^n$, so there's a lot of room for the function to misbehave. By adding noise, we create a large cloud of points that cluster around a smooth surface close to the training set. As a result, the algorithm tends to fit a relatively smooth surface to the points.

Is a smooth surface what we want? When we look for generalization, we expect similar stimuli to give similar responses. In other words, inputs that are close together should give similar outputs, which is the same as saying the surface should be smooth.

A better justification for adding random noise to data is provided by Holmström and Koistinen [17], who also propose methods for choosing the variance of the noise.   ■

In addition to improving generalization, adding noise can be useful in another way. Since noise changes the function $f(\vec{x})$ that we want to minimize, it will also move the local minima. A small amount of noise changes the surface determined by $(\vec{x}, f(\vec{x}))$ only slightly. Hence deep minima will be affected only slightly. On the other hand, shallow minima may move significantly; in fact, some may disappear and new ones may appear. What effect does this have on minimization? It provides a way of escaping from local minima. Suppose the algorithm is close to a shallow local minimum. On the next iteration, the noise may have moved or destroyed that minimum. Since deep minima are more stable, they are still able to trap the algorithm.

Example 13.5   Weight Decay

Weight decay was introduced in (11.36) (p. 483) as a way of improving generalization by making it difficult for parameters to get large. Unfortunately, the justification was completely ad hoc. A probabilistic approach offers a less ad hoc justification.

Let $\vec{\pi}$ denote the entire set of parameters—both $\vec{w}$ and $\vec{\theta}$. The statement that π_i is more likely than not to be small can be viewed as a statement about the prior probability distribution for π_i. Thus we can think of $\pi_i \in \mathbb{R}$ as a random variable. What is its pdf? Since a normal with mean 0 and some variance σ^2 makes calculations easy, favors small values of $|\pi_i|$, and is a commonly occurring distribution, we'll use it. Furthermore, we'll make $\pi_1, \ldots, \pi_n$ independent. It follows that the pdf of $\vec{\pi}$ is

$$f(\vec{x}) = C \prod_{i=1}^{n} \exp(-x_i^2/2\sigma^2) = C \exp(-\vec{x} \cdot \vec{x}/2\sigma^2) \qquad (13.7)$$

for some constant $C = C(n, \sigma)$.

From the talk about priors in the previous paragraph, you may have guessed that we're moving in a Bayesian direction. We've only stated and proved Bayes' Theorem for finite probability spaces. There's also a version for infinite spaces in which pdfs play the same role that probabilities do in the finite case, but we won't prove it. Because it's more familiar, let's phrase the discussion in terms of Pr—the finite case—rather than pdf—the infinite case. The training set $\mathcal{T}$ consists of a collection of pairs $(\vec{P_i}, R_i)$ of patterns and desired responses. Let $\mathcal{P}$ be the list of patterns and $\mathcal{R}$ the list of responses. We want $\Pr(\vec{\pi} \mid \mathcal{P} \wedge \mathcal{R})$. In the following, C_i is something independent of $\vec{\pi}$.

$$
\begin{aligned}
\Pr(\vec{\pi} \mid \mathcal{P} \wedge \mathcal{R}) &= C_1 \Pr(\mathcal{P} \wedge \mathcal{R} \mid \vec{\pi}) \Pr(\vec{\pi}) && \text{by Bayes' Theorem} \\
&= C_1 \Pr(\mathcal{R} \mid \mathcal{P} \wedge \vec{\pi}) \Pr(\mathcal{P} \mid \vec{\pi}) \Pr(\vec{\pi}) \\
&= C_2 \Pr(\mathcal{R} \mid \mathcal{P} \wedge \vec{\pi}) \Pr(\vec{\pi})
\end{aligned}
$$

because the patterns are given and do not depend on how we select the parameters $\vec{\pi}$.

When we put $\vec{P_i}$ into a net with parameters $\vec{\pi}$, we get some output R'_i. Somehow we need to compute the probability of R_i given that we predicted R'_i. Insisting on perfect prediction would mean that the only possible outcome for correct $\vec{\pi}$ is $R'_i = R_i$. This is unrealistic because nets are usually not perfect classifiers. It's also not useful for minimization algorithms because there's nothing continuous to climb on. If we imagine that there are random sorts of errors in either the approximation R'_i or in the original data R_i or in both, then it's reasonable to assume a gaussian distribution for $R'_i - R_i$ with mean $\vec{0}$. For simplicity, let's assume the $R'_i - R_i$ are independent and all have the same variance τ^2. We can now do the same sort of manipulations as in (13.7).

Translating all this into the neural net terminology found on page 472 tells us that the pdf of the parameters given the training set is proportional to e^{-s} where

$$ s = \frac{1}{2\tau^2} \sum_{p=1}^{P} \sum_{\gamma \in O} \left(o_\gamma(p) - o^*_\gamma(p) \right)^2 + \frac{1}{2\sigma^2} \left(\sum_{i,j} w_{i,j}^2 + \sum_i \theta_i^2 \right). \qquad (13.8) $$

To maximize e^{-s}, we minimize (13.8). This is equivalent to minimizing (11.36) (p. 483).

To complete the discussion, we need to describe how σ and τ (or simply λ) are determined. Here's where the approach runs into some problems. Various methods have been advocated and the problem is still being researched. ∎

Exercises

13.2.A. Why might adding noise to the training data improve generalization? convergence?

13.2.B. What is weight decay?

13.2.1. Suppose you're given a feedforward network and training data. You train the net obtaining $\vec{w}$ and $\vec{\theta}$. The values obtained may depend heavily on how the parameters are initialized since you expect to converge to one of many local minima, not to the global minimum. Imagine estimating the generalization error rate using cross-validation. To do so, you must train the network many times using different training sets. In addition to the problem caused by the choice of random starts, there's the problem that training is a time-consuming process.

There are various ways you might initialize the parameters for cross-validation training Here are some possibilities:

(i) Use a random start.

(ii) Use the same start used to train the net you're cross-validating.

(iii) Start them at the values of $\vec{w}$ and $\vec{\theta}$ of the net you're cross-validating, perhaps with a random (gaussian?) variable added to each component.

(a) What can you see as potential strengths of each approach? Potential weaknesses?

*(b) Can you suggest another approach that might be promising? If so, what is it and why do you think it is promising?

(c) What method are you inclined to prefer and why? (There is no one right answer to this.)

(d) What sort of experiments might you perform to determine which method seems preferable?

13.2.2. Fill in the missing steps in the paragraph containing (13.8).

13.2.3. Imagine a feedforward net with a single output vertex that is designed to produce outputs in the range $(0, 1)$. Introduce a 0/1-valued random variables X_γ and interpret o_γ as the probability that $X_\gamma = 1$ given the input. The training data consists of input/output pairs in which the outputs are 0/1-valued corresponding to "false" and "true." A given input may appear many times in the training data, not always with the same outputs. (See page 472 for net notation.)

(a) Explain how you could try to use such a neural net to produce the probability that a patient has a particular disease given the patient's symptoms.

(b) Show that the probability of the outputs in the training data given the training inputs and the nets is

$$\prod_{p=1}^{P} o_\gamma(p)^{o_\gamma^*(p)} (1 - o_\gamma(p))^{1 - o_\gamma^*(p)}.$$

(c) Suppose that the training set contains $N(\vec{x}, r)$ copies of the input pattern $\vec{x}$ with desired output $o_\gamma^* = r$ where $r \in \{0, 1\}$. Show that we should minimize

$$-\sum_{\vec{x}} \left(N(\vec{x}, 1) \log(o_\gamma(\vec{x})) + N(\vec{x}, 0) \log(1 - o_\gamma(\vec{x})) \right),$$

where $o_\gamma(\vec{x})$ is the output the net produces when the input is $\vec{x}$.

(d) Show that the minimum is achieved when

$$o_\gamma(\vec{x}) = \frac{N(\vec{x}, 1)}{N(\vec{x}, 0) + N(\vec{x}, 1)},$$

provided parameters can be chosen so that the net produces such outputs $o_\gamma(\vec{x})$.

Further discussion of this use of entropy as an error measure can be found in [30].

*13.2.4. Suppose that there are several output vertices O, all of which output non-negative numbers. In some situations people want to interpret the vector of outputs as an indication that the correct answer is γ for each $\gamma \in O$. One way of doing this is by setting $\Pr(\gamma) = o_\gamma / \sum_{\delta \in O} o_\delta$. Discuss adapting the ideas in the previous exercise to this situation.

Comments on Large Feedforward Nets

> *As yet, there seem to be few principles or methodologies for designing the specific connectivity patterns in these networks. All network designs in the literature seem to have been rather ad hoc constructions for specific experiments. This is a major inadequacy of the discipline.*
>
> —J. Stephen Judd (1990)

It's been estimated that a human brain has about 10^{11} neurons ("nodes" in neural net terminology) and about 10^4 connections (edges) per neuron. To approach anything like the number and connectivity of the neurons in the human brain, we must build *much* larger nets. Researchers have discovered that building large neural nets ("scaling up") presents great difficulties. This is a recurrent problem in AI. To cite just three examples:

- In the early days, it was thought that simple search was the key to AI. Unfortunately, search time expands explosively with problem size. Extensive knowledge of the problem domain is often needed to reduce the search to a manageable size.

- Another early hope for AI was simple logic approaches. Again, scaling up led to excessively slow programs. The reasoning domain was structured by such methods as semantic nets and defeasible logic in an attempt to reduce running time.

- Planning, which we haven't discussed, also becomes bogged down as problem size increases. Structuring the planning space so that work on strategy (large-scale steps) precedes work on tactics (refinements of the large steps into smaller ones) has helped.

Why is it so difficult to train large nets? In the first place, the number of parameters is quadratic in the number of vertices: If we increase the number of vertices by a factor of k, the number of possible edges increases by a factor of about k^2. Hence the number of parameters will increase by a factor of about k^2. In the second place, the number of steps required to reach an acceptable local minimum usually increases rapidly with the dimension of the parameter space. Improved minimization techniques could allow somewhat larger nets, but much more than that is needed. At this time, no one knows what to do. Experiments are being carried out with a variety of methods based on analogy and experience rather than theory. Some general design principles should emerge in the near future. A clear theoretical understanding will take much longer. What has been achieved is based largely on clever ad hoc ideas, many or all of which may be supplanted by future research. Here are some approaches that are being explored. The list is incomplete and some topics overlap.

- **Incremental Growth**: Start with a small network. Train it. Add one or more vertices in strategic positions. Continue training, perhaps inhibiting change in many of the old parameters. (Some may even be frozen.) This has an effect similar to working in a space with fewer parameters—an inhibited parameter counts for less than a free one. If the net has reached some reasonable partial approximation to a solution, the old parameters should be at or near useful values. As a result, freezing or slowing them should be okay.

- **Incremental Training**: This is closely related to incremental growth. Here's one such strategy. Train a small network N_1 on T. Suppose it responds well on $T_1 \subset T$ and poorly on $T_2 = T - T_1$. Train another network N_2 to do well on T_2 and then train a third network N_{12} to decide between using the output of N_1 and using the output of N_2. Network N_{12}'s decision would probably be based on distinguishing between T_1 and T_2. Taken together, N_1, N_2, and N_{12} provide a network that does well on all of T. The nets N_2 and N_{12} could be constructed iteratively.

- **Structure**: Structure the network based on the nature of the problem it is facing. A traditional example is vision where the inputs form a two-dimensional array. Hidden units are connected to blocks of adjacent inputs and may have inhibitory connectors to each other. These, in turn, feed to other hidden units, still preserving some of the metric structure of the inputs. With this hierarchical structure, the number of edges grows much more slowly than the square of the number of vertices and the outputs from one level provide highly informative inputs for the next level.

- **Modularity**: Split the job into subtasks, train nets on the subtasks, and then build a new net that uses the outputs of the trained nets as inputs. Finding subtasks can be difficult, especially since we probably need more than just a few if we want to obtain very large improvements. We've already discussed one obvious division into two subtasks: See the discussion of cleaning up inputs on page 488. The method described above in "incremental training" is a form of automated modularity.

- **Biology**: Considerable structure is present in the neural connections of brains—most are to nearby neurons, but there are also organized long-distance connections. A large number of synaptic connections (edges) and neurons (vertices) die in the brains of babies. A better understanding of the structure of brains and Nature's "pruning" technique may lead to useful tools for designing neural nets.

- **Tool Combining:** Don't use neural nets for large-scale problems. Instead, use them for preliminary processing of input and then use their outputs as input to some reasoning-based approach. This is called a *hybrid system* because it uses more than one knowledge manipulation technique. Some cognitive scientists argue that there is support for this approach in brains: Timing information implies that relatively few "machine cycles" occur between a brain's reception of input and the initiation of processing on a more conscious level. Neural nets may be the best devices for the initial low-level processing, but other approaches may be more appropriate for the more conscious type of processing. Even this low-level processing is more complex than what we can achieve artificially, so we would still need to improve neural nets.

All these topics are areas of current research.

Exercises

13.2.C. What are some of the difficulties in scaling up neural nets?

13.2.D. What are some avenues of approach for the problem of building large nets?

The Storage Capacity of Hopfield Nets

Let's return to Theorem 11.1 (p. 433), for which we gave a heuristic partial proof in Example 11.1 (p. 436). At that time, we were faced with the problem of showing that $P_t(k)$ and

$$(n-1)P_t(k) + \sum_{i:\, i \neq t} \sum_{l:\, l \neq k} P_i(l) P_t(l) P_i(k) \qquad (13.9)$$

have the same sign, where $\vec{P}(k)$ is an n-long pattern of ± 1's and $1 \leq k \leq s$. This cannot be guaranteed to always be true; however, it can be shown that it's probably true for a case selected at random. This is the reason for the "almost certainly" in the statement of Theorem 11.1. Theorem 11.1 must be regarded as a probabilistic result—if the components in the patterns are selected independently, then with high probability $\pm \vec{P}(k)$ will be an energy minimum and points not too far away will be in its basin of attraction.

With this probabilistic approach, we can use the Central Limit Theorem or something similar.

Let the $P_i(j)$'s be independent random variables that are as likely to be $+1$ as -1. By the definition of the $P_i(j)$'s each term in the summation in (13.9) is a random variable $X_{i,l}$ that is equally likely to be $+1$ or -1. Now let's cheat and apply the Central Limit Theorem (p. 513) to the sum. Why is this cheating? We don't know that the terms in the sum are mutually independent. In fact, there's a slight dependency, but it is small enough that it shouldn't distort the results too much.

To avoid confusion with the n in (13.9), imagine replacing the n in Theorem 12.3 (p. 513) by m. That sum contains $m = (n-1)(s-1) \approx ns$ terms. Since $\mathrm{E}(X_{i,l}) = 0$ and $\mathrm{E}(X_{i,l}^2) = 1$, the variance of each term is 1. Thus $\mu(m) = 0$ and $\sigma(m) = \sqrt{m} \approx \sqrt{ns}$ in Theorem 12.3. By the theorem,

$$\Pr(|Y_m| > C) \approx \frac{2}{(2\pi)^{1/2}} \int_C^\infty e^{-t^2/2} dt,$$

a function that is quite small when C is large. Since the magnitude of the summation in (13.9) is $m|Y|$, it's likely that (13.9) and $P_t(k)$ will have the same sign if $m \approx \sqrt{ns}$ is small compared to n; that is, if s is small compared to n. Unfortunately, this is not good enough. This result must hold for *all* n components of $\vec{P}(k)$ and also for all k. Even though each individual event has a high probability, the probability of *all* components of *all* $\vec{P}$'s working out correctly may be low.

We need a stronger result than the Central Limit Theorem. Instead of (12.9), we need

$$\lim_{m \to \infty} \left(\frac{\Pr(Y_m \le y(m))}{(2\pi)^{-1/2} \int_{-\infty}^{y(m)} e^{-t^2/2} dt} \right) = 1 \tag{13.10}$$

even if $y(m) \to -\infty$ as long as $|y(m)|$ does not grow too fast. (You should be able to easily see that (12.9) and (13.10) are the same when $y(m)$ is constant.) Such "large deviation" results exist. We'll simply use (13.10) as if everything is okay—after all, we've already cheated once by pretending the $X_{i,l}$'s are independent. According to Exercise 13.2.5, the integral in (13.10) can be bounded above by $e^{-y(m)^2/2}/|y(m)|$. Suppose n/s is large and let

$$y(m) = \frac{-(n-2)}{\sqrt{m}} \approx \sqrt{\frac{n}{s}} \quad \text{where } n/s \text{ is large.}$$

Using (13.10) and arguments like those in the previous paragraph, you should be able to show that the probability the summation in (13.9) will cause a sign reversal is less than $\exp(-y(m)^2/2) \approx e^{-n/2s}$.

Assuming the sign-reversal events are independent, the probability that the sign of (13.9) will be correct for *all values* of t and k is at least

$$\left(1 - e^{-n/2s}\right)^{ns} \approx \exp\left(-e^{-n/2s} ns\right), \tag{13.11}$$

where we used $(1-\delta) \approx e^{-\delta}$. How large can we make s and still have (13.11) close to 1? Experimentation or experience may suggest trying $s = n/C \ln n$. Then (13.11) becomes $\exp(-n^{2-C/2}/C \ln n)$. Hence we take $C = 4$.

Since our argument was based on independence assumptions that were only approximately correct, it isn't mathematically rigorous. The conclusion we reached—that $s = n/4 \ln n$ is a critical value—is correct. But a rigorous proof requires additional mathematical background and involves more work than this heuristic derivation.

Since each pattern contains n components, each of which is ± 1, it can be thought of as n bits. All the patterns together consist of $ns \approx n^2/4 \ln n$ bits. On the other hand, an n-node Hopfield net has about $n^2/2$ weights. Consequently, it uses about $2 \ln n$ weights per bit of pattern. Thus a Hopfield net requires many more bits to store the patterns than it would take to simply list the patterns.

Exercises

13.2.5. Prove that when $x > 0$,

$$\int_{-\infty}^{-x} e^{-t^2/2}\, dt = \int_x^\infty e^{-t^2/2}\, dt < \frac{e^{-x^2}}{x}.$$

Hint. Change variables in the second integral so the limits are 0 and ∞.

13.2.6. The estimate of $2 \ln n$ weights per component stored in a Hopfield net is actually rather bleak: A weight is a real number and so contains an infinite number of bits of information since it is an infinite decimal.

(a) Show that our weights are actually integers of magnitude less than s.

(b) Conclude that we are using about $c(\log n)^2$ bits to store one bit.

*(c) Perhaps we can round off our weights. Explore the possibility that it may be sufficient to keep just some of the high-order bits of a weight.

13.3   Heuristic and Partial Search

For this section, you'll need to review the definitions and concepts associated with search trees and heuristic search that were introduced in Sections 2.3 (p. 44) and 2.5 (p. 60). You'll also need to understand the notion of *expanding a vertex*; that is, putting all the vertex's sons on the list $\mathcal{L}$ of vertices to be considered. Best-first heuristic search always removes the vertex v of least heuristic cost $C(v)$ from $\mathcal{L}$, and if v is not a goal, expands it.

Beware of Small Heuristics

Recall that a heuristic with $h \leq h^*$ is called admissible and that an admissible heuristic guarantees that best-first heuristic search finds a least-cost goal. (See Theorem 2.4 (p. 62).) So far, admissibility appears to be a good feature for a heuristic to have. But you'll soon see that focusing on admissible heuristics could be a bad strategy.

To simplify the discussion, let's consider a highly structured search problem. Imagine a maze that is a tree. The tree maze has a branching factor of $b > 2$ at the root and $b - 1$ elsewhere. The goal is one vertex of the tree maze, and we'll denote its depth by N. As with most mazes, we must start at the root and walk to the goal and we may traverse edges of the maze in either direction. The cost of the walk is the number of edges it contains. Equivalently, each edge has unit length and the cost of the walk is its length.

The tree maze is not the search tree. (You may want to review the construction of search trees.) Because we can traverse edges in both directions in the maze, each maze vertex corresponds to infinitely many search tree vertices. The search tree will have branching factor b because we can descend along any of the $b - 1$ tree maze edges or go up toward the root. In the search tree, depth equals cost since depth is simply the number of maze edges traversed, counting repetition.

Of all the vertices in the search tree that correspond to the goal vertex, one of them, say z, corresponds to traversing the maze without backtracking. It will be at depth N in the search tree, just as it is in the maze. All other goal vertices involve backtracking in the maze and so have depth greater than N.

The following theorem contains a completely unrealistic model for the heuristic cost. It and its proof are here as preparation for a more realistic theorem.

Theorem 13.1

Let the search space be the tree maze described above. Suppose that $h = \lambda h^*$ for some constant $\lambda > 0$ and that best-first heuristic search is used. We have the following conclusions:

(a) If $\lambda < 1$, the number of vertices expanded is exponential in N. This result is true even if the search keeps track of all vertices it visits in the maze and never expands the same vertex twice.

(b) If $\lambda \geq 1$, the search expands only those vertices on the path to the least-cost goal z.

In other words, $\lambda < 1$ is bad and $\lambda \geq 1$ is good as far as search time is concerned.

Proof: Let's start with $\lambda \geq 1$. Suppose that we are at some vertex v in the search tree at depth d and are on the path to the least cost goal vertex z. Let $\{v_1, \ldots, v_b\}$ be the sons of v and let v_1 be on the path to z. From the definitions we have

$$g(v_i) = d+1, \quad h^*(v_1) = N-d-1, \quad \text{and} \quad h^*(v_j) = N-d+1 \quad \text{for } j \neq 1,$$

where the last follows because, if we move to the vertex in the maze that corresponds to v_j, we must then backtrack to the vertex corresponding to v before proceeding to the goal. Since $h = \lambda h^*$, the heuristic costs are

$$C(v_1) = d+1+\lambda(N-d-1) = \lambda N - (\lambda-1)(d+1) \qquad (13.12)$$

and

$$C(v_j) = d+1+\lambda(N-d+1) = C(v_1) + 2\lambda \quad \text{for } j \neq 1.$$

Thus, we would always move to v_1 rather than to a v_j with $j \neq 1$. Of course, it's possible that some time later in our search, we may reach a vertex v' at depth $d'-1$ on the path to z and that v_j may look more attractive than any of the sons of v'. We must show that this can't happen. By (13.12), the correct son v_1' of v' has heuristic cost

$$C(v_1') = \lambda N - (\lambda-1)(d'+1) = C(v_j) - (\lambda-1)(d'-d) + 2\lambda.$$

Since $\lambda - 1 \geq 0$ and $d' - d > 0$,

$$C(v_1') \leq C(v_j) + 2\lambda < C(v_j) \quad \text{for } j \neq 1. \qquad (13.13)$$

In other words, an incorrect son of v always has higher cost that a vertex v_1' on the correct path. This completes the proof for $\lambda \geq 1$.

Now suppose $\lambda < 1$. The first inequality in (13.13) breaks down because $\lambda - 1 < 0$. When $d' - d$ is sufficiently large, we can have

$$-(\lambda-1)(d'-d) + 2\lambda > 0$$

and so $C(v_1') > C(v_j)$. Thus a best-first algorithm would examine the sons of v_j before those of v_1'. We need to see how far this goes.

Note that $C(z) = N$ and all vertices between the root and z have cost at most N. Thus we must determine what vertices can be reached along a path whose vertices have heuristic cost at most N. Since $C(u) \geq g(u)$, which is the depth of u, no vertices with depth exceeding N are expanded. Since the number of vertices at depth d is b^d, the number of vertices expanded is at most

$$\sum_{b<N} b^d = \frac{1-b^N}{1-b} < b^N$$

by the formula for the sum of a geometric series.

To complete the proof, we'll show that the number expanded is exponential in N even when repeated maze vertices are rejected. This rejection corresponds to prohibiting backtracking in the maze. Elimination of backtracking changes the search tree by eliminating one edge out of each vertex except the root. Thus, there are now $b(b-1)^{d-1}$ vertices at depth d instead of b^d. Of course g, h^*, and h are unchanged.

Let w be a vertex in the search tree at a depth d. Since its distance from a goal vertex is at most $d + N$,

$$C(w) \le d + \lambda(d + N) = \lambda N + (1 + \lambda)d. \tag{13.14}$$

Thus, $C(w) < N$ whenever $d < \frac{\lambda N}{1+\lambda}$, and so all vertices with depth less than $\frac{\lambda N}{1+\lambda}$ will be expanded unless they have been reached by backtracking. It follows that more than a^{N-1} vertices will be expanded where

$$a = (b-1)^{\lambda/(\lambda+1)} > 1.$$

This proves that the number of vertices expanded is at least exponential in the depth of the goal z. ∎

As already mentioned, the model of search on which Theorem 13.1 was based is too simple: Since h is a *heuristic* function, it shouldn't just be a function of h^*—for then we could probably recover h^* from h. (In the model, $h^* = h/\lambda$.) Some randomness is needed to model the fact that h is a heuristic function. What happens to the conclusions in Theorem 13.1? The bad result (a) still holds and the good result (b) becomes weaker.

Theorem 13.2   Slowly and Rapidly Growing Heuristics

Let the search space be the tree maze described above. Suppose that there is some constant $\lambda > 0$ and some nondecreasing function $\theta > 0$ such that

$$\lim_{x \to \infty} \theta(x)/x = 0 \quad \text{and} \quad \big|h(v) - \lambda h^*(v)\big| < \theta\big(h^*(v)\big).$$

Let best-first heuristic search be used, and suppose that it keeps track of all vertices it visits in the maze and never expands the same vertex twice. We have the following conclusions:

(a) If $\lambda < 1$, the number of vertices expanded is exponential in N.

(b) If $\lambda \ge 1$, there are constants A and B such that the search expands at most $ANB^{\theta(2N)}$ vertices before reaching the goal.

It would be useful to have additional results such as the following:

- **Probabilistic results**: Relax the bound θ to a probabilistic statement and make the conclusion probabilistic.

- **A converse result**: Under appropriate assumptions, at least $ca^{\psi(N)}$ vertices are expanded for some constants a and c and some function ψ.

- **More general search trees**: Both the shape of the search tree and the form of the cost function are highly restricted.

Proof: The proof of (a) is essentially the same as that given in Theorem 13.1; however, we need to take into account the fact that h is not simply λh^*. The greatest cost on the path to z is at least $N - \theta(N)$. Suppose that the depth of w is $d < N$. Since a vertex at depth d is at most $d + N$ from the goal, (13.14) becomes

$$C(w) < d + \lambda(d + N) + \theta(d + N) \leq (1 + \lambda)d + \lambda N + \theta(2N).$$

Thus $C(w) < N - \theta(N)$ whenever

$$d < \frac{1 - \lambda - \theta(N)/N - 2\theta(2N)/2N}{1 + \lambda} N.$$

Since $\lambda < 1$ and $\theta(x)/x \to 0$, the numerator is bounded away from zero for large enough N. The proof of (a) now continues as in the previous theorem.

The proof of (b) uses the same idea as the proof of (a) with two exceptions. First, we need inequalities in the reverse direction. Second, the reverse inequalities are most easily obtained by noting when the path first departs from the path to z.

The greatest cost on the path to z is less than $\lambda N + \theta(N)$. Suppose that the depth of w is $d < N$ and that the path to w departs from the path to z starting at the root. Since there is no backtracking, w is a distance $d + N$ from the goal in the maze tree. Hence

$$C(w) > d + \lambda(d + N) - \theta(d + N) \geq (1 + \lambda)d + \lambda N - \theta(2N).$$

Since w will not be expanded whenever $C(w) \geq \lambda N + \theta(N)$, the greatest possible depth for w is given by the solution to

$$\lambda N + \theta(N) = (1 + \lambda)d + \lambda N - \theta(2N),$$

which is

$$d = \frac{\theta(2N) + \theta(N)}{1 + \lambda} \leq \frac{2\theta(2N)}{1 + \lambda}. \tag{13.15}$$

It follows that the number of such w is at most $AB^{\theta(2N)}$ for some constants A and B.

This takes care of the vertices on paths that leave the path to z at the root. Suppose a path leaves the path to z at a vertex v at depth d'. Regarding v as a new root and measuring depth from there, we see that (13.15) still applies provided N is replaced by $N - d'$. Hence there are at most $AB^{\theta(2(N-d'))} \leq AB^{\theta(2N)}$ such vertices. Summing over the N values of d' gives (b). ∎

What lessons can be drawn from the discussion, the theorem, and its likely extensions? Here are some.

- To control the number of nodes expanded, avoid any h that underestimates h^*. You might do this by replacing a small h with Ch for a sufficiently large constant h.

- Assuming a sort of converse to Theorem 13.2(b), the number of nodes expanded will grow rapidly unless h is nearly a function of h^*; that is, unless $|h - \lambda h^*|$ is small compared to h^* for some $\lambda \geq 1$.

- Since this limitation on h is often unrealistically stringent, you should probably look for other ways to reduce branching. The most likely candidate is a better search space. Trying to improve the search space may be more profitable than trying to improve the heuristic.

Partial Search

In Section 2.6, we saw that time constraints frequently make it impossible to exhaustively examine a search tree. We must then use a partial search and rely on a heuristic to evaluate vertices of the search tree. As we'll discuss below, we frequently use a partial search to select a "move"—a son of the search tree root. The search may be terminated by either

- the search algorithm when it decides that the expected additional gains don't justify the expected additional time or

- an agent external to the search algorithm, which must then be an anytime algorithm.

We'll avoid the problem of when to stop searching by assuming that an outside agent terminates the search.

Although heuristic and partial search methods are closely related, the search literature makes it appear otherwise. In heuristic search we usually look for heuristic cost functions that minimize search time when we use the strategy of expanding least-cost vertices first. As shown in Theorem 13.2, such heuristics need not be unbiased, and it may be best if they exceed actual costs. In partial search we usually look for unbiased heuristics and we speak of gain rather than cost. We'll follow that approach here. Changing cost to gain is a minor issue since one can be taken to be the negative of the other. The sort of heuristic we should use is not clear because Theorem 13.2 suggests it may be best to underestimate gain. How can this be resolved? The key lies in the nature of the search problem. Heuristic search research frequently deals with problems in which the cost can be viewed as the length of a path, so the depth of the goal vertex is crucial. In contrast, partial search research frequently deals with problems in which we're looking for the largest "pot of gold" at the end of the search, so distance is irrelevant. For example, what matters most in a game is whether we win or lose, not how many moves are required. In this case, an estimate of the probability of winning could be used as a heuristic gain.

Two-person games will provide the model for partial search in this section. Games with complete information were discussed Section 2.7 (p. 76) as a form of AND/OR tree. Recall that in game trees

- each vertex is a state of the game,
- the sons of the root correspond to my possible moves,
- for any particular son v of the root, the sons of v correspond to my opponent's possible moves if I choose v, and
- this alternation between my possible moves and my opponent's continues on down through the tree.

Players may or may not have complete control over their moves. For example, the control is incomplete when there is an element of chance, as in card games. An extreme case of lack of control arises when my opponent is "Nature." In that case, my opponent doesn't make a choice with the idea of winning the game, and all I can do is assign a probability distribution to my opponent's possible moves. In a game like backgammon, it may be best to introduce three players by making the rolls of the dice into moves by Nature.

Suppose we have a list $\mathcal{L}$, each entry being a vertex to consider together with information about the estimated gain if it is chosen. When it's time to make a decision, the information at these vertices must somehow be propagated back to the root so that we can make a choice among the various sons of the root. If the information about gain were exact, it would be clear what to do:

- A vertex at which it's my turn receives the maximum of its sons' gains.
- If there's no randomness in my opponent's move, a vertex at which it's her turn receives the minimum of its sons' gains.
- A vertex at which it's Nature's turn receives an expected value based on the probability of Nature's various move choices.

In partial search the gain is not exact, so it's unclear what to do. It's also unclear which vertices to expand to gain the most improvement in information. Thus a partial search algorithm contains three components:

- a heuristic function for estimating gain,
- a strategy for deciding what vertex to expand next, and
- a method for propagating heuristic information backward toward the root.

Let's assume that gain estimates are available. The expansion strategy and propagation method are closely related because an understanding of what makes the propagated information unreliable will indicate what vertex should be expanded next.

Example 13.6   Computing a Heuristic Maximum

Suppose that $h_1, \ldots, h_n$ are unbiased heuristic estimates of $h_1^*, \ldots, h_n^*$. What should our heuristic estimate of $\max(h_i^*)$ be? At first glance, it seems that the obvious choice, $\max(h_i)$, must be correct. That's not so.

Suppose that the h_i^* all equal h^* and that the h_i are independent samples from a distribution with mean h^*. Ideally, we'd like some function h of the h_i whose expected value is h^*. After all, if the h_i are unbiased estimators of h^*, that doesn't seem like too much to ask of h.

The expected value of $\max(h_i)$ will exceed h^*. In fact, it can be shown that

$$
\begin{aligned}
\mathrm{E}\big(\max(h_1, \ldots, h_n)\big) &= \int_{-\infty}^{\infty} x \big(F(x)^n\big)' \, dx \\
&= h^* + \int_{-\infty}^{\infty} F(x)(1 - F(x)^{n-1}) \, dx > h^*,
\end{aligned}
\tag{13.16}
$$

where $F(x)$ is the distribution function for each of the h_i's. (We won't prove this.) Computing (13.16) is difficult unless $F(x)$ is simple. To illustrate, suppose that f is uniform on $[0, 1]$; that is, $f(x) = 1$ if $0 \leq x \leq 1$ and $f(x) = 0$ otherwise. Thus $F(x) = x$ for $0 \leq x \leq 1$. Using the first integral in (13.16), you should be able to easily show that $\mathrm{E}(\max(h_1, \ldots, h_n)) = \frac{n}{n+1}$, in contrast to $\mathrm{E}(h_i) = \frac{1}{2}$. Clearly, the maximum is far from being an unbiased estimator.

Why not simply let h equal the average of the h_i's? In fact, the average is an unbiased estimator of the maximum *if all the h_i's come from distributions with the same mean*. We don't even need any independence assumptions! Unfortunately, we only have the h_i's—we don't know that the underlying distributions have the same mean. If they don't, averaging is bad. You should be able to see why this is so.

Some researchers have looked at using alternatives to the maximum. As far as I know, all that's available are some empirical results using ad hoc alternatives. ■

Since we use expectation for a move by Nature, that part of the propagation doesn't suffer from the problems that plague propagating maxima. In fact, if h_i are independent unbiased estimates of gain, p_i is the probability that alternative i occurs and σ_i^2 is the variance of h_i. Then the expectation estimate is $\sum p_i h_i$ and its variance is $\sum p_i \sigma_i^2$. For the remainder of this section we'll discuss maxima.

Making a choice: Rather than propagating the maximum to the root, we want to choose a move at the root. In search graph terms, this corresponds to selecting an action at the root. Thus, instead of estimating a maximum, we need to decide which son offers the maximum gain. Let's begin with this problem.

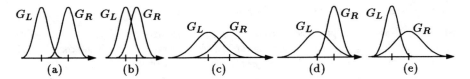

Figure 13.4 In each figure the true gains satisfy $G_{L}* < G_R^*$ and are indicated by tic marks on the horizontal axes. We want to estimate the maximum true gain (G_L^*) from the two heuristic gains, G_L and G_R. Each heuristic gain is an independent random variable whose density function is shown in the figure. See the text for a discussion of these five examples. (**Note:** I've used gaussians here merely as a convenient curve—I don't intend to imply that the distributions are gaussian.)

Figure 13.4 contains five examples. The caption explains the notation. It's important to remember that we don't actually "see" the subscripts on G_L and G_R—we have two estimated gains, but we don't know which is associated with the larger true gain. Since we choose the move associated with the larger heuristic, an incorrect decision will be made if $G_L > G_R$. This is highly unlikely in (a). Figures (b)–(e) were all constructed to have $\Pr(G_L > G_R) \approx$ 13%. Thus they all have the same probability of error; however, the expected loss of gain varies. It equals the amount lost times the probability of loss; that is, $(G_R^* - G_L^*)\Pr(G_L > G_R)$. If we decide to collect further information in either (d) or (e), it's usually more effective to try to improve the estimate with the higher variance. You should be able to convince yourself of this. In (b) and (c) it doesn't matter whether we work on G_R or G_L.

Computing a maximum: This differs from the previous situation because we want a good estimate for the maximum, not a good guess for which choice produces the maximum. In (a), (b), and (d), the larger of G_L and G_R is likely to be a rather good estimate for G_R^*. In cases (c) and (e) the larger of G_L and G_R is likely to be only a fair estimate for G_R^*. You should be able to explain this. Since we don't "see" the subscripts, we can't distinguish selecting G_L in (d) from selecting G_R in (e). Hence, a proposal for gathering further information in cases (c) and (e) must apparently also include (d).

The cold, cruel world intrudes: While the previous discussion may be interesting, it's based on some highly unrealistic assumptions:

- **Independence:** It's quite likely that G_L and G_R will be dependent: Their heuristics are based on similar situations in the search space. The more similar the situation, the more likely it is that what throws one estimate off will throw the other off, too.

- **Unbiased estimates:** As we saw in Example 13.6, producing an unbiased estimate of a maximum is difficult.

- **Known variance:** The recommendations for expansion require knowledge of variances.

Here are two ways we might gain some information about variances.

- If deeper exploration of the tree causes relatively large changes in the estimates propagated back to a vertex v, the variance is likely to be rather large. Conversely, if the changes are small, the variance is probably rather small. When changes are small, we call the vertex *quiescent*. The observations concerning Figure 13.4 lead to the rule of thumb that a nonquiescent vertex should often be expanded deeper than a quiescent one.

- Another way we might obtain variance information is by comparing different heuristic estimates for the gain in the position. If they vary widely, it's reasonable to assume our estimate has a high variance.

The Future of Search in AI

In the early years of AI, many researchers believed that powerful general search techniques would play an important role. But, as we saw in Chapter 2, the time required for a tree search usually grows exponentially with the size of the problem. What can be done? Clever search algorithms can reduce the rate of exponential growth, but their use only forestalls the computational difficulties for a little while. As a result, AI researchers turned their attention to other methods of reducing computation.

One method is to make extensive use of knowledge about the particular area. This approach has produced a variety of expert systems, but it suffers from a serious drawback: Since an understanding of the knowledge must be built into the system, construction of an expert system is time-consuming; moreover, the system is able to function only in one very circumscribed area such as chess playing, diagnosis of bacterial diseases, or prospecting for oil.

Another method is to design systems that somehow manage to reduce search time as they develop. Is this possible? It must be, since one such system already exists—human beings.

There is evidence that minimization (which is a type of search) also grows exponentially with problem size. This latter fact has hampered attempts to "scale up" neural nets.

Exercises

13.3.A. What is an admissible heuristic and why might it be a bad choice?

13.3.B. If you're planning to design a heuristic search, why might it be more profitable to attempt to reduce the search space than to attempt to improve the heuristic?

13.3.C. What are three important components of a partial search algorithm?

13.3.D. Why is it not clear how to compute an unbiased estimate of the maximum of $\max(h_1^*, \ldots, h_n^*)$ given unbiased estimates of the h_i^*'s?

13.3.1. Prove (13.16). You may assume that $\lim_{x \to -\infty} x F(x) = 0$ and $\lim_{x \to \infty} x(1 - F(x)) = 0$.

13.3.2. Suppose that h_1 and h_2 are sampled from independent uniform distributions where h_1 is uniform on $[-a, a]$ and the density function for h_2 is that for h_1 shifted leftward by $2ra$, where $r \geq 0$. Let $h = \max(h_1, h_2)$.

   (a) Show that we must have $h_2 < h_1$ whenever $r > 1$ and that the distribution function for h is then the same as the distribution function for h_1.

   (b) Show that when $0 \leq r \leq 1$ we have $E(h) = a(1 - r)^3/3$.

13.3.3. Do Exercise 2.7.3 (p. 78).

Notes

Decision Trees

Quinlan introduced his Iterative Dichotomizer 3 (ID3) for building decision trees in the context of chess playing in 1979. The method has become quite popular and led to the development of various extensions and modifications (as in [24]) and to connections with neural networks (as in [9]). The definitive book on decision trees in the statistical literature is [2], where the "CART" method is discussed. Tsoi and Pearson [29] briefly discuss and then compare ID3 and CART.

Chapter 9 of [2] develops a more sophisticated approach to evaluating a decision tree based on the errors it makes. A correct classification is a gain and an incorrect is a loss; however, the amount of gain or loss depends on the seriousness of the situation—diagnosing cancer is more important than diagnosing the flu.

Fisher's test and its generalizations (p. 529) provide another possible method for evaluating the goodness of a split either for growing or pruning a decision tree. Since the test provides a probability, some people may be more comfortable using it.

Neural Nets

One of several methods for constructing large networks by splitting up the classification work is discussed in [18].

Two methods for adding nodes are "cascade-correlation" [12] and "adaptive resonance theory" (ART) [4] or [5]. The other side of adding nodes is pruning them, just as we did with decision trees—we make them too big and then trim. Various researchers have worked on this. See [21] and [3] for two approaches.

In his thesis Lee [22] discusses some ideas for the general problem of adjusting the structure of neural networks. Such growing and pruning is akin to hill climbing in the space of neural nets. The space could be searched in other ways, such as with "genetic algorithms."

A more rigorous and thorough analysis of the storage capacity of Hopfield nets can be found in the literature. See [20] or [16, Ch. 2].

People are interested in the question of what neural nets can learn and how difficult such learning must be. The subject is still in its infancy. Judd's thesis [19] is an example of research in this area.

I've not discussed the connections between Bayesian methods and neural nets. Buntine and Weigand [3] discuss Bayesian methods for training nets. Richard and Lippmann [25] view things from a different perspective: Neural networks often estimate Bayesian posterior probability (the probability of various classifications of the input given the training data). With such an interpretation, the outputs could then be used as inputs to probabilistic reasoning or decision making systems.

The pattern recognition aspects of neural nets are treated in various places. A somewhat advanced discussion is the special journal issue [15]. A large amount of material is appearing on higher level aspects of connectionist systems and hybrid systems. One source of information is the ongoing series of research papers [1].

The books mentioned here and in Chapter 11 just scratch the surface of the neural net literature. To mention three more, the text [13] and the more advanced books [6] and [28] contain interesting material, some of which I've mentioned in passing.

Search

The topics in the search section need much more research.

Chenoweth and Davis's general theorem on rapidly growing heuristics [7] motivated my discussion of the subject. Various people have worked on the problem of automatically developing a better search space; for example, Ernst and Goldstein [11]. Nau and others ([8], [23]) have experimented with alternatives to "max" in game trees. Russell and Wefald [26] have begun studying the problems of propagating heuristic information from a more global perspective. DeGroot's text [10] provides an introduction to optimal decision theory.

A search topic I've not mentioned is *tabu search*. It's a heuristic method designed to allow escape from local optima. Glover [14] provides an introduction. I'm not aware of any theoretical results.

References

1. J. A. Barnden (series ed.) *Advances in Connectionist and Neural Computation Theory*, Ablex, Norwood, NJ. At present the series includes Vol. 1 *High-Level Connectionist Models* (1991), Vol. 2 *Analogical Connections* (1994), and Vol. 3 *Analogy, Metaphor, and Reminding* (1994).

2. L. Breiman, J. Friedman, R. Olshen, and C. Stone, *Classification and Regression Trees*, Wadsworth, Monterey, CA (1984).

3. W. L. Buntine and A. S. Weigand, Bayesian back-propagation, *Complex Systems* 5 (1991) 603–643.

4. G. A. Carpenter and S. Grossberg, A massively parallel architecture for a self-organizing neural pattern recognition machine, *Computer Vision, Graphics and Image Processing* 37 (1987) 54–115.

5. G. A. Carpenter and S. Grossberg, The ART of adaptive pattern recognition by a self-organizing neural network, *Computer* 21:3 (March, 1988) 77–88.

6. Y. Chauvin and D. E. Rumelhart (eds.), *Back-propagation: Theory, Architectures and Applications*, Lawrence Erlbaum Associates, Hillsdale, NJ (1994).

7. S. V. Chenoweth and H. W. Davis, High-performance A* search using rapidly growing heuristics, *IJCAI-91*, Morgan Kaufmann, San Mateo, CA (1991) 198–203. ("IJCAI" stands for *International Joint Conference on Artificial Intelligence*.)

8. P.-C. Chi and D. S. Nau, Comparison of the minimax and product back-up rules in a variety of games. In L. Kanal and V. Kumar (eds.), *Search in Artificial Intelligence*, Springer-Verlag, Berlin (1988) 450–471.

9. K. J. Cios and N. Liu, A machine learning method for generation of a neural network architecture: A continuous ID3 algorithm, *IEEE Trans. on Neural Networks* **3** (1992) 280–290.

10. M. H. DeGroot, *Optimal Statistical Decisions*, McGraw-Hill, New York (1970).

11. G. W. Ernst and M. M. Goldstein, Mechanical discovery of classes of problem-solving strategies, *JACM* **29** (1982) 1–23.

12. S. E. Fahlman and C. Lebiere, The cascade-correlation learning architecture, *NIPS* **2**, Morgan Kaufmann, San Mateo, CA (1990) 534–532. ("NIPS" stands for *Advances in Neural Information Processing Systems.*)

13. S. I. Gallant, *Neural Network Learning and Expert Systems*, MIT Press, Cambridge, MA (1993).

14. F. Glover, Tabu search: A tutorial, *Interfaces* **20**:4 (Aug. 1990) 74–94.

15. I. Guyon and P. S. P. Wang (eds.), *International Journal of Pattern Recognition and Artificial Intelligence*, Vol. 7 no. 4 (1993). Reprinted as *Advances in Pattern Recognition Systems Using Neural Network Technologies*, World Scientific, Singapore (1993).

16. J. A. Hertz, A. Krogh, and R. G. Palmer, *Introduction to the Theory of Neural Computation*, Addison-Wesley, Reading, MA (1991).

17. L. Holmström and P. Koistinen, Using additive noise in back-propagation training, *IEEE Trans. on Neural Networks* **3** (1992) 24–38.

18. M. I. Jordan and R. A. Jacobs, Hierarchical mixtures of experts and the EM algorithm, *Neural Computation* **6** (1994) 181–214.

19. J. S. Judd, *Neural Network Design and the Complexity of Learning*, MIT Press, Cambridge, MA (1990).

20. J. Komlós and R. Paturi, Convergence results in an associative memeory model, *Neural Networks* **1** (1988) 239–250.

21. Y. Le Cun, J. S. Denker, and S. A. Solla, Optimal brain damage, *NIPS* **2**, Morgan Kaufmann, San Mateo, CA (1990) 598–605. ("NIPS" stands for *Advances in Neural Information Processing Systems.*)

22. T.-C. Lee, *Structure Level Adaptation for Artificial Neural Networks*, Kluwer, Boston (1991).

23. D. S. Nau, Pathology of game trees revisited, and an alternative to mini-maxing, *Artificial Intelligence* **21** (1983) 221–244.

24. J. R. Quinlan, *C4.5: Programs for Machine Learning*, Morgan Kaufmann, San Mateo, CA (1993).

25. M. D. Richard and R. P. Lippmann, Neural network classifiers estimate Bayesian a posteriori probabilities, *Neural Computation* **3** (1991) 461–483.

26. S. Russell and E. Wefald, *Do the Right Thing: Studies in Limited Rationality*, MIT Press, Cambridge, MA (1991).

27. C. J. Stone, Admissible selection of an accurate and parsimonious normal linear regression model, *Annals of Statistics* **9** (1981) 475–485.

28. J. G. Taylor (ed.), *Mathematical Approaches to Neural Networks*, North-Holland, Amsterdam (1993).

29. A. C. Tsoi and R. A. Pearson, Comparison of three classification techniques, CART, C4.5 and multi-layer perceptrons, *NIPS* **3**, Morgan Kaufmann, San Mateo, CA (1991) 963–969. ("NIPS" stands for *Advances in Neural Information Processing Systems*.)

30. A. van Ooyen and B. Nienhuis, Improving the convergence of the back-propagation algorithm, *Neural Networks* **5** (1992) 465–471.

14

Last Things

A little inaccuracy sometimes saves tons of explanation.

—Saki (1924)

Introduction

Final chapters usually tie up loose ends, but this one creates them: It begins discussions on a variety of topics in AI. Since they're almost independent, you can pick and choose pretty much at will. Here's a brief guide by sections.

1. I discuss optimization again. Two additional methods are presented: genetic algorithms and hidden Markov models (HMMs). Reserachers have applied genetic algorithms to a variety of problems. The less important HMMs are used primarily in speech processing.

2. Learning is the process of modifying the knowledge base as new information becomes available. I discuss symbolic learning nonmathematically and briefly introduce the less important but highly mathematical topic of learning theory.

3. I give a brief, nonmathematical introduction to planning. On one level, planning is simply a form of reasoning; on another, it's a separate area because of its concern with time and with modifying reality.

4. This section has a brief nonmathematical introduction to natural language in both written and spoken forms.

5. The book draws to a close with a very brief discussion of robotics, especially vision and motion planning. Although mathematics is important in these areas, I don't have the space or inclination for a mathematical discussion.

Prerequisites: Although material from a variety of previous chapters is alluded to, deep familiarity with it is unnecessary. The main reference is to Chapter 11 in the discussion of optimization.

14.1   Optimization Again

Optimization is the process of selecting something "desirable" from a set of "possibilities." Some definitions of "desirable" are

- Best: This is called *global optimization*.
- Better than its neighbors: This is called *local optimization*.
- Good enough: This is called *satisficing*.

Search methods depend on the definition of desirable as well as on the structure of the set of possibilities. We frequently have a complicated function $f : D \to \mathbb{R}$, where D is the set of possibilities and the best choices are those for which f is largest (or smallest).

Why do search methods work? Let's review some methods of search that we've discussed in previous chapters.

- Decision trees (Chapter 2) give rise to sequences. In the simplest case, a sequence of T's and F's describes a path from the root to a leaf l and f gives the value at l. In Section 13.3 we briefly discussed partial tree search—how to look for good paths without exploring the entire tree. Partial search extends those paths that are potentially the most promising because of either

  (a) larger estimated values of f at leaves reached by extending the path or

  (b) greater uncertainty in the estimates.

  If we want to view this as exploring D, we need to enlarge D to include paths from the root to internal vertices. The problem of partial search is then twofold: How should f be defined for nonleaf vertices and how should this extended f be used? See Section 13.3 (p. 570) for some further discussion.

- Hopfield nets were discussed in Section 11.1. D is finite—it's just a list of the values assigned to the vertices of the net. Unlike partial search, Algorithm 11.1 (p. 434) starts at a particular point in D and chooses a better nearby point. "Nearby" here means differing in the value assigned to precisely one of the vertices in the net.

- We discussed $f : \mathbb{R}^n \to \mathbb{R}$ in connection with neural nets in Section 11.4. In this case, D is infinite and so it's impossible to explore all of D. The various hill-climbing methods in Section 11.4 conduct partial searches under the assumption that the surface determined by plotting $f(\vec{x})$ against $\vec{x}$ is reasonably smooth.

What makes it reasonable to conduct a partial search?

Where we are in D lets us obtain information about other nearby places in D. Based on this information, we either favor or shun these places:

- In partial tree search, $f(v)$ gives information about the best we can do by looking at leaves reached through v. If this is good, we expand v; if it is not, we abandon v.

- In Hopfield nets, we simply look around and make a single change that improves our situation.

- For functions $f : \mathbb{R}^n \to \mathbb{R}$, smoothness of the surface allows us to use the shape of the surface at $\vec{x}$ to estimate f at points near $\vec{x}$.

For other examples, see simulated annealing (Example 11.9 (p. 469)) and genetic algorithms (discussed below). Unfortunately this information requirement is vague and insufficient. No one has been able to formulate a more precise and/or stronger requirement that can be checked with reasonable effort.

The previous paragraph deals with *local* properties—being at $x \in D$ should provide information about f at certain other nearby values. *Global* properties are also important and hill-climbing methods are usually faced with a tradeoff problem between time spent on local and global aspects of the problem:

- In partial tree search, it may happen that another path leads to better leaves because $f(v)$ is only an *estimate* of how well we can do on paths through v. To avoid this, we might explore many of the more promising paths.

- In both the Hopfield net algorithm and gradient descent methods it is possible to get stuck in a local minimum because the algorithm is exploring only one region of D. This is the problem of escaping from local minima.

Escaping from Local Minima

Suppose we want to minimize a function $f : D \to \mathbb{R}$ and have an algorithm for attempting to do so. After using the algorithm, we find a point m such that $f(x) > f(m)$ for all neighbors x of m. There may be some point y further from m for which $f(y)$ is much smaller than $f(m)$. In other words, m is a local minimum of f but not a global minimum. We'd like to find y, but we have no hope of doing so unless we can escape from the local minimum m. Here are some ways to modify an algorithm to provide escape routes.

- **Multiple Starts:** Restart the algorithm, using a (random) starting value that's near the present position. If the starting value doesn't depend on the present position, you'd just be doing global exploration.

Instead of starting over, you might change the rules of the game:

- **Smoother Functions:** By sacrificing some accuracy, we might replace f with a function f^* that has fewer local minima. It's often very difficult to find such a function that can be computed in a reasonable time. If you're able to find an f^*, you could start your search with it and then gradually shift to f as a local minimum is reached. For example, use the function $\lambda f + (1 - \lambda)f^*$, starting with $\lambda = 0$ and increasing λ to 1 in a series of steps.

- **Delete Minima:** Having found m, continue the algorithm using the function $f(x) + g(x; m)$ instead. The function g should be designed to "cancel out" the minimum at m without disturbing the function too much. When $x \in \mathbb{R}^n$, a possible choice is $g(x; m) = C \exp(-A|x-m|^2)$. The constant $A > 0$ controls how local the modification is—the larger A is, the more local the modification. The constant $C > 0$ controls how strong the modification is—the larger C is, the more the function is distorted near m. When you encounter a local minimum using the modified f, you do two things. First continue minimization using the original f so you find a true local minimum. Then escape from the new local minimum by adding another g term to f.

- **Bigger Neighborhoods:** Suppose that D is finite and you look at points that are near the current point. When at a local minimum m, redefine the notion of nearby points in D (neighbors of m) so that m has more neighbors. With a big enough neighborhood you're bound to succeed, but larger neighborhoods usually make the algorithm look longer for its next step. As a result, implementations of this approach usually depend heavily on the nature of f.

Instead of changing the rules, you could allow some uphill movement. Here are two methods that imitate physical processes.

- **Simulated Annealing:** Suppose D is finite and you look at the nearby points There's a probability of moving to a neighbor even if it's worse. The chance of doing so decreases with running time and with how much worse the neighbor is. See Example 11.9 (p. 469) for more information on using this for discrete minimization.

- **Momentum:** Suppose $D = \mathbb{R}^n$. A marble rolling on a surface changes direction slowly because of momentum. The marble can roll uphill by converting some of its kinetic energy to potential energy. It eventually comes to rest because friction dissipates its energy. This can be simulated with $f(x)$ being the height of the surface at x. See Example 11.6 (p. 461) for further discussion.

Genetic Algorithms

It may be easier to evolve virtual entities with intelligent behavior than to design and build them.

—Karl Sims (1994)

The variety, complexity, and abilities of living organisms demonstrate that evolution is a flexible method for satisficing—obtaining solutions that are good enough. The idea behind genetic algorithms is simple—mimic some of the processes that occur in evolution in hopes of reaping similar benefits. To do that we need a (simplified) understanding of the processes that make evolution possible.

Living organisms store information on chromosomes, each of which is a linear array of many genes. These are built from an "alphabet" consisting of four chemical "letters" that geneticists denote by A, G, P, and T. The genes are used to produce chemicals that create the organism, often through complex interactions. Interactions among individuals and between individuals and the environment determine the number of offspring (if any) an individual produces. This ability to produce offspring is called *fitness*. Genetic information is modified in a variety of ways. The most important are sexual reproduction (combining genes from two organisms), mutation (modifying a gene), and crossover (exchanging portions of two chromosomes).

Here's how we might use a genetic algorithm to help us design a neural net. The fitness of a net is some decreasing function of its error rate. If nets of different complexity are allowed, fitness is also a decreasing function of complexity. If we already know the structure we want the net to have, the complexity is fixed and the genes describe the numerical parameters. On the other hand, if the structure is to be determined, the complexity is variable and the genes describe the net's structure. Fixed structure is simpler than variable because it's far from obvious how we should describe structural information "genetically." However, it's not even clear how we should encode numerical parameters such as edge weights. Although finding good genetic descriptions is the most important part of designing a genetic algorithm, we won't discuss that here.

Let's look at genetic algorithms from an algorithmic viewpoint. What is the function we want to maximize? In the simplest case, its domain consists of all n-long sequences of zeros and ones. In other words, each "organism" is described by a single, fixed-length chromosome whose alphabet has two letters. Each organism can be thought of as the current step in a search. The function maps the domain to a real number—the fitness. Our goal is to maximize its value. A genetic algorithm can be described as follows.

Algorithm 14.1   Genetic Algorithm

The following algorithm seeks local maxima for a function from n-long sequences to the real numbers:

1. **Start:** Construct a collection (the population) of domain elements (the individuals).

2. **Search (global part):** By comparing the function values (fitnesses), determine how many offspring, if any, each organism will have in the next generation. This allows us to pursue the most promising searches in parallel: It decides which "organisms" to focus on, and all those organisms are studied together.

3. **Search (local part):** This modifies organisms to produce others which are, in some sense, near them.

   - Carry out sexual reproduction. This can be accomplished in various ways. One possibility is crossover: If $\vec{a}$ and $\vec{b}$ are the sequences of the two parents, then possible sequences for the offspring are

   $$a_1,\ldots,a_k,b_{k+1},\ldots,b_n \text{ and } b_1,\ldots,b_k,a_{k+1},\ldots,a_n$$

   Another possibility is recombining chromosomes: Partition the n-long sequences into "chromosomes" and choose the offspring's ith chromosome to be the ith chromosome of one of its parents. This is a type of multiple crossover.

   - Before, after, or during reproduction, modify individuals. The simplest method is by randomly changing some sequence elements (mutation).

   This step is vague because there are a variety of ways to carry out reproduction and modification.

4. **Iterate:** Either decide to terminate the algorithm or go to Step 2 using the new population.

Random number generators are used in Step 3 and often is Step 2.

The various modifications that occur in Step 3 are a form of random search in the neighborhoods of high-fitness domain elements. For such a search procedure to work, we must have reason to expect that modifications of good elements have a reasonable chance of producing better elements. A "reasonable chance" can be quite small; for example, a mutation in the biological world has a very low probability of producing a more fit individual.

The algorithm pursues many possibilities in parallel, with more time spent on the more promising organisms because they reproduce more often. This is like a best-first search except that more than one alternative is pursued simultaneously. Since such extensive parallelism runs slowly on a serial computer, why not simply implement a best-first search? The use of sexual reproduction in Step 3 allows the possibility that good results from different locations may combine to provide a significant improvement. Best-first search doesn't allow that.

Since the success of a search algorithm depends on how well the algorithm is suited to the function being maximized or minimized, we should ask, "What sorts of functions on sequences are good for genetic algorithms?" One answer is provided in the preceding paragraph: It should be possible to combine parts of two good solutions to obtain a better solution. For other answers, we need to look at genetic algorithms mathematically. Holland's original work [28] contains two basic mathematical observations, one of which we'll discuss.

A *schema* (plural: *schemata*) is a genetic pattern; for example, it might be the pattern $S = 01?1??1$ in a 7-long sequence with the alphabet $\{0,1\}$. The symbol ? indicates an irrelevant position so the example refers to all sequences with a 0 in position 1 and a 1 in positions 2, 4, and 7. Thus, S could be thought of as the set

$$\{0101001, 0101011, 0101101, 0101111, 0111001, 0111011, 0111101, 0111111\}$$

Since four positions matter for this schema, we say that it contains four genes.

Holland's first basic observation is that a schema that causes above-average fitness tends to increase at an exponential rate until it becomes a significant portion of the population. Let's prove this. Assume that:

• There are no recombinations or alterations of genes.

• Initially, a fraction f of the population possesses the schema S and those individuals reproduce at some fixed rate r.

• All individuals lacking S reproduce at some fixed rate $s < r$.

Let N be starting size of the population. After t generations, we have Nfr^t individuals possessing the schema and $N(1-f)s^t$ of the other individuals. Thus, after t generations, the fraction of individuals with the schema is

$$\frac{Nfr^t}{Nfr^t + N(1-f)s^t} = \frac{fr^t}{s^t + f(r^t - s^t)} = \frac{f(r/s)^t}{1 + f((r/s)^t - 1)}.$$

When f is small and t is not too large, $f(r/s)^t$ will be small and so the above fraction will be about $f(r/s)^t$. In other words, the fraction of the population having the schema increases exponentially at a rate r/s.

What happens when other schemata affect reproduction rates? Instead of r and s, we have a whole range of reproductive rates to consider. Suppose S improves fitness; that is, if two organisms O_1 and O_2 differ only in that O_1 contains S, then O_1 has a higher reproductive rate than O_2. In this case, it's still possible to derive an exponential growth result.

What happens when mutations and crossovers occur?

Let $p_m(S)$ and $p_c(S)$ be the probabilities that schema S is destroyed by mutation and crossover, respectively, and suppose that these are independent and small. The probability that the schema in an individual is *not* destroyed is $(1 - p_m(S))(1 - p_c(S))$. Thus M individuals with the schema produce only $Mr(1 - p_m(S))(1 - p_c(S))$ like individuals instead of Mr. Of course, some of the general population might mutate or crossover to produce the schema. Such events are usually very rare and can usually be neglected. You should now be able to continue as in the previous paragraph and conclude that we have an exponential growth rate of about $\rho = (r/s)(1 - p_m(S))(1 - p_c(S))$. This is less than r/s. There's a tradeoff here:

> In effect, exponential growth of favorable schemata consolidates genetic information that is known to be good. Mutation, crossover, and other genetic modifications are necessary for exploring the space to find good genetic information, but they interfere with consolidation.

Thus we want relatively high mutation and crossover rates (to ensure rapid exploration) together with relatively small values for $p_m(S)$ and $p_c(S)$ (to make ρ large). How can we minimize this conflict?

Suppose the alphabet is $\{0, 1\}$ and let $n(S)$ be the number of zeros and ones in S. For example, $n(01?1??1) = 4$. Let p_m be the probability that a particular gene mutates and suppose that the probability of mutation is independent for different positions in the gene sequence. You should be able to show that

$$1 - p_m(S) = (1 - p_m)^{n(S)}.$$

Hence, the smaller $n(S)$ is, the better the mutation conflict is resolved.

$*$ $*$ $*$ Stop and think about this! $*$ $*$ $*$

Let $l(S)$ be the number of genes between the first and last non-? gene in the schema S. For example, $l(\ldots??01?1??1??\ldots) = 6$. Let p_c be the probability that a crossover occurs at a particular point, and suppose that probabilities of crossovers are independent. Then

$$1 - p_c(S) = (1 - p_c)^{l(S)}.$$

Hence, the smaller $l(S)$ is, the better the crossover conflict is resolved.

In summary, we want a schema to contain relatively few genes (the value of $n(S)$) and to be relatively short (the value of $l(S)$). Furthermore, for the schemata concept to be meaningful, the design should be modular. What does this mean? Simply that the fitness of a gene pattern is roughly determined by some sort of fitness measure on the good schemata that it contains. That's about it for mathematically based practical insights. Like neural network design, genetic algorithm implementation is still more of an art than a science. Some design heuristics can be found in the literature. It's quite possible that future mathematical research will provide other insights in this difficult area.

*Hidden Markov Models

The application of hidden Markov models (HMM) in AI is rather limited. They've been used in speech analysis and, more recently, in other areas of language learning. Since they're easily explained, we'll do so here in case you run across them later.

Let $\mathcal{X}$ be a finite set. Let $X_1, X_2, \ldots,$ be a sequence of random variables that lie in $\mathcal{X}$ and suppose that

$$\Pr(X_n \mid X_1, X_2, \ldots, X_{n-1}) = \Pr(X_n \mid X_{n-1}) \text{ for } n > 1; \qquad (14.1)$$

that is, the probability distribution for X_n depends only on its immediate predecessor. In this case the sequence is called a (stationary) Markov chain and a simple computation with probabilities gives

$$\Pr(X_1, X_2, \ldots, X_n) = \Pr(X_1) \prod_{k=2}^{n} \Pr(X_k \mid X_{k-1}).$$

If a sequence of random variables associated with a situation of interest is assumed to be a Markov chain, we speak of a Markov model for the situation.

A *hidden Markov model* (HMM) is a situation in which the random variables X_i in (14.1) are not observable. Instead, we see other random variables Y_i where the Y_i are in some finite set $\mathcal{Y}$ and

$$\Pr\big(Y_k \mid (X_1, X_2, \ldots, X_n) \wedge (Y_1, \ldots, Y_{k-1})\big) = \Pr(Y_k \mid X_k)$$

for all n and k with $n \geq k \geq 1$. In other words, the probability distribution for the kth observed random variable Y_k depends only on the value of the kth hidden random variable X_k. Since we observe the Y_i's, we need to know how to compute their probabilities. A simple probability computation gives

$$
\begin{aligned}
\Pr(Y_1, \ldots, Y_n) &= \sum_{X_1, \ldots, X_n} \Pr\big((Y_1, \ldots, Y_n) \wedge (X_1, \ldots, X_n)\big) \\
&= \sum_{X_1, \ldots, X_n} \Pr(Y_1, \ldots, Y_n \mid X_1, \ldots, X_n) \, \Pr(X_1, \ldots, X_n) \\
&= \sum_{X_1, \ldots, X_n} \left(\prod_{k=1}^{n} \Pr\big(Y_k \mid (X_1, \ldots, X_n) \wedge (Y_1, \ldots, Y_{k-1})\big) \right) \\
&\qquad\qquad \times \left(\Pr(X_1) \prod_{k=2}^{n} \Pr(X_k \mid X_{k-1}) \right) \\
&= \sum_{X_1, \ldots, X_n} \Pr(Y_1 \mid X_1) \, \Pr(X_1) \\
&\qquad\qquad \times \prod_{k=2}^{n} \Pr(Y_k \mid X_k) \, \Pr(X_k \mid X_{k-1}).
\end{aligned}
$$

The last equation involves summing $|\mathcal{X}|^n$ products, which is computationally intractable for large n. There is a recursive formula that reduces the work to a polynomial in n. It can be found in discussions of HMMs.

In most applications $\Pr(Y_k \mid X_k)$ and $\Pr(X_k \mid X_{k-1})$ are treated as unknown probabilities that should be chosen to maximize $\Pr(Y_1, \ldots, Y_n)$, the probability of the observed sequence. Since $\mathcal{X}$ and $\mathcal{Y}$ are finite, we have a finite number of nonnegative unknowns to determine. We also have the constraints $\sum_{X'} \Pr(X' \mid X) = 1$ and $\sum_Y \Pr(Y \mid X) = 1$. In other words, we have a *constrained* optimization problem. Various methods have been proposed for finding probabilities that give a (local) maximum for $\Pr(Y_1, \ldots, Y_n)$. They require computing the gradient, which can be done in polynomial time in n by a method based on the recursive formula mentioned earlier.

14.2   Learning

> *The ability to learn, to adapt, to modify behavior is*
> *an inalienable component of human intelligence.*
> *How can we build truly artificially intelligent*
> *machines that are not capable of self-improvement?*
>
> —Jaime G. Carbonell (1990)

> *A wise man changes his mind; a fool never will.*
>
> —Spanish proverb

The term "learning" denotes a useful modification of a knowledge base by an organism or a computer program. In previous chapters, symbolic knowledge bases were static and subsymbolic pattern classifiers had rather limited learning: We viewed classifiers as proceeding from a state of no knowledge to a state in which they learned a specific situation, namely the particular set of patterns and classifications they were supposed to learn. How can we program continuing education? In other words, how can we design a system that is able to modify itself to learn additional material without forgetting what it already knows?

Some ideas for growing larger neural networks that were discussed on page 567 can be used to design nets with more general learning abilities. Similar ideas can be applied to decision trees. Genetic algorithms are another tool for growing subsymbolic classifiers as the learning tasks emerge. Quite a bit of research is being done in these areas, but we'll discuss only symbolic learning methods.

How knowledge is represented and manipulated determines the strengths and weaknesses of any learning system. Most symbolic representations are based on variants and/or extensions of the methods discussed in Chapter 6. The more expressive the representation and the more varied the possible manipulations, the greater the *potential* power of the system and, unfortunately, the greater the likelihood of computational bottlenecks. We've run into this again and again—Gödel's incompleteness results for logic, the intractability of general Bayesian nets, and the difficulty of training large neural nets. Theory and experiment both tell us that completely general reasoning is impossible (Gödel) and even somewhat restricted reasoning is too difficult (NP-hardness theorems and neural net results). Hence, compromises are inevitable. Humans are an existence proof that reasonably general learning systems are possible. Unfortunately, the proof doesn't provide an algorithm for constructing non-biological systems.

Let's look at the different types of learning. Imagine a machine translation program that queried the user for information whenever a new word was encountered. The program would then store the information supplied by the user in its data base. This is an example of *rote learning*, the simplest form of learning. More interesting forms of learning require manipulation of the data—i.e., reasoning. Researchers distinguish two types of such learning:

- **Deductive**: Also called *analytic learning*, this includes all conclusions that are a consequence of information in the knowledge base. A classic example of this is mathematics.

- **Inductive**: Also called *synthetic learning*, this includes all conclusions that rely on some assumptions beyond the information in the knowledge base. A classic example of this is the formation of scientific theories.

Truth maintenance systems are closely associated with learning. These are methods for resolving conflicts that arise when new information is incorporated into a knowledge base.

The previous discussion dealt with *how* to learn. Instead, we might ask *what* can be learned. Work on this question is called *learning theory* and is heavily mathematical.

Inductive Learning

"Induction" has many meanings. There is the notion of mathematical induction, which is closely related to recursion—see page 40. From the point of view of logic, mathematical induction is deductive reasoning because the conclusions must be true if the hypotheses are true. *This is not the sort of induction we're interested in here.*

In logic (and learning and reasoning) *induction* refers to a process that reaches conclusions that aren't necessarily true. The main forms of induction are generalization and analogy.

- **Generalization**: In this type of reasoning we seek a statement α such that some known facts are specializations of α. Preschool children use generalization to learn some rules of grammar. Sometimes their generalizations are wrong and so they make errors, as in "I swimmed in the pool."

- **Analogical Reasoning**: Suppose we know that A entails B and we observe that A and A' are similar. If B and B' are similar in the same manner, then analogical reasoning tells us that it is reasonable to conclude that A' entails B'. When first learning how to do some type of mathematical word problem (e.g., percentage problems or calculus minimization problems), most people consciously use analogical reasoning. Choosing the proper analogy can be difficult. With time and experience, some people develop a subconscious knack for choosing the right analogies. Other people never do.

Although inductive reasoning methods may fail, they're often the best we have.

How can we program generalization? Suppose that we have rules of the form $\alpha_i \rightarrow \beta$ and we want to construct the *minimum generalization* $\alpha \rightarrow \beta$. A generalization is a statement $\alpha \rightarrow \beta$ such that, whenever some α_i is true, α is also true; that is, $\alpha_i \rightarrow \alpha$ for all i. A generalization $\alpha \rightarrow \beta$ is minimum if, whenever $\alpha_i \rightarrow \alpha'$ for all i, we also have $\alpha \rightarrow \alpha'$. Some conditions—at present unspecified—are imposed on the structure of the α_i's, α, and α'. You should be able to convince yourself that this is a reasonable definition.

*   *   *   Stop and think about this!   *   *   *

Depending on the conditions imposed on the form of α, there is no guarantee that a generalization, much less a minimum one, exists.

If no conditions are imposed on α, we could simply let it be disjunction of the α_i's. What we have then is equivalent to the set of original statements; that is,

$$\Big((\alpha_1 \rightarrow \beta) \wedge (\alpha_2 \rightarrow \beta) \wedge \cdots \Big) \equiv \Big((\alpha_1 \vee \alpha_2 \vee \cdots) \rightarrow \beta\Big). \qquad (14.2)$$

(Prove it!) Hence $\alpha \rightarrow \beta$ is simply a more compact representation of the original formulas—not a generalization.

One popular method of generalization is the *version space* approach. The α_i's and α are required to be conjunctions of literals. In other words, α_i lists some conditions such that β is true whenever all those conditions hold. Let $\mathcal{A}_i$ be the set of literals whose conjunction produces α_i, let $\mathcal{A}$ be the intersection of the sets $\mathcal{A}_i$, and let α be the conjunction of the literals in $\mathcal{A}$. We claim that $\alpha \rightarrow \beta$ is the minimum version space generalization. The proof is left as an exercise. There are at least two objections to the preceding:

- Why use conjunctions of literals? As (14.2) shows, we need to impose some condition. It's often natural to phrase specific rules such as $\alpha_i \rightarrow \beta$ in the form "If A and B and $\cdots$, then C." The constraint that α also be a conjunction of literals is needed if we want to iterate generalization.

- The resulting set $\mathcal{A}$ may not contain enough things for a reasonable statement; indeed, it may even be empty. We need some way of deciding when induction has gone too far—a reality check, so to speak.

How can we program analogy? To do that, we need to be able to measure when two things A and A' are similar, or we need to be told that A is like A'. One way of measuring similarity is by looking at attributes of A and A' or by seeing if there is a common generalization of A and A' that is not "too" general. Using hierarchical systems, discussed briefly in Section 6.5 (p. 234), to represent data may facilitate this sort of reasoning.

Since the early 1980s, *explanation-based learning* (EBL) has become increasingly popular as a learning method. It attempts to explain how conclusions are reached. To do so, EBL progams must obtain a fairly deep understanding of the material being learned. EBL requires very few examples to learn concepts. The knowledge it produces can be used to reduce search time in reasoning and to automatically provide explanations for how an expert system reached a conclusion. *Case-based reasoning* is a form of EBL that facilitates the building of explanations. In this method, the reasoning program studies solved problems ("cases") that are similar to a given problem and uses their solutions to propose a solution for the current problem. This requires analogical reasoning. The method learns from both success and failure. The solved cases provide a ready explanation for the proposed solution. Advocates claim that CBR and EBL are similar to much human problem solving.

Inductive learning, then, is the process of attempting to infer the important aspects of the knowledge so that it can be applied to new problems. Stated this way, inductive learning appears to be a form of pattern recognition. It is.

Truth Maintenance Systems

The main problem with inductive learning is that the learner may come to believe something that conflicts with other information. Resolving such conflicts gracefully and efficiently is the goal of truth maintenance systems.

The simplest approach may be to use a nonmonotonic logic in which all inductive conclusions are capable of being retracted when faced with a contradictory conclusion. We discussed two approaches to this in Chapter 6, default reasoning and defeasible reasoning. Difficulties arise when we want to choose between two contradictory conclusions. The default approach is silent on this issue while the defeasible approach favors the conclusion with the less general hypothesis. Unfortunately, this is insufficient for resolving conflicts like

Dog-like animals are safe to pet.

and

Strange animals are unsafe to pet.

Little Skipper obtained the former by generalization from the observation of pets, while she was told the latter by her parents. What should Little Skipper decide about a strange dog? To answer such questions, we need to know how a piece of information was obtained. Such knowledge not only facilitates conflict resolution; it also helps to isolate faulty assumptions or reasoning. There are a variety of approaches to truth maintenance.

Learning Theory

Here's a rough idea of one approach to developing a mathematical theory of learning. Imagine a collection $\mathcal{H}$ of possible things to be learned; for example, a collection of logic functions. We call $\mathcal{H}$ the *hypothesis space* and call the elements of $\mathcal{H}$ *concepts*. Given data, we want to learn which concept each comes from. Can we write a program that will be able to nearly determine the correct concept for every element of $\mathcal{H}$? Can the program run in a reasonable time? Here's a complicated mathematical definition:

Definition 14.1   Probably Approximately Correct Learning

Let $\mathcal{E}$ be a probability space and R a set. The hypothesis space is some set $\mathcal{H}$ of random variables on $\mathcal{E}$ taking values in R. Let $\vec{v} \in \mathcal{E}^m$ and $\vec{r} \in R^m$. Suppose we have an algorithm whose input is a sequence $(v_1, r_1), \ldots, (v_m, r_m)$ and whose output is a concept $L \in \mathcal{H}$. For $C \in \mathcal{H}$, let $L_C(\vec{v})$ denote the output of the algorithm when the input is $(v_1, C(v_1)), \ldots, (v_m, C(v_m))$. The idea is that L_C should be close to C if we have enough data. This is expressed as follows.

We say that the algorithm is a *probably approximately correct*, or *pac*, learning algorithm for $\mathcal{H}$ if the following is true. For every positive δ and ϵ, there is a positive integer $m = m(\delta, \epsilon)$ such that, for *every* $C \in \mathcal{H}$,

$$\mathrm{Pr}_m\left(\mathrm{Pr}(L_C(\vec{v}) \neq C) > \epsilon\right) < \delta, \tag{14.3}$$

where Pr is on $\mathcal{E}$ and Pr_m is defined on $\mathcal{E}^m$ by $\mathrm{Pr}_m(\vec{v}) = \prod_{i=1}^{m} \mathrm{Pr}(v_i)$. (See Example 7.16 (p. 286).) We say the algorithm is *efficient* if

- the running time of the algorithm is bounded by a polynomial in m and

- $m(\delta, \epsilon)$ is bounded by a polynomial in $\log(\delta^{-1})$ and ϵ^{-1}.

This is worse than the ϵ-δ definitions from calculus! What does it mean?

*   *   *   Stop and think about this!   *   *   *

$\mathrm{Pr}(L_C(\vec{v}) \neq C)$ is the probability that the target concept C and the random variable produced by the algorithm disagree at points in $\mathcal{E}$; that is, it is the *error rate* when $L_C(\vec{v})$ is used in place of C. Thus the expression in the large parentheses in (14.3) simply says that the error rate exceeds ϵ. Pr_m applies to $\vec{v}$ and simply means that the components are chosen independently at random. Hence (14.3) says that if $\vec{v}$ is chosen randomly, the probability that the error rate exceeds ϵ is less than δ. I won't justify the definition of efficiency.

If $\mathcal{E}$ is finite, the definition is useless. Here's a sketch of the reason. Since knowing $C(v)$ for all $v \in R$ determines the random variable C, it's easy to imagine an algorithm that gives $L_C(\vec{v}) = C$ if every element of $\mathcal{E}$ appears in

$\vec{v}$. The probability that e does not appear in $\vec{v}$ is $(1 - \Pr(e))^m$. Hence the probability that every e appears is at least

$$1 - \sum_{e \in \mathcal{E}} (1 - \Pr(e))^m \geq 1 - |\mathcal{E}| a^m,$$

where $a = \max(1 - \Pr(e))$. It then follows that

$$\Pr_m \Big(\Pr(L_C(\vec{v}) \neq C) > 0 \Big) \leq |\mathcal{E}| a^m,$$

and so we may take $m > \big(\log(\delta/|\mathcal{E}|) / \log a \big)$, which is linear in $\log(\delta^{-1})$ and independent of ϵ. Should we be concerned about finite $\mathcal{E}$? Yes. Many pattern classifiers have inputs taken from $\mathcal{E} = \{0,1\}^n$. One method for dealing with the finite case is by embedding it in an infinite situation. For example, any particular propositional logic formula is finite, but the set of *all* such formulas contains formulas of arbitrary size.

To return to the logic functions alluded to earlier, imagine a collection $\mathcal{P}$ of propositional letters. We have a probability space because some assignments of truth values are more likely to be observed than others. An element of $\mathcal{E}$ is an assignment of T/F values to the letters in $\mathcal{P}$. A random variable in $\mathcal{H}$ is a formula on $\mathcal{P}$ and its value at any point in $\mathcal{E}$ is the truth value of the formula given the assignments of truth values to $\mathcal{P}$. Thus $R = \{\text{T}, \text{F}\}$. For example, if we have assigned meanings to the letters, a random variable might be a formula that expresses the condition that something is alive ("It takes nourishment and it reproduces and ...").

The pac learning concept has been criticized on several grounds. Here are some:

- As just discussed, Definition 14.1 is useless in finite probability spaces, but learners are only faced with finite situations.

- In the real world, we are frequently learning several interacting concepts simultaneously, but pac learning deals with single-concept learning.

- Learning theory looks only at the reactions of the learner (the value of $L_C(\vec{v})$) rather than at concept formation, which is the heart of learning.

We won't discuss these issues here. You might like to debate them in class.

Learning theory hasn't produced any useful learning algorithms. Instead, it's helping researchers to better understand what's doable—some things are not efficiently pac-learnable. Developments in this area may make it necessary for researchers who are attempting to develop practical learning systems to revise their concepts and/or goals.

14.3 Planning

We discussed symbolic reasoning in Chapters 3–6, but we paid little attention to a special form of reasoning called *planning*. To plan, we must analyze the present situation and devise a strategy for achieving a goal. As the strategy is carried out, new information may cause the planner to modify the strategy. For example, suppose I am visiting San Diego and want to attend a meeting that is being held in Los Angeles. I also want to give some books to an L.A. colleague who will be attending the meeting. Since flying is fast, I plan to fly to Los Angeles and then take ground transportation from the airport. Because the books are heavy, I decide to ship them by UPS instead of carrying them in my luggage. The first step in my plan is to find out what ground transportation to use. While seeking the latter information, I discover that the meeting is far from the airport. Since this makes the ground transportation both time-consuming and expensive, I consider modifying my plan. I decide it is faster and perhaps cheaper to rent a car and drive the eighty miles from San Diego to the meeting. I also modify my plan for shipping the books since I can easily carry them in the car. After calling some companies, I choose the best deal and reserve a car. Having made my plans, I call a colleague in New York who is also planning to attend the meeting. After discussing transportation with her, I decide to fly in and share the car she plans to rent at the airport. I cancel my car reservation, make an airline reservation, make a shuttle reservation for getting to the San Diego airport and arrange to ship the books by UPS. This scenario illustrates various features of planning, some of which are:

- **Top-Down Approach:** I worked on the general plan before attending to details such as making reservations. Furthermore, the order in which reservations were made differs from the order in which they will be used—I reserved the flight before the shuttle.

- **Interaction:** My travel and book transporting plans interacted. Sometimes this interaction can set up interferences that a planning system must resolve. A famous example is the Sussman anomaly. Suppose we can move one block at a time and can move only the top block in a pile. We are given the two piles

$$\begin{array}{|c|}\hline A \\\hline B \\\hline\end{array} \text{ and } \boxed{C}$$

and have two goals: we want C on A and we want want B on C. Attempting to achieve either goal separately interferes with the other; however, they can be achieved by "cooperative" action. (Think about why these claims are true.)

- **New Knowledge**: Actions produce reactions in the world and so lead to new knowledge that can significantly affect a plan. First I changed my plan because of new knowledge that I sought concerning the relative locations of the meeting and the airport. Then I changed it because of knowledge gained incidentally by discussing the meeting with my colleague.

- **Time and Change**: Planning leads to actions in the real world and deals with situations where time is important—two issues we haven't dealt with. We've just alluded to some problems on page 190 and mentioned temporal logics in passing.

Planning has been of interest to AI researchers for many years. Considerable progess has been made, but much remains to be done.

14.4 Language and Speech

> *Marvin Minsky has characterized machine translation*
> *as the most typical and hardest AI problem ....*
> *[It] promises, but will it ever deliver?*
>
> —Klaus K. Obermeier (1994)

Natural-language understanding (NLU) and machine translation (MT) were early goals of AI research. By the mid-1960s, the research was generally considered a failure. In 1971, Winograd's SHRDLU program proved that natural-language understanding is possible if the domain is sufficiently restricted. Today, a variety of programs exist for processing natural language and speech. They are still far from achieving the capabilities of a typical seven-year-old child.

Processing natural language presents a variety of problems. Here are some major ones.

- **Grammatical Errors**: Conversation is filled with grammatical errors, which we ignore or correct through the use of common sense and context. Some grammatical errors, such as misplaced modifiers, are so common that they're more like ambiguities.

- **Ambiguity**: Resolving ambiguity is a central problem in natural-language processing. A simple, commonly used example of ambiguity is "Time flies." It may be a statement about temporal phenomena or it may be a command to make temporal measurements of certain insects. Grammatical errors, such as misplaced modifiers and pronouns and phrases having uncertain antecedents, often cause ambiguities. Ambiguities at one level can often be resolved at a higher level; for example, ambiguities in speech sounds could be resolved by syntactic constraints while syntactic ambiguities could be resolved by semantic constraints.

- **Common Sense:** Common sense plays an important part in our use of natural language. It helps us resolve ambiguity—you just did that in the first part of this sentence by deciding that "it" did not refer to "our use of natural language." Common sense overrode the grammatical rule that a pronoun's antecedent is normally the last preceding noun phrase. Researchers have had limited success incorporating common sense in software.

- **Context:** Common sense is a form of context. In addition, each natural-language usage has its own context, which is not always easily described. "I'll kill them" has different meanings when spoken by

  - a mystery writer about characters in a novel she's writing,

  - a teenager who has just been frustrated by parents,

  - a gangster in reference to gang members who became informers, and

  - a comic referring to the audience he's about to entertain.

- **Noise:** The noise level is usually fairly low in written language, and "iz usely easu two over come." In contrast, speech often has a high level of noise. For example, people frequently carry on conversations in a room where many other people are also conversing.

Computer languages and compilers deal with these problems by defining them out of existence: An unambiguous grammar is defined and any errors in using it are the programmer's problem—the compiler needn't try to resolve them.

Understanding Language

A complete understanding of natural language would presumably lead to the ability to manipulate the information content. This is a difficult problem: Commercial natural-language interfaces can usually do a limited amount of manipulation. More is required to extract information from news articles. Programs that do this are currently rather limited in scope. The need to manipulate information content has hampered the development of learning systems having sophisticated knowledge bases.

Although some approaches to NLP (natural-language processing) are unsophisticated, others make use of (a) concepts from linguistics, (b) lexicons that incorporate syntactic, semantic, and commonsense information about word usage, and (c) statistical information.

Syntax is associated with a grammar, which is something that provides structural information about what constitutes a legal sentence. This information is often phrased in terms of rules, frames, or networks, which were discussed in Chapter 6. A typical rule for English might look like

{sentence} $\longrightarrow$ {noun pharase} {verb phrase} {noun phrase} except that more information is needed. For example, the first noun phrase is a subject and so cannot contain pronouns such as "them." For another, the subject and verb must both be either singular or plural. These sorts of constraints can be incorporated by expanding the grammar in various ways. There are more subtle constraints as well. For example, the verb "to feel" requires animate subjects. These constraints require basic semantic information about the vocabulary.

A parser is any program that uses a knowledge base of syntactic information to determine the grammatical structure of a string of words. There will often be more than one syntactic interpretation of a phrase or sentence. This ambiguity can sometimes be resolved by an augmented parser that is able to handle limited contextual information and some basic semantics (meaning). More extensive knowledge of the real world and/or the context in which the ambiguity appears is often required. Hence the parsing and understanding levels of a sophisticated natural-language program must interact.

Why is it that people are so adept at language understanding and computers are, comparatively, so inept? We don't know; however, it appears that human brains contain an innate specialization for language. If we knew how this specialization worked, the problem of language processing would probably make a great leap forward. Similar comments apply to vision.

Recognizing Speech

One of the simplest speech-recognition tasks is developing a system to recognize a limited set of syllables spoken by a particular individual in a nearly noise-free environment. One of the hardest is developing a system to transcribe someone's speech in a situation (e.g., a party) where many conversations are going on simultaneously. Researchers have progressed well beyond the former and are still far from the latter. Current speaker-independent speech-recognition systems with restricted vocabularies perform fairly well.

There are commercial programs that do a reasonable job of transcribing clearly enunciated speech. How is this done?

Since pitch, inflection, duration, and so forth vary from one person to another, a model used for speech must contain some parameters that are adjusted to the situation and speaker in some manner. Hidden Markov models and neural nets are two methods that are used. Since such models are only as good as their inputs, preprocessing is used in various degrees to extract relevant information from the acoustic data. This is not enough. Such systems can only provide lists of likely transcriptions. Syntactic and some semantic information must be used to resolve the ambiguities. One possible approach is to make crude estimates of the probabilities of various phonemes (basic units of speech) in the auditory processing. This information would be combined with linguistic requirements and common sense by the rest of the system to produce a transcription.

Humans behave in a similar manner. It's been estimated that we often acquire only 70% of a conversation by direct auditory input. The "blanks" are filled in by our understanding of what is being said. This critical interaction with natural-language understanding is usually subconscious. It becomes conscious only when we are faced with a particularly difficult situation or when our subconscious choice is obviously wrong.

To read the speech recognition literature, you'll find that some knowledge of probability, statistics, and signal processing is useful.

Machine Translation

A variety of machine translation (MT) programs exist. Some even provide real-time translation of speech. Current MT programs have rather limited vocabularies and/or require human intervention to "clean up" the translation. Some approaches to machine translation attempt to go from the source langauge to the target language without "understanding" the text. In contrast, "interlingua" approaches are based on understanding the text.

14.5   Robotics

> *[C]omputer vision made very little progress. However, simplifying the scenes to be viewed, essentially to a world of blocks with uniformly colored faces, did allow for rapid progress. ... An intellectual trap had been sprung, and it ensnared computer vision researchers for the next twenty years.*
>
> —Rodney A. Brooks (1992)

Most researchers classify robotics as a part of AI, but some classify it as a separate discipline. It's a major research area with an extensive literature. The mathematics used in robotics falls primarily into two categories: (a) applications and adaptations of methods commonly used in other areas of AI and (b) applications specific to robotics, such as the geometric theory used in some aspects of vision research. It seems better to leave these to more specialized textbooks than to attempt to include them in an already long text. Consequently, our discussion will be exceedingly brief.

What is robotics? It includes everything from simple assembly-line automation to the self-aware, mobile machines of science fiction. The ultimate goal of robotics is the creation of *autonomous* devices that, when told *what* to do in everday speech, have the "mental" ability to plan *how* to carry out the task in an ordinary environment and have the "physical" ability to implement the plan. Constructing such devices requires background in many areas, including the following:

- **Engineering**: In addition to producing off-the-shelf computer hardware, engineering is needed to design equipment for sensory input, locomotion, and manipulation.

- **Sensory-Input Processing**: Vision is the most studied input modality. Robots might also use active "vision" such as radar and sonar which automatically provide distance and speed information. Robots may also have tactile input based on pressure sensors. Not only must a robot extract information from sensory input, it must combine input from multiple sensors.

- **Planning**: Given a task to accomplish, the robot must decide *how* to do it. This requires the general-purpose planning discussed in Section 14.3. It also requires "motion planning"—how to get all or part of the robot from here to there. Because it's needed even for simple robots such as assembly-line arms and because it presents its own peculiar difficulties, motion planning is a major area of robotics research distinct from general planning.

- **Speech Processing and NLU**: Getting from auditory input to a task description requires speech processing and some understanding of natural language. Of course, we might design a robot that accepted only limited, easily understood input.

- **Learning**: To function efficiently, a robot should be able to learn from its previous planning and execution efforts. It should also be able to learn from instruction such as "Keep off the grass" or "Here's how to use a hammer."

In addition, a robot must be able to quickly perceive and react to unexpected danger. This brings to the forefront the tradeoff between thinking longer and acting quickly illustrated in Figure 1.1 (p. 15) but largely ignored since then. Robotics cannot afford to ignore it.

Vision

A considerable portion of the human nervous system is associated with vision. Significant processing begins before the nerve impulses even leave the eye. With such warning signs, we might expect vision to be a difficult problem. And so it is. A variety of mathematical techniques have been brought to bear on the problem. They've been applied at all stages; for example,

- detecting edges in early processing,
- obtaining shape information from the location of edges,
- recognizing an object regardless of distance and orientation.

The richness and the paucity of input data are two central difficulties in vision. Visual input provides a large amount of data that contains extensive information. This makes it essential to have efficient methods for extracting the feature information needed for the task at hand. Unfortunately, visual input is also poor in data: It seldom contains this needed information in easily accessible form and may lack information needed to resolve ambiguities. For example, boundaries of objects must be inferred indirectly and the three-dimensional nature of the world must be reconstructed from two-dimensional images. As in natural-language processing, vision processing suffers from ambiguity at all levels of processing.

Apparent progress was made by limiting attention to artificial worlds of simple shapes. This reduced input richness and also made it much easier to extract the information needed to reconstruct the three-dimensional scene.

These early systems often took whatever information was available from a single picture of the scene and tried to recover as much information about the scene as possible. In a newer approach, termed *active vision*, the observer moves and processes the data with a task in mind and may also move to obtain additional input. Task orientation allows us to ignore some of the richness while motion provides data that makes it easier to extract certain information. To contrast the two approaches:

- The goal of the early method, "general vision," is to completely reconstruct the scene.
- The goal of active vision is to obtain task-specific information about the scene.

Because the goal of the active-vision approach is more limited, it is able to deal with more complicated (i.e., realistic) scenes.

A general-purpose robot is likely to combine active-vision methods with active probes such as sonar and coherent radiation. Since the robot controls the active probes, it can more easily gain information from them than it can from passive illumination.

Motion Planning

Motion planning could be defined as the process of deciding what movements
a robot should carry out to achieve a task. This would include long-range
planning, such as getting a box of paper from the stockroom and delivering
it to the secretarial pool. It would also include shorter-range planning such
as extracting the box of paper from behind a pile of other supplies. Longer-
range tasks may involve general planning, but shorter-range tasks often belong
entirely to the field of motion planning. Such planning must take into account
the ways in which the robot can move, the need to avoid collisions, and the
physical properties of the world (e.g., gravity).

Action may begin while a motion plan is incomplete. The plan is filled
in and modified as the task is being carried out. One reason for this may be
a lack of information; e.g., we may not know if a door is locked until we try
to open it. Another is the complexity of the planning problem—it may be
more efficient to plan only the initial stages in detail and adapt the plan as
additional information is available. Another reason is the inaccuracy of robot
motion. For example, if a manipulator must be moved 100 cm with an error of
0.1 cm in its final position and the accuracy of motion is only 1%, corrective
motion will be needed as the manipulator approaches its target.

The motion-planning problem lies in a continuous space: The rate at
which a joint moves can be viewed as a real number; a path from point A to
point B that avoids obstacles can be any one of an infinite number of possible
"curves." (Curves includes paths consisting of sequences of line segments.)
One of the problems in motion planning is to transform the problem from one
with an infinite number of choices to one with a finite number.

Motion planners use computational geometry, search, and other areas of
mathematics.

Notes

From time to time the Association for Computing Machinery (ACM) and the
Institute of Electrical and Electronics Engineers (IEEE) devote issues of their
nonspecialist journals to survey articles written by experts in various fields.
Since the purpose of such issues is to keep members informed, you should find
many of the articles readable and informative. For recent special issues on
AI see [35] and [39]. The roughly annual *Advances in Computers*, edited by
Yovits and intended for a general computer science audience, frequently has
one or more articles on some aspect of AI. As in Chapter 1, I recommend the
texts by Ginsberg [23] and by Russell and Norvig [46].

Blackboard systems are a type of AI system that I haven't even mentioned. The basic idea is that a collection of problem solvers post problems to a "blackboard" and wait for solutions. At the same time, solvers work on posted problems to which they are suited and post solutions. The book [16] is a tutorial on the subject.

Optimization

John Holland introduced the concept of genetic algorithms and published the first book on the subject in 1975 [28]. Additional books began to appear about fifteen years later. These include [17, 18, 24, 33, 36]. A brief introduction is provided in the paper [7] and in Chapter 25 of Winston's text [54]. Another source of introductory material is the selected papers in [10]. Harp and Samad [27] discuss applications to neural nets.

I mentioned that Holland made two mathematical observations. His second deals with the connection between results from statistical decision theory, randomness in genetic algorithms, and exponential growth of good schemata. This interplay suggests that such exponential growth is just what is needed, but it does not say that the rate is correct.

Two introductions to hidden Markov models are Rabiner's paper [43] and Charniak's text [12, pp. 39–73]. The HMMs I introduced are called *first*-order HMMs. In a *j*th-order HMM, the probability distribution of X_k depends on the j previous X values. This general HMM can be reduced to a first-order HMM.

Yuille and Kosowsky [56] review some algorithms that have migrated from physics to neural net optimization.

I imagine you've been convinced by now that optimization is difficult. If not, look at one of the journals on the subject: either the long-standing *Journal of Optimization Theory and Applications* or the more recent *Journal of Global Optimization*.

Learning

Although machine learning was considered important in the early years of AI research, results were disappointing and interest shifted elsewhere. Interest revived in the 1980s and many believe that solid progress is now being made. Many introductory AI texts have some discussion of machine learning. The two collections of papers [9] and [50] have relatively little overlap and the former concludes with an extensive introductory bibliography. Carbonell [11] has edited a collection of papers that were solicited with the goal of providing an introduction to the various approaches to machine learning. An overview of some research in robot learning is presented in [14].

Prieditis [42] edited a volume on various approaches to analogical reasoning. For explanation-based learning, see Chapter 5 of [9] or [11]. For case-based reasoning, see Chapter 7 of [9], the book [32], or the book [45].

For a good introduction to truth maintenance in the context of practical problem solving, see the text by Forbus and de Kleer [21].

Since it takes time to learn and since too much knowledge can slow a system down, it's important to know *when* to learn as well as *how* to learn. Minton [38] explores this problem.

Valiant introduced the notion of pac learning in 1984 [52]. Anthony and Biggs [6], Kearns and Vazirani [31], and Natarajan [41] have written texts on learning theory. You may also find the doctoral thesis [48] interesting, especially the first three or four chapters. It's possible to regard pac learning as a form of pattern recognition. As such, it belongs in one of Chapters 10–13. Wolpert's article [55] is written from this viewpoint.

From time to time, *Machine Learning* and other AI journals have informative special issues that are more up to date than the books I've mentioned.

Planning

Although planning is an active research area, there are relatively few books on this subject. You might read Chapter 14 of [23] or Part IV of [46]. A variety of papers are collected in [3]. Allen et al. [4] and Dean and Wellman [19] have written books on planning. The latter discusses the problem of planning under uncertainty and touches on issues related to robotics. Planning plays an important role in robotics and considerable research has been done on robot motion planning. Constraint satisfaction plays a role in planning and other areas. See Tsang's book [51] for more information about constraint programming.

Language

The collection [53] provides a good introduction to the speech-recognition literature. Although now somewhat dated, [25] is still a good source of information. Good introductory texts to mainstream NLP are available: Allen's [2] is based on Lisp and Covington's [15] on Prolog. The book by Gazdar and Mellish [22] is available in both Lisp and Prolog versions. Charniak's book [12] takes a different approach, focusing on statistical approaches, particularly hidden Markov models. Neural net approaches are discussed by Miikkulainen [37] and in the collection [44].

Jackendoff [30] presents a lively defense of the thesis that language ability is innate.

Robotics

As noted in Section 5.4 (p. 191), combining data from several sources is a major theoretical problem. It plays at least a minor role in many areas of AI and is of critical importance in robotic sensory input and in sophisticated learning. See [1] for a discussion of various methods for dealing with quantitative data.

Many books have been written on robotics and I'm not very familiar with the literature. The two areas with a large amount of literature are vision and motion planning.

Schilling's introductory book [49] deals mainly with motion planning. Latombe [34] also treats motion planning. Murray, Li, and Sastry [40] discuss manipulation, which is a form of motion planning.

Some books on vision are those by Chen, Pau, and Wang [13], Faugeras [20], Haralick and Shapiro [26], Horn [29], and Sarkar and Boyer [47]. The book edited by Aloimonos [5] and the papers in [8] deal with active vision.

References

1. M. A. Abidi and R. C. Gonzalez (eds.), *Data Fusion in Robotics and Machine Intelligence*, Academic Press, Boston (1992).

2. J. Allen, *Natural Language Understanding*, 2d ed., Benjamin/Cummings, Redwood City, CA (1994).

3. J. Allen, J. Hendler, and A. Tate, *Readings in Planning*, Morgan Kaufmann, San Mateo, CA (1990).

4. J. Allen, H. A. Kautz, R. N. Pelavinand, and J. D. Tenenberg, *Reasoning About Plans*, Morgan Kaufmann, San Mateo, CA (1991).

5. Y. Aloimonos (ed.), *Active Perception*, Lawrence Erlbaum Associates, Hillsdale, NJ (1993).

6. M. Anthony and N. Biggs, *Computational Learning Theory: An Introduction*, Cambridge University Press, Cambridge, England (1992).

7. D. Beasley, D. R. Bull, and R. R. Martin, An overview of genetic algorithms: Part I, Fundamentals; Part II, Research topics, *University Computing* **4** (1993) 58–69, 170–181.

8. A. Blake and A. Yuille (eds.), *Active Vision*, MIT Press, Cambridge, MA (1992).

9. B. G. Buchanan and D. C. Wilkins (eds.), *Readings in Knowledge Acquisition and Learning: Automating the Construction and Improvement of Expert Systems*, Morgan Kaufmann, San Mateo, CA (1993).

10. B. P. Buckles and F. E. Petry (eds.) *Genetic Algorithms*, Computer Society Press, Los Alamitos, CA (1992).

11. J. G. Carbonell (ed.) *Machine Learning: Paradigms and Methods*, MIT Press, Cambridge, MA (1990).

12. E. Charniak, *Statistical Language Learning*, MIT Press, Cambridge, MA (1993).

13. C. H. Chen, L. F. Pau, and P. S. P. Wang (eds.), *Handbook of Pattern Recognition and Computer Vision*, World Scientific, Singapore (1993).

14. J. H. Connell and S. Mahadevan (eds.), *Robot Learning*, Kluwer, Boston (1993).

15. M. A. Covington, *Natural Language Processing for Prolog Programmers*, Prentice Hall, Englewood Cliffs, NJ (1994).

16. I. D. Craig, *Blackboard Systems*, Ablex, Norwood, NJ (1994).

17. Y. Davidor, *Genetic Algorithms and Robotics: A Heuristic Strategy for Optimization*, World Scientific, Singapore (1991).

18. L. Davis (ed.), *Handbook of Genetic Algorithms*, Van Nostrand Reinhold, New York (1991).

19. T. L. Dean and M. P. Wellman, *Planning and Control*, Morgan Kaufmann, San Mateo, CA (1991).

20. O. D. Faugeras, *Three-Dimensional Computer Vision*, MIT Press, Cambridge, MA (1993).

21. K. J. Forbus and J. de Kleer, *Building Problem Solvers*, MIT Press, Cambridge, MA (1993).

22. G. Gazdar and C. Mellish, *Natural Language Processing in ****: An Introduction to Computational Linguistics*, Addison-Wesley, Reading, MA (1989). (Replace **** by Lisp or Prolog.)

23. M. Ginsberg, *Essentials of Artificial Intelligence*, Morgan Kaufmann, San Mateo, CA (1993).

24. D. E. Goldberg, *Genetic Algorithms in Search, Optimization, and Machine Learning*, Addison-Wesley, Reading, MA (1989).

25. B. J. Grosz, K. S. Jones, and B. L. Webber (eds.), *Readings in Natural Language Processing*, Morgan Kaufmann, San Mateo, CA (1986).

26. R. M. Haralick and L. G. Shapiro, *Computer and Robot Vision* (2 vols.), Addison-Wesley, Reading, MA (1992, 1993).

27. S. A. Harp and T. Samad, Genetic synthesis of neural network architecture. In [18] 202–221.

28. J. Holland, *Adaptation in Natural and Artificial Systems*, University of Michigan Press, Ann Arbor, MI (1975, 1992).
The only difference between the original and the 1992 reprinting is a brief chapter on what happened in the intervening 17 years.

29. B. K. P. Horn, *Robot Vision*, MIT Press, Cambridge, MA (1986).

30. R. Jackendoff, *Patterns in the Mind. Language and Human Nature*, Basic Books, New York (1994).

31. M. J. Kearns and U. V. Vazirani, *Introduction to Computational Learning Theory*, MIT Press, Cambridge, MA (1994).

32. J. L. Kolodner, *Case Based Reasoning*, Morgan Kaufmann, San Mateo, CA (1992). A nice additional feature is a lengthy appendix with brief information on a wide variety of case-based reasoning systems.

33. J. R. Koza, *Genetic Programming: On the Programming of Computers by Means of Natural Selection*, MIT Press, Cambridge, MA (1992). This massive (819 pp.) book concentrates on implementation.

34. J.-C. Latombe, *Robot Motion Planning*, Kluwer, Boston (1991).

35. C. C. Liu (ed.) *Proc. IEEE* **80** no. 5 (May 1992). A special issue on "knowledge-based systems in electric power systems."

36. Z. Michalewicz, *Genetic Algorithms + Data Structures = Evolution Programs*, Springer-Verlag, Berlin (1992).

37. R. Miikkulainen, *Subsymbolic Natural Language Processing: An Integrated Model of Scripts, Lexicon and Memory*, MIT Press, Cambridge, MA (1993).

38. S. Minton, *Learning Search Control Knowledge: An Explanation Based Approach*, Kluwer, Boston (1988).

39. T. Munakata (ed.) *Communications of the ACM* **37** no. 3 (March 1994). The editor's goal is to provide "an overview of the everyday, practical uses of AI technology."

40. R. M. Murray, Z. Li, and S. S. Sastry, *A Mathematical Introduction to Robotic Manipulation*, CRC Press, Boca Raton, FL (1993).

41. B. K. Natarajan, *Machine Learning: A Theoretical Approach*, Morgan Kaufmann, San Mateo, CA (1991).

42. A. Prieditis (ed.), *Analogica*, Morgan Kaufmann, San Mateo, CA (1988).

43. L R. Rabiner, A tutorial on hidden Markov models and selected applications in speech recognition, *IEEE Proceedings* **77** (1989) 257–286. Reprinted in [53] 267–295.

44. R. G. Reilly and N. E. Sharkey (eds.) *Connectionist Approaches to Natural Language Processing*, Lawrence Erlbaum Associates, Hillsdale, NJ (1992).

45. C. K. Riesbeck and R. C. Schank, *Inside Case-Based Reasoning*, Lawrence Erlbaum Associates, Hillsdale, NJ (1989).

46. S. Russell and P. Norvig, *Artificial Intelligence. A Modern Approach*, Prentice Hall, Englewood Cliffs, NJ (1994).

47. S. Sarkar and K. L. Boyer, *Computer Perceptual Organization in Computer Vision*, World Scientific, Singapore (1994).

48. R. E. Schapire, *The Design and Analysis of Efficient Learning Algorithms*, MIT Press, Cambridge, MA (1992).

49. R. J. Schilling, *Fundamentals of Robotics. Analysis and Control*, Prentice Hall, Englewood Cliffs, NJ (1990).

50. J. W. Shavlik and T. G. Dietterich (eds.), *Readings in Machine Learning*, Morgan Kaufmann, San Mateo, CA (1990).

51. E. Tsang, *Foundations of Constraint Satisfaction*, Academic Press, San Diego (1993).

52. L. G. Valiant, A theory of the learnable, *Comm. ACM* **27** (1984) 1134-1142.

53. A. Waibel and K.-F. Lee (eds.), *Readings in Speech Recognition*, Morgan Kaufmann, San Mateo, CA (1990).

54. P. H. Winston, *Artificial Intelligence*, 3d ed., Addison-Wesley, Reading, MA (1992).

55. D. H. Wolpert, The relationship between PAC, the statistical physics framework, the Bayesian framework, and the VC framework. In D. H. Wolpert (ed.), *The Mathematics of Generalization: Proceedings of the SFI/CNLS Workshop on Formal Approaches to Supervised Learning*, Addison-Wesley, Reading, MA (1995) 117-214.

56. A. L. Yuille and J. J. Kosowsky, Statistical physics algorithms that converge, *Neural Computation* **6** (1994) 341-356.

Subject Index

The style of a page number indicates the nature of the text material:

the style *123* indicates a definition,
the style 123 indicates a brief mention and
the style 123 indicates an extended discussion.

A

abductive inference, 315–317, 319, 345–350, 352.
 bipartite net, 320–328.
acyclic (DAG) digraph, *303.*
acyclic digraph (DAG), *303*, *471.*
adaptive resonance theory (ART), 581.
admissible heuristic, *62.*
algorithm (general),
 anytime, *70.*
 greedy, *550.*
 hill-climbing, 455.
 interruptible, *70.*
 probably approximately correct (pac), *599.*
 recursive, *40.*
algorithm (specific),
 alpha-beta pruning, 80.
 backpropagation, 475.
 breadth-first search, 49–50.
 decision tree, 548–561.
 defeasible reasoning, 247–249.
 depth-first search, 53.
 fuzzy without chaining, 375–379.
 gradient method, 457–460.
 Hopfield net, 434.
 Horn clause resolution, 152.

algorithm (specific) (*continued*):
 irredundant cover, 325.
 iterative-deepening search, 56–57, 65.
 normal default theory, 222.
 Prolog, 120–128.
 Quickprop, *464–465.*
 resolution, 146.
 single-layer perceptron, 451.
 singly connected Bayesian net, 330–331.
 Skolemization, 156.
 unification, 159.
almost surely, *435.*
alpha-beta pruning, 78–81.
alternative hypothesis, *524.*
Amdahl's Law, *51.*
analytic learning, *596.*
AND/OR tree, 76–81.
annealing schedule, *469.*
annealing, simulated, *469.*
antecedent, *374.*
anytime algorithm, *70.*
Arrow Impossibility Theorem, 193–196.
ART (adaptive resonance theory), 581.
associative binary operation, *100.*

Author Index

Since all entries in the references at the end of a chapter are cited in the text, reference section page numbers aren't included in the index. An italicized name is fictional The style of a page number indicates the nature of the text material:

the style 123 indicates a quotation,
the style 123 indicates a citation or reference in text and
the style *123* indicates a biographical entry.

A

Aarts, E., 492.
Abidi, M.A., 197, 611.
Acton, F.S., 454.
Akyürek, A., 28.
Albert, L., 180.
Alcott, A.B., 11.
Alexander, S.M., 192.
Alice, 185.
Allen, J., 25, 610.
Aloimonds, J., 25.
Aloimonos, Y., 611.
Anantharaman, T., 23.
Andersen, S.K., 351.
Anderson, J.A., 25.
Anthony, M., 610.
Antoniou, G., 133.
Apocrypha, 48.
Appelt, D.E., 351.
Armstrong, J.S., 422.
Arrow, K.J., 197.
Artin, E., 139.
Asai, K., 400.
Azencott, R., 492.

B

Babbage, C., 415.
Bacchus, F., 196.
Baker-Ward, L., 6.
Baldi, P., 492.
Barnden, J.A., 581.
Barr, A., 25.
Barrow, J., 180.
Bartley, W.W., 113, 119.
Bayes, T., 280.
Beasley, D., 609.
Becker, S., 464, 476.
Behara, M., 542.
Bench-Capon, T.J.M., 253.
Bender, E.A., 26, 83.
Bernstein, J., 27.
Bhatnagar, R., 189.
Bienenstock, E., 454, 542, 547.
Biggs, N., 610.
Billings, J., 191.
Blake, A., 611.
Bobrow, D.G., 13.
Bohr, N., 96.
Bolc, L., 83.
Bonissone, P.P., 351.
Borenstein, N.S., 491, 517.

IEEE COMPUTER SOCIETY
50 YEARS OF SERVICE • 1946-1996

IEEE Computer Society

The IEEE Computer Society advances the theory and practice of computer science and engineering, promotes the exchange of technical information among 100,000 members worldwide, and provides a wide range of services to members and nonmembers.

Membership

All members receive the monthly magazine *Computer*, discounts, and opportunities to serve (all activities are led by volunteer members). Membership is open to all IEEE members, affiliate society members, and others interested in the computer field.

Publications and Activities

Computer Society On-Line: Provides electronic access to abstracts and tables of contents from society periodicals and conference proceedings, plus information on membership and volunteer activities. To access, telnet to the Internet address info.computer.org (user i.d.: guest). The web address is http://www.computer.org.

Computer magazine: An authoritative, easy-to-read magazine containing tutorial and in-depth articles on topics across the computer field, plus news, conferences, calendar, interviews, and product reviews.

Periodicals: The society publishes 10 magazines and seven research transactions.

Conference proceedings, tutorial texts, and standards documents: The Computer Society Press publishes more than 100 titles every year.

Standards working groups: Over 200 of these groups produce IEEE standards used throughout the industrial world.

Technical committees: Over 29 TCs publish newsletters, provide interaction with peers in specialty areas, and directly influence standards, conferences, and education.

Conferences/Education: The society holds about 100 conferences each year and sponsors many educational activities, including computing science accreditation.

Chapters: Regular and student chapters worldwide provide the opportunity to interact with colleagues, hear technical experts, and serve the local professional community.

IEEE Computer Society Press Publications

CS Press publishes, promotes, and distributes original and reprint computer science and engineering texts. Original books consist of 100 percent original material; reprint books contain a carefully selected group of previously published papers with accompanying original introductory and explanatory text.

Submission of proposals: For guidelines on preparing CS Press books, write to Manager, Press Product Development, IEEE Computer Society Press, P.O. Box 3014, 10662 Los Vaqueros Circle, Los Alamitos, CA 90720-1264, or telephone (714) 821-8380.

10/30/95

DATE DUE

~~JUL 1 4 2003~~ MAR 23			
'JUL 1 4 2003			